The Effect

D1263732

The Effect
An Introduction to Research Design and Causality

Nick Huntington-Klein

CRC Press
Taylor & Francis Group
Boca Raton London New York

CRC Press is an imprint of the
Taylor & Francis Group, an **informa** business

A CHAPMAN & HALL BOOK

First edition published 2022
by CRC Press
6000 Broken Sound Parkway NW, Suite 300, Boca Raton, FL 33487-2742

and by CRC Press
2 Park Square, Milton Park, Abingdon, Oxon, OX14 4RN

© 2022 Nick Huntington-Klein

CRC Press is an imprint of Taylor & Francis Group, LLC

Reasonable efforts have been made to publish reliable data and information, but the author and publisher cannot assume responsibility for the validity of all materials or the consequences of their use. The authors and publishers have attempted to trace the copyright holders of all material reproduced in this publication and apologize to copyright holders if permission to publish in this form has not been obtained. If any copyright material has not been acknowledged please write and let us know so we may rectify in any future reprint.

Except as permitted under U.S. Copyright Law, no part of this book may be reprinted, reproduced, transmitted, or utilized in any form by any electronic, mechanical, or other means, now known or hereafter invented, including photocopying, microfilming, and recording, or in any information storage or retrieval system, without written permission from the publishers.

For permission to photocopy or use material electronically from this work, access www.copyright.com or contact the Copyright Clearance Center, Inc. (CCC), 222 Rosewood Drive, Danvers, MA 01923, 978-750-8400. For works that are not available on CCC please contact mpkbookspermissions@tandf.co.uk

Trademark notice: Product or corporate names may be trademarks or registered trademarks and are used only for identification and explanation without intent to infringe.

Library of Congress Cataloging-in-Publication Data
[Insert LoC Data here when available]

ISBN: 978-1-032-12745-3 (hbk)
ISBN: 978-1-032-12578-7 (pbk)
ISBN: 978-1-003-22605-5 (ebk)

DOI: 10.1201/9781003226055

Publisher's note: This book has been prepared from camera-ready copy provided by the authors.

Thanks to Scott Cunningham for encouraging me to write this book in the first place. Thanks to Joan for being a good enough sleeper to turn midnight-to-3A.M. into writing time. Thanks to Spike for everything.

Contents

List of Figures

List of Tables

Introduction

0.1 Welcome!

THE EFFECT is a textbook all about causal inference, specifically causal inference done with observational data. We want to know whether X causes Y, and by how much, but we can't or don't want to run an experiment. How can we design a research study to answer that question? A tricky task, one that I've had more than one researcher tell me to my face was impossible and not worth trying.[1]

And that's what this book is for. In this book I'll cover what a causal research question even is, and how we can do the hard work of answering that causal research question once we have it.

I'll do that while scaling far back on equations and proofs. There's absolutely a technical element to causal inference, and we'll get to some of that in this book. But when you talk to people who actually do causal research, they think of this stuff intuitively first, not mathematically. They talk about assumptions about the real world and whether they're reasonable, and what the *story* is behind the data. After they've got that settled, *then* they worry about equations and statistical properties. Designing good research and proving (or even understanding)

[1] I think they were just trying to get out of it.

statistical theorems are separate tasks. I think they should be introduced in that order.

IN THE FIRST PART OF THE BOOK, *The Design of Research*, I'll go over the concept of *identification*—the process of figuring out what part of the data has your answer in it, so you can start to work on digging it out. This will require us to use what we know about how the world works to learn just a little more. You'll come out of the first half with an idea of *what it is you need to do* to answer a research question—what your research design is! Or, if you prefer, how you can tell whether to believe causal claims that you've seen other people make. What is it that *they* needed to do to support that claim, and did they do it?

The first part of the book is great. You're gonna love it. I've read a lot of causal inference books at this point, and I don't think there's anything quite like it out there. My mom loved the first part of this book, and she is allergic to statistics.

THE SECOND PART OF THE BOOK, *The Toolbox*, is more technical. In *The Toolbox* I go over the standard set of tools that someone doing causal inference is likely to reach for. Some of these are statistical tools like regression. Others are common research designs that have turned out to be handy in answering lots of research questions, like difference-in-differences.

Of course, I say it's more technical, but the emphasis is still heavily on intuition. I'm never going to try to sell you on a method by proving mathematically that it works. Instead, my goal is to get you to understand what these methods are trying to do, why they're useful, and when they can be used. Then, I want to help you learn how to carry out these methods, which in the 2020s means how to code them up in R, Stata, or Python. Plenty of code examples in the second part of this book.

I want you to come out of the second part of this book feeling competent—ready to implement these methods and understand what's going on when they're used. With a little work, I think you can be.

I'M DEFINITELY BIASED, BUT I THINK THIS BOOK IS A LOT OF FUN. It was fun to write, and I think it will be about as fun to read as a causal inference textbook can be. And it will take you on a tour of all kinds of methods and all kinds of research. Causal inference is a mutt of a field, with important contributions from medicine, from epidemiology, from economics, from sociology, from political science, from finance, from data science, and so on and so on. My home is in economics, so you know where I'm coming from, but I can guarantee I'll be visiting all of these fields in the course of this book. I hope you'll come with me.

0.2   Note to Students

WHEN I WAS IN COLLEGE, I was introduced to the very barest of causal inference methods. But even those felt to me like a real kind of power. These tools and ideas are the kinds of things that, if you use them right, can turn you from a *consumer* of knowledge into a *producer*. You can find the answer to questions nobody else has the answer to. You can figure out how the world really works on your own. I think that's pretty darn cool!

That's the kind of power I want you to keep in mind as you read this book. The point I want to make clear in every page is this: the methods in this book are not blunt instruments to be whacked against some data until an answer emerges.[2] They're designed to be used *as an extension of the researcher's understanding about the world.* They take what we know and tell us how we can learn more, and what assumptions we need to make to do so.

So don't treat causal inference as a technical task. There are technical elements, and you'll need to do some technical work, but that's not the main point. Think of it as a reasoning task. What do you know? What theory can you rely on? And how can you use that to turn data from something confusing and jumbled into something useful and insightful?

0.3   Note to Educators

I'M A PROFESSOR. People have tried to sell me textbooks, so I know how it is. Every textbook is new, and different, and really provides real-world examples that students can latch onto, not like those stuffy *old* books that kids hate! Then you read it and it's the exact same thing as the other books, but with different stock photos and *New Yorker* cartoons in the margins this time.

This book really is different in a few major ways, though. I promise! For one, no stock photos. But the differences aren't just cosmetic, they're structural. I suspect you're either going to think this book is the perfect teaching tool you've been waiting for all these years, or you're going to think it's completely wrong-headed and focusing entirely on the wrong things.[3]

What we end up with is a book that is, for its subject, a fairly easy read without cutting back on rigor or breadth. The

[2] Good thing they're not—if these methods could just be applied blindly without real understanding, then before you finished reading this book someone would have just written a computer program that would do all this causal inference stuff for you, and you'd have wasted your time.

[3] Not to mention all the stuff I left out. Apologies if your favorite causal or statistical method didn't make it in. Every cut from the outline was agony, I assure you.

difficulty level is such that it is most appropriate for an undergraduate causality, observational methods, or applied econometrics course. Depending on the program, it could also be used in masters-level versions of those courses. In conjunction with more technical materials you may also find it useful reading for PhD courses. Readings from the first part of the book, *The Design of Research*, would also be acceptable for high school statistics classes that want to discuss causality.

There are many ways to use the book, but the way I organize the course myself is to spend the first third of my time discussing the concept of identification and how to work with causal diagrams to figure out identification. Then, the latter part of the course goes into specific methods, with plenty of opportunities to read and replicate existing research that uses those methods. Assignments and video materials are available on the textbook website at theeffectbook.net.

What makes this book so different then?

THE FIRST POINT OF DIFFERENCE IS THE LEVEL OF MATHEMATICS. Compared to the existing crop of causal inference textbooks (and certainly to the existing crop of econometrics textbooks), this one is very light on equations. In my experience, even among students who are good at math and can answer mathematical questions, it's a pretty slim portion of students who understand a method *because* of an equation.[4] But if you know what a method is trying to do, then when you do get to the equations they take on meaning beyond just being a homework problem to conquer.

[4] But the students who *do* gain real understanding from equations, myself included, are more likely to become professors, and thus a problem...

The priorities in this book place a conceptual understanding of research design far, far above anything else. The second priority is the ability to *implement*. That means writing code to perform these methods so you can see first-hand what they do and how they behave. The upside: if it works, students will understand what they're doing and how to do it. Plus, I can introduce more advanced and up-to-date methods than a typical textbook that would expect to have to lay out the whole mathematical foundation.

A downside—and it is a real downside—is that this won't prepare students to write proofs in a graduate statistical methods course, or develop their own estimators. However, I suspect that's not where a lot of students, even the ones going into research, are heading anyway.

THE SECOND POINT OF DIFFERENCE IS THE THEORETICAL APPROACH TO CAUSALITY. This book focuses heavily not just on causal inference methods but also causal inference concepts. As such, it has a theoretical underpinning for those concepts!

There are two main theoretical frameworks to choose from in causal inference. One is potential outcomes, associated primarily with Donald Rubin, and the other is the structural causal model/causal diagram framework associated with Judea Pearl.

I make two potentially controversial choices here. The first is to omit almost entirely the potential outcomes framework. The logic of potential outcomes certainly makes its way into the book several times, but I never introduce the model formally. Why? Because the stuff that potential outcomes is great at—clarifying the "missing-data" problem, handling treatment-effect averages, expressing ignorability conditions—I either don't do, or I do in ways I think are more intuitive for students. I've taught potential outcomes to undergraduates before. The intuition is helpful; the math is a barrier. I take what I like!

So I'm largely using the causal diagram framework. The second controversial choice is to use what I think of as "causal diagrams lite." No do-calculus, and I do some things that are helpful for clarity but not part of the formal causal diagram setup, like occasionally including functional form terms on the diagram.

Both of these choices mean that there will be some additional work to do for students who want to continue on to advanced study of these methods. But hopefully they'll understand what they're *trying* to do extremely well. I hope you'll agree with me that, while the things I've left out are valuable and worth knowing in the long run, it's the right call to leave them for later. Better to learn *one* thing well than *two* things poorly.

0.4   Note to the Interested Passerby

THE MARK OF A TRULY EXCELLENT TEXTBOOK is when someone chooses to read it even when it's not assigned. A truly, truly excellent textbook is one that someone just wants to sit down and read the whole way through. If you do this, please let me know. My ego doesn't need the boost, but it does *want* the boost.

While I'm sure I'm leaving some people out, I can imagine three kinds of people who might be likely to read this outside of a classroom. And for those three kinds of people, I have some reading recommendations.

For the data scientist or business analyst with little background in causality who wants to answer causal questions: glad you're here! This book will be taking a rather different approach to data analysis than you're probably used to. For the most part, data science and business analytics are both fields that are first *data-driven*.[5] You look for patterns in the data and see what it tells you. Your goal, usually, is to make some prediction or measurement with the data.

Causal research, on the other hand, is *theory-driven*. You start with what you know and use that to interpret the data. Your goal is to use the data to uncover some broader truth about the processes and laws that generated the data.

Coming into this book, you not only are going to learn some new methods but also a whole new approach to thinking about research! Using a whole different mental framework is tough to do—I find it very difficult when I try to go in the other direction to read data science results, for example. For you, the key chapters of the book are going to be 2 and 5. Maybe even read those ones a couple times until you really *get* them. Once you do, the rest is methods. Given your background you can pick those up in a snap. Open your mind and step inside.

For the non-researcher who wants to understand how causal inference works or get better at interpreting and evaluating studies that use causal inference: this book is conveniently arranged in such a way that you can learn what you need without getting in over your head. Chapters 1 through 9 will give you a look into what studies that use causal inference, as a whole, are *trying to do*. It also gives you a leg up in determining how studies (or people) who make causal claims can support, or fail to support, the claims they're making. Even as a non-researcher, you'll become fully capable of drawing your own causal diagrams (Chapter 7), and then thinking about *what must be done* to support (identify) a causal claim (Chapters 8—9). Then you can ask yourself if they did that or not! If you find the prospect of reading about math terrifying, you can probably get away with skipping Chapters 3-4, but do give them a shot and see how far you get before skipping to Chapter 5.

The second part of the book may also be handy for you. If you're not planning to do your own statistical research, you're not going to need to read any chapter all the way through. But if there's a study out there you want to be able to understand that uses one of the designs in these chapters, you can look through the "How Does It Work?" section at the beginning of

[5] Not always! But usually.

many of the *Toolbox* chapters to see what that design is trying to actually do. And if you wanna get real fancy and interpret some of their results more directly, the "How Is It Performed?" sections will help with that too.

For the researcher with experience in causal inference who wants to review the standard methods or get a better sense of how they work, the second part of this book, *The Toolbox*, is perfect for you. Each chapter on a standard causal inference design is split in three parts—a "How Does It Work?" section that will refresh you on the concepts and theory underlying the design, a "How Is It Performed?" section that probably treads the closest to the econometrics-textbook introduction to these methods you've already seen, and a "How the Pros Do It" section that tracks modern tweaks, concerns, and fixes that you probably want to know about. And while we're at it, those "How the Pros Do It" sections are perfect **for the researcher who learned methods long ago and needs to get up to speed on recent developments.**

That said, I'd also recommend checking out the material from the first half of the book on causal diagrams.[6] The marginal value of learning causal diagrams for someone already trained in potential outcomes isn't enormous in my opinion, but it's still a powerful tool to have in your belt. And they really *are* magical little things when it comes to teaching. If you teach, you may find yourself adding causal diagrams in class even if you don't teach methods and statistics. I always used them in my Economics of the Education System class to help students understand empirical papers above their heads.

[6] And Chapter 5 on Identification, too—you already know this stuff but I think the chapter turned out really well; you might pick up a few new perspectives on identification.

0.5 Thoughts

WRITING THIS BOOK HAS BEEN A WILD TIME, in every sense of the word. I started this thing in February 2020, when my kid was six months old and almost exactly one month before the coronavirus pandemic hit the United States. Now, as I finish the book, my kid is calling every animal "cat" and, well, the pandemic isn't exactly over, but I did just get my second vaccine shot on Friday.

Having a book to write has been a distraction, a passion project, and a way of keeping time during this weird and very tense year. If you're reading this in the future and have no idea what I'm talking about, I'm sure there are many history books about 2020 you can read. Or, by the standards of your time,

2020 isn't even that notably weird, in which case I hope you're doing okay and am shocked you have the time to read causal inference textbooks.

You can also consider this book a strange and minor out-growth of the economics community on Twitter (and the other academic communities it overlaps with). Not only did I learn quite a lot about causal inference from Twitter, but it's just a great intellectual environment to be in, and contributing to it has been a great motivator. There's a real rush in sharing some of your teaching materials and having a thousand people tell you they like them. So I spent a year writing a book. I guess flattery will get you anything.

Part I

The Design of
Research

1
Designing Research

1.1 I Have a Question

HOW DOES THE WORLD WORK? That's the big question, isn't it? And it's one for which our answers will always be ever so slightly incomplete.

That incompleteness is a curse, but also a blessing. Sure, we'll never know everything perfectly.[1] But that also means that we'll always have questions that need answering. For a certain kind of person, at least, answering those questions sounds like a pretty good way to spend your time on Earth. Maybe you're that kind of person. I certainly am.

A RESEARCH QUESTION is a question that you have that you plan to answer, or at least try to answer, by doing research. Simple as that. Or, rather, as difficult as that. A good research question is well-defined, answerable, and understandable—those can be hard to figure out! We'll talk more about this in Chapter 2.

For one example, let's say that our research question is "does adding an additional highway lane reduce traffic?"

[1] Quantum mechanics is perhaps the most precise scientific field in existence, with accurate measurements and predictions out to more than a dozen decimal places. But even then, "accurate to fourteen decimal places" isn't the same thing as "accurate."

DOI: 10.1201/9781003226055-1

That's a question about how the world works. Traffic is, unfortunately, a part of the world. And it's something we could likely figure out with some research!

What kind of research? Well-designed research is research capable of answering the question it's trying to answer. That seems simple but it actually requires quite a lot of thought and effort.

And that's the real trick.

How can you do research in such a way that, when you're done, you have an answer to your research question?

That's what this book is about.

1.2   Empirical Research

"RESEARCH CAPABLE OF ANSWERING THE QUESTION IT'S TRYING TO ANSWER" could mean a lot of things, of course.

There are many kinds of research. You could look in books to see what people have already had to say about your question ("what do the traffic experts say about the effects of an additional highway lane?"). You could philosophically reason your way around the question ("if I assume people try to minimize their commute times, how would I expect them to respond to an additional highway lane?"). These are all forms of research.

This book will focus on *empirical* research and, specifically, *quantitative empirical research*.

Empirical research is any research that uses structured observations from the real world to attempt to answer questions. So instead of trying to *reason* our way through what drivers would do if given an additional highway lane, we try to *observe* the choices that drivers take. Perhaps we interview drivers about how they make decisions. Or maybe we get a big data set of traffic violations, or of traffic flow numbers on highways.

Empirical research. Research that uses observations from the real world in a structured way.

Quantitative empirical research is just empirical research that uses quantitative measurements (numbers, usually). More data sets, fewer interviews.

QUANTITATIVE EMPIRICAL RESEARCH, like any kind of research, can be tricky! Measurements are hard to take precisely or interpret accurately. Statistics is a difficult field.

One particularly sticky problem with quantitative empirical research is that the numbers that we observe often don't tell us exactly what we want to know.

After all, we might want to study the impact of additional lanes by comparing two-lane highways to three-lane highways. But we probably aren't *actually* interested in how much traffic there is on three-lane highways and on two-lane highways. We're *probably* interested in whether we can make traffic go down by turning a two-lane highway into a three-lane highway! But as much as we want them to, the numbers we have don't actually tell us that right away. All we have are two-lane highways and three-lane highways. We don't have a "what if" highway that tells us how much traffic there *would have been* if we'd made *that* two-lane highway one lane wider.

This problem constitutes a major headache for us researchers. If the numbers we have don't actually answer the research question we have, what can we do?

Well, it turns out that, if you do it right, you often *can* figure out how to collect the right numbers, or do the right things to those numbers, to get an actual answer to our question. But it doesn't come free. We have to carefully design the right kind of analysis that will answer our question.

1.3   Why Research Needs a Design

WHY IS IT SO IMPORTANT for research to be properly designed? One way we can think about this is by looking at what happens when it's not.

Let's take our highways and traffic example. How might we go about researching an answer to this question? Our first pass might be to just compare traffic patterns on highways with more lanes against traffic patterns on highways with fewer lanes.

Seems reasonable. But then you do it, and it turns out that more lanes seem to go along with more traffic! Surprising. However, why do those highways have more lanes in the first place? It might be that the busiest routes tend to be the ones that get expanded, and so it's no surprise that more lanes are associated with more traffic! Sure, maybe additional lanes *do* lead to more traffic.[2] But it takes research design to know that our first-pass analysis wasn't right and to figure out what to do instead.

A LACK OF SOLID RESEARCH DESIGN can be seen in the results, as well. Have you ever noticed, for example, how the studies you read about nutrition in the news can't seem to make up their mind? When I was a kid in the 90s, high-carb, low-fat food was what you were supposed to eat, and frozen yogurt and bagels both counted as pretty darn good for you. None of that

[2] Transportation researchers generally say that more lanes at least leads to more driving! See for example Milam et al. (2017).

is considered true today. And is a glass of wine a night good for you or not? Or coffee? Or butter versus margarine? Or sugar versus corn syrup?[3] The underlying truth about what food is good to eat can't possibly be changing that much, but the scientific results sure do!

Some of this we can blame on the news hyping up studies beyond reason, or misinterpreting them completely. But some of this comes down to a lot of nutrition studies not having research designs that allow them to answer the question "what food will make you healthier?" Different studies seem to give different answers to "what food will make you healthier?" because they're not actually answering that question in the first place, even if they claim to! $2 + 2$ only has one answer,[4] but if you're actually calculating something entirely different from $2 + 2$, you might well come back with an answer of 6, or 1, or -52. Then you wake up to a news headline reading that scientists have determined that $2 + 2 = -52$.

Nutrition is a good field to pick on here because it isn't really the fault of the nutrition researchers themselves. Nutrition just happens to be a topic that makes good research design really elusive.[5] So you end up with a field with shaky research design. And what does that give us? Inconsistent results that people have unfortunately learned not to pay all that much attention to today, because they know it might change tomorrow.

RESEARCH DESIGN IS HARD, and just because you want to answer a question doesn't mean there's necessarily a straightforward way of doing it. But the worst that could happen is that we'd figure out that the answer will be difficult to get. Then, at least, we'll know.

The best that could happen is that we *can* answer our question. And we do. And then we win a Nobel prize.

1.4   In This Book

THIS BOOK IS DESIGNED TO DO A FEW THINGS.

In the first half, it aims to teach you the principles of research design. Specifically, it will go into the ways that you can build an answerable research question, and then think about what kind of quantitative empirical research you could perform in order to answer that question. What would you need to measure? How could you be sure your method would actually answer your question?

[3] Ever noticed that some candy and soda brands advertise their use of "real sugar" as though not being corn syrup makes it a health food? This is only marginally related to what we're talking about here, but boy do I hate it a lot.

[4] In a typical system, anyway. Advanced mathematics gets up to some hijinks.

[5] It's really hard to accurately measure what people eat, it's really hard to pick apart the effect of one food from all the other stuff people eat, it's really hard to separate out the effect of the food from the effect of the stuff that made you choose to eat the food, and so on and so on...

Then, in the second half, it introduces you to some of the "toolbox" methods for causal research designs using observational data (i.e., answering a causal research question without running an experiment). These methods are very commonly used in modern research as they tend to be widely applicable in answering a wide range of research questions, and the assumptions they rely on are well-understood.

Hopefully, you can walk away from this book confident in your ability to craft a research project, figure out what kind of data you need to answer your research question, and figure out what calculations you need to perform on your data.

2
Research Questions

2.1 *What Is a Research Question?*

COMING UP WITH A QUESTION IS EASY. Just ask any five-year-old and they can provide you with dozens. Coming up with a good research question is much harder.

What's the difference? The difference, at least in the case of quantitative empirical research, is that a research question is a question *that can be answered*, and for which having that answer will *improve your understanding of how the world works*.

Those are both a little abstract. Let's take them one at a time.

WHAT DOES IT MEAN to have a question *that can be answered*? It means that it's possible for there to be some set of evidence in the world that, if you found that evidence, your question would have a believable answer. So for example, "what is the best James Bond movie?" can't really be answered.[1] No matter what evidence you find, "best" is ambiguous enough that you can't even imagine the evidence that would settle the question for you. You could get every person on Earth to agree it's

[1] Although, having seen exactly two James Bond movies myself, I certainly hope it's not either of them.

DOI: 10.1201/9781003226055-2

Moonraker and that still wouldn't necessarily settle the question.

On the other hand, "which era of Bond movies had the highest ticket sales?" can definitely be answered. You look at the ticket sales and see when they were highest. Evidence can tell you the answer to this question.

So WE HAVE A QUESTION THAT CAN BE ANSWERED. But does it *improve our understanding of how the world works?* What this means is that the research question, once answered, should tell you about something broader than itself. It should inform *theory* in some way. Theory doesn't have to be something as important as the theory of gravity or the theory of evolution. It could even be something as generic as "bread costs more today than last year because bread prices have been generally increasing over time." Theory just means that there's a *why* or a *because* lurking around somewhere. Even hydrogen is a theory—it says that we see material like water behaving a certain way *because* there's a kind of atom that behaves in certain ways and has a certain structure.

Take germ theory, for example. Germ theory says that microorganisms like bacteria and viruses can cause disease. This explains *why* we have diseases, and also *why* disease can spread from one person to another. We don't call it "a theory" because we're uncertain about whether it's true.[2] We call it a theory because it tells us *why*.

A good research question *takes us from theory to hypothesis*, where a hypothesis is a specific statement about what we will observe in the world, like "people who wash their hands will get sick less often." That is, a research question should be something that, if you answer it, helps improve your *why* explanation. Put another way, you can ask "if I find result X, what can I do with it? Does that inform how policy should be created? Does it change how I think the world functions?" Great research questions often come from the theory themselves—the line of thinking being "if this is my explanation of how the world works, then what should I observe in the world? Do I observe it?"

This is easy to miss! Let's keep working with germ theory as an example. We might ponder germ theory for a while and think "Hey, I wonder how small the smallest microorganism is." That's a research question that we could answer with the right evidence, and it's *related* to germ theory, and would be kind of neat to know. However, learning the answer to this question wouldn't really tell us anything new about why we have diseases or why

Theory. An explanation of why the things we observe happen, or otherwise generalizes what we observe in one situation to another situation.

[2] Theories that are almost certainly correct, like germ theory, aren't any more or less a theory than theories that are almost certainly incorrect, like the theory that the Egyptians were able to build the pyramids because they had help from aliens.

Hypothesis. A specific statement about what you will observe in the data.

disease can spread from one person to another.[3] Maybe it helps us understand some *other* theory better. That would make our small-microorganism question a better research question for that other theory than for germ theory.[4]

So does asking "which era of Bond movies had the highest ticket sales?" improve our understanding of how the world works? Maybe, for the right theory. Maybe we have a theory that says that action movies were generally at their most popular in the 1980s. Asking about Bond sales over time might tell us a little more information about whether that theory is an accurate explanation of ticket sales.

LET'S WALK THROUGH AN EXAMPLE, starting with our theory. Let's say we have a theory that your curiosity as an adult is harmed by exposure to passive entertainment like TV and movies. Regardless of whether this is actually true or not, it still qualifies as a theory—it explains why we might see certain levels of curiosity in adults.

A natural research question here is "does watching a lot of TV as a child dull your curiosity as an adult?"

Let's check our two conditions for research questions. Could we answer this question? Yes! The data necessary to answer this question might be hard to come by, but we can at least conceive of it existing. If we randomized a bunch of kids to watch different amounts of TV, and then followed them to adulthood and measured their curiosity, that would be some pretty convincing evidence on our research question.[5]

Second, does this research question tell us about how the world works? Yes! If we answered this question, that would pretty clearly inform our theory. If we answered our research question with "no, watching a lot of TV as a child does not dull your curiosity as an adult," then it would be a pretty hard sell to explain adult curiosity by saying it's because of passive entertainment. The research question does help us figure out if the theory is any good.

A good test for whether a research question informs theory is to imagine that you find an unexpected result, and then wonder whether it would make you change your understanding of the world. Let's say that instead of answering the research question "does watching a lot of TV as a child dull your curiosity as an adult?" we use the research question "do kids who watch lots of Sesame Street tend to have lower levels of curiosity later?" We do our research and find that, actually, kids who watch Sesame Street have *higher* levels of curiosity! Uh oh. With this new

[3] Or at least I don't think so... I'm not a biologist.

[4] And even if it doesn't help us understand any theory or broaden our understanding, it might still be worth pursuing just because it's kinda neat. Nothing wrong with curiosity for its own sake.

[5] Of course, in the real world, very few research questions are ever answered conclusively. Even after this experiment we'd wonder if the results would be the same in a different decade, or country, or if we chose different lengths of TV watching. But even if we couldn't answer the question *conclusively*, we'd still be be providing evidence that unambiguously *informs* our answer to the question.

information, do we have to change our theory? Well, we hem and we haw, and we think about how fond we were of that original theory. And we explain away the Sesame Street result by noting that Sesame Street might be different from most kinds of TV, and also that we just looked at which kids *did* watch Sesame Street, not whether Sesame Street is actually responsible for their curiosity—maybe kids who are more curious in the first place choose Sesame Street.

This ability to see a bad result and still hold on to the original theory tells us that the research question wasn't very good, at least not for this theory.[6] A really good research question, once answered, should be hard to explain away just because it's inconvenient.

So there we have it—"does watching a lot of TV as a child dull your curiosity as an adult?" is a good research question that could be answered with the right data, and would inform our understanding of the world if answered. Granted, the process of actually *answering* that question is another hurdle.[7] But at the very least we know that the question itself is good, even if the answer is elusive.

2.2   Why Start with a Question?

THIS SOUNDS HARD. WHY BOTHER? We have a bunch of data at our fingertips. In fact, we're awash in it. There's data everywhere. So why not skip the hard part of deriving a research question from a theory and instead just see what sorts of patterns are in the data?

Well, you could. In fact, a lot of people do. This is called "data mining," and there are people who do that very thing, and manage to do it quite well. They go to the data, look for patterns, and report back. You find a lot of this in the field of data science,[8] but data mining can be done any time you have some data. Just look at the data, see what's in there, and work backwards.

So, sounds good, right? Well, data mining is well and good, but it turns out to be very good at some things, and very bad at others.

The kinds of things that data mining is good at are in *finding patterns* and in *making predictions under stability*.[9] The kinds of things that data mining is less good at are in *improving our understanding*, or in other words *helping improve theory*. It also has a tendency to find *false positives* if you aren't careful.

[6] This means if the Sesame Street study had turned out in favor of our theory we shouldn't have *increased* our confidence either.

[7] Good luck running that big experiment.

[8] Although it's certainly not the only thing data scientists do. They also make a lot of money.
Okay, fine, and sometimes they work from theory too. Most tools designed for data mining can be used to work with theory if applied properly. At that point it's no longer data mining. It may still be data science though.

[9] What do I mean by "under stability"? I mean that the process giving us the data doesn't change. If I roll a six-sided die a thousand times, data mining would be great at predicting that the probability of a 1 is 1/6. But if I then switch to a twenty-sided die, that data mining prediction will be bad. It will still predict a 1/6 chance of a 1 until it gets a lot more data. Probability theory, on the other hand, will properly predict the switch to a 1/20 chance immediately.

FINDING PATTERNS AND MAKING PREDICTIONS ARE VERY VALUABLE. And we probably do want to rely on some sort of data mining for these tasks. After all, there's no way we can really theorize about every possible pattern that could be in the data and think to check it. Doing something that just asks *what* we see rather than *why* is the right angle to take there. Plus, sometimes seeing patterns in data can give us ideas for research questions that we can examine further in other data sources.

It's also probably the best angle to take when we don't care about why! If I don't care why the stock market goes up or down, and I just want to predict if it will or not so I know whether to buy or sell, then data mining may well be the way to go.

But outside of those realms?

WHY DOES DATA MINING HAVE DIFFICULTY HELPING THEORY? There are a few main reasons.

One of the reasons is that data mining, by definition, focuses on what's in the data, not *why* it's in the data. In other words, it's fantastic at revealing correlations—patterns in the data of how variables we've observed have varied together in the past—but the correlations it uncovers may have little to do with causality, or an understanding of *why* those variables move together.

To introduce an example that will pop up a few times in this book, someone using data mining to try to understand ice cream sales may well notice that the proportion of people who wear shorts is a fantastic predictor of ice cream sales. But shorts-wearing isn't *why* people buy ice cream. They buy ice cream and wear shorts because it's hot. But to a data miner, the shorts/ice cream connection is pretty compelling! After all, shorts can be a great way of *predicting* ice cream eating, even if there's no "why" there.

However, if what we're really interested in isn't predicting ice cream but *explaining why* people eat ice cream, it's pretty tempting at that point to try to invent a story to justify why shorts might actually be the reason people eat ice cream. In the case of ice cream and shorts we can tell that's ridiculous, but it's a lot harder when we don't actually know what's ridiculous and what's not ahead of time.

For example, we'd love to know what causes children to act aggressively. That sounds really important! A data mining exercise here might look through all of the things kids do or are exposed to, and check whether any of them are associated with higher levels of aggression. Maybe kids who play a lot of video games are more likely to be aggressive. So... are the video games responsible? Maybe, maybe not.[10] Data mining is well-equipped to find the relationship but poorly-equipped to tell us *why* that relationship is there. Hopefully someone doesn't get worried and ban all the video games before the researcher can carefully explain the distinction.

Another reason is that, because it's so focused on the data, data mining doesn't really deal in *abstraction*. For example, take a look at a chair. How do you know it's a chair? Well, it's probably got some legs, maybe a back, definitely a flat-ish seating area, and it's clearly designed for sitting in. This is our "chair theory"—we theorize that there are these objects called chairs that have certain chair-like properties, united in the ability to sit on them some distance off the ground. The chair you're looking at now is one example of chair theory.[11]

[11] Were Plato not already dead I'm sure this paragraph would kill him.

But what's actually *in the data?* There's no "chair" in the data. There's just a flat bit and some straight-up-and-down bits underneath the flat bit. Data mining would be great at noticing that it sees a lot of flat bits on top of straight-up-and-down bits, but it would not be good at developing "chair theory" for us because it would miss *why* we keep seeing that arrangement— because it allows us to sit on it. A data miner would never guess that the four-legged chair has anything to do with, say, a bean-bag chair, which has no straight-up-and-down bits at all.

FALSE POSITIVES are another reason why data mining can be a dangerous. Take the video games and aggression example. Okay, sure, maybe the video games aren't *why* the kids are aggressive, but we still found the relationship. Surely there's *something* there.

Well, maybe, and maybe not. Data mining means looking in the data to see what's there. And there's a lot of stuff to look at! If you check, say, a hundred variables and see if they're related to aggression, *something* is going to pop up as looking related, just by random chance. That random relationship is unlikely to pop up again if you tried another sample. It's only in the sample you have by random chance, which is what makes it a "false positive."

That's one major danger in proceeding in your work without starting with a solid research question. Without a disciplined research question, there's no reason not to just check everything. *Something* is going to pop up as related by random chance if you check enough stuff. It takes a really well-behaved researcher to not pretend that's exactly what they'd been looking for from the start, and to fill in some reason why the 100th relationship they checked makes perfect sense and supports some theory.

There are ways of avoiding false positives while doing data mining—this is something they worry about a lot in data science and have a lot of tools for.[12] But if you're just sort of trawling through a data set to see what you see, you're likely to end up with a whole lot of false positives mixed in with the real positives. You'll have no way of telling one from another.

THAT SAID, DATA MINING ISN'T ALL BAD. There's no way we could possibly *think up* every interesting theory to test. Plenty of theories come from looking at the data in the first place, noticing a pattern, and wondering why the pattern is showing up, or whether the pattern is even real.

The drug Viagra, for example, was initially being tested as a blood pressure medication. The researchers testing it out to see if it worked to lower blood pressure happened to notice, uh, its other effects.

They've done data mining there—instead of coming to the data with a theory, they noticed an interesting pattern in the data.

Of course, the responsible thing to do at that point is to not just take the pattern as given. *That's* where the real problem of data mining is. Instead, they took the interesting pattern they noticed and looked to see if it held up in *other* data—if it replicated—before being sure that the pattern they noticed was real and explained how the drug worked.

Data mining isn't bad. It's just bad as a *final step* if you're trying to explain the world. It can still work as a source of ideas. And heck, maybe it can help you earn a hojillion dollars like Viagra did, too.

[12] Some examples are "cross-validation" and "training and testing sets." If this interests you, read more about data science. You'll get some data science in this book but not much.

2.3 Where Do Research Questions Come From?

RESEARCH QUESTIONS CAN COME FROM LOTS OF PLACES. Mostly, curiosity. We want to know how the world works, and that naturally leads to questions!

There are two steps in this process: thinking about theory, and coming up with a research question. Either one can come first.

Perhaps it begins with theory: "I think this is how the world works" or "I wonder if this is how the world works"—that's your theory. This could be anything from "I think people make the decisions they do because they follow incentives" to "I think plants survive without eating because they collect energy from the sun" to "I think CD sales are down because people stream music now instead."

With the theory in place, the process continues with our hypothesis: "if this is how the world works, what would I expect to see in the world?" Our above theories might lead to the research questions "will students work harder in school if you pay them for good grades?" or "will plants die if you store them in a dark room?" or "are CDs more popular now in areas with bad Internet connections?" These research questions tell us a *hypothesis to test* such that the result of that test *tells us something about the theory.*

The question might come first. "Will students work harder in school if you pay them for good grades?" we might ask. Then, we might wonder why we came up with such an idea in the first place. Probably because we think students respond to incentives. Or we might wonder, if we answered the question, what sense we might make of it, leading us back to our theory. If you can't figure out why you would ask the question, it may not be a great research question. Or at least you'd have a hard time getting anyone to care about the answer once you had it.

LET'S BE HONEST, sometimes research questions also come from *opportunity*.

Have a neat data set? Think about what data is available to you and whether any related research questions or theories come to mind.[13]

Or, perhaps you've learned about something unusual or interesting that has happened in the world. Maybe you've learned that a few school districts have decided to try paying their students for good grades. When you hear about something like this, you might ask "what research questions would this allow me to answer?" and from there you have a research question, and from there a theory!

[13] Try to do this after understanding what is in the data but before actually analyzing the data, unless your goal is data mining.

2.4   How Do You Know if You've Got a Good One?

YOU'VE FOLLOWED THE PROCESS. You have a research question in mind. You know it can be answered with data, and you're pretty sure that if you get the answer to it, it will help you learn how the world works.

But is it really a good one? Just a few things to check before you get too far into the process:

1. **Consider Potential Results.** A good way to double-check
   the relationship between your research question and your theory is to *consider the potential answers you might get.* Then,
   imagine what kind of sense you'd make of that result, or what
   conclusion you would draw. Let's say you find that students
   do tend to work harder in school when they're paid for good
   grades. What would this tell us about how students respond
   to incentives? Or let's say you find that students *don't* work
   harder when they're paid. What would *that* tell us about
   how students respond to incentives? If you can't say something interesting about your potential results, that probably means your research question and your theory aren't as
   closely linked as you think! Let's say we *do* find that kids
   who happen to play video games are more aggressive. Can
   we take that result and claim that video games are a cause of
   aggression? Not really, for the reasons we've discussed previously. So maybe that research question really isn't linked
   to that theory very well.

2. **Consider Feasibility.** A research question should be a
   question that can be answered using the right data, if the
   right data is available. But *is* the right data available? If
   answering your research question is *possible* but requires following millions of people repeatedly for decades, or trying to
   measure something that's really hard to measure accurately,
   like trying to get people to remember what they had for lunch
   three years ago, or getting access to the private finances of
   thousands of unwilling people, then that research question
   might not be feasible. While sometimes you can get around
   these problems with a clever design, you might want to consider going back to the drawing board.

3. **Consider Scale.** What kind of resources and time can you
   dedicate to answering the research question? Given a lifetime
   of effort and considerable resources, you might be able to
   tackle massive questions like "What causes some countries to
   become rich and others poor?" Given the confines of, say, a
   term paper, you could take some wild swings at that question,
   but you're likely to do a much more thorough job answering
   questions with a lot less complexity.

4. **Consider Design.** A research question can be great on its
   own, but it can only be so interesting without an answer. So,
   an important part of evaluating whether you have a workable

research question is figuring out if there's a reasonable research design you can use to answer it. Figuring out whether you do have a reasonable research design is the topic of the rest of this book.

5. **Keep It Simple!** Answering any research question can be difficult. Don't make it even harder on yourself by biting off more than you can chew. A common mistake is to bundle a bunch of research questions into one. "What are the determinants of social mobility?" I.e., how someone can move from one social class to another throughout their lifetime. There are *many* determinants of social mobility. You're unlikely to answer that question well. Instead try "Is birth location a determinant of social mobility?" For another example, how about the question "How was the medium of painting affected by the Italian renaissance?" In a million ways! You'll get lost and do a poor job on a bunch of minor pieces instead of getting at the whole. Instead maybe "What similar characteristics are there among the countries that adopted the use of perspective in painting most quickly?"

So, consider feasibility, scale, and design. Keep it simple, and think about whether the results you might likely see would tell you anything interesting about the world. After all, learning something interesting and new about the world is our goal!

3

Describing Variables

3.1   Descriptions of Variables

THIS CHAPTER WILL BE ALL ABOUT HOW TO DESCRIBE A VARI-
ABLE. That seems like an odd goal.[1] The opening to this book
was all about setting up research questions and how empirical
research can help us understand the world. And we jet right
from that into describing variables? What gives?

It turns out that empirical research questions really come
down entirely to describing the density distributions of statistical variables. That's, well, that's really all that quantitative
empirical research is. Sorry.

Maybe that's the wrong approach for me to take. Perhaps
I should say that all the interesting empirical research findings you've ever heard about—in physics, sociology, biology,
medicine, economics, political science, and so on— can be all
connected by a single thread. That thread is laid delicately on
top of a mass of probability. The shape it takes as it lies is
the density of a statistical variable, tying together all empirical
knowledge, throughout the universe, forever.

[1] And what does it even
mean? We'll get there.

DOI: 10.1201/9781003226055-3

Is that better? Am I at the top of *The New York Times* nonfiction bestseller list yet?

Look, in order to make any sense of data we have to know how to take some observations and describe them. The way we do that is by describing the types of variables we have and the distributions they take. Part of that description will be in the form of describing how different variables interact with each other. That will be Chapter 4. In this chapter, we'll be describing variables all on their own. It will be less interesting than Chapter 4, but, I am sorry to say, more important.

A *variable*, in the context of empirical research, is a bunch of observations of the same measurement—the monthly incomes of 433 South Africans, the number of business mergers in France in each year from 1984—2014, the psychological "neuroticism" score from interviews with 744 children, the color of 532 flowers, the top headline from 2,348 consecutive days of *The Washington Post*. Successfully *describing a variable* means being able to take those observations and clearly explain what was observed without making someone look through all 744 neuroticism scores themselves. Trickier than it sounds.

> **Variable.** A set of observations of the same thing.

3.2   Types of Variables

THE FIRST STEP in figuring out how to describe a variable is figuring out what kind of variable it is.

While there are always exceptions, in general the most common kinds of variables you will encounter are:

Continuous Variables. Continuous variables are variables that could take any value (perhaps within some range). For example, the monthly income of a South African would be a continuous variable. It could be 20,000 ZAR,[2] or it could be 34,123.32 ZAR, or anything in between, or from 0 to infinity. There's no such thing as "the next highest value," since the variable changes, well, continuously. 20,000 ZAR isn't followed by 20,001 ZAR, because 20,000.5 ZAR is between them. And before you get there you have to go through 20,000.25 ZAR, and 20,000.10 ZAR, and so on.

> [2] ZAR is the South African rand, the official currency in South Africa.

Count Variables. Count variables are those that, well, count something. Perhaps how many times something happened or how many of something there are. The number of business mergers in France in a given year is an example of a count variable. Count variables can't be negative, and they certainly can't take fractional values. They can be a little tougher to deal with than continuous variables. Sometimes, if a count variable takes many different values, it acts a lot like a continuous variable and so researchers often treat them as continuous.

Ordinal Variables. Ordinal variables are variables where some values are "more" and others are "less," but there's not necessarily a rule as to how *much* more "more" is. A "neuroticism" score with the options "low levels of neuroticism," "medium levels of neuroticism," and "high levels of neuroticism" would be an example of an ordinal variable. High is higher than low, but how much higher? It's not clear. We don't even know if the difference between "low" and "medium" is the same as the difference between "medium" and "high." Another example of an ordinal variable that might make this clear is "final completeted level of schooling" with options like "elementary school," "middle school," "high school," and "college." Sure, completing high school means you got more schooling than people who completed middle school. But how much more? Is that... two more school? That's not really how it works. It's just "more." So that's an ordinal variable.

Categorical Variables. Categorical variables are variables recording which category an observation is in—simple enough! The color of a flower is an example of a categorical variable. Is the flower white, orange, or red? None of those options is "more" than the others; they're just different. Categorical variables are very common in social science research, where lots of things we're interested in, like religious affiliation, race, or geographic location, are better described as categories than as numbers.

A special version of categorical variables are *binary variables*, which are categorical variables that only take two values. Often, these values are "yes" and "no." That is, "Were you ever in the military?" Yes or no. "Was this animal given the medicine?" Yes or no. Binary variables are handy because they're a little easier to deal with than categorical variables, because they're useful in asking about the effects of treatments (Did you get the treatment? Yes or no) and also because categorical variables can be turned into a series of binary variables. Instead of our religious affiliation variable being categorical with options for "Christian," "Jewish," "Muslim," etc., we could have a bunch of binary variables—"Are you Christian?" Yes or no. "Are you Jewish?" Yes or no. Why would we want that? As you'll find out throughout this book, it just happens to be kind of convenient. Plus it allows for things like someone in your data being both Christian *and* Jewish.

Qualitative Variables Qualitative variables are a sort of catch-all category for everything else. They aren't numeric in nature, but also they're not categorical. The text of a *Washington Post* headline is an example of a qualitative variable. These can be very tricky to work with and describe, as these kinds of variables tend to contain a lot of detail that resists boiling-down-and-summarizing. Often, in order to summarize these variables, they get turned into one of the other variable types above first. For example, instead of trying to describe the *Washington Post* headlines as a whole, perhaps asking first "how many times is a president referred to in this headline?"—a count variable—and summarizing that instead.

3.3   The Distribution

ONCE WE HAVE AN IDEA OF WHAT KIND OF VARIABLE WE'RE DEALING WITH, the next step is to look at the *distribution* of that variable.

A variable's *distribution* is a description of *how often different values occur.* That's it! So, for example, the distribution of a coin flip is that it will be heads 50% of the time and tails 50% of the time. Or, the distribution of "the number of limbs a person has" is that it will be 4 most often, and each of 0, 1, 2, 3, and 5+ will occur less often.

Distribution. A description of the probability that each possible value of a variable will occur.

When it comes to categorical or ordinal variables, the variable's distribution can be described by simply giving the percentage of observations that are in each category or value. The full distribution can be shown in a frequency table or bar graph, which just shows the percentage of the sample or population that has each value.

Variable	N	Percent
Primary Degree Type Awarded	7424	
... Less than Two-Year Degree	3495	47.1%
... Two-Year Degree	1647	22.2%
... Four-Year or More	2282	30.7%

Data from College Scorecard

Table 3.1: Distribution of Kinds of Degrees US Colleges Award

Frequency table. A table that shows the proportion of the time that a variable takes a given value. **A bar graph** is a graphical representation of a frequency table.

These tables tell you all you need to know. From Table 3.1 we can see that, of the 7,424 colleges in our data,[3] 3,495 of them (47.1%) predominantly grant degrees that take less than two years to complete, 1,647 of them (22.2%) predominantly grant degrees that take two years to complete, and 2,282 of

[3] Generally in statistics, as well as in this book, "N" means "number of observations."

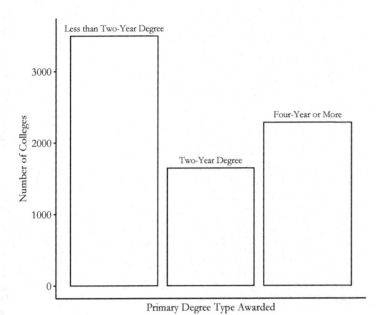

them (30.7%) predominantly grant degrees that take four years or more to complete or advanced degrees. Figure 3.1 shows the exact same information in graph format.

There are only so many possibilities to consider, and the table and graph each show you how often each of these possibilities comes up. Once you've done that, you've fully described the distribution of the variable. There's literally no more information in this variable to show you! If we wanted to show more detail (maybe *which majors* each college tends to specialize in) we'd need a different data source with a different variable.

CONTINUOUS VARIABLES ARE A LITTLE TRICKIER. We can't just do a frequency table for continuous variables since it's unlikely that more than one observation takes any specific value. Sure, one person's 24,201 ZAR income is very close to someone else's 24,202 ZAR. But they're not the same and so wouldn't take the same spot on a frequency table or bar chart.

For continuous variables, distributions are described not by the probability that the variable takes a given value, but by the probability that the variable takes a value *close* to that one.

One common way of expressing the distribution of a continuous variable is with a *histogram*. A histogram carves up the potential range of the data into bins, and shows the proportion of observations that fall into each bin. It's the exact same thing as the frequency table or graph we used for the categorical variable, except that the categories are ranges of the variable rather than the full list of values it could take.

Histogram. A graph showing the proportion of the time that a variable falls into a given range between two values.

For example, Figure 3.2 shows the distribution of the earnings of college graduates a few years after they graduate, with one observation per college per graduating class ("cohorts"). We can see that there are over 20,000 college cohorts whose graduates make between $20,000 and $40,000 per year. There are a smaller number—about 4,000—making on average $10,000 to $20,000. For a very tiny number of college cohorts, the cohort average is between $80,000 and $160,000. There are so few between $160,000 and $320,000 that you can't even really see them on the graph.

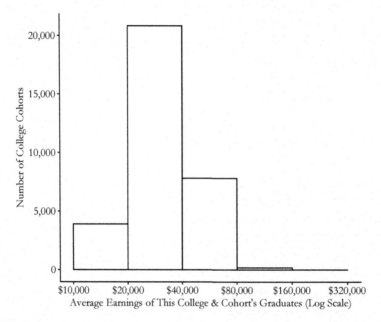

WITH A CONTINUOUS VARIABLE WE CAN GO ONE STEP FUR-THER than a histogram all the way to a *density*.[4] A density shows what would happen to a histogram if the bins got narrower and narrower, as you can see in Figure 3.3.[5]

When we have a density plot, we can describe the probability of being in a given range of the variable by seeing how large the area underneath the distribution is. For example, Figure 3.4 shows the distribution of earnings and has shaded the area between $40,000 and $50,000. That area, relative to the size of the area under the *entire* distribution curve, is the probability of being between $40,000 and $50,000. That particular shaded area makes up about 16% of the area underneath the curve, and so 16% of all cohorts have average earnings between $40,000 and $50,000.[6]

[4] Density distribution if you're graphing it, probability density if you're describing it.

[5] This only works because we have enough observations to fill in each bin. Otherwise it looks a lot choppier. Formally we need the bins to get narrower *and* the number of observations to get bigger.

[6] Calculus-heads will recognize all of this as taking integrals of the density curve. Did you know there's calculus hidden inside statistics? The things your professor won't tell you until it's too late to drop the class. We won't be doing calculus in this book, though.

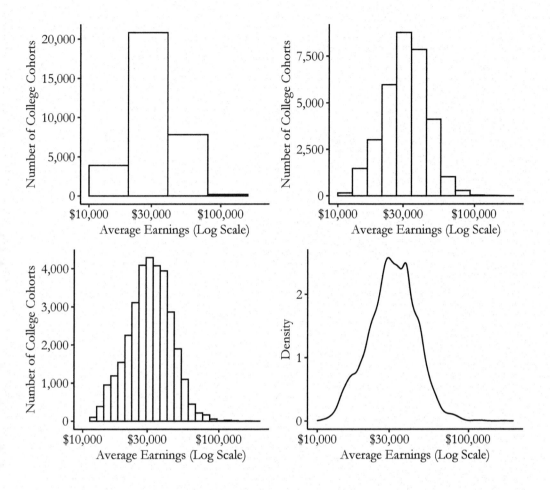

Figure 3.3: Distribution of Average Earnings across US College Cohorts

AND THAT'S IT. Once you have the distribution of the variable, that's really all you can say about it.[7] After all, what do we have? For each possible value the variable *could* take, we know how likely that outcome, or at least an outcome like it, *is*. What else could you say about a variable?

Of course, in many cases these distributions are a little too detailed to show in full. Sure, for categorical variables with only a few categories we can easily show the full frequency table. But for any sort of continuous variable, even if we show someone the density plot, it's going to be difficult to take all that information in.

So, what can we do with the distribution to make it easier to understand? We pick a few key characteristics of it and tell you about them. In other words, we summarize it.

[7] Until you start to incorporate how it relates to *other* variables, as we'll talk about in Chapter 4.

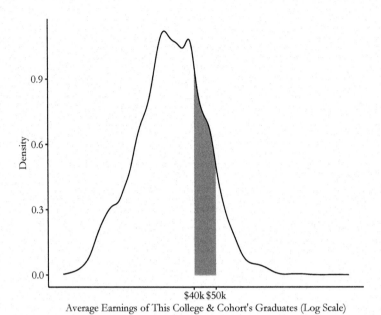

Figure 3.4: Shaded Distribution of Earnings across US College Cohorts

3.4   Summarizing the Distribution

ONCE WE HAVE THE VARIABLE'S DISTRIBUTION, we can turn our attention to *summarizing* that variable. The whole distribution might be a bit too much information for us to make any use of, especially for continuous variables. So our goal is to pick ways to take the *entire* distribution and produce a few numbers that describe that distribution pretty well.

Probably the most well-known example of a single number that tries to describe an entire distribution is the *mean*. The mean is what you get if you add up all the observations you have and then divide by the number of observations. So if you have 2, 5, 5, and 6, the mean is $(2 + 5 + 5 + 6)/4 = 18/4 = 4.5$.

A little more formally, what the mean does is it takes each value you might get, scales it by *how likely you are to get it,* and then adds it all up. And so on! What does the distribution look like for our data set of 2, 5, 5, and 6? Our frequency table is shown in Table 3.2.

Variable	N	Percent
Observed.Values	4	
... 2	1	25%
... 5	2	50%
... 6	1	25%

Table 3.2: Distribution of a Variable

Table 3.2 gives the distribution of our variable. Now we can calculate the mean. Again, this scales each value by how likely we are to get it. In 2, 5, 5, and 6, we get 5 half the time, so we count *half of five*. We get 2 a quarter of the time, so we count *a quarter of 2*.

Okay, so, since 2 only shows up 25% (1/4) of the time, we only count 1/4 of 2 to get .5. Next, 5 shows up 50% (1/2) of the time, so we count half of 5 and get 2.5. We see 6 shows up 25% (1/4) of the time as well, so we scale 6 by 1/4 and get 1.5. Add it all up to get our mean of $.5 + 2.5 + 1.5 = 4.5$.

So what the mean is *actually doing* is looking at the *distribution* of the variable and summarizing it, boiling it down to a single number. What is that number? The mean is supposed to represent a central tendency of the data—it's in the middle of what you might get. More specifically, it tries to produce a representative *value*. If the variable is "how many dollars this slot machine pays out" with a mean of $4.50, and it costs $4.50 to play, then if you played the slot machine a bunch of times you'd break even exactly.

SOMETIMES IT PAYS TO BE MORE DIRECT. We can certainly use the mean to describe a distribution, and in this book we will, many times. If the goal is to describe the distribution to someone, why bother doing a calculation of the mean when we could just tell people about the distribution itself?

The Xth percentile of a variable's distribution is the value for which X% of the observations are less. So for example, if you lined up 100 people by height, if the person in line with 5 people in front of them is 5 foot 4 inches tall, then 5 foot 4 inches tall is the 5th percentile.

Xth percentile. The value for which X percent of the variable's observations are smaller.

We can see percentiles on our distribution graphs. Figure 3.5 shows our distribution of college cohort earnings from before. We started shading in the left part of the distribution and kept going until we'd shaded in 5% of the area underneath the curve. The point on the *x*-axis where we stopped, $16,400, is the 5th percentile.

We can actually describe the entire distribution perfectly this way. What's the 1st percentile? Okay, now what's the 2nd? And so on.[8] Pretty soon we'll have mapped out the entire distribution by just shading in a little more each time. So percentiles are a fairly direct way of describing a distribution.

There are a few percentiles that deserve special mention.

The first is the *median*, or the 50th percentile. This is the person right in the middle—half the sample is taller than them,

[8] Okay, fine, you can't actually get it *perfectly* unless you also do the 1.00001th percentile and the 1.00002nd and and the infinite percentiles between those two and so on. But you know what I mean.

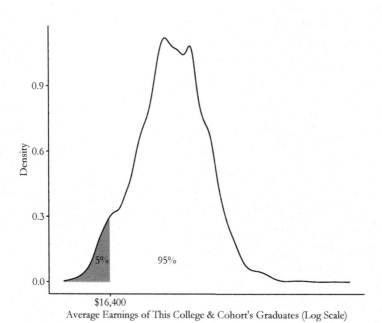

Figure 3.5: Distribution of Average Earnings across US College Cohorts

half the sample is shorter. Like the mean, the median is measuring a central tendency of the data. Instead of trying to produce a representative *value*, like the mean does, the median gives a representative *observation*.

For example, say you're looking at the wealth of 10,000 people, one of whom is Amazon founder Jeff Bezos. The mean says "hmm... sure, most people don't have much wealth, but once in a while you're Jeff Bezos and that makes up for it. The mean is very high." But the median says "Jeff Bezos isn't very representative of the rest of the people. He's going to count exactly the same as everyone else. The median is relatively low."[9]

For this reason, the median is generally used over the mean when you want to describe what a *typical observation* looks like, or when you have a variable that is highly skewed, with a few *really big* observations, like with wealth and Jeff Bezos. The mean wealth of that room might be $15,000,000, but that's almost all Jeff Bezos and doesn't really represent anyone else. As soon as he walks out of the room, the mean drops like a stone. The mean is very sensitive to Jeff! But the median might be closer to $90,000, a fairly typical net worth for an American family,[10] and it would stay pretty much exactly the same if Jeff left the room. The median is great for stuff like this![11]

The other two percentiles to focus on are the *minimum*, or the 0th percentile, and the *maximum*, or the 100th percentile. These are the lowest and highest values the variable takes. They're

[9] So why use the mean at all? It has a lot of nice properties too! For reasons too technical to go into here, it's usually much easier to work with mathematically and pops up in all sorts of statistical calculations beyond just describing a variable. Besides, sometimes you *do* want the representative value, not the representative observation. A place for everything.

[10] Yes, really. Start saving, kids.

[11] Sometimes the mean can work well with skewed data if it's transformed first. A common approach is to take the logarithm of the data, which reins in those really big observations. We'll talk about this more in the Theoretical Distributions section.

handy because they show you the kinds of values that the variable produces. The minimum and maximum height of a large group of people would tell you something about how tall or short humans can possibly be, for example.

Another nice thing about the minimum and maximum is that we can take the difference between them to get the *range*. The range is one way of seeing how much a variable *varies*. If the maximum and minimum are very far apart, as they would be for wealth with Jeff Bezos in the room, you know the variable can take a very wide range of values. If the maximum and minimum are close together, as they might be for "number of eyes you have," you know the range of values that a variable can take is fairly small.

Range. The difference between the maximum value of a variable and the minimum value.

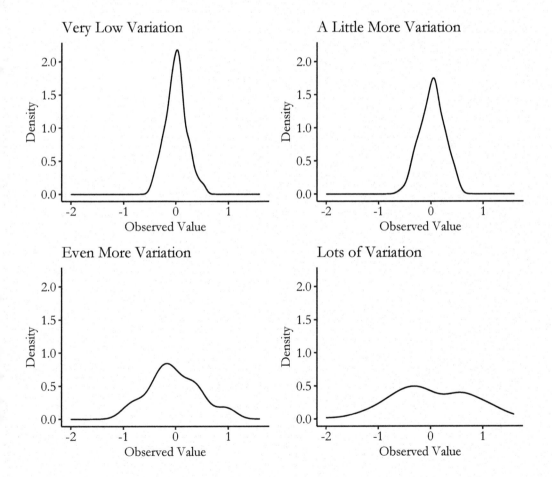

Figure 3.6: Four Variables with Different Levels of Variation

SOME VARIABLES VARY A LITTLE, OTHERS VARY A LOT. For example, take "the number of children a person has." For many people, this is zero. The mean, for people in their thirties perhaps, is somewhere around 2. Some people have lots and lots of children, though. A few rare women have ten or more. There are men in the world with dozens of children. The number of children someone has can vary quite a bit.

Compare that to "the number of eyes a person has." For a small number of people, this might be 0, or 1, or maybe even 3. But the vast majority of people have two eyes. The number of eyes a person has varies fairly little.

These two variables—number of children and number of eyes— have similar means and medians, but they are clearly very different kinds of variables. We need ways to describe *variation* in addition to central tendencies or percentiles.

Variation. How a variable changes in value from observation to observation.

The way that variation shows up in a distribution graph is in how *wide* the distribution is. If the distribution is tall and skinny, then all of the observations are scrunched in very close to the mean. Low variation. If it's flat and wide, then there are a lot of observations in those fat "tails" on the left and right that are far away from the mean. High variation! See Figure 3.6 as an example. The distributions with more area "piled in the middle" have little variation—not a lot of area far from that middle point! The distributions with less "piled in the middle" and more in the "tails" on either side have plenty of observations far away from the middle.

There are quite a few ways to describe variation. Some of them, like the mean, focus on *values*, and others, like the median and percentiles, focus on *observations*.

Variation around the mean. The difference between the value of an observation and the mean.

VARIANCE IS A MEASURE OF VARIATION that focuses on values and is derived from the mean. To calculate the variance in a sample of observations of our data, we:

1. Find the mean. If our data is 2, 5, 5, 6, we get $(2 + 5 + 5 + 6)/4 = 18/4 = 4.5$.

2. Subtract the mean from each value. This turns our 2, 5, 5, 6 into $-2.5, .5, .5, 1.5$. This is our *variation around the mean*.

3. Square each of these values.[12] So now we have 6.25, .25, .25, and 2.25.

4. Add them up! We have $6.25 + .25 + .25 + 2.25 = 9$.

5. Divide by the number of observations minus 1.[13] So our sample variance is $9/(4 - 1) = 3$.

The bigger the variance is, the more variation there is in that variable. How does this work? Well, notice in steps 4 and 5 previously that we're sort of taking a mean. But the thing we're taking the mean *of* is squared variation around the actual mean. So any observations that are far from the mean get squared—making them even bigger and count for more in our mean! In this way we get a sense of how far from the mean our data is, on average.

One downside of the variance is that it's a little hard to interpret, since it is in "squared units." For example, the variance of the college cohort earnings variable is 153,287,962. 153,287,962... dollars squared? I'm not entirely sure what to make of that. So we often convert the variance into the *standard deviation* by taking the square root of it to get us back to our original units. The standard deviation of the college cohort earnings variable is $\sqrt{153,287,962} = 12,380.95$. And that 12,380.95 we can think of in dollars, like the original variable!

If we know that the mean is $33,348.62, and we see a particular college cohort with average earnings of $38,000, we know that that cohort is $(38,000 - 33,348.62)/12,380.95 = .376$, or *37.6% of a standard deviation above the mean.* This lets us know not just how much money that cohort earned ($38,000), and how far above the mean they are ($38,000 − $33,348.62 = $4,651.38), but how unusual that is relative to the amount of variation we typically see (37.6% of a standard deviation).

Figuring out how much variation one standard deviation is can be kind of tricky, and largely just takes practice and intuition. But a graph can help. Figure 3.7 shows how far to the left and right you have to go to find a one standard-deviation distance. It just so happens that 32.7% of the sample is under the curve between the "Mean − 1 SD" line and the "Mean" line, and another 35.5% of the sample is between the "Mean" line and the "Mean + 1 SD" line. In this case, more than 60% of the sample is closer than a single standard deviation.

[12] Why square? Why not something else? We could do something else! The kind of thing I'm calculating here for the variance is called a *moment* of the distribution. The mean is the first moment, and the variance is the second moment (this step squares), telling us about variation. The third moment (this step cubes) tells us about how lopsided the distribution is to one side or the other. The fourth moment (to the fourth power) tells us how much of the distribution is out in the "tails" (in the left and right edges). The third and fourth moments, after being "standardized," are called skewness and kurtosis.

[13] Why minus 1? Well, we estimated that mean already, and in different samples we might get slightly different calculations for the mean, in ways that are related to the variance. That introduces a little bias into the calculation. Specifically, if we divide by the number of observations n, we'll be off by $(n − 1)/n$. So we divide by $1/(n − 1)$ instead of $1/n$, in effect multiplying by $n/(n − 1)$ and getting rid of that bias!

So how weird is being one standard deviation away from the mean? Well, roughly a third of people are between you and the average. Make of that what you will.

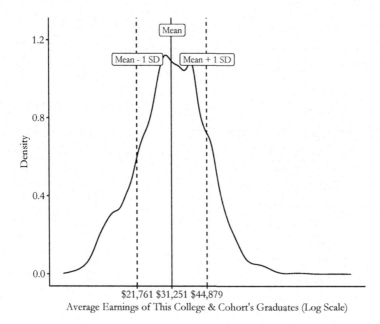

Figure 3.7: Distribution of Average Earnings across US College Cohorts

WE CAN ALSO COMPARE PERCENTILES to see how much a variable varies.

This is actually quite a straightforward process. All we have to do is pick a percentile above the median, and a percentile below the median, and see how different they are. That's it!

We've already discussed the range, which gives the distance between the biggest observation and the smallest. But the range can be very sensitive to really big observations—the range of wealth is very different depending on whether Jeff Bezos is in the room. So it's not a great measure.

Instead, the most common percentile-based measure of variation you'll tend to see is the *interquartile range*, or IQR.[14] This gives the difference between the 75th percentile and the 25th percentile. The IQR is handy for a few reasons. First, you know that the value given by the IQR covers exactly half of your sample. So for the half of your sample closest to the median, the IQR gives you a good sense of *how* close to the median they are. Second, unlike the variance, the IQR isn't very strongly affected by big tail observations. So, as always, it's a good way of representing observations rather than values.

[14] Another one you might see is the "90/10" ratio, or the 90th percentile divided by the 10th percentile. This is a measure commonly used in studies of inequality, to give a sense of *just how different* the top and bottom of the distribution are.

Figure 3.8 shows where the IQR comes from on a distribution. In this case, the 25th and 75th percentiles are at $21,761 and $44,879, respectively, giving us an IQR of $44,879 - 21,761 = 23,118$. So the 50% of the cohorts nearest the median have a range of average incomes of $23,118.

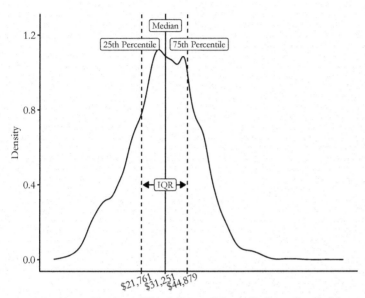

Figure 3.8: Distribution of Average Earnings across US College Cohorts

BEYOND THE VARIATION there are of course a million other things we could describe about a distribution. I will cover only one of them here, and that's the *skew*.

Skew describes how the distribution *leans* to one side or the other. For example, let's talk about annual income. Most people have an income in a relatively narrow range—somewhere between $0 and, say, $150,000. But there are some people—and a fair number of them, actually, who have *enormous* incomes, *way* bigger than $150,000, perhaps in the millions or tens or hundreds of millions.

So for annual income, the *right tail*—the part of the distribution on the right edge—is very big. Figure 3.9 shows what I'm talking about. Most of the weight is down near 0, but there are people with millions of dollars in income making the right tail of the distribution stretch way far out. The same isn't true on the left side—at least in this data, we're not seeing people with negative incomes.

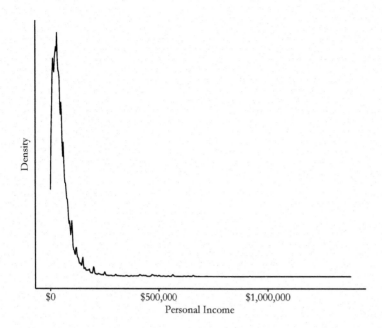

Figure 3.9: Distribution of Personal Income in 2018 American Community Survey

We say that distributions like this one, with a heavy right tail but no big left tail to speak of, has a "right skew" since it has a lot of right-tailed observations. Similarly, a distribution with lots of observations in the left tail would have a left skew. A distribution with similar tails on both sides is *symmetric*.

Right-skewed variables pop up all the time in social science. Basically anything that's unequally distributed, like income, will have a lot of people with relatively little, and a few people with a lot, and the people with a lot have *a lot*.

Skew can be an important feature of a distribution to describe. It can also give us problems if we're working with means and variances, since those really-huge values will affect any measure that tries to represent values.

One way of handling skew in data is by *transforming* the data. If we apply some function to the data that shrinks the impact of those really-big observations, the mean and variance work better. A common transformation in this case is the *log* transformation, where we take the natural logarithm of the data and use that instead. This can make the data much better-behaved.[15]

Figure 3.10 shows that once we take the log of income, there's still a bit of a tail remaining (on the left this time!), but in general we have a roughly symmetric distribution that our mean will work a lot better with.[16]

[15] The "natural" logarithm uses a logarithmic base of e. This is so common in statistics that if you just see "log" without any detail on what the base is, you can assume it's base-e.

[16] How about downsides of a log transformation? It can't handle negative values, or even 0 values, since $\log(0)$ is undefined. There are in fact a lot of 0 incomes in this data that aren't graphed. If we wanted to include them, we couldn't use a log transformation. If you have negative values things get a lot trickier, but if it's just 0s worrying you (and often it is), you might want to try the inverse hyperbolic sine transform (asinh), as that's similar to log for large values but sets 0s to 0.

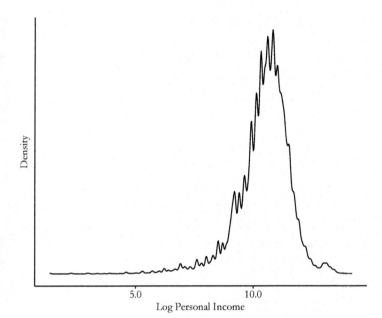

Figure 3.10: Distribution of Logged Personal Income in 2018 American Community Survey

One reason the natural log transformation is so popular is that the transformed variable has an easy interpretation. A log increase of, let's say, .01, translates to *approximately* a .01 × 100 = 1% increase in the original variable. So an increase in log income from 10.00 to 10.02 in log terms means a (10.02 − 10.00) = .02 ≈ 2% increase in income itself. This approximation works pretty well for small increases like .01 or .02, but it starts to break down for bigger increases like, say, .2. Anything above .1/10% or so and you should avoid the approximation.[17]

There might be trouble brewing if you take the log and it *still* looks skewed. This can be the case when you have *fat tails*, i.e., observations far away from the mean are very common. When you have fat tails on one side but not the other, this can make your data very difficult to work with indeed. I'll cover this a little bit all the way at the end of the book in Chapter 22.

[17] For bigger increases, an increase of p in the log is actually equivalent to a $(e^p - 1) \times 100\%$ increase in the variable. Or, a $X\%$ increase in the variable is actually equivalent to a $\log(1 + X)$ log increase. The approximation works because $e^p - 1$ and p are very close together for small values of p.

3.5   Theoretical Distributions

THE DIFFERENCE BETWEEN REALITY AND THE TRUTH is that reality is always with you, but no matter how far you walk, truth is still on the horizon. So it's much more convenient to squint, shrug, go "eh, close enough," and head home.

Statistics makes a very clear distinction between the *truth* and the data we've collected. But isn't the data truly what we've collected? Well, sure, but what it is supposed to *represent* is some broader truth beyond that.

Let's say you want to understand the average age at which children learn to share toys. So you interview 1,000 parents about when their kids started doing that. You calculate a mean and get that kids in your sample start to share easily around 4.2 years old.

Of course, what you *actually* have is that *the 1,000 kids in your sample* started to share easily around 4.2 years old. And you didn't set out to learn something about those 1,000 kids, right? You set out to learn something about kids in general! So the *true* average age at which kids in general start to share is one thing, and the average age you calculated in your data is another.

That's the whole point of doing statistics. We can never check every kid who ever existed on the age they started sharing. So given the real data we actually have, *what can we say about that true number?*

Figuring this out will require us to think about how data behaves under different versions of the truth. If the truth is that kids learn to share on average at 3.8 years of age, what kind of data does that generate? If the truth is that kids learn to share at 5 years of age, what kind of data does that generate? We'll need to pair our observed distributions, the ones we've been talking about so far in this chapter, with *theoretical distributions* of how data behaves under different versions of the truth.

SOME QUICK NOTATION before we get much further.

If you've read any sort of statistics before, you may be familiar with symbols like β, μ, $\hat{\mu}$, $\bar{x}$. You may have memorized what means what. But it turns out you probably don't have to, as there's an order to all this madness. What do these all mean?

English/Latin letters represent *data*. So x might be a variable of actual observed data. That's our 1,000 surveys with parents about their kids' sharing ages.

Modifications of English/Latin letters represent *calculations* done with real data. A common way to indicate "mean" is to use a bar on top of the letter. So $\bar{x}$ is the mean of x we calculated in our data. That's the 4.2 we calculated from our survey.

Greek letters represent *the truth*.[18,19] We don't know what actual values these take, but we can make assumptions. Certain Greek letters are commonly used for certain kinds of truth—μ commonly indicates some sort of mean,[20] σ the standard devi-

Theoretical distribution. A distribution based on theoretical assumptions about the truth, rather than derived purely from data.

[18] So we have a system where English is the down-and-dirty real world and Greek is the lofty and perfect truth. Who designed this?

[19] It's worth pointing out that a Greek letter in a *bad model* might not necessarily be "true." But it is meant to *imply* the truth, even if it doesn't actually get there.

[20] Technically, it represents "expected values" more often than means, but we won't be going much into that in this book.

ation, ρ for correlation, β for regression coefficients, ε for "error terms" (we'll get there), and so on. But the important thing is that Greek letters represent the truth.

Modifications of Greek letters represent *our estimate of the truth.* We don't know what the truth is, but we can make our best guess of it. That guess may be good, or bad, or completely misguided, but it's our guess nonetheless. The most common way to represent "my guess" is to put a "hat" on top of the Greek letter. So $\hat{\mu}$ is "my estimate of what I think μ is." If the way that I plan to estimate μ is by taking the mean of x, then I would say $\hat{\mu} = \bar{x}$.

THE THEORETICAL DISTRIBUTION IS WHAT GENERATED YOUR DATA. That's actually a good way to think about theoretical distributions. They're the distribution of *all* the data, even the data you didn't actually collect, and maybe could never actually collect! If you could collect literally an infinite number of observations, their distribution would be the theoretical distribution.[21]

This fact tells us a few things. First, it tells us why we're interested in the theoretical distribution in the first place. Because that's where we get our data from! If we want to learn about the average age children share at, the place *that data comes from* is the theoretical distribution. So if we want to know the value of that number beyond the data we actually have, we have to use that data to claw our way back to the theoretical distribution. Only then will we know something really interesting!

Remember, we don't really care about the mean in our observed data, $\bar{x}$. We care about the *true average for everyone,* μ! The reason we bother gathering data in the first place is because it will let us make an *estimate* $\hat{\mu}$ about what the theoretical distribution it came from is like.

The second thing this "infinite observations" fact tells us is that the more observations we have, the better a job our observed data will do at matching that theoretical distribution. One observation isn't likely to do us much. But an infinite number would get the theoretical distribution exactly! Somewhere in the middle is going to have to be good enough. And the bigger our number of observations gets, the gooder-enougher we become.

This can be seen in Figure 3.11. No matter how many observations we have, the solid-line theoretical distribution always stays the same, of course. But while we do a pretty bad job at describing that distribution with only ten observations, by the

[21] In statistics this is known as the "limiting distribution" because in the *limit* (i.e., as the number of observations approaches infinity) it's what you get. (Psst, that's another calculus thing. The calculus is coming for you!)

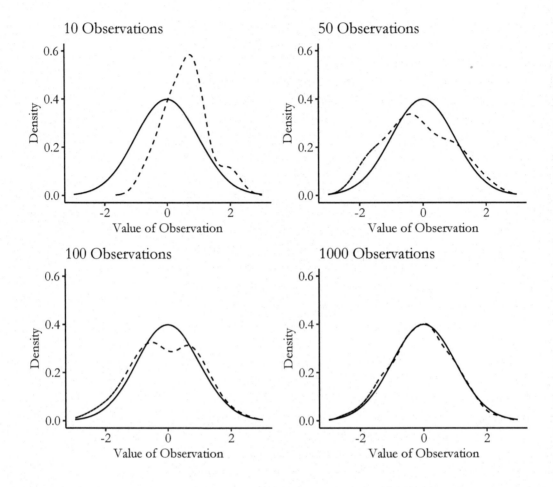

Figure 3.11: Trying to Match the Theoretical Distribution

time we're up to 100 we're doing a lot better. And by 1,000 we've got it pretty good! That's not to say that 1,000 is always "big enough to be just like the theoretical distribution." But here it worked pretty well.

This means as we get more and more observations, we're going to do a better and better job of getting an observed distribution that matches the theoretical one that we sampled the data from. Since that's the distribution we're interested in, that's a good thing! We just need to make sure to have plenty of observations.[22]

THERE ARE INFINITE DIFFERENT THEORETICAL DISTRIBUTIONS, but some pop up in applied work often. There are some well-known distributions that are applied over and over again. If we think that our data follows one of these distributions we're in luck, because it means we can use that theoretical distribution to do a lot of work for us!

[22] Although to be clear, the theoretical distribution we'll get is *the one our data came from*. This may not be the one we're actually interested in! Say our children-sharing data came only from children who go to daycare. Then, the distribution we'll get as we get more and more observations is the distribution *of daycare-going children*. If we're interested in *all* children, we won't get the result for all children no matter how many daycare-going children we survey.

I will cover only two that are especially important to know about in applied social science work, which are both depicted in Figure 3.12. There are many, many more I am leaving out: uniform distribution, Poisson, binomial, gamma, beta, and so on and so on. If you are interested, you may want to check out a more purely statistics-oriented book, like any of the eight zillion books I find when I search "Introduction to Statistics" on Amazon. I'm sure they're all nearly as good as my book. Nearly.

The first to cover is the *normal* distribution. The normal distribution is *symmetric* (i.e., the left and right tails are the same size, there's no skew/lean). The normal distribution often shows up when describing things that face real, physical restrictions, like height or intelligence.

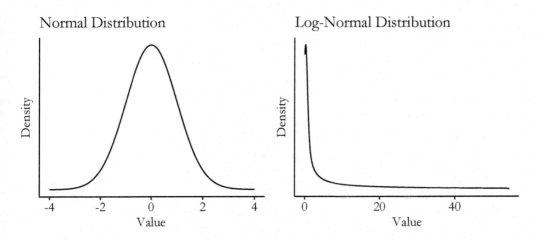

The normal distribution also pops up a lot when looking at aggregated values.[23] Income might have a strong right skew, but if we take the *mean* of income in one sample of data, aggregating across observations, and then again in another sample of data, aggregating across different observations, and then again and again in different samples, *the distribution of the mean across different samples* would have a normal distribution.

The normal distribution technically has *infinite range*, meaning that every value is possible, even if unlikely. This means that any variable for which some values are impossible (like how height can't be negative) technically can't be normal. But if the approximation is very good, we tend to let that slide. One reason we let that slide is that the normal distribution has fairly *thin tails*—observations far from the mean are extremely

Figure 3.12:   Normal and Log-Normal Distributions

[23] This is due to something called the "central limit theorem." Really, read one of those eight zillion Introduction to Statistics textbooks I talked about! Interesting stuff.

unlikely. Notice how quickly the distribution goes to basically 0 in Figure 3.12. So sure, maybe saying that height follows a normal distribution means you're technically saying that negative heights are possible. But you're saying it's possible with a .00000001% chance, so that's close enough to count.

The second is a bit of a cheat, and it's the log-normal distribution. The log-normal distribution has a heavy right skew, but once we take the logarithm of it, it turns out to be a normal distribution! How handy.

The log-normal is a very convenient version of a skewed distribution, since we can take all the skew out of it by just applying a logarithm. Heavily skewed data comes up all the time in the real world. Anything that's unequally distributed and that doesn't have a maximum possible value is generally skewed (income, wealth...) as well as many things that tend to be "winner take all" or have some super-big hits (number of song downloads, number of hours logged in a certain video game, company sizes...). Notice how much of the weight is scrunched over to the left to make room for a very tiny number of really huge observations on the right. That's skew for you!

When we see a skewed distribution, we tend to hope it's log-normal for convenience reasons. However, there are of course many other skewed distributions out there. Skewed distributions with fat tails can be difficult to work with and can take specialized tools. So it's a good idea, after taking the log of a skewed variable, to look at its distribution to confirm that it does indeed look normal.[24] If it doesn't you may be wading into deep waters!

How CAN WE USE our empirical data to learn about the theoretical distribution?

Remember, our real reason for looking at and describing our variables is because we want to get a better idea of the theoretical distribution. We're not really interested in the values of the variable in our sample, we're interested in *using* our sample to find out about how the variable behaves in general.

We can, if we like, take our sample and look at its distribution (as well as its mean, standard deviation, and so on), figure that's the best guess we have as to what the theoretical distribution looks like, and go from there.[25] Of course, we know that's imperfect. Would we really believe that the distribution we happened to get in some data really represents the true theoretical distribution?

[24] One common family of skewed distributions that are difficult to work with are "power distributions," which pop up when you have data where big values are fairly common, like in the stock market, where days with huge upswings and downswings happen with some regularity. These can be tricky to work with—many theoretical power distributions don't even have well-defined means or variances.

[25] Or, if we're using Bayesian statistics, we can take our observed distribution and combine it with our best guess for the theoretical distribution before we collected our data.

One thing that is a bit easier to do is to learn *whether certain theoretical distributions are unlikely.* Maybe we can't figure out exactly what the theoretical distribution *is*, but maybe we can rule some stuff out.

How can we figure out how likely a certain theoretical distribution is? We follow these steps:

1. Choose some description of the theoretical distribution—its mean, its median, its standard deviation, etc. Let's use the mean as an example.

2. Use the properties of the theoretical distribution and your sample size to find the theoretical distribution *of that description in random samples*—means generally follow a normal distribution, and the standard deviation of that normal distribution is smaller for bigger sample sizes.

3. Make that same description of your observed data—so now we have the distribution of our theoretical mean, and we have the actual observed mean.

4. Use the theoretical distribution of that description to find out how unlikely it would be to get the data you got—if the theoretical distribution of the mean we're looking at has mean 1 and standard deviation 2, and our observed mean is 1.5, we're asking "how likely is it that we'd get a 1.5 or more from a normal distribution with mean 1 and standard deviation 2?"

5. If it's really unlikely, then you probably started with the wrong theoretical distribution, and can rule it out. If we're doing statistical significance testing, we might say that our observed mean is "statistically significantly different from" the mean of the theoretical distribution we started with.

Let's walk through an example.

Say we're interested in how many points basketball players make in each game. You collect data on 100 basketball players. Your observed data doesn't look particularly well-behaved and doesn't look like any sort of theoretical distribution you've heard of before. But you calculate its mean and standard deviation and find a mean of 102 with a standard deviation of 30.

Following step 1, you ask "could this have come from a distribution with a mean μ of 90?"—notice we haven't said anything here about it being from a normal distribution, or log-normal, or anything else. We are going to try to rule out distributions with means of 90, that's all. Also notice we called the mean μ, a Greek letter (the truth!). We want to know if we can rule out that $\mu = 90$ is the truth.

Then, for step 2, we want to get the distribution of that description. As mentioned earlier in this chapter, means are generally distributed normally, centered around the theoretical mean itself. The standard deviation of the mean's distribution is just $\sigma/\sqrt{N}$, or the standard deviation of the overall distribution divided by $\sqrt{N}$, where N is the number of observations in the sample.

What's going on here? Well, what this is saying is: if you survey N basketball players and take the mean, and then survey another N basketball players, and then another N, and then another N, and so on, you'll get a different mean each time. If we take the mean from each sample as its own variable, the distribution of that variable will be normal, with a mean of the true theoretical mean, and a standard deviation of $\sigma/\sqrt{N}$. The $/\sqrt{N}$ is because the more players we survey each time, the more likely it is that we'll get very very close to the true mean. A mean of 10 basketball players could give you a result very far from the mean. But a mean of 1,000 basketball players is pretty likely to give you a mean close to the theoretical mean, and thus the smaller standard deviation.

σ is a Greek letter, and so of course that's the true standard deviation, which we don't know. But our best estimate of it is the standard deviation of 30 we got in our data. So we say that the mean is distributed normally with a mean of $\mu = 90$, and a standard deviation of $\hat{\sigma}/\sqrt{N} = 30/\sqrt{100} = 3$ (remember, $\hat{\sigma}$ means "our estimate of the true σ," which is just the standard deviation we got in our observed data).

Now for step 3, we get the same calculation in our observed data. As above, the mean in the observed data is 102.

Moving on to step 4, we can ask "how likely is it to get a 102, or even more, from a normal distribution with mean 90 and standard deviation 3?"[26] More precisely, we are generally interested in how likely it is to get a 102 *or something even farther away*, which could be more or less than μ.[27] So 102 is 12 away from 90, and thus we're interested in how likely it is to get a mean of 102 or more, or 78 or less (since 78 is also 12 away from 90).

By looking at the percentiles of the normal-with-mean-90-and-standard-deviation-3 distribution, we can determine that 78

[26] 3 and not 30? Remember, this is the standard deviation of the sampling distribution of the mean, not the standard deviation of the variable itself.

[27] This is a "two-sided test." A "one-sided test" would just ask how likely it would be to get 102 *or more*.

is not even the 1st percentile, it's more like the .004th percentile. So there's only a .004% chance of getting a mean of 78 or less in a 100-player sample if the true mean is 90 and the standard deviation is 30. Similarly, 102 is the 99.996th percentile, so again there's only a .004% chance of getting a mean of 102 or more in a 100-player sample if the true mean is 90 and the standard deviation is 30.

For step 5, we add those up and say that there's only a .008% chance of getting something as far off as 102 or more if the true mean is 90 and the true standard deviation is 30. That's pretty darn unlikely. So this data very likely did not come from a distribution with a mean of 90.

We could also frame all of this in terms of *hypothesis testing*. Following the same steps, we can say:

Step 1: Our *null hypothesis* is that the mean is 90. Our *alternative hypothesis* is that the mean is not 90.

Step 2: Pick a test statistic with a known distribution. Means are distributed normally, so we might use a Z-statistic, which is for describing points on normal distributions.

Step 3: Get that same test statistic in the data.

Step 4: Using the known distribution of the test statistic, calculate how likely it is to get your data's test statistic, or something even more extreme (*p*-value).

Step 5: Determine whether we can reject the null hypothesis. This comes down to your threshold (α)—how unlikely does your data's test statistic need to be for you to reject the null? A common number is 5%.[28] If that's your threshold, then if Step 4 gave you a lower *p*-value than your α threshold, then that's too unlikely for you, and you can reject the null hypothesis that the mean is 90.

[28] The choice of 5% is completely arbitrary and yet widely applied. Like your keyboard's QWERTY layout.

This section describes what hypothesis testing actually is and what it's for. Statistical significance is not a marker of being *correct* or *important*. It's just a marker of being able to *reject some theoretical distribution you've chosen*. That can certainly be interesting. At the very least, we've narrowed down the likely list of ways that our data could have been generated. And that's the whole point!

4
Describing Relationships

4.1 What Is a Relationship?

FOR MOST RESEARCH QUESTIONS, we are not just interested in the distribution of a single variable.[1] Instead, we are interested in the *relationship* we see in the data between two or more variables.

What does it mean for two variables to have a relationship? The relationship between two variables shows you *what learning about one variable tells you about the other*.

For example, take height and age among children. Generally, the older a child is, the taller they are. So, learning that one child is thirteen and another is six will give you a pretty good guess as to which of the two children is taller.

We can call the relationship between height and age *positive*, meaning that for higher values of one of the variables, we expect to see higher values of the other, too (more age is associated with more height). There are also negative relationships, where higher values of one tend to go along with lower values of the other (more age is associated with less crying). There are also null relationships where the variables have nothing to

[1] Get lost, Chapter 3, nobody likes you.

DOI: 10.1201/9781003226055-4

do with each other (older children aren't any more or less likely to live in France than younger children). All kinds of other relationships are positive sometimes and negative other times, or *really* positive at first and then only slightly positive later. Or perhaps one of the variables is categorical and there's not really a "higher" or "lower," just "different" (older children are more likely to use a bike for transportation than younger children). Lots of options here.

THE GOAL IN THIS CHAPTER is to figure out how to describe the relationship between two variables, so that we can accurately relay what we see in the data about our research question, which, once again, very likely has to do with the relationship between two variables. Once we know how to describe the relationship we see *in the data*, we can work in the rest of the book to make sure that the relationship we've described does indeed answer our research question.

Throughout this chapter, we're going to use some example data from a study by Emily Oster,[2] who used the National Health and Nutrition Examination Survey. Her research question was: do the health benefits of recommended medications look better than they actually are because already-otherwise-healthy people are more likely to follow the recommendations?

To study this question, she looked at vitamin E supplements, which were only recommended for a brief period of time. She then answers her research question by examining the relationship between taking vitamin E, other indicators of caring about your health like not smoking, and outcomes like mortality, and how those relationships change before, during, and after the time vitamin E was recommended.[3]

We can start off with an example of a very straightforward way of showing the relationship between two continuous variables, which is a scatterplot, as shown in Figure 4.1. Scatterplots simply show you every data point there is to see. They can be handy for getting a good look at the data and trying to visualize from them what kind of relationship the two variables have. Does the data tend to slope up? Does it slope down a lot? Or slope down just a little like in Figure 4.1? Or go up and down?

A scatterplot is a basic way to show *all* the information about a relationship between two continuous variables, like the density plots were for a single continuous variable in Chapter 3.[4] And they're usually a great place to start describing a relationship.

[2] Emily Oster. Health recommendations and selection in health behaviors. *American Economic Review: Insights*, 2(2):143–60, 2020b.

Scatterplot. A graph that plots every data point for an *x*-axis variable and a *y*-axis variable.

[3] In this chapter, I'll add some analyses that weren't exactly in the original study but are in the same spirit, wherever it helps explain how to describe relationships. It's almost like she had other purposes for her study besides providing good examples for my textbook. Rude if you ask me.

[4] Unlike density plots, though, they tend to get very hard to read if you have a lot of data. That's why I only used 150 observations for that graph, not all of them.

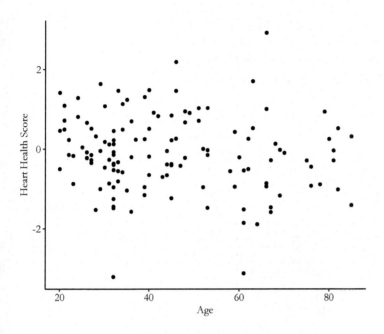

Figure 4.1: Age and Heart Health, 150 Observations

Scatterplots imply two things beyond what they actually show. One is bad, and one is good. The bad one is that it's very tempting to look at a relationship in a scatterplot and assume that it means that the x-axis causes the y-axis. Even if we know that's not true, it's very tempting. The good one is that it encourages us to use the scatterplot to imagine other ways of describing the relationship that might give us the information we want in a more digestible way. That's what the rest of this chapter is about.

4.2   Conditional Distributions

CHAPTER 3 WAS ALL ABOUT DESCRIBING the distributions of variables. However, the distributions in those chapters were what are called *unconditional* distributions.[5]

A *conditional* distribution is the distribution of one variable *given the value of another variable.*

Let's start with a more basic version—conditional probability. The probability that someone is a woman is roughly 50%. But the probability that someone *who is named Sarah* is a woman is much higher than 50%. You can also say "*among all Sarahs,* what proportion are women?" We would say that this is the "probability that someone is a woman conditional on being named Sarah."

[5] These are also called "marginal" distributions, but I really dislike this term, as I think it sounds like the opposite of what it means. **Conditional distribution.** The distribution of a variable conditional on another variable taking a certain value.

Learning that someone is named Sarah changes the probability that we can place on them being a woman. Conditional distributions work the same way, except that this time, instead of just a single probability changing, an entire distribution changes.

Take Figure 4.2 for example. In this graph, we look at the distribution of how much vitamin E someone takes, among people who take any. We then split out the distribution by whether someone has engaged in vigorous exercise in the last month.

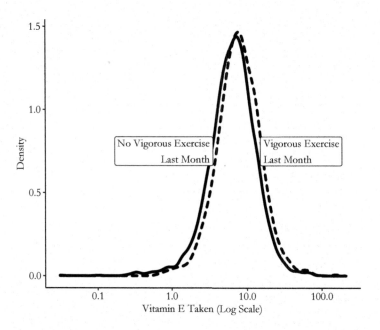

Figure 4.2: Distribution of Amount of vitamin E Taken by Exercise Level

We can see a small deviation in the distribution for those who exercise and those who don't.[6] In particular, those who exercise vigorously take larger doses of vitamin E when they take it. The distribution is different between exercisers and non-exercisers, telling us that vitamin E and exercise are *related* to each other in this data.

THE EXAMPLE I'VE GIVEN is for a continuous variable, but it works just as well for a categorical variable. Instead of looking at how large the doses are, let's look at whether someone takes vitamin E at all! Oster's hypothesis is that people who take vitamin E at all should be more likely to do other healthy things like exercise, because both are driven by how health-conscious you are.

Figure 4.3 shows an example of this. The distribution of whether you take vitamin E or not is shown twice here, once for those who currently smoke, and one for those who don't smoke.

[6] It doesn't look enormous, but this is actually how a lot of fairly prominent differences look in the social sciences. That rightward shift can be deceptively larger than it looks!

The distributions are clearly different, with a higher proportion taking vitamin E in the non-smoking crowd, exactly what Oster would expect.

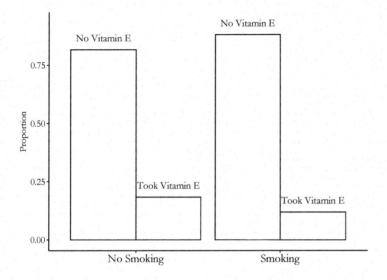

Figure 4.3: Distribution of Whether vitamin E is Taken by Whether you Smoke

4.3   Conditional Means

WITH THE CONCEPT OF A CONDITIONAL DISTRIBUTION UNDER OUR BELT, it should be clear that we can then calculate *any* feature of that distribution conditional on the value of another variable. What's the 95th percentile of vitamin E taking overall and for smokers? What's the median? What's the standard deviation of mortality for people who take 90th-percentile levels of vitamin E, and for people who take 10th-percentile levels?

While all those possibilities remain floating in the air, we will focus on the conditional mean. Given a certain value of X, what do I expect the mean of Y to be?[7]

Once we have the conditional mean, we can describe the relationship between the two variables fairly well. If the mean of Y is higher conditional on a higher value of X, then Y and X are positively related. Going further, we can map out all the conditional means of Y for each value of X, giving us the full picture on how the mean of one variable is related to the values of the other.

IN SOME CASES, THIS IS EASY TO CALCULATE. If the variable you are conditioning on is discrete (or categorical), you can just calculate the mean for all observations with that value.

Conditional mean. The mean of one variable given that another variable takes a certain value.

[7] Why the mean? One reason is that the mean behaves a bit better in small samples, and once we start looking at things separately by specific values of X, samples get small. Another reason is that it helps us weight prediction errors and and so figure out how to minimize those errors. It's just handy.

See Figure 4.4, for example, which shows the proportion taking vitamin E conditional on whether the observations are from before vitamin E was recommended, during recommendation, or after.[8] I just took all the observations in the data from before the recommendation and calculated the proportion who took vitamin E. Then I did the same for the data during the recommendation, and after the recommendation.

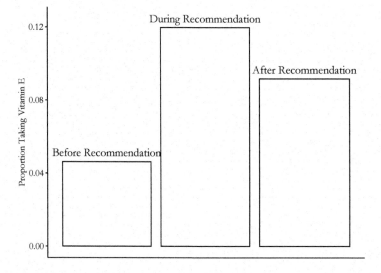

Figure 4.4: Proportion Taking Vitamin E Before It Was Recommended, During, and After

Figure 4.4 shows the relationship between the taking of vitamin E and the timing of the recommendation. We can see that the relationship between the taking of vitamin E and the recommendation being in place is positive (the proportion taking vitamin E is higher during the recommendation time). We also see that the relationship between vitamin E and *time* is at first positive (increasing as the recommendation goes into effect) and then negative (decreasing as the recommendation is removed).

THINGS GET A LITTLE MORE COMPLEX when you are conditioning on a continuous variable. After all, I can't give you the proportion taking vitamin E among those making \$84,325 per year because there's unlikely to be more than one person with that exact number. For lots of numbers we'd have no data at all.

There are two approaches we can take here. One approach is to use a *range* of values for the variable we're conditioning on rather than a single value. Another is to use some sort of shape or line to fill in those gaps with no observations.

Let's focus first on using a range of values. Table 4.1 shows the proportion of people taking vitamin E conditional on body mass index (BMI). Since BMI is continuous, I've cut it up into ten equally-sized ranges (bins) and calculated the proportion taking vitamin E within each of those ranges. Cutting the data into bins to take a conditional mean isn't actually done that often in real research, but it gives a good intuitive sense of what we're trying to do when we use other methods later.

BMI Bin	Proportion Taking Vitamin E
(11.6,20.6]	0.133
(20.6,29.5]	0.159
(29.5,38.4]	0.171
(38.4,47.3]	0.178
(47.3,56.2]	0.203
(56.2,65.1]	0.243
(65.1,74]	0.067
(74,83]	0.143

Table 4.1: Proportion Taking Vitamin E by Range of Body Mass Index Values

Those same ranges can be graphed, as in Figure 4.5. The flat lines reflect that we are assigning the same mean to every observation in that range of BMI values. They show the mean conditional on being in that BMI bin. We see from this that BMI has a positive relationship with taking vitamin E up until the 70+ ranges, at which point the conditional mean drops.

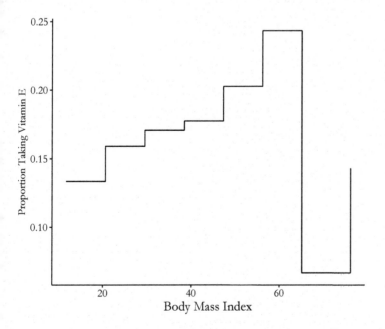

Figure 4.5: Proportion Taking Vitamin E by Range of Body Mass Index Values

OF COURSE, WHILE THIS APPROACH IS SIMPLE AND ILLUS-TRATIVE, IT'S ALSO FAIRLY ARBITRARY. I picked the use of ten bins (as opposed to nine, or eleven, or...) out of nowhere. It's also arbitrary to use evenly-sized bins; no real reason I had to do that. Plus, it's rather choppy. Do I really think that if someone is at the very top end of their bin, they're more like someone at the bottom of their bin than like the person at the very bottom end of the next bin?

Instead, we can use a range of X values to get conditional means of Y using *local means*. That is, to calculate the conditional mean of Y at a value of, say, $X = 2.5$, we take the mean of Y for all observations with X values *near* 2.5. There are different choices to make here—how close do you have to be? Do we count you equally if you're *very* close vs. *kind of* close?

Local mean. The mean of a variable Y calculated using only observations over a short range of another variable X.

A common way to do this kind of thing is with a LOESS curve,[9] also known as LOWESS.[10] LOESS provides a local prediction, which it gets by fitting a different shape for each value on the X axis, with the estimation of that shape weighting very-close observations more than kind-of close observations. The end result is nice and smooth.

[9] "Locally Estimated Scatterplot Smoothing"

[10] Depending on who you ask, LOESS and LOWESS might be the exact same thing, or might have slight differences in how they estimate their local prediction, with either name referring to either of the local-prediction variants.

Figure 4.6 shows the LOESS curve for the proportion taking vitamin E and BMI.

Figure 4.6: Proportion Taking Vitamin E by BMI with a LOESS Curve

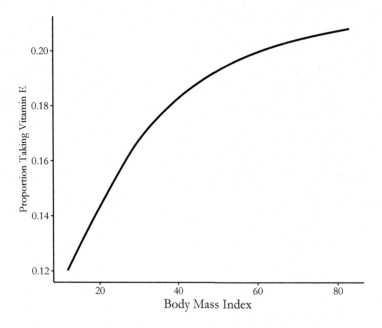

From Figure 4.6 we can see a clear relationship, with higher values of BMI being associated with more people taking vitamin E. The relationship is very strong at first, but then flattens out a bit, although it remains positive.[11] It got there by just

LOESS. A curve that uses local averages to smooth out the relationship between two variables.

[11] Why doesn't this dip down at the end like Figure 4.5? There are very, very few observations in those really-high BMI bins. LOESS doesn't let that tiny number of observations pull it way down, and so sort of ignores them in a way that Figure 4.5 doesn't.

calculating the proportion of people taking vitamin E among those who have BMIs in a certain range, with "a certain range" moving along to the right only a bit at a time while it constructed its conditional means.

4.4   Line-fitting

SHOWING THE MEAN OF Y AMONG LOCAL VALUES OF X IS VALUABLE, and can produce a highly detailed picture of the relationship between X and Y. But it also has limitations. There still might be gaps in your data it has trouble filling in, for one. Also, it can be hard sometimes to concisely describe the relationship you see.[12]

Enter the concept of *line-fitting,* also known as *regression.*[13]

Instead of thinking locally and producing estimates of the mean of Y conditional on values of X, we can assume that the underlying relationship between Y and X can be represented by some sort of *shape*. In basic forms of regression, that shape is a straight line. For example, the line

$$Y = 3 + 4X \qquad (4.1)$$

tells us that the mean of Y conditional on, say, $X = 5$ is $3 + 4(5) = 23$. It also tells us that the mean of Y conditional on a given value of X would be 4 higher if you instead made it conditional on a value of X one unit higher.

In Figure 4.7, we repeat the vitamin E/BMI relationship from before but now have a straight line fit to it. That particular straight line has a slope of .002, telling us that you are .2 percentage points more likely to take a vitamin E supplement than someone with a BMI one unit lower than you.

This approach has some real benefits. For one, it gives us the conditional mean of Y for *any* value of X we can think of, even if we don't have data for that specific value.[14] Also, it lets us very cleanly describe the relationship between Y and X. If the slope coefficient on X (.002 in the vitamin E/BMI regression) is positive, then X and Y are positively related. If it's negative, they're negatively related.

Those are pragmatic upsides for using a fitted line. There are more upsides in statistical terms in using a line-fitting procedure to estimate the relationship. Since the line is estimated using *all* the data, rather than just local data, the results are more precise. Also, the line can be easily extended to include more than one variable (more on that in the next section).

[12] Not to mention, it can be difficult, although certainly not impossible, to do what we do in the Conditional Conditional Means section with those methods.

[13] These two concepts are not the exact same thing, really. But they're close enough in most applications. Also, while I repeatedly mention conditional means in this section, there are versions of line-fitting that give conditional medians or percentiles or what-have-you as well.

Regression. The practice of fitting a shape, usually a line, to describe the relationship between two variables.

[14] Although if we don't have data anywhere near that value, we probably shouldn't be trying to get the conditional mean there.

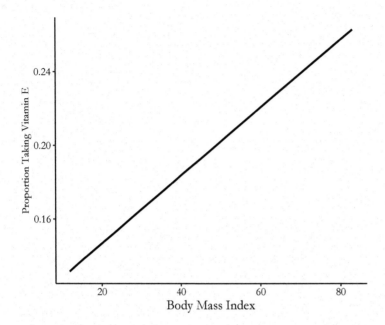

Figure 4.7: Proportion Taking Vitamin E by BMI with a Fitted Straight Line

There is a downside as well, of course. The biggest downside is that fitting a line requires us to *fit a line*. We need to pick what kind of shape the relationship is—a straight line? A curved line? A line that wobbles up and down and up and down? The line-fitting procedure will pick the best version of the shape we give it. But if the shape is all wrong to start with, our estimate of the conditional mean will be all wrong. Imagine trying to describe the relationship in Figure 4.6 using a straight line!

The weakness here isn't necessarily that straight lines aren't always correct—line-fitting procedures will let us use curvy lines. But we have to be aware ahead of time that a curvy line is the right thing to use, and then pick which kind of curvy line it is ahead of time.

That weakness is, naturally, set against the positives, which are strong enough that line-fitting is an extremely common practice across all applied statistical fields. So, then, how do we do it?

ORDINARY LEAST SQUARES (OLS) IS THE MOST WELL-KNOWN APPLICATION OF LINE-FITTING. OLS picks the line that gives the lowest *sum of squared residuals*. A residual is the difference between an observation's actual value and the conditional mean assigned by the line.[15]

Take that $Y = 3+4X$ line I described earlier. We determined that the conditional mean of Y when $X = 5$ was $3 + 4(5) = 23$. But what if we see someone in the data with $X = 5$ and $Y = 25$? Well then their *residual* is $25 - 23 = 2$. OLS takes that number,

Ordinary least squares. A regression method that uses a straight line and minimizes the sum of squared residuals.

[15] Or if you prefer, the difference between the actual value and the prediction.

squares it into a 4, then adds up all the predictions across all your data. Then it picks the values of β_0 and β_1 in the line $Y = \beta_0 + \beta_1 X$ that make that sum of squared residuals as small as possible, as in Figure 4.8.

Let's fit a line to four points
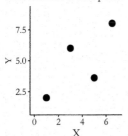

Add the OLS line
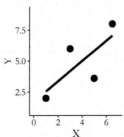

Figure 4.8: Fitting an OLS Line to Four Points

Residuals are from point to line
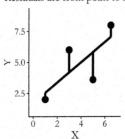

Goal: minimize squared residuals
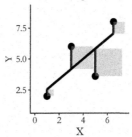

How does it do this?[16] It takes advantage of information about how the two variables move together or apart, encoded in the *covariance*.

If you recall the variance from Chapter 3, you'll remember that to calculate the variance of X, we: (a) subtracted the mean of X from X, (b) squared the result, (c) added up the result across all the observations, and (d) divided by the sample size minus one. The resulting variance shows how much a variable actually varies.

The covariance is the exact same thing, except that in step (a) you subtract the mean from *two* separate variables, and in step (b) you multiply the result from one variable by the result from the other. The resulting covariance shows how much two variables move together or apart. If they tend to be above average at the same time or below average at the same time, then multiplying one by the other will produce a positive result for most observations, increasing the covariance. If they have nothing to do with each other, then multiplying one by the other will give a positive result about half the time and a negative

[16] Calculus, for one. But besides that.
Covariance. A measurement of how much two variables vary with each other, as opposed to how much a single variable varies as in the variance. Technically, the average of the summed products of de-meaned variables.

result the other half, canceling out in step (c) and give you a covariance of 0.

How does OLS use covariance to get the relationship between Y and X? It just takes the covariance and divides it by the variance of X, i.e., $cov(X,Y)/var(X)$. That's it![17] This is roughly saying "of all the variation in X, how much of it varies along with Y?"[18] Then, once it has its slope, it picks an intercept for the line that makes the mean of the residuals (not the squared residuals) 0, i.e., the conditional mean is at least right on average.

The result from OLS is then a line with an intercept and a slope like $Y = 3 + 4X$. You can plug in a value of X to get the conditional mean of Y. And, crucially, you can describe the relationship between the variables using the slope. Since the line has $4X$ in it, we can say that a one-unit increase in X is associated with a four-unit increase in Y.

Sometimes we may find it useful to rescale the OLS result. This brings us to the concept of *correlation*. Correlation, specifically Pearson's correlation coefficient, takes this exact concept and just rescales it, multiplying the OLS slope by the standard deviation of X and dividing it by the standard deviation of Y. This is the same as taking the covariance between X and Y and dividing by both the standard deviation of X and the standard deviation of Y.

The correlation coefficient also relies on this concept of fitting a straight line. It just reports the result a little differently. We lose the ability to interpret the slope in terms of the units of X and Y.[19] However, we gain the ability to more easily tell how strong the relationship is. The correlation coefficient can only range from -1 to 1, and the interpretation is the same no matter what units the original variables were in. The closer to -1 it is, the more strongly the variables move in opposite directions (downward slope). The closer to 1 it is, the more strongly the variables move in the same direction (upward slope).

How about for vitamin E and BMI? OLS estimates the line

$$VitaminE = \beta_0 + \beta_1 BMI \qquad (4.2)$$

and selects the best-fit values of β_1 and β_2 to give us

$$VitaminE = .110 + .002BMI \qquad (4.3)$$

So for a one-unit increase in BMI we'd expect a .002 increase in the conditional mean of vitamin E. Since vitamin E is a binary

[17] For the two-variable version. We'll get to more complex ones in a bit.

[18] The sheer intuitive nature of this calculation might give a clue as to why we focus on minimizing the sum of squared residuals rather than, say, the residuals to the fourth power, or the product, or the sum of the absolute values. OLS gets some flak in some statistical circles for being restrictive, or for some of its assumptions. But the way that it seems to pop up everywhere and be linked to everything—it's the π of multivariate statistical methods, if you ask me. I could write a whole extra chapter just on cool stuff going on under the hood of OLS. Look at me, starstruck over a ratio.

[19] Why? Well, the slope of a straight line tells you the change in units-of-Y-per-units-of-X. You can read that "per" as "divided by." When we multiply by the standard deviation of X, that's in units of X, so the units cancel out with the per-units-of-X, leaving us with just units-of-Y. Then when we divide by the standard deviation of Y, that's in units of Y, canceling out with units-of-Y and leaving us without any units. **Correlation.** A measurement of how two variables vary linearly together or apart, scaled to be between -1 and 1.

variable, we can think of a .002 increase in conditional mean as being a .2 percentage point increase in the proportion of people taking vitamin E.

Then, since the standard deviation of taking vitamin E is .369 and the standard deviation of BMI is 6.543, the Pearson correlation between the two is $.002 \times 6.543/.369 = .355$.

SOMETIMES BEING STRAIGHT IS INSUFFICIENT. OLS fits a straight line, but many sets of variables do not have a straight-line relationship! In fact, as shown in Figure 4.6, our vitamin E/BMI relationship is one of them. What to do?

Two heroes come to our rescue.[20]

The first of them is apparently also the villain, OLS. Turns out OLS doesn't actually have to fit a *straight* line. Haha, gotcha. It just needs to fit a line that is "linear in the coefficients," meaning that the slope coefficients don't have to do anything wilder than just being multiplied by a variable.

Asking it to estimate the β values in $Y = \beta_0 + \beta_1 X$ is fine, as before. But so is $Y = \beta_0 + \beta_1 X + \beta_2 X^2$—not a straight line! Or $Y = \beta_0 + \beta_1 \ln(X)$—also not a straight line! And so on. What would be something that's *not* linear in coefficients? That would be something like $Y = \beta_0 + X_1^\beta$ or $Y = \frac{\beta_0}{1+\beta_1 X}$.

So that scary-looking curved line in Figure 4.6? Not a problem, actually. As long as we take a look at our data beforehand to see what kind of shape makes sense (do we need a squared term for a parabola? Do we need a log term to rise quickly and then level out?), we can mimic that shape. For Figure 4.6 we could probably do with $Y = \beta_0 + \beta_1 X + \beta_2 X^2$ to get the nice flexibility of the LOESS with the OLS bonuses of having fit a shape.

The second hero is "nonlinear regression" which can take many, many forms. Often it is of the form $Y = F(\beta_0 + \beta_1 X)$ where $F()$ is... some function, depending on what you're doing.

Nonlinear regression is commonly used when Y can only take a limited number of values. For example, we've been using all kinds of line-fitting approaches for the relationship between vitamin E and BMI, but vitamin E is *binary*—you take it or you don't. A straight line like OLS will give us something that doesn't really represent the true relationship—straight lines increase gradually, but something binary jumps from "no" to "yes" all at once! Even a line that obeys the curve like $VitaminE = \beta_0 + \beta_1 BMI + \beta_2 BMI^2$ will be a bit misleading. Even if we think about the dependent variable as the *probability* of taking vitamin E, which *can* change gradually like

Regression slope coefficient. The linear relationship between two variables, estimated by regression. A one-unit change in one variable is associated with a (coefficient)-unit change in the other.

[20] These are two heroes that will not really receive the attention necessary in this book, which in general covers regression just enough to get to the research design. See a little more in Chapter 13, or check out a more dedicated book on regression like Bailey's *Real Econometrics*.

a straight line, follow that line out far enough and eventually you'll predict that people with really high BMIs are more than 100% likely to use vitamin E, and people with really low BMIs are less than 0% likely. Uh-oh.

You can solve this by using an $F()$ that doesn't go above 100% or below 0%, like a "probit" or "logit" function. I'll cover more on these in Chapter 13.

There are many other functions you could use, of course, for all kinds of different Y variables and the values they can take. I won't be spending much time on them in this book, but do be aware that they're out there, and they represent another important way of fitting a (non-straight) line.

4.5   Conditional Conditional Means, a.k.a. "Controlling for a Variable"

LET US ENTER THE LAND OF THE UNEXPLAINED. By which I mean the residuals.

When you get the mean of Y conditional on X, no matter how you actually do it, you're splitting each observation into two parts—the part *explained by X* (the conditional mean), and the part *not explained by X* (the residual). If the mean of Y conditional on $X = 5$ is 10, and we get an observation with $X = 5$ and $Y = 13$, then the prediction is 10 and the residual is $13 - 10$. Figure 4.9 shows how we can distinguish the conditional mean from the residual.

Residual.   The difference between the actual and predicted values of an observation.

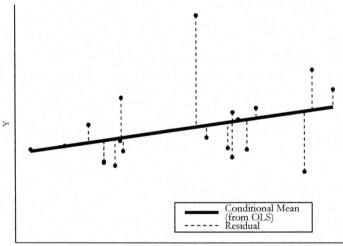

Figure 4.9:   An OLS Line and its Residuals

It might seem like those residuals are just little nuisances or failures, the parts we couldn't predict. But it turns out there's a little magic in there. Because we can also think of the residual as *the part of Y that has nothing to do with X*. After all, if the conditional mean is 10 and the actual value is 13, then X can only be responsible for the 10. The extra 3 must be because of some other part of the data generating process.

Why would we want that? It turns out there are a number of uses for the residual. Just off the bat, perhaps we don't just want to know the variation in vitamin E alone. Maybe what we want is to know how much variation there is in vitamin E-taking that *isn't explained by BMI*. Looking at the residuals from Figure 4.7 would answer exactly that question.

Things get real interesting when we look at the residuals of two variables at once.

WHAT IF WE TAKE THE EXPLAINED PART OUT OF TWO DIFFERENT VARIABLES? Let's expand our analysis to include a third variable. Let's keep it simple with Y, X, and Z. So, what do we do?[21]

1. Get the mean of Y conditional on Z.

2. Subtract out that conditional mean to get the residual of Y. Call this Y^R.

3. Get the mean of X conditional on Z.

4. Subtract out that conditional mean to get the residual of X. Call this X^R.

5. Describe the relationship between Y^R and X^R.

Now, since Y^R and X^R have had the parts of Y and X that can be explained with Z *removed*, the relationship we see between Y^R and X^R is *the part of the relationship between Y and X that is not explained by Z*.

In other words, were getting the *Mean of Y conditional on X* all conditional on Z. We're *washing out the part of the X/Y relationship that is explained by Z*.

In doing this, we are taking out all the variation related to Z, in effect not allowing Z to vary. This is why we call this process "controlling for" Z (although "adjusting for" Z might be a little more accurate).

Let's take our ice cream and shorts example. We see that days where more people eat ice cream also tend to be days where more

[21] This particular set of calculations, when applied to linear regression, is known as the Frisch-Waugh-Lovell theorem and doesn't apply precisely to regression approaches that are nonlinear in parameters, like logit or probit as previously described. However, for those regressions the concept is still the same.

Controlling for a variable. Removing all the variation associated with that variable from all the other variables.

people wear shorts. But we also know that the temperature outside affects both of these things.

If we really want to know if ice cream-eating affects shorts-wearing, we would want to know *how much of a relationship is there between ice cream and shorts that isn't explained by temperature?* So we would get the mean of ice cream conditional on temperature, and then take the residual, getting only the variation in ice cream that has nothing to do with temperature. Then we would take the mean of shorts-wearing conditional on temperature, and take the residual, getting only the variation in shorts-wearing that has nothing to do with temperature. Finally, we get the mean of the shorts-wearing residual conditional on the ice cream residual. If the shorts mean doesn't change much conditional on different values of ice cream eating, then the entire relationship was just explained by heat! If there's still a strong relationship there, maybe we do have something.

THE EASIEST WAY TO TAKE CONDITIONAL CONDITIONAL MEANS IS WITH REGRESSION. Regression allows us to control for a variable by simply adding it to the equation. Now we have "multivariate" regression. So instead of

$$Y = \beta_0 + \beta_1 X \tag{4.4}$$

we just use

$$Y = \beta_0 + \beta_1 X + \beta_2 Z \tag{4.5}$$

and voila, the OLS estimate for β_1 will automatically go through the steps of removing the conditional means and analyzing the relationship between Y^R and X^R.

Even better, we can do things conditional on *more than one variable.* So we could add W and do...

$$Y = \beta_0 + \beta_1 X + \beta_2 Z + \beta_3 W \tag{4.6}$$

and now the β_1 that OLS picks will give us the relationship between Y and X conditional on *both* Z and W.

Let's take a quick look at how this might affect our vitamin E/BMI relationship. Some variables that might be related to both taking vitamin E and to BMI are gender and age. So let's add those two variables to our regression and see what we get.

Before, with only BMI, we estimated

$$VitaminE = .110 + .002BMI \tag{4.7}$$

Now, with BMI, gender, and age, we get

$$VitaminE = -.006 + .001BMI + .002Age + .016Female \quad (4.8)$$

The effect of BMI has changed a bit, from .002 to .001, telling us that some of the relationship we saw between BMI and vitamin E was explained by age and/or gender. We also see that older people are more likely to take vitamin E—for each additional year of age we expect the proportion taking vitamin E to go up by .2 percentage points. Women are also more likely than men to take the supplement. A one-unit increase in "Female" (i.e., going from 0—a man—to 1—a woman) is associated with an increased proportion taking vitamin E of 1.6 percentage points.[22]

So how does regression do this? Put your mental-visualization glasses on.

One way is mathematically. If you happen to know a little linear algebra (and if you don't, you can skip straight to the next paragraph), the formula for multivariate OLS is $(A'A)^{-1}A'Y$, where A is a matrix of all the variables other than Y, including the X we're interested in. In other words, it washes out the influence of all the non-X variables on the X/Y relationship by dividing out a bunch of covariances.

Another way is graphically. If you can think of a two-variable OLS line $Y = \beta_0 + \beta_1 X$ as being a line, you can think of a three-variable OLS line as a *plane* in 3-D space (or with four variables, in 4-D space, and so on). We can visualize this by looking at each of the three sides of that 3-D image one at a time.

Figure 4.10 shows the X-Y axis on the top-left. Then to the right you can see the Z-Y axis, and below the Z-X axis. The coordinates are flipped on the Z-X axis—even though we're getting the mean of X conditional on Z here, I've put X on the x-axis to be consistent with the X-Y graph. The upward slope on the Z-Y and Z-X axes shows that Z is explaining part of both X and Y, and that we could take that explanation out to focus on the residuals.

Then, in Figure 4.11, we flatten out those explanations. The upward slopes get flattened out, moving the X and Y points with them. You can see how subtracting out the parts explained by Z literally leaves the X/Y relationship no part of Z to hold on to! Z has been flatlined in both directions, providing no additional "lift" to the points in the X/Y graph. What's still there on X/Y is there without Z.

[22] Of course, OLS by itself doesn't know *which* variable is the treatment. So the Age effect "controls for" BMI and Female, and the Female effect controls for BMI and Age. However, we don't want to get too wrapped up in interpreting the coefficients on controls, as we generally haven't put much work into identifying their effects (see Chapter 5).

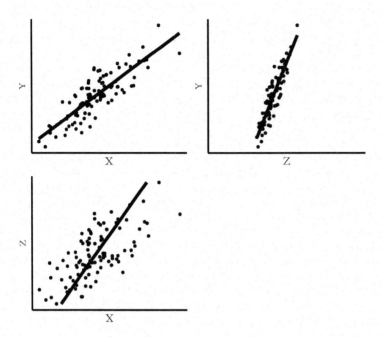

Figure 4.10: A Three-Variable Regression from All Three Dimensions

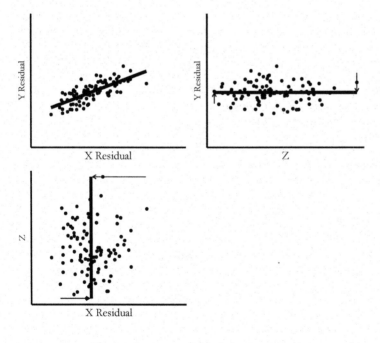

Figure 4.11: A Three-Variable Regression from All Three Dimensions After Removing the Variation Explained by Z

4.6   What We're Not Covering

In the previous chapter, on describing variables, we did a pretty good job covering a lot of what you'd want to know when describing a variable. This chapter, however, leaves out a whole lot more.

This is largely for reasons of focus. This book is about research design. Once you've got research design pinned down, there are certainly a lot of statistical issues you need to deal with at that point. But things like specific probability distributions (normal vs. log-normal vs. t vs. Poisson vs. a million others we didn't cover), functional form (OLS vs. probit/logit vs. many others), or standard errors and hypothesis testing can be a distraction when thinking about the broad strokes of how you're going to answer your research question.

In one case, omission is less for focus and more to cover it more appropriately later. Notice how I introduced the Oster paper as being all about how the relationship between vitamin E and health indicators changed over time... but then I never showed how the BMI relationship changed over time? There are a number of research designs that have to do with *how a relationship changes* in different settings.[23] However, a proper treatment of this will have to wait until Part II of the book.

[23] Controlling for time would not achieve this. Controlling for time would remove the part of the relationship explained by time, but would not show how the relationship changes over time.

To be clear, you want to know all this stuff. And I will cover it more in this book in Chapter 13, and many of the other Part II chapters. You can also check out a more traditional econometrics book like Bailey's *Real Econometrics* or Wooldridge's *Introductory Econometrics*. But for observational data, most of the time these are things to consider *after* you have your design and plan to take that design to actual data.

For now, I want you think about *what you want to do* with your data—what kinds of descriptions of variables your research design requires, what kinds of relationships, what kinds of conditional means and conditional conditional means. Figure out how you want your data to *move*. Figure out the journey you're going to take first; you can pack your bags when it's actually time to leave.

4.7   Relationships in Software

In this section, I'll show you how to calculate or graph the relationship between variables in three different languages: R, Stata, and Python.

These code chunks may rely on *packages* that you have to install. Anywhere you see `library(X)` or `X::` in R or `import X` or `from X import` in Python, that's a package **X** that will need to be installed if it isn't already installed. You can do this with `install.packages('X')` in R, or using a package manager like **pip** or **conda** in Python. In Stata, packages don't need to be loaded each time they're used like in R or Python, so I'll always specify in the code example if there's a package that might need to be installed. In all three languages, you only have to install each package once, and then you can load it and use it as many times as you want.

The data sets for all the examples in this book can be found in the **causaldata** package, which I've made available for all three languages. Do `install.packages('causaldata')` in R, `ssc install causaldata` in Stata, or `pip install causaldata` (if using **pip**) in Python.

So let's do those code examples! The Oster data, while free to download, would require special permissions to redistribute. Instead, I will be using data from Mroz (1987),[24] which is a data set of women's labor force participation and earnings from 1975.

In each of these languages, I'm going to:

1. Load in the data

2. Draw a scatterplot between log women's earnings and log other earnings in the household,[25] among women who work

3. Get the conditional mean of women's earnings by whether they attended college

4. Get the conditional mean of women's earnings by different bins of other household earnings

5. Draw the LOESS and linear regression curves of the mean of log women's earnings conditional on the log amount of other earnings in the household

6. Run a linear regression of log women's earnings on log other earnings in the household, by itself and including controls for college attendance and the number of children under five in the household

[24] Thomas A Mroz. The sensitivity of an empirical model of married women's hours of work to economic and statistical assumptions. *Econometrica*, 55(4): 765–799, 1987.

[25] Why am I using the log of earnings for most of these steps? Think carefully about what we learned about logarithms in Chapter 3.

```
# R CODE
library(tidyverse); library(modelsummary)

df <- causaldata::Mroz %>%
    # Keep just working women
    filter(lfp == TRUE) %>%
    # Get unlogged earnings %>%
    mutate(earn = exp(lwg))
```

```
10   # 1. Draw a scatterplot
11   ggplot(df, aes(x = inc, y = earn)) +
12       geom_point() +
13       # Use a log scale for both axes
14       # We'll get warnings as it drops the 0s, that's ok
15       scale_x_log10() + scale_y_log10()
16
17   # 2. Get the conditional mean by college attendance
18   df %>%
19       # wc is the college variable
20       group_by(wc) %>%
21       # Functions besides mean could be used here to get other conditionals
22       summarize(earn = mean(earn))
23
24   # 3. Get the conditional mean by bins
25   df %>%
26       # use cut() to cut the variable into 10 bins
27       mutate(inc_cut = cut(inc, 10)) %>%
28       group_by(inc_cut) %>%
29       summarize(earn = mean(earn))
30
31   # 4. Draw the LOESS and linear regression curves
32   ggplot(df, aes(x = inc, y = earn)) +
33       geom_point() +
34       # geom_smooth by default draws a LOESS; we don't want standard errors
35       geom_smooth(se = FALSE) +
36       scale_x_log10() + scale_y_log10()
37       # Linear regression needs a 'lm' method
38       ggplot(df, aes(x = inc, y = earn)) +
39       geom_point() +
40       geom_smooth(method = 'lm', se = FALSE) +
41       scale_x_log10() + scale_y_log10()
42
43   # 5. Run a linear regression, by itself and including controls
44   model1 <- lm(lwg ~ log(inc), data = df)
45   # k5 is number of kids under 5 in the house
46   model2 <- lm(lwg ~ log(inc) + wc + k5, data = df)
47   # And make a nice table
48   msummary(list(model1, model2))
```

```
1    * STATA CODE
2    * Don't forget to install causaldata with ssc install causaldata
3    * if you haven't yet.
4    causaldata Mroz.dta, use clear download
5    * Keep just working women
6    keep if lfp == 1
7    * Get unlogged earnings
8    g earn = exp(lwg)
9    * Drop negative other earnings
10   drop if inc < 0
11
12   * 1. Draw a scatterplot
13   twoway scatter inc earn, yscale(log) xscale(log)
14
15   * 2. Get the conditional mean college attendance
16   table wc, c(mean earn)
17
18   * 3. Get the conditional mean by bins
19   * Create the cut variable with ten groupings
20   egen inc_cut = cut(inc), group(10) label
```

```
21 | table inc_cut, c(mean earn)
22 |
23 | * 4. Draw the LOESS and linear regression curves
24 | * Create the logs manually for the fitted lines
25 | g loginc = log(inc)
26 | twoway scatter loginc lwg || lowess loginc lwg
27 | twoway scatter loginc lwg || lfit loginc lwg
28 |
29 | * 5. Run a linear regression, by itself and including controls
30 | reg lwg loginc
31 | reg lwg loginc wc k5
```

```
 1 | # PYTHON CODE
 2 | import pandas as pd
 3 | import numpy as np
 4 | import statsmodels.formula.api as sm
 5 | import matplotlib.pyplot as plt
 6 | import seaborn as sns
 7 | from causaldata import Mroz
 8 |
 9 | # Read in data
10 | dt = Mroz.load_pandas().data
11 | # Keep just working women
12 | dt = dt[dt['lfp'] == True]
13 | # Create unlogged earnings
14 | dt.loc[:,'earn'] = dt['lwg'].apply('exp')
15 |
16 | # 1. Draw a scatterplot
17 | sns.scatterplot(x = 'inc',
18 | y = 'earn',
19 | data = dt).set(xscale="log", yscale="log")
20 | # The .set() gives us log scale axes
21 |
22 | # 2. Get the conditional mean by college attendance
23 | # wc is the college variable
24 | dt.groupby('wc')[['earn']].mean()
25 |
26 | # 3. Get the conditional mean by bins
27 | # Use cut to get 10 bins
28 | dt.loc[:, 'inc_bin'] = pd.cut(dt['inc'],10)
29 | dt.groupby('inc_bin')[['earn']].mean()
30 |
31 | # 4. Draw the LOESS and linear regression curves
32 | # Do log beforehand for these axes
33 | dt.loc[:,'linc'] = dt['inc'].apply('log')
34 | sns.regplot(x = 'linc',
35 |             y = 'lwg',
36 |             data = dt,
37 |             lowess = True)
38 | sns.regplot(x = 'linc',
39 |             y = 'lwg',
40 |             data = dt,
41 |             ci = None)
42 |
43 | # 5. Run a linear regression, by itself and including controls
44 | m1 = sm.ols(formula = 'lwg ~ linc', data = dt).fit()
45 | print(m1.summary())
46 | # k5 is number of kids under 5 in the house
47 | m2 = sm.ols(formula = 'lwg ~ linc + wc + k5', data = dt).fit()
48 | print(m2.summary())
```

5

Identification

5.1   The Data Generating Process

ONE WAY TO THINK ABOUT SCIENCE GENERALLY is that scientists believe that there are regular laws that govern the way the universe works.

These laws are an example of a "data generating process." The laws work behind the scenes, doing what they do whether we know about them or not. We can't see them directly, but we do see the *data* that result from them. We can see that if you let go of a ball, it drops to the ground. That's our observation, our data. *Gravity* is a part of the data generating process for that ball. That's the underlying law.

Data generating process. The set of underlying laws that determine how the data we observe is created.

We have a pretty good idea how gravity works. Basically,[1] if we have two objects that are a distance r apart, one with a mass of m_1 and another with a mass of m_2, then the force F pulling them together is

[1] This "basically" means "don't write me angry emails about quantum gravity or whatever."

$$F = G\frac{m_1 m_2}{r^2} \qquad (5.1)$$

where G is the "gravitational constant." That equation is a *physical law*. It describes a process working behind the scenes

DOI: 10.1201/9781003226055-5

that determines how objects move, whether we know about it or not.

In addition to thinking that there *are* underlying laws like this, scientists also believe that we can learn what these laws are from empirical observation. We didn't always know about the gravity equation—Isaac Newton had to figure it out. But before we had Newton's equation, we had data. Lots and lots of data about objects moving around, specifically data about planets moving around in orbits.

Science says "we can see these planets moving around. We know there must be *some* law explaining *why* they move around the way they do. I bet we can use what we *see* to figure out *what that law is*." And we could! Or at least Newton could. Smart guy. Given the observation that planets have elliptical orbits, Newton showed that Equation 5.1 could generate these orbits.[2]

[2] Edmond Halley, of comet fame, played a role here, too.

We can infer how the world works from our data because things like Equation 5.1 are part of the *data generating process*, or DGP. If there are two planets flying around in space, their movement is *actually determined* by that equation, and then we *observe* their movement as data. If we didn't know the equation, then with enough observation of the actual movement, and enough understanding about the other parts of the DGP so we can block out the effects of things like momentum, we might be able to figure it out and identify the effect of gravity.

DGPs in the social sciences are generally not as well-behaved and precise as the ones in the physical sciences. Regardless, if we believe that observational data comes from at least somewhat regular laws, we are saying there's a DGP.[3]

THE TRICK TO DATA GENERATING PROCESSES (DGPs) is that there are really two parts to them—the parts we know, and the parts we don't. The parts we don't know are what we're hoping to learn about with our research. Learning about gravity lets us fill in a part of our DGP for how data about movement is created.

[3] If we don't believe there are some sort of underlying regular-ish laws, there's little point in doing social science, or at least little point in thinking about it as a science. You don't have to look hard to find physical science fans who disqualify social science from being a science on this basis. Are they right? It's an interesting question. But a question for a philosophy of science book on a good day or a YouTube comments section on a bad day, not this book.

The parts we already know about are just as important, though. The DGP combines *everything* we already know about a topic and its underlying laws. Then, we can use what we *do* know to learn new things. If he were starting from nothing, Newton would have no chance of figuring out gravity. Sure, we see planets acting as though they're moving towards each other. But maybe the planets move like that because of magic? That's not it. But we can only rule that out because of what we already know—the data doesn't do it for us. Or maybe it's just how momentum works? Again, that's not enough on its own, but we need to know a whole lot about momentum to be able to *tell* that it's not enough. We know about forces, momentum, and velocity. With those in place, what can we learn about gravity?

So! We can use what we know to figure out as much as we can about the data generating process. This will allow us to figure out reasons why we see the data we see, so we can focus just on the parts we're interested in. If we want to learn about gravity, we need to be sure that our data is telling us about gravity, not about momentum, or speed, or magic.

LET'S GENERATE SOME DATA. A good way to think about how data generating processes (DGPs) can help with research is to cheat a little and make some data where we know the data generating process for sure.

Simulated data. Data created using random number generators with a data generating process chosen by the researcher.

In the world that we're crafting, these will be the laws:

1. Income is log-normally distributed (see Chapter 3)

2. Being brown-haired gives you a 10% income boost

3. 20% of people are naturally brown-haired

4. Having a college degree gives you a 20% income boost

5. 30% of people have college degrees

6. 40% of people who don't have brown hair or a college degree will choose to dye their hair brown[4]

Let's say that we have some data that has been generated from these laws, but we have *no idea* what the laws are. Now let's say we're interested in the effect of being brown-haired on your income. We might start by looking at the distribution of income by whether you are brown-haired or not, as in Figure 5.1, or by just looking at average income by hair color, as in Table 5.1.

[4] The mathematical representation of these laws is:
$Pr(College) = .3$
$Pr(BrownHair) = .2 + .8 \times .4 \times (College == 0)$
$log(Income) = .1 \times BrownHair + .2 \times College + \varepsilon$
where ε is normally distributed.

Hair	Log Income
Brown	5.111
Other Color	5.095

Table 5.1: Mean Income by Hair Color

Raw data. Data that has not been adjusted in any way.

What do we see in the raw data? We see that people with brown hair earn a little more. It's hard to see in the figure, but in the table we can tell that they earn about .01 more in log

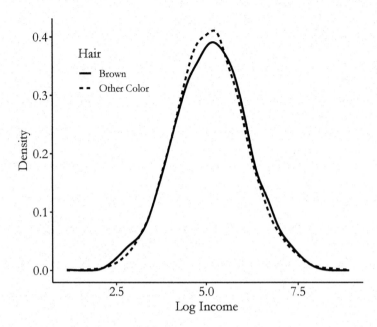

Figure 5.1: Distribution of Log Income by Hair Color

terms, which means they earn about 1% more than people with other hair colors (see Chapter 3).

Uh-oh, that's the wrong answer! We happen to know for a fact that brown hair gives you a 10% pay bump. But here in the data we see a 1% pay bump.

Where can we go from there in order to get the right answer? Not really anywhere. We have College information in our data, but without having any idea of how it fits into the DGP, we have no idea how to use it.

Now let's bring in some of the knowledge that we have. Specifically let's imagine we know everything about the data generating process *except* the effect of brown hair on income.

If we know that it's only non-college people who are dying their hair, and that College gives you a bump, we have some alternate explanations for our data. We don't see much effect of brown hair because a whole lot of non-college people have brown hair, but those people don't get the College wage bump, so it looks like brown hair doesn't do much for you even though it does.

Knowing about the DGP also lets us figure out what we need to do to the data to get the right answer. In this DGP, we can notice that *among* college students, nobody is dying their hair, and so there's no reason we can see why brown hair and income might be related *except* for brown hair giving you an income boost.

So in Table 5.2 we limit things just to college students. Now we see that brown-haired students get a bump of 13%. That's not exactly 10% (randomness is a pest sometimes), but it's a lot closer. And if we ran this little study a thousand times, on average we'd get 10% on the nose.

Hair	Log Income
Brown	5.34
Other Color	5.208

Table 5.2: Mean Income by Hair Color Among College Students

HOW THE HECK did we do that? We got the right answer (or close enough) in the end. We know that getting the right answer involves using information about the data generating process (DGP), and we did that. But what were we actually trying to accomplish by using that information, and how did we know how to use it? At this point I just sort of told you what to do. Not too helpful unless you've got me added as a contact in your phone.[5]

We can split this into two ideas, which we will cover in the next two sections.

The first is the idea of *looking for variation*. The data generating process shows us all the different processes working behind the scenes that give us our data. But we're only interested in part of that variation—in this case, it turned out to be the variation in income by hair color *just among college students*. How can we find the variation we need and focus just on that?

The second is the idea of *identification*. How can we use the data generating process to be sure that the variation we're digging out is the right variation? Figuring out what sorts of problems in the data you need to clear away—like how we noticed that the non-college students dying their hair was giving us problems—is the process of identification.

[5] If you must call me with questions about identification, please limit it to business hours.

Variation. How a variable changes from observation to observation.

5.2 Where's Your Variation?

IN A MOMENT I'm going to show you a graph, Figure 5.2. This scatterplot graph shows, on the x-axis, the price of an avocado, and on the y-axis, the total number of avocados sold.[6] Each point on the graph is the price and quantity of avocados sold in the state of California in a single week between January 2015 and March 2018.

[6] Price is on the x-axis rather than the y for the purposes of aiding statistical thinking, and possibly also for the purpose of enraging several economists.

I'm going to ask you to indulge me and spend a good long while looking at the graph. While you're doing that, I want you to ask yourself a few questions:

1. What conclusion can you draw from Figure 5.2? Think literally, as though you were a robot. Try not to see anything that isn't really there.

2. Now, think about the relationship between avocado prices and quantities more broadly. What kinds of research questions might we have about this relationship? For example, we might wonder what effect a 10% increase in price might have on the number of avocados people want to buy. Can you think of another research question?

3. Can you answer your research question from Step 2 using Figure 5.2? Why or why not?

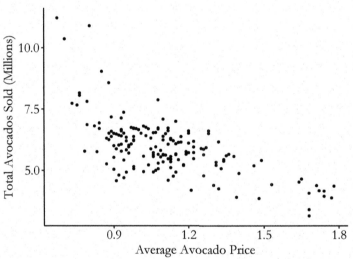

Figure 5.2: Weekly Sales of Avocados in California, Jan 2015-March 2018

Data from Hass Avocado Board
c/o https://www.kaggle.com/neuromusic/avocado-prices/

All right—hopefully you've been able to come up with some answers to those questions. What do we have?

First, what do we literally see on Figure 5.2? We see that there's clearly a negative relationship. What this tells us is that *avocado sales tend to be lower in weeks where the price of avocados is high.* Or, perhaps, *prices tend to be higher in weeks where fewer avocados are sold.* What we can see in the graph is the *covariation* or *correlation* between price and quantity of avocados.

Negative relationship. When one variable is higher than normal, the other is generally lower than normal.

That's more or less it. That's all. We might be tempted to say something like *an increase in the price of avocados drives down sales.* But that's not actually on the graph! Or we might want to say that an increase in prices makes people demand fewer avocados, but that's not on there either—any good economist will tell you that both supply *and* demand are part of what generates market prices and quantities, and we've got no way of pulling *just* the demand parts out yet. All we know is that quantities tend to be high when prices low. We have no idea yet why that is.

Second, what research questions might we have about the relationship between price and quantity? There are a lot of things we could say here. Perhaps we're interested in how price-sensitive consumers are—what is the effect of a price increase on the number of avocados people buy? Or heck, maybe we're thinking about this the wrong way—maybe we should ask what is the effect of the number of avocados brought to market on the price that sellers choose to charge?[7]

So, third, can we answer any of these questions by looking at the graph? Unfortunately, no.[8] The graph shows the *covariation* of price and quantity—how they move together or apart. But these variables move around for all sorts of reasons! Consider a simplified version of the graph, Figure 5.3. Figure 5.3 has only two points, from consecutive weeks.

[7] There are, of course, many other questions to ask! What is the effect of the price on the number of avocados brought to market? What is the effect of the quantity sold one week on the number of avocados people will want to buy the next week? What is the effect of the price on the number of avocados that are brought to market but never sell? Each of these questions would require us to dig out a different part of the variation.

[8] Unless your research question really was about what the correlation is. That question you can certainly answer. It's not a bad question, either.

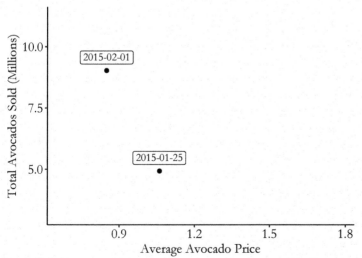

Figure 5.3: Weekly Sales of Avocados in California, Jan 2015-Feb 2015

Data from Hass Avocado Board
c/o https://www.kaggle.com/neuromusic/avocado-prices/

We can still see a negative relationship, which is the same as we saw in Figure 5.2. But *why* did price drop and quantity rise from January to February that year? Is it because a drop in price made people buy more? Is it because the market was flooded with avocados so people wouldn't pay as much for them? Is it because the high price in January made suppliers bring way more avocados to market in February?

It's probably a little bit of all of them. Variables move around for all sorts of reasons. Those reasons would be reflected in the data generating process. But when we have a research question in mind, we are usually only interested in one of those reasons.

SO THE TASK AHEAD OF US IS: *how can we find the variation in the data that answers our question?* The data as a whole is too messy—it varies for all sorts of reasons. But somewhere inside the data, *our* reason for variation is hiding. How can we get it out?

We have to ask *what is the variation that we want to find?* If we want to figure out what the effect of the price is on how many avocados people want to buy, then we want variation in *people buying avocados* (rather than people selling them) that is driven by *changes in the price* (rather than, say, avocados becoming less popular).[9]

As we discussed in the previous section, we're going to be hopeless at doing this if we don't know anything about the data generating process (DGP). We need to *use what we know* about the data generating process to *learn a little more*. So let's make up some facts about where this data came from. Let's imagine for a second that we know for a fact that at the beginning of each month, avocado suppliers make a plan for what avocado prices will be each week in that month, and never change their plans until the next month.

If that's true, then the "suppliers set prices" and "suppliers set quantities" explanations only matter *between* months. The variation in price and quantity *from week to week in the same month* will isolate variation in *people buying avocados* and get rid of variation from *people selling avocados*. Further, because the price is set by the sellers, the variation in quantity we're looking at can only be driven by *changes in the price*.

Our ability to find this answer is entirely based on that assumption we made about sellers making their choices between months. The reason I've made this particular assumption is that it helps us isolate (identify) only the variation on the part of consumers, conveniently getting rid of variation on the part

[9] Advanced readers may notice both the similarities and differences between this approach and the "ideal experiment" in Angrist & Pischke (2008).

of sellers and letting us just look at buyers.[10] This assumption, which works so much magic for us, is entirely a fiction made up for the convenience of writing this book. Hopefully you will not find all assumptions made for the purposes of digging out variation to be convenient fictions. Some probably are.

By tossing out any variation between months, we're digging through explanations that rely on that variation and tossing them out. Since sellers only change behavior between months (given our assumption), that explanation gets tossed out when we get rid of between-month variation, leaving us only with buyer behavior.

If we just look at changes *within* months, as in Figure 5.4, we can see that there's still a negative relationship. Note that *for each of the months*, there's a negative relationship, ignoring any differences between the months. So, given the data and what we know about how sellers operate, an increase in price does reduce how many avocados people want to buy.

[10] Would it be possible to identify what we want without this weird assumption? Sure! It's kind of tricky though. Outcomes that are a combination of two separate systems working together—like buyers and sellers—are tricky to work with. But researchers manage to isolate variation in just one side, often using "instrumental variables," which will be discussed more in Chapter 19.

Figure 5.4: Weekly Sales of Avocados in California, Feb 2015-Mar 2015

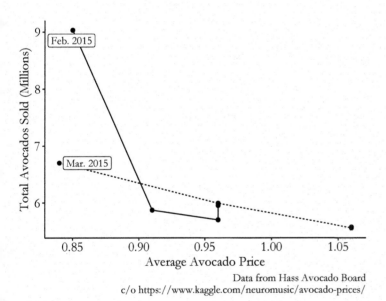

Data from Hass Avocado Board
c/o https://www.kaggle.com/neuromusic/avocado-prices/

The task of figuring out how to answer our research question is really the task of figuring out *where your variation is.* It's unlikely that the variation in the raw data answers the question you're really interested in. So where is the variation that *does* answer your question? How can you find it and dig it out? What variation needs to be removed to unearth the good stuff underneath?

That process—finding where the variation you're interested in is lurking and isolating *just that part* so you know that you're answering your research question—is called identification.

5.3   Identification

ABEL AND ANNIE VASQUEZ woke up Tuesday morning to find that their faithful dog Rex was not at the foot of their bed as normal. Distraught, they ran through the house, calling his name, whistling, offering treats. Finally, one of them opened the front door, and there was Rex, sitting on the grass, wagging his tail. They brought him back inside.

"I keep telling you to latch the doggie door," said Abel.

"He never uses that door anyway. I bet he jumps out that open window in the basement," said Annie.

"Sure," said Abel, rolling his eyes. That night, he made sure to latch the doggie door.

Abel and Annie Vasquez woke up Wednesday morning. The sun was warm, the smells of the automatic coffee machine had already reached their bedroom, and Rex was gone.

It didn't take them long to find him this time, out on the lawn again, rolling in dew. A little patch of fur was getting rubbed raw—however he was getting out, it was clearly the same way every night.

That night, they latched the doggie door and double-bolted the back door. Thursday morning, Rex was gone.

Thursday night, they latched the doggie door, double-bolted the back door, and closed the blinds. Friday, Rex was gone.

Friday night, they latched the doggie door, double-bolted the back, closed the blinds, and blocked the air vent. Saturday morning Rex had eaten two butterflies before they found him outside.

"I've had enough of this," said Abel. That night they locked and latched every door, closed every window, boarded up the chimney, sealed up the half-inch crack between the boards in the garage, plugged the drain in the tub, and gave Rex a very stern talking-to.

Sunday morning, Annie came downstairs to find Abel, distraught, sitting on the stoop and looking at his dog, who had once again escaped the house.

"I don't understand... I closed off every possible way he could get out. Every crack, every crevice, every hole. Every possible exit was blocked. But there he is!"

Annie drank her coffee. "Oh, I opened the basement window back up before I went to bed. Told you that's how he was getting out."

IDENTIFICATION IS THE PROCESS of figuring out what part of the variation in your data answers your research question. It's called identification because we've ensured that our calculation *identifies* a single theoretical mechanism of interest. In other words, we've gotten past all the misleading clues and identified our culprit.

A research question takes us from theory to hypothesis, making sure that the hypothesis we're testing will actually tell us something about the theory. Identification takes us from hypothesis to the data, making sure that we have a way of testing that hypothesis in the data, and not accidentally testing some other hypothesis instead.

Take Abel and Annie's case. What do we see in the actual data? We see, on different nights, different parts of the house being sealed off, and we also see that Rex makes it outside each night. Annie's research question is whether the basement window being open is what lets Rex out.

We—and Abel—know that there are plenty of ways to get out of the house. So this data alone doesn't answer Annie's research question. As Abel closes up more and more stuff, he is tossing out some of the variation. Every time he does, the number of places Rex could get out besides the window shrinks, so there are fewer and fewer ways that Annie could be *wrong* about the window, but there are still other explanations for how the dog got out.

But once we've closed up literally every other possible avenue out of the house, and the window is all that's left, there's only one conclusion to come to: the window being open is what let the dog out. This observation has *identified* an answer to our question, simply by closing off every other possible explanation.

The process of closing off other possible explanations is also how we identify the answers to research questions in empirical studies.

Consider the avocado example in the previous section. Just like Abel and Annie considered all the ways their dog might be able to get out, we asked *what are all the ways that prices and quantities might be related?* Once we have an idea of what our data generating process (DGP) is, i.e., what are all the different ways that prices and quantities might be related, we will have a good idea of what work we need to do. Just like Abel and Annie closed off every possible avenue except the basement window, we closed off undesirable explanations by getting rid of between-month variation driven by sellers. The only way for price to affect quantity at that point is through the consumers.

Identification requires statistical procedures in order to properly get rid of the kinds of variation we don't want. But just as important, it relies on *theory* and *assumptions* about the way that the world works in order to figure out what those undesirable explanations are, and which statistical procedures are necessary. Specifically, it relies on theory and assumptions about what the DGP looks like. We need to make a claim about what we *already know* in order to have any hopes of learning something new. In the avocado example, we used our knowledge about how markets work to realize that sellers might be setting price in response to the quantity—an alternate explanation we need to deal with. Then, we used an assumption about how sellers set prices to figure out how we can block out this alternate explanation by looking within-month just like Abel blocked up the chimney by using wood and nails.

It doesn't instill a lot of confidence to acknowledge that we have to be making some assumptions in order to identify our answers, does it? But it is necessary. Imagine if Abel and Annie tried to figure out how the dog got out without any assumptions or knowledge about the data generating process. The list of alternate explanations they'd have to deal with would include "the dog can teleport," "the universe is a hologram," and "our walls have invisible holes only dogs can see." They'd never come to a conclusion, and they'd be at serious risk of becoming philosophers.[11]

SO THEN THAT'S OUR GOAL. If we want to identify the part of our data that gives the answer to our research question, we must:

1. Using theory, paint the most accurate picture possible of what the data generating process looks like

2. Use that data generating process to figure out the reasons our data might look the way it does that *don't* answer our research question

3. Find ways to block out those alternate reasons and so dig out the variation we need

This process is a lot more difficult than just "look at the data and see what it says." But if we don't go the extra mile of

[11] Granted, while we can't avoid making assumptions we can try to avoid making unreasonable ones. "Dogs can't teleport" seems fine. "Sellers make all choices at the beginning of the month and never change" is a bit iffy, at least without further information.

following these steps, we can end up with confusing, inconsistent, or just plain wrong results. Let's see what happens when we don't take identification quite seriously enough.

5.4   Alcohol and Mortality

IF YOU ARE THE TYPE to pay attention to news articles with headlines beginning "A New Study Says..." you may find yourself deeply confused about the health effects of alcohol. Sometimes the study says that a few drinks a week is even healthier than none at all. Other studies find that alcohol is unsafe at any level. Or maybe it's just wine that's good for you?[12]

Let's take a look at one major study on the effects of alcohol on "all cause mortality"—basically, your chances of dying sooner.[13] This study has more than 200 authors and studied the relationship between drinking and outcomes like mortality and cardiovascular disease among nearly $600,000$ people. It was published in 2018 in the prominent medical journal *The Lancet*, and it has been used in discussions of how to set medical drinking guidelines.[14]

The study is very well-regarded! What it found is that there did not appear to be a benefit of small amounts of drinking. Further, it found that the amount of alcohol it took to start noticing increased risk for serious outcomes like mortality was at about 100 grams of alcohol per week, which is about a drink per day, and below current guidelines in some countries. Figure 5.5 shows the relationship they found between weekly alcohol consumption and the chances of mortality. Mortality starts to rise at around 100 grams of alcohol per week, and goes up sharply from there.

We know from previous sections in this chapter that in order to make sense of the data we need to think carefully about the data generating process.

What is the data generating process? What leads us to observe people drinking? What leads us to observe them dying? What reasons might there be for us to see an association between alcohol and mortality?

Pause here for a moment and try to list five things that would be a cause of someone drinking, or at least be related to drinking. Then, try to list five things that would be a cause of someone dying. Bonus if the same things end up on both lists.

DID YOU COME UP with some determinants of drinking and mortality? There are certainly plenty of options. For example, a

[12] This same phenomenon plays out not just with alcohol but with pretty much every kind of food you can think of—chocolate, eggs, wheat. Great for us one month, deadly the next.

[13] Angela M Wood, Stephen Kaptoge, Adam S Butterworth, and 239 more. Risk thresholds for alcohol consumption: Combined analysis of individual-participant data for 599,912 current drinkers in 83 prospective studies. *The Lancet*, 391 (10129):1513–1523, April 2018. DOI: 10.1016/S0140-6736(18)30134-X.

[14] Jason Connor and Wayne Hall. Thresholds for safer alcohol use might need lowering. *The Lancet*, 391(10129): 1460–1461, 2018.

Hazard ratio or **odds ratio**. Not something we'll be focusing a lot on in this book, but basically a *proportional* effect. So a 1.25 means you're multiplying your hazard rate or odds (as appropriate) by 1.25.

Determinants of Y. The set of variables that are part of the data generating process when Y is the outcome variable.

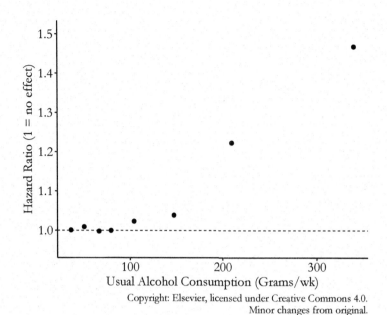

Figure 5.5:   Alcohol Consumption   and   Mortality from Wood et al. (2018).

Copyright: Elsevier, licensed under Creative Commons 4.0.
Minor changes from original.

tendency to take more risks might lead someone to drink more, and may cause other things like smoking. So risk-taking is a cause of drinking, and because of that, smoking is related to drinking too. As for mortality, any one of a long list of bad-health indicators could cause mortality. So could something like smoking or risk-taking. Or just working in a dangerous job!

You can see pretty immediately from this that anything that ends up on both lists is automatically going to be an "alternate explanation" for the results. Smoking, for instance. If smokers are more likely to drink, and smoking increases mortality, then one reason for the relationship between drinking and mortality might just be that smokers tend to drink, and also die earlier because of their smoking. Anything else that ends up on both lists is going to give us an alternate explanation.

There's something else you may not have thought of—how about people who don't drink at all? If drinking is very bad for you, surely non-drinkers would have very low mortality rates? Maybe, but keep the data generating process in mind. Why don't they drink? One reason people choose not to drink at all is because their health is too poor to handle it. Another reason is if they are recovering alcoholics. In these cases, we may actually see worse mortality for non-drinkers, but that relationship is almost certainly not because not-drinking is bad for them. That's an alternate explanation too.

I'm going to bug you once again. Now that we've thought about it some more, I want you to pause and try to think of a few more reasons why we might see a relationship between alcohol and mortality other than alcohol being the cause of mortality. There's no quiz,[15] but this is a very good habit to get into, both for the purposes of understanding the content in this book and for the purpose of trying to make sense of new studies and findings you read about.

THANKFULLY, THE AUTHORS of the Woods et al. study did manage to deal with some of these alternate explanations. They were putting some thought into what the data generating process looks like.

You may have noticed, for one, that Figure 5.5 actually doesn't contain non-drinkers. They've been left out of the study because of the too-sick-to-drink and ex-alcoholic alternate explanations they want to block out. This is one reason why the study doesn't find a positive effect of a little alcohol while other studies do—some of those other studies leave in the non-drinkers (oops!).

The authors also use statistical adjustment to account for some other alternate explanations. They adjust for smoking, which was one thing we were concerned about, as well as age, gender, and a few health indicators like body mass index (BMI) and history of diabetes.

SO THEY'RE GOOD, RIGHT? Well, not necessarily. Just because they've been able to account for some of the alternate explanations doesn't mean they've accounted for all of them. After all, we mentioned risk-taking above. They adjusted for smoking, but surely risk-taking is going to matter for other reasons, and that's not something they can easily measure. They took out non-drinkers who might not drink because they're sick or ex-alcoholics. But that means they also took out non-drinkers who are neither of those things. And what if some very sick people just choose to drink less rather than not at all? There's plenty more to think about, some of which you might have even noticed yourself while reading this chapter, that isn't in their study. They clearly spent some time thinking about the data generating process. But they might not have spent enough—or maybe they did spend enough but just weren't realistically able to account for everything.

There are still some alternate explanations for their results that they couldn't address. So it might be a little premature to take these results, despite the hundreds of thousands of people they examined, and use them to conclude that we have now identified the effects of alcohol on mortality.

[15] Unless you're reading this book for a class and there is a quiz.

If it feels like they did their part in addressing some of the alternate explanations and what's left over feels trivial, keep in mind that these alternate explanations can lead us to very strange conclusions. Chris Auld read the study,[16] and then took the same methods as in the original paper and used them to "prove" that drinking more causes you to become a man (see Figure 5.6).

[16] Chris Auld. Breaking news! https://twitter.com/Chris_Auld/status/1035230771957485568, 2018. Accessed: 2020-02-20.

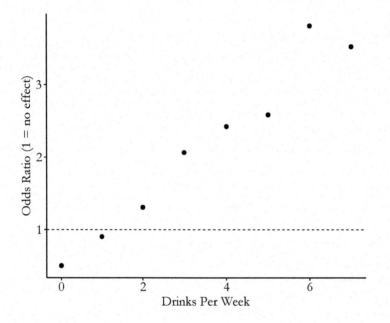

Figure 5.6: Alcohol Consumption and Being a Man

The idea that alcohol turns you into a male seems ridiculous, and in contrast it is pretty darn plausible that alcohol does really increase mortality. But if these methods can give us Figure 5.6, then even if there is really an effect of alcohol on mortality, how close do we think Figure 5.5 is to identifying that effect?

Is it, then, maybe a little concerning how much certainty there seems to be in the response to the study, and others like it, from the media and from medical authorities? The authors of this study have without a doubt found a very interesting relationship, and have addressed some alternate explanations for it. But have they *identified* the answer to the research question you're really interested in—how drinking causes your health to change? That's the question you would want to ask when making policy decisions like setting medical alcohol guidelines. Have they addressed all the necessary context? Is it even possible to do that?

5.5   *Context and Omniscience*

ONE THING THAT I HOPE this chapter has made clear is this: for most research questions, especially any questions that have a causal element to them, understanding context is incredibly important. If you don't understand where your data came from, there's no way that you'll be able to block alternative explanations and identify the answer to your question.

In the words of Joshua Angrist and Alan Krueger (2001),[17,18]

> Here the challenges are not primarily technical in the sense of requiring new theorems or estimators. Rather, progress comes from detailed institutional knowledge and the careful investigation and quantification of the forces at work in a particular setting. Of course, such endeavors are not really new. They have always been at the heart of good empirical research.

They were specifically discussing identification using instrumental variables (see Chapter 19), but it's true generally. Before attempting to answer a research question empirically, be sure to put in plenty of time understanding the context that the data came from and how things work there. Fill in as much of the data generating process as you can! That's going to be important both for interpreting your results properly, and for ensuring that there isn't some equally-plausible explanation for your data that you can't block out.

This means that there's a lot of power in general understanding. So how should you start out designing your research project? By learning a lot of context. Read books about the environment you're studying. Look at the documents describing how a policy was implemented. Read accounts and talk to people about how things actually work. Make sure you get the details right, because the details all show up in the part of the data generating process you're supposed to know.

This is either exciting news for you, or extraordinarily frustrating, depending on what kind of person you are.[19]

IT'S PROBABLY HEALTHY to see this necessary reliance on context as both a great opportunity and also as a revelation about our limitations.

It's a great opportunity because it means you get to learn a lot of stuff. And learning a lot of stuff is fun. If we didn't like learning we probably wouldn't be interested in doing research in the first place. Also, learning about context may well be where we got the idea for the research question in the first place. As discussed in Chapter 2, a great way to come up with a research

[17] Joshua D. Angrist and Alan B. Krueger. Instrumental variables and the search for identification: From supply and demand to natural experiments. *Journal of Economic Perspectives*, 15(4):69–85, December 2001.

[18] Copyright American Economic Association; reproduced with permission of the *Journal of Economic Perspectives*.

[19] I will cop to finding this frustrating. What, you mean I have to figure out what *reality* is like to learn things about reality? I could prove way more fun things if we all just agreed the world should be how it is in my head.

question is by studying an environment or setting and realizing that there's something important about it that you don't know. If that's what you're doing, by the time you have a research question the contextual-learning part will already be done.

That learning may also help you figure out how to block off alternate explanations. Someone with a research question about gravity who has no idea about physics might be able to guess that magic isn't responsible for planetary movement. But someone who *really* studies how planets move will be able to do the appropriate calculations to show that it's not just momentum explaining how they move—you need gravity, too.

HOWEVER, THE NEED FOR CONTEXT ALSO REVEALS OUR LIMITATIONS. If we can only *really* answer our research questions when we understand the context very well, then our only successful projects will tend to be in areas that we already mostly understand. Boring! What about those wild, out-there frontiers where nobody understands anything? Surely it's not a good idea to just ignore them. After all, even if we will have trouble pinning down an exact answer, some questions, like the effects of alcohol on mortality, are too important to ignore completely.

We shouldn't ignore those out-there exciting realms for research. But the need for context means that we should be prepared for our results to be more questionable, more prone to alternate explanations. More "exploratory." This is just plain intellectual humility, and it's a good thing. Knowing less means we can prove less. Forgetting that can lead us to make wild claims that end up being wrong when we know more.

You can see this pretty much any time anything *new* is introduced to society—novels, trains, universal voting, television, the Internet, video games. When these things are new, we don't know how they fit into the data generating process since we haven't seen them produce any data. But that doesn't stop people from producing wild claims about how they will save/destroy all of society. And there's never a shortage of questionably performed research proving both the devotees and the skeptics correct, only for it all to be overturned as we learn more about how these things work.

EVEN AT A SMALLER SCALE, the need for more context can be stymieing. Forget trying to study the effects of flying cars or space aliens. How can we even have enough context to study the effect of avocado prices on avocados sold? You can spend years studying the topic and become the world's leading expert, and you'll still never know *everything* about it. In order to truly be able to see all the alternative explanations for your data and truly understand all the underlying tantalizing-but-invisible laws, it's almost like you need to be omniscient. How can you be sure there aren't any holes left in your research?

Well... you can't!

Sorry.

No research project is perfect. All we can hope to do is:

- Learn what we can about the context so that we don't miss any hugely important part of the data generating process

- Be careful to acknowledge what assumptions we're making, and think about how they might be wrong

- Try to spot gaps in our knowledge about the data generating process, and make some realistic guesses about what *might* be in that gap

- Not aim for perfection, but aim for getting *as close as we can*

If we try to be omniscient, we'll always fail. So there's no point in that. But there is a point in trying to be useful. For that we simply need to learn what we can, work carefully, and try to make our errors small ones.

6

Causal Diagrams

6.1 Causality

IN THIS CHAPTER, we are going to discuss *causal diagrams*, which are a way of drawing a graph that represents a data generating process. We are going to be using causal diagrams in the rest of the book. We will use these diagrams to work out our research designs and figure out what exactly we should do with our data, and how we can identify the answers to our research questions.

Before we get to the *diagrams* part of causal diagrams, let's talk about the *causal*. We are going to use causal diagrams to think about research because it will help us address causality. What does it mean for something to be causal? What is causality? We can say "X causes Y," but what do we specifically mean by that?

For that matter, why worry so much about causality? Because many research questions we are interested in are *causal* in nature. We don't want to know if countries with higher minimum wages have less poverty, we want to know if *raising the minimum wage reduces poverty*. We don't want want to know

DOI: 10.1201/9781003226055-6

if people who take a popular common-cold-shortening medicine get better, we want to know if *the medicine made them get better more quickly*. We don't want to know if the central bank cutting interest rates was shortly followed by a recession, we want to know if *the interest rate cut caused the recession*.

Okay, maybe we do want to know those other things too.[1] But a lot of the time, even non-causal analyses and research questions have a causal question lurking underneath. How many times have you seen a headline like "getting less than eight hours of sleep a night is linked to early death!"? The actual study itself may have made no causal claims. And "linked to" doesn't technically say anything about causality. But because "does a lack of sleep kill you?" (causal) is so much more interesting a question than "do people who die early also tend to sleep less?" (non-causal), we the readers apply wishful thinking to incorrectly interpret the result as causal.[2]

THERE ARE LOTS OF WORDS that are generally taken to imply causality, as well as words that describe relationships without implying causality. There are also some "weasel words" that don't technically *say* anything about causality but clearly really want you to *hear* it.

What are some of these words?

We can say that X causes Y by saying: X causes Y, X affects Y, the effect of X on Y, X increases/decreases Y, X changes Y, X leads to Y, X determines Y, X triggers Y, X improves Y, X is responsible for Y, and so on...

We can say that X and Y are related without implying causality by saying X and Y: are associated, are correlated, are related, tend to occur together, tend not to occur together, go together, and so on...

If some weaselly writer (certainly not *us*) doesn't want to *say* causality but does want the reader to *hear* it, they might say: X is linked to Y, X is followed by Y, X has ramifications for Y, X predicts Y, people who X are more likely to Y, Y happens as X happens, and many others.

Knowing these terms can help you interpret what scientific studies are really saying, and when someone might be trying to pull one over on you.

Also, looking at these words might help us figure out what exactly causality is. Look at the causal phrases. They have *direction*. They tell us that X is *doing something* to Y. In contrast, the clearly non-causal terms don't even need to specify which of X and Y goes first. They just talk about these two

[2] And in case you're wondering, yes. The author of this book *does* get far angrier at this stuff than at actual problems.

variables and how they work together. The weasel terms are so weaselly specifically because they're written in a way that implies that direction from X to Y, even if they're not literally claiming a causal relationship.[3]

This idea, that X is *doing something* to Y, is going to help us think about what causality *is*.

THERE IS NO SINGLE DEFINITION FOR CAUSALITY. But a good way to think about it, at least for the purposes of this book, is this:

We can say that X causes Y if, were we to intervene and *change* the value of X, then the distribution of Y would *also* change as a result.

Let's take that definition for a spin.

Let's say that penicillin *causes* bacteria to die. So, imagine there's a bunch of bacteria in a petri dish without any penicillin on it. The bacteria is alive. If we intervene and put some penicillin in the dish (changing the value of X, where X is "is the bacteria exposed to penicillin"), then the bacteria will die (changing the value of Y, "is the bacteria alive," as a result). So far so good.

This definition lets us distinguish between correlation and causation. For example, we can observe that the number of people who wear shorts is much higher on days when people eat ice cream. However, if we were to intervene and swap out someone's pants for shorts, would it make them more likely to eat ice cream? Probably not! So this is a non-causal relationship.

We can even use this definition to link far-off variables. For example, surely the price of cigarettes by itself has no causal effect on your health. But if we were to intervene and raise the price of cigarettes, that would likely reduce the number of cigarettes smoked. So the price of cigarettes causes cigarette smoking (to go down). Also, if we were to intervene and reduce the number of cigarettes smoked, that would cause your health (to improve). So the price of cigarettes causes cigarette smoking causes health. In sum, the price of cigarettes causes health.[4]

One addendum we will need is that we'll still say X causes Y even if changing X doesn't *always* change Y, but just changes the *probability* that Y occurs. As I said earlier, it changes the *distribution* of Y (see Chapter 3).[5] For example, it's pretty obvious that turning on a light switch causes the light to come on, right? The light is off and so is the switch. Then we intervene to turn the switch on (change X), and the light comes on as well (Y changes too). However, sometimes the light is burned

[3] The key is that all the weasel phrases are equally true if you swap the positions of X and Y, even though swapping them would really change how we'd interpret the claim. "Aspirin is linked to headaches" is no more or less true than "headaches are linked to aspirin." But you can *hear* the notion that aspirin causes your headache in the first one, which is false. You can get away with it without technically lying because it's true that they're "linked."

Cause. X causes Y if, by intervening to change X, the distribution of Y changes as a result.

[4] Note also that simply saying one thing causes another doesn't necessarily imply *which way*. By itself, "cigarettes cause health" could mean "cigarettes improve health" or "cigarettes harm health." If you want to be more specific, you gotta say so!

[5] The definition I'm using does imply a question of whether variables that can't actually be manipulated— race, for example—can causally affect anything. There's a lively debate here, but I think it's valid to say they can. We can't practically manipulate race, but it's reasonable to ask something like "what would have happened differently if this person's race had been different?" which is a kind of *theoretical* manipulation.

out. So on occasion, turning the switch on does nothing. Regardless, we'd still say that "flipping a light switch" causes "the light comes on."[6]

This last part turns out to be very important in social science, where very little is ever for certain. Does buying a child a copy of *Alice in Wonderland* cause them to read it? Not always! Some kids won't read it no matter what, and some kids would manage to read it on their own without you buying it for them. But in general, buying children copies of *Alice in Wonderland* increases the *probability* that a child reads it, and so we'd say that buying them the book causes them to read it.[7]

NOW THAT WE HAVE a basic idea of what causality *is*, let's see what we can do with it. How can we use our ideas about what variables have causal relationships to address our research questions?

After all, what do we know so far?

- For many research questions, in order to identify an answer to them we need to have an idea of the data generating process (Chapter 5).

- If we can think of some variables as causing others, then the causal relationships between them must be *a part of* that data generating process. If X causes Y, then X must be a part of what generates our observations of Y.

So then, how can we represent the data generating process and the causal relationships within it so that we can figure out how to identify the answer to our research question?

6.2   Causal Diagrams

A *causal diagram* is a graphical representation of a data generating process (DGP). Causal diagrams were developed in the mid-1990s by the computer scientist Judea Pearl,[8] who was trying to develop a way for artificial intelligence to think about causality. He wanted reasoning about DGPs and causality to be so easy that a computer could do it. And he succeeded![9] Pearl brought us a whole new way of thinking about causality that's far easier to approach than what came before.

A causal diagram contains only two things:

1. The variables in the DGP, each represented by a *node* on the diagram

[6] A very small percentage of you may notice that this differs from the legal concept of causality, which requires something to actually happen. The law does funny things to a person's brain. More charitably, our concept of causality would be pragmatically difficult to apply to the law because then everyone would be causing each other harm all the time. "Your honor, the respondent played a loud sound that increased my client's chances of tripping and falling by .05%. He didn't actually trip, but we demand .05% of a million dollars."

[7] There are other interesting results that come out of this definition. A well-known example by Judea Pearl comes from sailing. Surely, a wind from the west causes the sailboat to go east. However, the sailor will respond to the wind by moving the sail, canceling out the effect so as to continue in a straight line. So even though we don't even see a correlation between wind and direction, we'd still say that there's a causal effect of wind on direction—the wind changed the distribution of direction. It just happens to be exactly balanced out by the causal effect of the wind on the sailor's decisions.

Causal diagram. A representation of a data generating process (DGP) including the variables in that DGP and the causal relationships between them.

[8] Judea Pearl. *Causality*. Cambridge University Press, Cambridge, MA, 2nd edition, 2009.

[9] If you are a graph theory buff—and let's be honest, who among us isn't—you'll recognize a lot of the concepts here. If you have no idea what graph theory is, don't worry, you'll be fine.

2. The causal relationships in the DGP, each represented by an *arrow* from the cause variable to the caused variable

Node. A variable on a causal diagram with arrows coming in and going out of it.

Let's start with a basic example. Let's say Brad has taken the last slice of cake.[10] You also want the last slice of cake, and offer to flip a coin for it. If it's heads, you get the cake. Brad agrees.

[10] *Totally* something Brad would do.

There are two variables we can observe: the outcome of the coin flip (which can be heads or tails), and whether you have cake (you have it or you don't).

There is one causal relationship—the outcome of the coin flip determines whether you get cake. We can add that to our graph with an arrow from one to the other: Coin → Cake.

CoinFlip ——————————————→ Cake

Figure 6.1: The Effect of a Coinflip on Cake

And that's it! There are only two relevant variables here— the coin flip that determines whether you get cake, and whether or not you get cake. They're both there, and we've accurately described the relationship between them.

NOTE A COUPLE OF THINGS HERE. First, each *variable* on the graph may take multiple values. There's only one node for "CoinFlip" here, rather than one node for "Heads" and another for "Tails."[11] Similarly, there's only one node for "Cake" rather than one for "You get cake" and another for "You don't get cake."

[11] This is different from how you might do things on, say, a flowchart, where each value of the variable sends you in a different direction.

Second, the arrow just tells us that one variable causes another. It doesn't say anything about whether that causal effect is positive or negative. All we have on the diagram is that the outcome of the coin flip affects your cake status.

Causal arrow. An arrow on a causal diagram indicating that the variable it's coming from causes the variable it's pointing at.

Third, we're focusing on the important stuff here. There are plenty of things that could happen at this point that could affect the outcome of the coin flip or whether you get cake. A gust of wind could affect the outcome of your coin flip. Or perhaps a masked stranger might steal the cake while you're busy coin-flipping and you'll end up with no cake regardless of the coin flip. But we can ignore these trivial possibilities and really focus on the important parts of the data generating process.[12]

[12] These are clear-cut examples, but can't the question of which parts of the DGP are "important enough" for inclusion be highly subjective and a topic for heated debate? Yes! Dang, research is hard.

Along the lines of "focusing on important stuff," causal diagrams are often drawn with a particular *outcome variable* in mind. This is done because it allows you to ignore anything that is *caused by* that outcome variable.[13] So in this graph, if we're thinking of Cake as the outcome variable, we don't need

[13] Most of the time. Sometimes you need to consider some of the things caused by the outcome variable, like when you're avoiding collider bias. See Chapter 8.

to worry about the fact that the Cake might give you a Stom-achAche later on. This is a relief—otherwise every single causal diagram we drew would basically have to contain the entire world in it.

OF COURSE, in most data generating processes, there will be more than two relevant variables.

So let's make our little cake scenario a bit more elaborate. Let's say that, first, Brad only agrees to the bet if he gets to keep the coin if he loses. And second, you know there's a chance that your friend Terry might happen to walk in the room. If they do, they'll feel bad if you lose your coin and so give you two coins to replace it.[14]

[14] They're a really great friend.

So now what are the relevant variables?

1. CoinFlip: as before

2. Cake: as before

3. Money: how much money you have left after keeping or losing your coin, and potentially being paid by Terry

4. TerryInRoom: whether or not Terry happens to show up

And what are the causal effects in the data generating process?

1. CoinFlip causes Cake and Money

2. Cake doesn't cause anything

3. Money doesn't cause anything

4. TerryInRoom causes Money

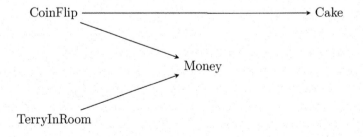

Figure 6.2:   Coin Flips, Cake, and How Great Terry Is

Now that our diagram is a little more complex, a few more things to note:

First, keep in mind that we had to think about the causal relationships between *all* the variables. One variable might cause multiple things (like CoinFlip), and other variables might be caused by multiple things (like Money).

Second, when one variable is caused by multiple things, the diagram doesn't tell us exactly how those things come together. Terry only gives us money if we lost the coin flip. So Money isn't so much "caused by Terry" as it is "caused by a combination of CoinFlip and Terry." But on the diagram we just have TerryIn-Room → Money and CoinFlip → Money. Keep in mind when looking at a diagram that there might be complex interactions between the causes.[15]

ONE LAST THING to keep in mind is that *all* (non-trivial) variables relevant to the data generating process should be included, *even if we can't measure or see them.* This pops up all the time in social science.

For example, let's say we're trying to figure out what variables cause people to get promotions at their job. One thing that probably causes you to get a promotion is if the clients at your job really like you. But as researchers we probably don't have access to data about clients liking you, and even if we did, how could we measure that accurately? So "clients liking you" would be an important part of the data generating process, but it would be an *unobserved* variable.

In this book, we will indicate unobserved variables as being a shade of gray. So now let's say that Terry doesn't join you in the room at random, but rather decides to come in based on their mood today. We have no way of measuring their mood, so that's unobservable.

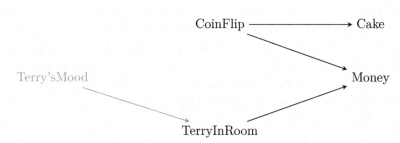

[15] At some points in the book, we'll simplify by putting some of those interactions on the graphs themselves so they're explicit. This technically breaks the causal diagram rules a bit, but it will make things much more clear. See for example the Moderators in Causal Diagrams section at the end of this chapter.

Unobserved/unmeasured variable. A variable that is relevant to the data generating process, but that the researcher either doesn't have access to or could never have access to.

Figure 6.3: Terry's Unobservable Mood

UNMEASURED VARIABLES SERVE TWO PURPOSES in causal diagrams. In addition to being key parts of the data generating process, they can sometimes fill in for variables that we know must *be there*, but we have no idea what they are.

In particular, let's say that we have two variables that are *correlated* but neither of them causes the other. To borrow a previous example, people are more likely to wear shorts on days they eat ice cream, but shorts don't cause you to eat ice cream and ice cream doesn't cause you to wear shorts.

So what do we do? There has to be something between them. Otherwise they wouldn't be correlated.[16] But we can't have an arrow from one to the other, since neither causes the other.

In these cases we imagine that there's some sort of *latent variable* causing both of them, and we can put that on the diagram, perhaps labeled something like "L1" or "U1" (and if we had another unobserved variable that might be "U2", and so on) that similarly indicates we have no idea what it is. Then we can have that "U1" variable cause both IceCream and Shorts.

Some approaches to drawing causal diagrams will opt to leave U1 out of the picture and instead just draw a double-headed arrow between Shorts and IceCream, like Shorts ↔ IceCream, or a no-headed line between them, like Shorts—IceCream. However, I think this approach is too easily confused for "Shorts and Ice-Cream both cause each other" and so will use the latent variable approach in this book. But if you look at materials elsewhere and see them doing this, that's what is happening.

6.3   The Real World: On an Omission Mission

SO FAR, causal diagrams seem pretty simple. And in a technical sense, they really are. As always seems to be the case, though, things get more complex when we take this neat theoretical object and expose it to the horror that is the real world.

Let's take a look at a causal diagram designed to address a classic question in causal inference in the social sciences: what is the effect of police presence on crime?

You would probably expect that sending additional police out into the streets, whether you think that's a good idea or not, would almost certainly have the effect of getting you *less crime*, right?

However, if you look in the data, as in Figure 6.5, additional police presence is consistently associated with *more crime*.[17]

TO UNTANGLE THIS PUZZLE, let's think about what a causal diagram for this data generating process might look like. A

[16] In this case, it's no mystery what's going on. We know that temperature causes both of them. But let's imagine we're extremely unfamiliar with both ice cream and shorts and say we don't know and so can't measure it.

Figure 6.4: Shorts and Ice Cream

Latent variable. An unobserved or unmeasurable variable, often representing a general concept. For example, in many labor economics models, a worker's "ability" at a job is a latent variable. It certainly matters, but is too broad and ill-defined to actually be measured. Things like "IQ score" may be measurable and represent *parts* of ability but aren't the same thing.

[17] Data is from Cornwell & Trumbull (1994) via Wooldridge (2016) and Shea (2018).

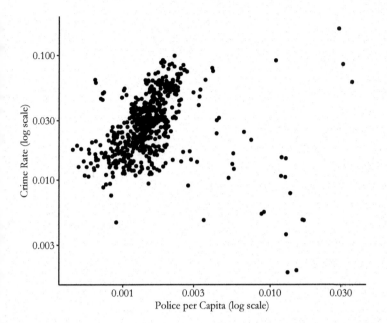

key part of what might be going on is that crime in *previous years* (what we'll call "lagged crime") is likely to cause police presence *now*. So the raw correlation is picking up the fact that high-crime areas get assigned lots of police.

Lag. The value of a variable from a previous time period.

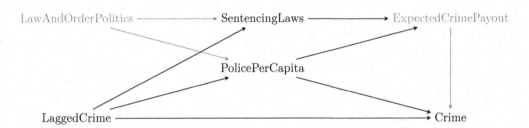

Figure 6.6: Police Presence and Crime

What do we have here? We've incorporated the effect that LaggedCrime has on PolicePerCapita and Crime. We've also added some other things that might be important. The unobservable ExpectedCrimePayout is how profitable a criminal might reasonably expect their crime to be before they commit it, which will be based on how many police there are (making them more likely to get caught), and the sentencing laws in their area (go to jail for longer if they are caught). Both sentencing laws and police per capita are also caused by LawAndOrderPolitics, an unmeasurable variable indicating how tough the political system in the local area wants to be on crime.

We can see very clearly the list of variables included in the diagram. We can also see the arrows that are between those variables. Hopefully, you can agree that those variables and arrows make sense.

HOWEVER, JUST AS IMPORTANT as what's on the diagram is what's *not* on the diagram! *Every variable and arrow that's not on the diagram is an assumption we're making.* In particular, for every variable not on the diagram, we're assuming that variable is not an important part of the data generating process. And for every arrow not on the diagram, we're assuming that the variables not connected by an arrow have no direct causal relationship between them.

What does this mean in the context of our police and crime graph? Some assumptions we've made:

Direct effect. X has a direct effect on Y if $X \rightarrow Y$ is on the graph.

1. LaggedCrime doesn't cause LawAndOrderPolitics

2. PovertyRate isn't a part of the data generating process

3. LaggedPolicePerCapita doesn't cause PolicePerCapita (or anything else for that matter)

4. RecentPopularCrimeMovie doesn't cause Crime

These are some hefty assumptions. And, likely, not all of them are true.

THERE'S A BALANCING ACT to walk when talking about causal diagrams. On one hand, we want to *omit* from the diagram every variable and arrow we can possibly get away with. The simpler the diagram is, the easier it is to understand, and the more likely it is that we'll be able to figure out how to identify the answer to our research question.

A common misstep that some people fall into when working with causal diagrams is to omit too little. They end up with a diagram with dozens of variables, every one of which seems to cause every other one.[18] But if you do that, you're going to let some variables or arrows that may not even be very important possibly ruin your chances of answering your research question of interest![19]

On the other hand, omitting things makes the model simpler. But the real world is complex. So in our quest for simplicity, we might end up leaving out something that's really important. Like, PovertyRate definitely belongs on this graph, right? How could it not? And if we proceed with our diagram while it's still missing, we may end up getting our identification wrong. You've

[18] In my experience, this tendency is strongest in Education departments. In their defense, they probably think this whole book is a bit hopeless.

[19] #4 on the list of omissions is a good example of one we are probably fine omitting. Sure, maybe a big heist movie coming out might inspire a couple of random crimes. But it seems unlikely to be all that important.

probably thought of some other important omissions from this graph on your own.

As discussed in Chapter 5, no study is ever perfect, and we'll never know everything about our data generating process. All we can try to do is build a diagram that's *useful*—it incorporates all the most important stuff while shedding as much extraneous baggage as it possibly can. We will discuss how to pick out just the most important bits when we don't know everything in Chapter 11.

IN FULL, if we want to say anything about our research question, we need to have some model in mind. It's not going to be perfect, but we'll try to put all the most important stuff on there, while still making it simple enough to use.

The willingness to make a stand and say "yes, we really *can* omit that"—that's what you need to answer your research question! It is truly necessary. Anybody trying to answer a research question who *won't* do that is still making those assumptions implicitly. They just aren't making clear what those assumptions are. Assumptions that are necessary to identify your answer are called "identifying assumptions."

That can be dangerous. That's one of the benefits of drawing these causal diagrams. They make very explicit what your assumptions are about the data generating process. That opens those assumptions up to scrutiny, which can be scary. But that scrutiny can help make our research stronger.

The causal diagram, once we have it in hand and are willing to believe it's close enough to accurate, also gives us an extremely powerful tool for answering our research question. But how can we do that?

Identifying assumption. An assumption you make about your causal diagram or data that, if incorrect, means that your approach to identifying the answer to your research question doesn't work.

6.4 Research Questions in Causal Diagrams

NOW THAT WE HAVE our causal diagram, we can use it to figure out how to identify the answer to our research question.

One way that the diagram can help us is in figuring out which parts of the variation in our data identify the answer and which parts don't.

Let's use our police-and-crime figure again. Even though it's not perfect, for a moment let's assume that it's correct, and see how we can use it to think about how to answer the question "does additional police presence reduce crime?"

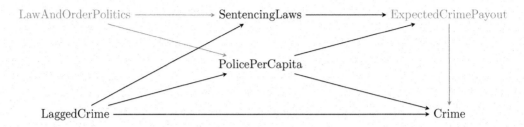

Looking at this diagram, we should be able to see which parts of the diagram answer our research question of interest. In particular, we should be interested in any parts of it that allow PolicePerCapita to cause Crime.

We have an obvious one here, PolicePerCapita → Crime. This is the *direct effect*. However, we also have PolicePerCapita → ExpectedCrimePayout → Crime, which is an *indirect effect*. Why might that count? Ask yourself—if the *reason that police presence reduces crime* is *because criminals commit less crime when they expect to be caught* (i.e., ExpectedCrimePayout is low), then isn't that an example of police presence reducing crime?[20]

So THE VARIATION in our data that answers our research question has to do with PolicePerCapita causing Crime, and to do with PolicePerCapita causing ExpectedCrimePayout, which then affects Crime. To identify our answer, we have to dig out that part of the variation and block out the alternative explanations.

We can see the alternative explanations on the graph pretty easily, too, which is a benefit of using the graph. One alternate explanation is that LaggedCrime causes both PolicePerCapita and Crime, and so PolicePerCapita and Crime might be related because LaggedCrime causes both to rise. LawAndOrderPolitics also affects both PolicePerCapita and (indirectly) Crime. So PolicePerCapita and Crime might be related just because LawAndOrderPolitics drives both of them.

If we are willing to believe this diagram, then the next steps are pretty clear. We need to get rid of the variation due to LaggedCrime and LawAndOrderPolitics in order to isolate just the variation we need. We'll start to talk more in depth about how to do this in Chapter 8.

Figure 6.7: Police Presence and Crime

Indirect effect. X has an indirect effect on Y if $X → Z → Y$ is on the graph for some Z. This implies that X has an effect on Y because it has an effect on something else, Z, which itself affects Y. You can think of this like dominos—the first domino doesn't directly cause the last domino to fall. But by affecting the dominos in between, it has an indirect effect.

[20] You might decide that no, it doesn't count. That's valid! Plenty of research questions concern *only* the direct effect and not any indirect effects. But make sure that's what you want.

6.5 Moderators in Causal Diagrams

THE FOLLOWING SECTION IS AN ODD ONE in that it addresses an aspect of causal diagrams that is likely to be confusing for seasoned researchers, but that new researchers might not have a problem with at all. That is: what about moderators?

Moderating variable. The effect of X on Y is *moderated* by Z if the effect is different depending on Z's value.

Moderators are variables that don't necessarily cause another variable (although they might do that too). Instead, they modify the *effect* of one variable on another. For example, consider the effect of a fertility drug (X) on the chances of getting pregnant (Y). The effect is *moderated* by the variable "having a uterus" (Z). If you don't have a uterus, the drug can't do much for you. But if you do have a uterus, it can increase your chances of conceiving.

But in our causal diagrams, we only have arrows going from one variable to another. We don't have arrows that go to *other arrows*, which it seems like you'd want to give a moderator effect.

TECHNICALLY, THOSE MODERATOR EFFECTS *are* on the diagram. That's because, in a causal diagram, we know that a variable is caused by all its incoming arrows, but *nothing* about the exact shape that influence takes, just like our example with Terry in Figure 6.2.

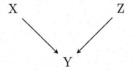

Figure 6.8: X and Z Cause Y, But in What Way?

For example, consider Figure 6.8. All we know from this diagram is that $X \rightarrow Y$ and $Z \rightarrow Y$.[21] However, this could be consistent with any of the following data generating processes:[22]

[21] Formally, all we know is that Y is *some causal function* of X and Z.

[22] All the numbers chosen for this list are arbitrary.

1. $Y = .2X + .3Z$

2. $Y = 4X + 3Z + 2Z^2$

3. $Y = 1.5X + 5Z + 3XZ$

4. $Y = 2X + 3XZ$

and infinitely many more. Note that in #3 on that list, Z has a moderating influence on the effect of X. If $Z = 1$, then a one-unit increase in X will increase Y by $1.5 + 3 = 4.5$. But if $Z = 2$, then a one-unit increase in X will increase Y by $1.5 + 3 \times 2 = 7.5$. It goes the other way, too, with X moderating the effect of Z— the effect of Z is different at $X = 3$ or $X = 5$. In #4, we go even further—Z has no direct effect on Y, but does moderate the effect of X.

Phew! We can handle moderating variables in a causal diagram. But while this satisfies a technical need for moderating variables, it's not a very *clear* way of describing your data generating process. After all, if you thought there was an important moderating effect, do you really think Figure 6.8 would get across that idea?

IN THIS BOOK, we will cheat a little bit for the sake of clarity, and include our moderated effects right on the diagram,[23] even though that's not technically the right way to do it.[24]

In this slightly more-relaxed approach to causal diagrams, we might reflect the data generating process #3 from our previous list as it's depicted in Figure 6.9. In this diagram, X and Z both affect Y directly. However, they also combine to form $X \times Z$, which affects Y as well, demonstrating that there is an interaction between X and Z, and each moderates the other's effect.

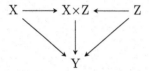

[23] I will also, on occasion, include other functional form restrictions on the diagram, like in Chapter 20.

[24] Please don't hurt me Judea Pearl.

Figure 6.9: X and Z Cause Y, Each Moderating the Other's Effect

We can represent data generating processes like #4 in our list above, where Z had no direct effect but moderated the effect of X, as in Figure 6.10. Here, Z only affects Y by changing the effect of X, reflected in the $X \times Z$ node.

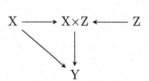

Figure 6.10: X Causes Y and Z Moderates That Effect

7

Drawing Causal Diagrams

7.1 Our Idea of the World

WE HAVE A BASIC IDEA NOW about how causal diagrams work. But that assumes we have a causal diagram to work with. How can we get one? We'll have to draw one ourselves.[1]

Causal diagrams represent the data generating process that got us our data. So, drawing our own causal diagram will come down to putting our idea of what the data generating process is onto paper (or a computer screen).

This can be tricky! It requires that we know as much as possible about that data generating process before getting started. So the first step in drawing a causal diagram is really to do some research. Lots and lots of research on your topic.

Once you've done that, you can combine what you've learned with your intuition, follow this chapter, and suddenly, hey, there it is—a causal diagram. One step closer to answering that research question.

[1] This chapter will be focused on how to design and put together a causal diagram. For the actual literal drawing portion of the show, I recommend either doing it by hand or using the website dagitty.net.

DOI: 10.1201/9781003226055-7

7.2 Thinking Through the Data Generating Process

HOW CAN WE POSSIBLY WRITE THE WORLD DOWN IN A GRAPH? After all, the world is very complex, and a graph is clean and simple. But we'll figure it out.

The key is to focus in as much as possible. Think about our research question and try to live in the world of that research question. We can't possibly put everything in the world on our graph, so we need to work hard to not fall for the trap of trying to do so. What we *can* do is try to put everything relevant to our research question on the graph.

All through the process, we're going to want to keep in mind that we're trying to make a graph that mimics the *data generating process* relevant to our research question. What leads us to observe the data we do? What causes the outcome? What causes the treatment?

It will probably help if we work with an example.

Let's walk through a study that I worked on.[2] That way, I can tell you exactly what sorts of things we were thinking about when considering what we thought the data generating process looked like. In this study, we were interested in the effect of *taking online courses* on *staying in college*, specifically in community college in Washington State.[3,4]

OUR FIRST TASK WILL BE THINKING THROUGH the list of relevant variables.

First off, just so we're clear, *what is a variable?* We've covered this in previous chapters, but it bears repeating since it's easy to forget when applying this stuff to the real world.

A *variable* on a causal diagram is a *measurement* we could take that could result in different values. So, for example, one of the variables relevant to our research question is "online class." Is the class you're taking online?—That's the measurement. It could be "yes," it could be "no"—those are the values. If we like we could call it "type of class" with the values "online class" and "face-to-face class." Notice that we don't have one variable for "online class" and another for "face-to-face class." That's because those aren't two separate variables, they're two separate values that the same variable could take.

So there's one relevant variable—online class. That's our treatment variable. We're interested in the effect of that variable. Another relevant variable will be "dropout"—did the student drop out of college since taking the class? This is our outcome variable. Treatment and outcome are always a good

[2] Nick Huntington-Klein, James Cowan, and Dan Goldhaber. Selection into Online Community College Courses and their Effects on Persistence. *Research in Higher Education*, 58(3): 244–269, 2017.

[3] The world is just dripping with context. The data generating process may well be different depending on where, when, or who you're talking about. The diagram in this chapter might look very different if the study were to be done in Oregon. Or Thailand!

[4] Is this a good research question? Sure! We can imagine randomly assigning students into online or face-to-face college classes and recording afterwards whether they stay in college or drop out, so it's a question we can answer with evidence. And learning the answer to this question would certainly help shape a theory about why people stay in school. If online classes lead to dropout, then something about the face-to-face experience might make people want to stick around.

Treatment variable. The variable we want to know the effect *of*. How does it affect the outcome?

Outcome variable. The variable we want to know the effect *on*. How does the treatment variable affect it?

place to start.

Variable list:

- OnlineClass

- Dropout

All right, now what else?

We want to include all variables relevant to the data generating process. That means any variable that has something to say about whether we observe online class-taking, or whether we observe dropout, or whether we observe them both together or apart. *Every variable that causes the treatment or outcome, or causes something that causes something that causes the treatment or outcome, or causes something that causes something that causes something...* is a good candidate for inclusion.[5]

What are some things that might cause people to take online classes? Well, different students have different preferences for or against online courses, so Preferences. Those preferences might be driven by background factors like Race, Gender, Age, and SocioeconomicStatus. Those same background factors might influence how much AvailableTime students have—time-pressed students may prefer online courses. And AvailableTime might be influenced by how many WorkHours the student is doing. You also need solid InternetAccess to take online courses.[6]

Now how about things that might cause people to drop out of community college? Some of the same background factors as before might be relevant, like Race, Gender, SES (socioeconomic status), and WorkHours. Your previous performance in school, Academics, is also likely to be a factor.

Now what does our list look like?

- OnlineClass

- Dropout

- Preferences

- Race

- Gender

- Age

- SES

- AvailableTime

[5] You may want to include some variables that *don't* fit that description too, if they happen to be closely *related* to the treatment or outcome, or to some of the other variables.

[6] Writing this during the coronavirus lockdown in 2020 gives me another idea for a *preeeetty* darn important cause of taking online classes. But of course this isn't relevant to the data generating process in the paper since the the data is from long ago—lockdown was not occurring at that time. So we leave it out. The diagram should be drawn to reflect the context the data comes from.

- WorkHours

- InternetAccess

- Academics

This is a good time to pause, look at our list, and think hard about whether there's anything important that we've left off. You may be able to think of a thing or two. But for now, let's leave it at that.

HOW CAN WE TELL if something is important enough to be included? I just mentioned we should think about whether there's anything important being left off. But how can we tell if a variable is important or not?

What it really comes down to is *how strong* we think the causal links out of that variable are.[7]

For example, the presence of QuietCafes in someone's area might encourage them to take an online course. A nice quiet place outside the house to do schoolwork. That may well be a real thing that encourages a few additional students to take online courses. But it's unlikely to really be a determining factor for too many students.

So yes, it's relevant, but it seems unlikely that it would have anything but a tiny effect, on average, on whether a student takes an online course. So we're probably okay leaving it out.

WITH OUR SET OF VARIABLES IN HAND, we must try to think about which variables *cause* which others.

Conveniently, we've already done most of the work here. When we were thinking about which variables to include, we were asking ourselves what variables might be out there that cause our treatment or outcome variables. So we already have an idea of what might cause those.

What's left is to think about how those variables might cause *each other*, or perhaps be *caused by* the treatment or outcome. We might also want to consider whether any of the variables are related but neither causes the other, in which case they must have some sort of common cause we can include.

We already have some causes of our treatment and outcome, as well as some other causes, from how we described the variables as we introduced them in the previous section:

- OnlineClass:[8] causes dropout

- Dropout

- Preferences: causes OnlineClass

[7] Or how strong the links *into* that variable are if it's a collider anywhere... that's a different chapter.

[8] This (especially in light of several demographic causes below) would be an excellent time to point out two things about this process: (1) There's no *sign* implied here—none of these causal arrows say whether a given cause makes dropout more or less likely, just that it changes the distribution in some way, and (2) the *mechanism* might be omitted— having the arrow doesn't say *why* one thing causes another, just that it does. In most cases you could drop another variable in the middle of the cause and caused variables, perhaps "School Spirit" or "Education Quality" in this case if you like, to explain the link. But often we don't have enough information to quantitatively study every mechanism, so we leave it out for a later study, or theorize outside the diagram about what we think the mechanism might be.

- Race: causes Dropout

- Gender: causes Dropout

- Age: causes Dropout

- SES: causes Dropout

- AvailableTime: causes OnlineClass, Dropout

- WorkHours

- InternetAccess: causes OnlineClass

- Academics: causes Dropout

How about which non-treatment and non-control variables cause each other? Age certainly causes SES, and all of the background variables (Race, Gender, Age, SES) affect AvailableTime and WorkHours. SES probably causes InternetAccess as well.

How about which variables are related to each other without there necessarily being a clear causal arrow in either direction? In this case, we add on common causes we can just call U1, U2, etc., that cause both variables.

Academics and SES are clearly correlated with Race and Gender, so we'll want to have some sort of common cause there. Academics might also be related to the kinds of employment someone has and so affect WorkHours. InternetAccess is also likely to be caused not just by SES, but also Location, which we've left out up to now. So Location might want to get in that list.

Now what do we have?

- OnlineClass: causes dropout

- Dropout

- Preferences: causes OnlineClass

- Race: causes Preferences, Dropout, AvailableTime, WorkHours, related to Academics, SES

- Gender: causes Preferences, Dropout, AvailableTime, WorkHours, related to Academics, SES

- Age: causes Preferences, Dropout, SES, AvailableTime, WorkHours

- SES: causes Preferences, Dropout, InternetAccess, Available-Time, WorkHours

- AvailableTime: causes OnlineClass, Dropout

- WorkHours: causes AvailableTime

- InternetAccess: causes OnlineClass

- Academics: causes Dropout, WorkHours, related to Race, Gender

- Location: causes InternetAccess, related to SES

And now that we have our list we can draw a diagram. With a list this long and this many causal arrows described, I can warn you it's going to be a little messy. Okay, very messy. That can happen. We'll be coming back to clean it up a bit later.

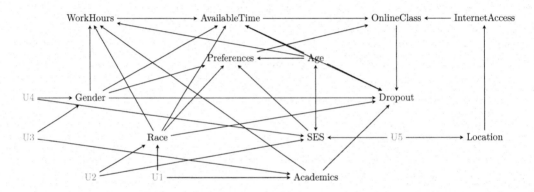

Figure 7.1: A Messy Diagram of the Effect of Online Classes on Dropout

AND THERE WE HAVE OUR DIAGRAM in Figure 7.1. Now that we do, it's a good time for *revision*. Looking at the diagram, what might be left off? What variables are likely to be relevant but are missing? What arrows should probably be there but aren't?

There are probably a lot of things we're missing here.[9] One big one is the specific community college being attended—some offer lots of online courses and some don't! So that's likely to be a big cause of OnlineClass, and caused by plenty of other things on the list, especially Location.

But that's not all. It's a good exercise to look at a causal diagram and think carefully about what's both important and missing.

[9] And don't worry... things weren't quite this simple in the actual paper.

7.3   Simplify

THE REAL WORLD IS COMPLEX. The true data generating process is too.

That leads us to a problem. The whole point of having a model like a causal diagram is to help us make sense of the data generating process and, eventually, figure out how we can use it to identify the answer to our research question.

But if the diagram we end up with looks like what we have in the previous section, we're going to be very hard-pressed to make any sort of sense of it. It would be handy if we could simplify it in some way.

Why is it important to simplify? Ultimately, the more complex a causal diagram is, the less helpful it is likely to be. Imagine you were asking someone directions to the next gas station and instead of saying "it's two exits north on the freeway, then next to the Wendy's" they handed you a giant atlas where each page is so intricately detailed that it only covers a single square mile. Sometimes less information is more information.

The trick will be to simplify where we can without getting *so* simple that our diagram no longer represents the true data generating process. It's a bit too far in the other direction to take the atlas away and just say the nearest gas station is "on Earth somewhere."

HOW CAN WE HIT THAT GOLDEN MEAN OF SIMPLE BUT NOT TOO SIMPLE? We can apply a few simple tests to see if there's any needless complexity in our diagram.

1. **Unimportance.** We've already discussed this one—if the arrows coming in and out of a variable are likely to be tiny and unimportant effects, we can probably remove the variable.

2. **Redundancy.** If there are any variables on the diagram that occupy the *same space*—that is, they have the arrows coming in and going out of them from/to the same variables—we can probably combine them and describe them together (this works even if there are arrows between some of the variables being grouped together).

3. **Mediators.** If one variable is *only* on the graph as a way for one variable to affect another (i.e. B in A → B → C where nothing else connects to B), then we can probably remove it and just have A → C directly.[10]

4. **Irrelevance.** This one will take some additional knowledge about causal paths from Chapter 8. Some variables are an important part of the data generating process but irrelevant to the research question at hand. If a variable isn't on any

[10] This is what I mentioned in an earlier footnote about the mechanism—*why* one variable causes another—being omitted.

path between the treatment and outcome variables, we can probably remove the variable.[11]

Can we apply these steps to our diagram in Figure 7.1? We've already done Unimportance and left a few variables off for that reason. Let's leave Irrelevance for now since we haven't gotten to Chapter 8 yet. Can we do Redundancy or Mediators?

We do have some variables that occupy the same space on the diagram and so might be redundant. In particular, Gender and Race have the exact same set of arrows coming in and going out.[12] So we can combine those in to one, which we can call Demographics.[13]

How about Mediators? We have a few here, the most prominent of which is Preferences. Instead of having Gender, Race, SES, and Age affect Preferences and then have Preferences affect OnlineClass, we can just have those four variables affect OnlineClass directly. Another one here is Location → InternetAccess → OnlineClass. We can chuck InternetAccess right out and lose nothing!

The last one is a bit less certain. WorkHours affects OnlineClass through AvailableTime. WorkHours and AvailableTime don't quite fall under Redundancy, since Academics affects WorkHours but not AvailableTime. And they don't quite fall under Mediators, because other variables besides WorkHours affect AvailableTime.

However, the other variables besides WorkHours that cause AvailableTime *also* cause WorkHours. So if we got rid of AvailableTime and just had WorkHours affect OnlineClass directly (Mediators), we'd still have all those same AvailableTime causes affecting WorkHours and wouldn't lose anything (Redundancy). The only sticky thing is that Academics doesn't cause AvailableTime. But that's fine in this case, because Academics *does* cause AvailableTime... through WorkHours! We do lose a little bit of information with this simplification because of the Academics variable, so we'd have to think carefully about whether we're okay with that.

Now we have the much-better-looking, although still slightly messy, causal diagram below:

While these steps can come in very handy, pay close attention to the use of "probably" in each of them. We can't just apply these blindly. Even if a variable is subject to one of these steps, we don't want to remove it if it's key for our research design or crucial for communicating what's going on.

[11] This also requires that we aren't planning to use the variable as an instrument (Chapter 19).

[12] Other than the U1, U2, etc. But for this purpose we can ignore those. The new U1, U2, etc. that we include in the diagram will just technically be a mix of the old ones.

[13] Combined variables like this can make diagrams nicer but make things slightly trickier when we get to controlling for things in later chapters. We have to remember that Demographics was made up of Gender and Race, and so if we want to control for Demographics we need to control for *both* Gender and Race.

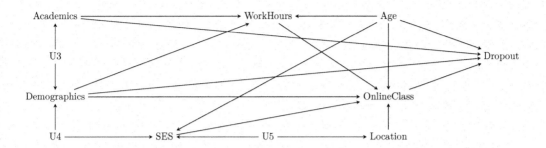

Figure 7.2: A Cleaner Diagram of the Effect of Online Classes on Dropout

For example, let's say we're interested in the effect of exercise on your lifespan. One reason exercise might lengthen your lifespan is because it raises your heart rate, and another reason is that it develops muscle. Heart rate and muscle development might be subject to the Mediator step—if the only arrows pointing to them are from exercise, and the only ones out are to lifespan, then we could eliminate both and just have exercise point to lifespan. But if we're interested in *why* exercise works (is it heart rate or is it muscle?) we'd have no hope of answering that question if heart rate and muscle development aren't actually on the diagram. So that would be a simplification too far.

To give another example, just a few paragraphs ago we eliminated InternetAccess because it was a mediator. But in the original study we're talking about, the fact that InternetAccess caused OnlineClass was crucial because it acted as an "instrument" (Chapter 19). If we did that simplification in our study there would be no study!

7.4   Avoiding Cycles

THERE IS ONE THING A CAUSAL DIAGRAM CANNOT ABIDE, and that is *cycles*.[14] That is, you shouldn't be able to start at one variable, follow down the path of the arrows, and end up back where you started.

There are two examples of graphs with cycles below in Figure 7.3. In the first one, you can go A → B → C → A. In the second, the variables cause each other, and so you can just go A → B → A.[15]

Why can't we have this? Because if we do, then a variable can *cause itself*, and suddenly we've lost all hope of ever isolating the cause of anything, since we can't separate the effect of B on A from the effect of A on B on A from the effect of B on A on B on A... and so on.

[14] The formal name for a causal diagram is *directed acyclic graph*. It's even right there in the name! Acyclic! No cycles!

[15] Some people, when drawing causal diagrams, will use a double-headed arrow like this to indicate that these two variables share a common cause, i.e. A ← U1 → B. But we don't do that in this book.

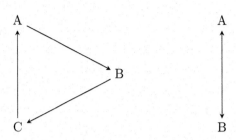

Figure 7.3: Two Causal Diagrams with Cycles

BUT HOLD ON A MINUTE. Surely there are plenty of real-world data generating processes with feedback loops like that. The rich get richer, objects have momentum, and if I punch you that makes you punch me, which makes me punch you.

Surely we're not just going to have to give up any time this happens?

Well, no. But that's because in the true data generating process there *can't* really be any cycles, if you think about it right. That's because of *time*.

Let's think about that punching feedback loop. It certainly seems like the diagram should look like Figure 7.4.

IPunchYou ⟷ YouPunchMe

Figure 7.4: A Diagram About Punching

But that's not quite right. After all, if I punch you, and you punch me back, that doesn't cause me to send the *first* punch—it can't, I already did it. It might, however, cause me to *punch again later*.

Let's pay attention to *when* these punches are thrown. As is common in statistical applications where time is a factor, let's refer to these time periods as t, $t+1$, $t+2$, and so on, where t is "some particular time," $t+1$ is "the time right after that," and so on. Now the diagram looks like Figure 7.5, and the cycle is gone. The diagram as is only has t and $t+1$, but we could keep going out to the right with $t+2$, $t+3$, and so on, if we wanted.

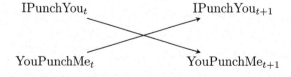

Figure 7.5: Cycles are Even Worse than Punching. But Also Don't Punch People.

Whenever we have a cycle in our diagram, we can get out of it by thinking about adding a time dimension. And it has to work—the cycles pop up because the arrows loop back on themselves. But time's arrow only moves in one direction.[16]

THERE'S ANOTHER WAY to break a cycle in a causal diagram.[17] If you can find a source of random variation for one of the variables in the cycle (say, with a randomized experiment), then if we just focus on the part of the variable driven by randomness, the effect can't loop back on itself. So if instead of waiting for you to punch me, I decide to punch you based on the outcome of a coin flip, then we still have $IPunchYou \rightarrow YouPunchMe$ in the diagram, but instead of $YouPunchMe \rightarrow IPunchYou$ we have $CoinFlip \rightarrow IPunchYou$. Now the cycle is broken.[18]

7.5   Getting Comfortable With Assumptions

IN THIS CHAPTER, I've emphasized that your causal diagram should be based as much as possible on real-world knowledge and prior research. But since we can't possibly know everything about every part of the data generating process,[19] it also contains a lot of assumptions.

That's both necessary and scary! Writing down a diagram like this means sticking your neck out. *This* causes *that*, you have to say. *That* doesn't cause *this*, or at least not enough to draw an arrow. *This other thing* isn't even worth including on the diagram. You think surely this will draw the pitchforks outside your door, or at least a slight disapproving glance from a professor.[20]

But in order to progress, the assumptions do have to be made. The quality of your research will hinge on how accurate those assumptions are.[21] So how can we get comfortable with the idea that we have to make assumptions, and how can we make those assumptions as accurate as possible?

THE CONVENIENT THING ABOUT EMPIRICAL WORK IS THAT ASSUMPTIONS ARE RARELY RIGHT OR WRONG. They're more on a scale of probably-false to probably-true.

After all, if we have to make an assumption, it's usually because there's a gap in our knowledge. There's no way to know for sure. Unlike a math problem it's not up to us to prove we're right, but rather to get a critical reader to think "okay, that sounds plausible. I buy it."

So it's not our job to *prove* we're right, at least not in the mathematical sense of "prove," but it *is* our job to get that critical reader to buy it.

[16] Time travel notwithstanding.

[17] This is a common approach when researchers think they have a cycle on their hands. Although for the reasons given in this section, maybe they shouldn't *actually* think of it as a cycle on the true diagram. But even with the approach of breaking the cycle up by introducing time, there are still some very cycle-like elements there to work through in identification.

[18] This approach still leaves you with some problems if you can't randomly determine *all* of the variable. See, for example, the discussion of social networks in Chapter 22.

[19] If we did, we wouldn't need to bother doing research, now would we?

[20] Which of these, you wonder, is worse?

[21] Plus, even great researchers make bad assumptions. Plenty of research goes back to look at old work and finds flaws in its assumptions, replacing them with better ones to progress the field. Heck, the economist Gary Becker was a genius, founded like four major subfields of economics, and probably would have won a second econ Nobel if you were allowed to do that. But pretty much everywhere he made a mark, people have spent decades showing how wrong his assumptions were.

That narrows down our work for us. For a given assumption, ask yourself: "Is this probably true? What evidence can I provide to push this away from *possible* and towards *probable?*"

Put yourself in the head of that critical reader. Why might they not believe that assumption? What evidence could they be shown to convince them? Then, produce whatever evidence you can.

For example, let's say you're drawing a diagram of whether door-knocking for a candidate actually increases votes for that candidate. On your diagram there's no arrow between "how much money the candidate has" and "having a door-knocking campaign."

A reader might think "hold on... surely candidates with more money can more easily afford a door-knocking campaign, right? There should be an arrow there." And to that you would look at whatever evidence you had. Do you have prior studies about the effects of candidate campaign coffers? Check them! Do you have data on the topic? Look in your data to see if there's a correlation between money and door-knocking. Neither of these things would truly prove that the arrow shouldn't be there, but they might help tip that reader's scales from skeptical to buying-it (and gives you an opportunity to find out that your assumption was, in fact, wrong so you can fix it).

THAT'S A LOT OF WHAT IT COMES DOWN TO. Think about whether our assumptions are reasonable, try to base them as much on well-established knowledge and prior research as possible, and if we think there's reason to be skeptical of them, ask what evidence would support the assumption and try to provide that evidence.

There are a few other approaches we can take that can help.

The first is just to get another set of eyes on it. It can be hard to be skeptical of your own assumptions—you made them, after all, so you probably think they're pretty reasonable. But maybe there are reasons to be skeptical that you didn't think of! Show your model to another person, especially a person who knows something about the setting or topic you're trying to make a causal diagram for. Or just describe some of the assumptions you made and see what they think. You might be surprised with what they are and are not okay with.

There are also some more formal tests you can do. One nice thing about causal diagrams is that they produce *testable implications* for us. That is, once we have the diagram written down, it will tell us some relationships that *should be* zero. And we can check those relationships in our actual data using basic correlations.[22] If they're not zero, something about our diagram is wrong! We'll talk more about these formal tests in Chapter 8.

[22] Formally, we should check more kinds of statistics than just correlations, like ones that allow for nonlinear relationships. See Chapter 4.

8

Causal Paths and Closing Back Doors

8.1 I Walk the Line

PREVIOUS CHAPTERS have emphasized the concept of the *causal diagram*. I, the author of this book, clearly have a thing for nodes and arrows.

The reason these diagrams can be so useful for applied researchers is that they make clear *the different reasons why* two variables might be related to each other, and also what we need to do to identify the answer to our research question.

Both of these benefits of causal diagrams come about by thinking carefully about the *paths* from one variable to another on a diagram.

These paths tell us *why*—why are two variables related? Which causes the other? That's the kind of crucial context we need to figure out our research question.

Those paths are what this chapter is all about.

WHAT IS A PATH? A path between two variables on a causal diagram is a description of the set of arrows and nodes you visit when "walking" from one variable to another.

Path. A path on a causal diagram is the set of arrows and nodes you pass when going from one variable to another.

DOI: 10.1201/9781003226055-8

For example, consider the below diagram Figure 8.1. We can observe in data that B and C are related. But *why* are they related?

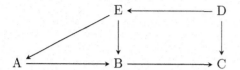

Figure 8.1:   An Example Causal Diagram for Path-Finding

One reason is that B causes C. After all, B → C is on the diagram—that's one *path* between B and C.

Another reason is that D causes both E and C, and E causes B. In other words, we have the path B ← E ← D → C. We can "walk" from B to E, and then onwards to D, and finally to C. We've taken a nice little stroll along the paths laid out on our diagram, and moved from one variable, B, to another, C.[1]

So why are B and C related? As mentioned, one reason is that B causes C. The path B → C is on the diagram. Another is that D causes C, and D also causes E, which causes B. In other words, B and C appear to move together because they're both caused by D.

If our research question is about the effect of B on C, then this second pathway—the one that D is responsible for, is another reason we would see B and C being related other than B → C. It's an alternate explanation for why B and C might be related, other than the explanation that answers our research question of whether (and how much) B causes C.

The paths can tell us the road we want to walk on, and also the road we want to avoid. They're crucial for figuring out our identification. So let's learn about them.

[1] There's a third way to go, too: B ← A ← E ← D → C.

8.2   Any Way You Like It

OUR FIRST TASK WILL BE FINDING ALL THE PATHS. That's right, all of them. We want to be able to write out every single path that *starts* with the treatment variable and *ends* with the outcome variable.

Why do we need to be sure to get every path? Because each path explains one way in which the treatment and outcome variables might be related. They're alternate explanations. For example, if we are looking at the effect of smoking on cancer, and one path is Smoking → Cancer, and another is Smoking ← Income → Cancer, then if you say "smoking causes cancer," then someone else could, quite reasonably, say "maybe, but also, being low-income can affect both whether you smoke and your health generally, so maybe it's just a statistical illusion!"

If you want to really show how much your treatment causes your outcome, you have to be able to *find* those alternate explanations so you can account for them in your research and identify *just* the explanation you're interested in.

Those alternate explanations live on the paths. If we want to destroy them, first we must know them.

SO HOW CAN WE FIND EVERY PATH FROM TREATMENT TO OUTCOME?[2] There might be lots of paths on a more complex graph. So how can we find them?

We can follow a few convenient steps:[3]

1. Start at the treatment variable

2. Follow one of the arrows coming in or out (either is fine!) of the treatment variable to find another variable

3. Then, follow one of the arrows coming in or out of *that* variable

4. Keep repeating step 3 until you either come to a variable you've already visited (A loop! That's not a path), or find the outcome variable (that's a path. Write it down)

5. Every time you either find a path or a loop, back up one and try a different arrow in/out until you've tried them all. Then, back up again and try all *those* arrows

6. Once you've tried all the ways out of the treatment variable and all the eventual paths, you've got all the paths!

Let's do a quick example with a simple diagram in Figure 8.2.

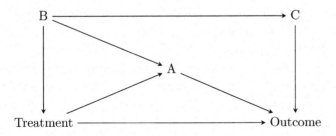

Let's follow the steps outlined above.[4]

- Start at Treatment.

[2] This method also works for finding paths between two variables that *aren't* treatment and outcome, which comes in handy in some cases, including later in this chapter when we test the diagram.

[3] Computer science enthusiasts may recognize this as a *greedy algorithm*.

Figure 8.2: Diagram that we will Look for Paths On

[4] And if this seems tedious (and it is), don't worry. Once you've done this to a few diagrams, you'll start to be able to spot all the paths without having to go through this slog.

- Let's follow an arrow. Let's go straight to Outcome.

- We've reached outcome so that's a path. Treatment → Outcome is a path.

- Back up to Treatment. Follow another arrow. This time to A.

- Now follow an arrow out of A. Let's go to Outcome. Done! Treatment → A → Outcome is a path.

- Back up to A. Take the other arrow out to B.

- Where can we go from here? Only to C without repeating a variable.

- And from C we can only go to Outcome. Done! Treatment → A ← B → C → Outcome is a path.

- Back up to B, but there's nowhere else to go.

- Back up to A, and nowhere else to go.

- Back up to Treatment. The only arrow left is B.

- From B we can go to A, and then on to Outcome. Done! Treatment ← B → A → Outcome is a path.

- Back up to A, then back up to B. Only path remaining is C, then Outcome. Done! Treatment ← B → C → Outcome is a path.

- Back up to C, nowhere to go, back up to B, nowhere else to go, back up to Treatment, nowhere else to go. We've exhausted all the possibilities. We're done!

The full list of paths is:

- Treatment → Outcome

- Treatment → A → Outcome

- Treatment → A ← B → C → Outcome

- Treatment ← B → A → Outcome

- Treatment ← B → C → Outcome

LET'S PRACTICE. The example diagram in Figure 8.3 has a few more moving parts so we can try to work through the full list of paths. Give it a shot yourself first—try to find all the paths you can, and then see how well you've matched up to the full list.

Figure 8.3 shows the relationship between drinking wine and your lifespan. There are a few reasons why you might expect people who drink more wine to live longer. Perhaps wine itself really does have a causal effect on your lifespan, for one. Also, the kinds of people who choose to drink wine might also live longer for other reasons, such as their baseline level of health or their income.[5] Also, maybe wine-drinking affects your penchant for taking other drugs (perhaps as a substitute?) and so affects your lifespan in that way. All of that is on the diagram.

[5] "Health" here is meant to imply a *baseline* level of health, i.e., how healthy you might be before drinking (or not drinking) wine.

Figure 8.3: The Effect of Wine on Lifespan

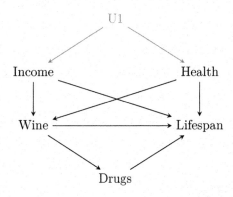

Let's follow our steps and see where it leads us.

- From Wine, we can go to Lifespan. Easy. Wine → Lifespan.

- Now back to Wine. The next arrow out is Drugs, which only goes to Lifespan. Wine → Drugs → Lifespan

- Back to Drugs, no other way out, so back to Wine. Next out let's go to Income.

- From Income we can go to Lifespan: Wine ← Income → Lifespan.

- Now back to Income. The only way out we haven't covered is U1. From there we have to go to Health. And from Health we've already visited. Wine so the only way is Lifespan. Wine ← Income ← U1 → Health → Lifespan.

- Back to Health, every other arrow out is somewhere we've been. Then back to U1, have to go back to Income, and

again we've already taken every arrow out already, so back
to Wine.

- The only arrow we haven't been out of yet is Health. So we
go to Health and can then go on to Lifespan. Wine ← Health
→ Lifespan.

- Back to Health. We can take the arrow to U1 out of Health,
which goes to Income, which goes to Lifespan. Wine ←
Health ← U1 → Income → Lifespan.

And there we have it. We've exhausted all the possible ways
we can get from Wine to Lifespan. The full list of paths is

- Wine → Lifespan

- Wine → Drugs → Lifespan

- Wine ← Income → Lifespan

- Wine ← Income ← U1 → Health → Lifespan

- Wine ← Health → Lifespan

- Wine ← Health ← U1 → Income → Lifespan

And each of those paths contains a story—a reason why we'd
see a relationship between Wine and Lifespan in data. Wine
can affect your drug-taking, which affects your lifespan (Wine →
Drugs → Lifespan). Or people who drink wine tend to be richer,
and richer people live longer (Wine ← Income → Lifespan).
Or people who drink Wine tend to be richer, richer people on
average have better health, and health improves your lifespan
(Wine ← Income ← U1 → Health → Lifespan). A story for
every path. But which stories matter to us?

8.3   Good Paths and Bad Paths, Front Doors and Back Doors

NOW THAT WE HAVE our list of paths, what can we do with
them?

The promise I've given you is that we can use these paths to
figure out how to answer your research question. And we can.
At this point, the next step in using the paths we've come up
with is to figure out just exactly how they relate to your research
question.

This brings us to the concept of *Good Paths* and *Bad Paths*.[6]

[6] In this chapter, I'll capitalize these terms to draw attention to them. That seems like it would get old real fast though, so I'll limit that practice to just this chapter.

In short, Good Paths are the reasons why the treatment and outcome variables are related *that you think should "count" for your research question.* Bad Paths are the paths that *shouldn't count,* in other words the alternate explanations.

Usually, this is as simple as "every path in which all the arrows face *away* from Treatment are Good Paths, and the rest are Bad Paths." Paths where all the arrows face *away* from Treatment are also known as *front door paths.* The rest would then be "back door paths." Paths with at least one arrow pointing towards Treatment are back door paths. So, usually, all the front door paths are good, and all the back door paths are bad.[7]

Taking Figure 8.3 as an example, there are two paths where all the arrows are going *away* from Wine: Wine → Lifespan and Wine → Drugs → Lifespan. These are all the ways in which a change in Wine would cause a change in Lifespan. So these are our front door paths. And if our research question is "does Wine cause Lifespan?" then these front door paths are the Good Paths, and the other, back door paths are bad ones.

IT MIGHT NOT BE QUITE AS SIMPLE AS THAT depending on what your research question is. For example, if instead of asking "does Wine cause Lifespan?" you want to know "other than how it might affect drug-taking, does Wine cause Lifespan?"

With this research question, the Wine → Drugs → Lifespan suddenly becomes a path we're *not* interested in, even though it's a front door. If we look at the raw data, one reason why Wine and Lifespan would be related is because of Drugs. But we are no longer interested in that part of the effect! We might at this point consider it a Bad Path.

Research questions like this, where the only Good Path is Treatment → Outcome, are looking for "direct effects." They are uninterested in "indirect effects" that take detours through variables like Drugs.

SO WE HAVE ALL OUR PATHS, and we've figured out which of them are Good and which are Bad. The key to identifying the answer to our research question, then, is to find a way to dig *just the Good Paths* out of the data without getting distracted by the Bad ones.

Good path. A causal path is Good if it is describing a reason why treatment and outcome are related that answers your research question.

Bad path. A causal path is bad if it is describing a reason why treatment and outcome are related that is unrelated to your research question, or is an alternate explanation of the data.

Front door path. A causal path where all the arrows point away from Treatment.

Back door path. A causal path where at least one arrow, somewhere along the line, is pointing back left towards the treatment variable.

[7] In Abel and Annie's case in Chapter 5, the Good, front door path was through the basement window, and all the other Bad Paths out of the house were back door paths.

Direct effect. The causal path Treatment → Outcome.

8.4   Open and Closed Paths

THOSE PESKY BAD PATHS can cause us some real trouble. But only as long as they're open for business. Open for... trouble business. In the business for trouble? Troubling business. Put something cool here.

The presence of a path on your diagram means that there is a relationship between the variables at the beginning and end of the path explained by the variables along the path. Wine → Drugs → Lifespan means that Wine and Lifespan will be related partly because of how Wine affects Drugs. Wine ← Income → Lifespan means that Wine and Lifespan will be related partly because of how Income affects both Wine and Lifespan.

However, this is only true if the path is *Open*. A path is Open if all of the variables along that paths are allowed to vary. It is instead Closed if at least one of the variables along that path has no variation.

What do I mean by this? Let's take Wine → Drugs → Lifespan as an example. If in our data set we have wine drinkers and non-wine drinkers, drug-users and non-drug users, and people with shorter and longer lifespans, then all the variables along this path have variation. They vary in value.

But what if we pick a data set in which *nobody uses drugs?* In this case, there's no variation in Drugs, and thus none of the relationship between Wine and Lifespan can possibly be driven by Drugs. The path has Closed.[8]

That's the basic idea—paths become Closed, and thus no longer a threat to our identification—if we can remove all the variation due to a variable along that path. Picking a sample in which there's no variation in that variable, as we did by picking a sample where nobody did drugs, is one way to do that, although it's not a particularly great way. There are other ways.

For now, let's just assume we have some way of controlling for a variable. In other words, we have some way of statistically adjusting our data so as to remove all the variation in a variable and thus closing a path with that variable on it. Holding that variable constant and un-varying, if you prefer. We talked about how to do this by estimating conditional conditional means in Chapter 4.[9] We could also do this by, as just mentioned, picking a data set where the variable doesn't vary. Or by picking a set of untreated observations with very similar (or exactly the same) values of that variable.[10]

If we can control for at least one variable on each of our Bad Paths without controlling for anything on one of our Good Paths, we have identified the answer to our research question.

And there it is. That's how you do it.

Open path. A causal path is Open if all of the variables along the path have variation in the data.
Closed path. A causal path is Closed if at least one of the variables along the path has no variation in the data.

[8] You can also imagine also picking a sample in which nobody drinks wine—all the paths with Wine (i.e., all the paths) would Close. We obviously wouldn't be able to study the effects of wine in that sample, since there aren't any wine drinkers to compare the non-wine-drinkers too. Another way to say that is that all the paths are Closed.

[9] Or adding it as a control variable to a regression, as will be discussed further in Chapter 13. But this is a specific application of "controlling"—we're speaking generally here.
[10] This last approach is called matching, and we will cover it much more thoroughly in Chapter 14.

LET'S TRY THIS OUR ON OUR WINE STUDY. As a recap, here's the diagram again in Figure 8.4, as well as our list of paths:

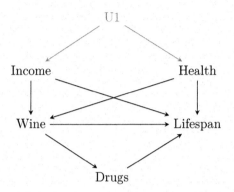

1. *Wine → Lifespan

2. *Wine → Drugs → Lifespan

3. Wine ← Income → Lifespan

4. Wine ← Income ← U1 → Health → Lifespan

5. Wine ← Health → Lifespan

6. Wine ← Health ← U1 → Income → Lifespan

If we're interested in the effect of Wine on Lifespan, we want all the ways in which Wine can cause Lifespan to change. We want all the Front Door Paths. We have two Front Door Paths, which means we have two Good Paths, marked with *. The rest are Bad Paths.

We can identify the answer to this research question by picking at least one variable along each Bad Path to control, without controlling for anything on a Good Path.

Let's start by controlling for Income. What does that get us? Let's turn all the Closed Paths gray to see. Income shows up in paths 3, 4, and 6, so those are all now Closed.

1. *Wine → Lifespan

2. *Wine → Drugs → Lifespan

3. Wine ← Income → Lifespan

4. Wine ← Income ← U1 → Health → Lifespan

5. Wine ← Health → Lifespan

6. Wine ← Health ← U1 → Income → Lifespan

Only one Bad Path to close! If we control for Health we can do it. Controlling for both Health and Income we get...

1. *Wine → Lifespan

2. *Wine → Drugs → Lifespan

3. Wine ← Income → Lifespan

4. Wine ← Income ← U1 → Health → Lifespan

5. Wine ← Health → Lifespan

6. Wine ← Health ← U1 → Income → Lifespan

So if we control for both Health and Income, any remaining relationship between Wine and Lifespan identifies the answer to our research question—what is the effect of Wine on Lifespan?[11]

THERE'S A WRINKLE: COLLIDERS. There's something I've been leaving out about this whole Open and Closed Paths thing. So far I've said that paths are Open as long as every variable along the path is allowed to vary, but removing the variation from a variable on the path (controlling/adjusting for it) Closes the path.

But that's not always true. Specifically, that's not true when there is a *collider* on the path. A variable is a *collider* on a particular path if, on that path, both arrows point at it.

For example, on the path Treatment ← A → B ← C → Outcome, B is a collider, since the arrows on either side of it on the path point towards it.[12]

When there's a collider on the path, that path is *Closed by default*. That is, it's Closed even if all of the variables on the path are allowed to vary.

Further, if you control for the collider, *the path Opens back up!* Thankfully, you can Close it back down again by controlling for another variable along the path.

Why does this happen? Why does it matter that both arrows are pointing at the variable?

You can think of it this way: the collider variable doesn't cause anything else on the path. It's just being *caused by* the variables to its left and right on the path. So if we're looking for alternate explanations of why Treatment and Outcome might be related, the collider shuts that alternate explanation down.

[11] Assuming our model was correct.

Collider. A variable is a collider on a path if the arrows on either side of it both point at it.

[12] Notice that being a collider has to do with where it is *on a particular path*. It's not a feature of the variable, it's a feature of the path. B might not be a collider on other paths it's a part of.

If the path were Treatment ← C → Outcome, without a collider, then one reason why Treatment and Outcome vary together is because C causes them both. But with a collider, Treatment ← A → B ← C → Outcome, C can affect Outcome, and C can affect B, but because B doesn't affect Treatment, C can no longer induce a relationship between Treatment and Outcome. B saved us.

Okay, so that's why a collider Closes a path by default. But why does controlling for the collider Open it back up? Because once you control for the collider, *the two variables pointing to the collider become related*,[13] and suddenly C can affect A and thus affect Treatment again. The alternate explanation returns.

Colliders seem like not a big deal. They're only a problem if you control for them. So don't control for them, right? Easy. Two problems with that. First, we need to figure out that a variable *is* a collider so we know not to control for it—if you aren't careful, colliders are often disguised as variables it feels like you *should* control for—after all, they do belong on the diagram, so seems reasonable to include it as a control, right? Second, one common way we control for colliders is by *selecting a sample*. Remember, picking a sample with no variation in Z is one way of controlling for Z. So if you do a study, say, of college students, then you're controlling for college attendance whether you want to or not. If college attendance is a collider... oops!

WHAT'S THE GOAL, THEN? If we want to identify the answer to our research question, what we have to do is *Close all the Bad Paths* while *leaving all the Good Paths Open*. Once we've done that, any remaining relationship between Treatment and Control can *only* be going through the Good Paths, and all of the Good Paths we want are included. This is exactly what we want.

This takes a careful consideration of the diagram and the paths along it. Which ones are Good? Which are Bad? Which are Closed by default because of colliders? Which are Open and need one of the variables along it controlled for so it can be Closed? Can we be careful not to Close any Good Paths?

This is the process of identification.

[13] Uh, why is that? For example, imagine your boss has a habit of ordering sandwiches for the whole office, but they always announce it after you've already ordered lunch. So we have the path Buy A Sandwich → Eat A Sandwich ← Gifted A Sandwich. Whether you've bought a sandwich is totally unrelated to whether your boss gives you one. But *if we know you Eat A Sandwich* (i.e., control for the collider), they *are* related—if you didn't Buy A Sandwich then you *must* have been Gifted A Sandwich. So now there's a link from Buy A Sandwich and Gifted A Sandwich, and an alternate explanation can occur.

8.5 *Using Paths to Test Your Diagram*

BEFORE WE MOVE ON, THERE'S ONE MORE NEAT TRICK WE CAN PLAY with causal paths.

Everything we've talked about so far has been about looking for paths between Treatment and Outcome. But that's not all you can do in a diagram. We can look for paths between *any* two variables.

Now, this might not directly help us answer our research question, which really only cares about paths between Treatment and Outcome. But what it *can* do is help us determine whether we have the right diagram in the first place.

Let's pick two variables on our diagram other than Treatment and Outcome. Let's call them A and B. The basic idea is this: list all of the paths between A and B. Then, do what you need to do to make sure they're *all* Closed.[14] Then, if A and B are still related to each other (as might be determined using any of the methods in Chapter 4), that means there must be some *other* path you didn't account for. Your diagram is deficient, and perhaps in an important way.

[14] In the lingo of causal diagrams this is known as "d-separation." The two variables are *separated* from each other by the control variables we've chosen.

LET'S DO A QUICK EXAMPLE, ONCE AGAIN TURNING TO WINE. Let's pick two variables on Figure 8.4. Let's try it with Income and Drugs. What are all the paths between Drugs and Income? Using our steps from before, we get:

- Drugs ← Wine ← Income

- Drugs ← Wine ← Health ← U1 → Income

- Drugs ← Wine ← Health → Lifespan ← Income

- Drugs ← Wine → Lifespan ← Income

- Drugs → Lifespan ← Income

- Drugs → Lifespan ← Wine ← Income

- Drugs → Lifespan ← Wine ← Health ← U1 → Income

- Drugs → Lifespan ← Health → Wine ← Income

- Drugs → Lifespan ← Health ← U1 → Income

That's a lot! But the list of Open Paths is much smaller, since Lifespan is a collider everywhere it turns up. The list of Open Paths is only:

- Drugs ← Wine ← Income

- Drugs ← Wine ← Health ← U1 → Income

If we can then control for Wine, both of these paths Close too. What we've learned is that our diagram effectively makes the claim that *there's no way that Income and Drugs are related to each other except through the amount of wine you drink.* That seems... unlikely. We can then check in the data to see if Income and Drugs *are* related after controlling for Wine. If they are (and they likely will be), our model is incomplete.

Tests like these, where we *expect* that a relationship should be zero because our diagram says there are no Open Paths, and we see whether it's actually zero, are called *placebo tests*. They get the name because, like a placebo drug, there shouldn't be anything there. But maybe there is anyway! We'll be seeing plenty of placebo tests throughout this book.

Placebo test. Finding a relationship that your causal diagram says should be zero, and checking whether it's actually zero.

FAILING A PLACEBO TEST is not the end of the world, thankfully. Yes, failing one of these tests does prove that your model is incorrect and incomplete. But let's be honest, we already knew that before we started. Models necessarily leave things out. The question isn't whether you've omitted anything, it's how important the omission is.

So then what's the point of these tests?

Well, even though failing a test doesn't mean the end of the world, it should at least lead you to reflect on whether the model can be improved. Even the process of setting up the test can be informative. Did you notice when looking at the original Figure 8.4 that it implied that Income only affected Drugs through Wine? Maybe, or maybe not. Regardless, the test definitely made that clear.

Also, even if we find a relationship, there are degrees of relationship. If you find a small but nonzero relationship that, according to the diagram, shouldn't be there, that might be a minor case for concern. But if there's an *enormous* and super-strong relationship that shouldn't be there, that's when you should really worry and maybe go back to the drawing board on your diagram.

8.6   Path Glossary

We've talked about a lot of different kinds of paths. Here are some reminders as to what these terms mean.

Good Path. A path that relates to your research question/a path you are trying to identify.

Bad Path. A path that is not related to your research question.

Front Door Path. A path where all the arrows point away from Treatment.

Back Door Path. A path where at least one of the arrows points towards Treatment.

Open Path. A path in which there is variation in all variables along the path (and no variation in any colliders on that path).

Closed Path. A path in which there is at least one variable with no variation (or a collider with variation).

Collider. A variable is a collider along a path if both arrows on either side of it point at it.

9
Finding Front Doors

9.1 Looking Ahead

THE PROSPECT OF IDENTIFYING the answer to your research
question by closing all back doors is actually kind of daunting.
Think about everything you have to do. You have to model the
data generating process, you have to list out all the paths, you
have to find a set of variables that close all the back doors, and
you have to measure and control for all of those variables.

Tough work! Especially that last part. Actually controlling
for everything is surprisingly tricky, especially in social science
where there are *so many things* that might matter that you're
almost certain to run into a variable you *have* to control for
but can't. Even if it's possible to measure, it's probably not
in *your* data. "Attitude towards risk," "Curiosity," "Customer
sentiment," "Intellectual ability"—these are all things that you
might well find sitting on a bad path but don't have access to
in your data.

Even worse, *because* there are so many variables that might
be on back doors, what are the odds you're actually going to

DOI: 10.1201/9781003226055-9

think of them all and include them in your causal diagram? Seems pretty likely that you'd miss one.[1]

SOUNDS BAD. SO WHAT TO DO? An alternate approach to identifying the answer to a research question is to, instead of actively closing back doors, find ways of isolating just Front Doors. If we can estimate the Front Doors directly, we don't need to worry about closing back doors.

How is this possible? This can work in two different ways: *either* find a setting in which only *some* of the variation in Treatment has back doors, but *some* of the variation has no back doors, or at least only has back doors you can close— natural experiments—*or* you can estimate individual arrows on the Front Door paths even if the overall effect isn't identified— the front door method.

For example, imagine trying to estimate the effect of Wealth on Lifespan. There are bound to be plenty of back door paths on that causal diagram, passing through all kinds of variables you don't have data on, like "Business Skill," "Willingness to Commit Robberies," and so on.

But how about wealth among people who play the lottery? Certainly, for most of the variation in wealth between people who play the lottery, they got it through means like working, inheritance, or buying assets, and for that variation in wealth all those same back doors are there. However, among people who buy lottery tickets, work, inheritance, and buying assets have *nothing* to do with whether you *win the lottery*. So if your research design focuses just on differences in wealth driven by lottery winnings among people who play the lottery—surprise! No back doors on that variation.[2]

9.2 *Trying to Push a String*

SO HOW CAN WE PICK OUT JUST THE VARIATION WE WANT? It all comes down to the fact that your treatment variable varies for different reasons. If A, B, and C are all causes of treatment, then A, B, and C are all reasons why treatment varies from one person to another. If you're lucky, some of those reasons lead to variation that has back doors, and other reasons don't. If you're even luckier, you can actually isolate just the part that doesn't have back doors.

The key idea here is that we can *partition the variation* in Treatment. By either selecting a particular sample or using certain approaches to statistical adjustment, we can throw out

[1] For this reason, you don't have to look hard to find researchers who would distrust *any* causal claim made using an approach of closing Bad Doors by controlling for things. Much of economics is this way. Those researchers would instead focus on the kinds of designs described in this chapter, and detailed much more fully in the second half of this book.

[2] Of course, you might also think that the effect of *lottery wealth* might be different from the effect of *wealth overall*. And you'd be correct—this analysis would give you only the effect of lottery wealth, not wealth overall, which might be your actual interest. Technically this analysis would give you a "local average treatment effect," which we'll discuss further in Chapter 10.

the part that *is* driven by all that nasty back door business, and leave ourselves to do analysis with just another part that *isn't*. Then we focus on the part that *isn't* so we don't have to worry about back doors.

PERHAPS THE CLEANEST APPLICATION OF THIS APPROACH is the randomized controlled experiment. In a randomized controlled experiment, the researcher actually steps in and assigns treatment (or the absence of treatment) to people, and watches the resulting differences in outcome.

You're probably familiar with the concept. You probably did a few in your science classes in middle school. You may even be familiar with the idea that randomized experiments are sometimes referred to as the "gold standard" of causal research designs.[3] But *why* do they work?

Experiments work because they create a form of variation in the treatment that has no back doors. If the treatment was assigned randomly, then for everyone in the experiment, variation in all the variables on all the back doors should be unrelated to whether they got the treatment or not. So all the back doors are closed!

Let's say we're interested in figuring out whether charter schools improve students' test scores more (or less) than traditional public schools, a hot-button issue in the United States and a frequent setting for isolating front doors.[4] There are a *whole lot* of variables that cause you to attend a charter school or not, with race, background, personality, location, and academic interest giving us only a few. No way we could control for enough things to close all those back doors. This is represented by the unobservable AllKindsaStuff in Figure 9.1.

Randomized controlled experiment. When someone has explicit control over who gets treatment and who doesn't, and assigns treatment in a random way.

[3] Is that actually true? Randomized experiments have their pros and cons compared to research using observational data. The identification is, obviously, much easier and more believable with a randomized experiment. On the other hand, people might not behave as they normally would when they're part of an experiment. Or the logistical limitations of recruiting participants may mean that experimental samples aren't great representations of the wider population. Plus, samples tend to be smaller in experiments. There are no easy answers.

[4] Julia Chabrier, Sarah Cohodes, and Philip Oreopoulos. What can we learn from charter school lotteries? *Journal of Economic Perspectives*, 30(3): 57–84, 2016.

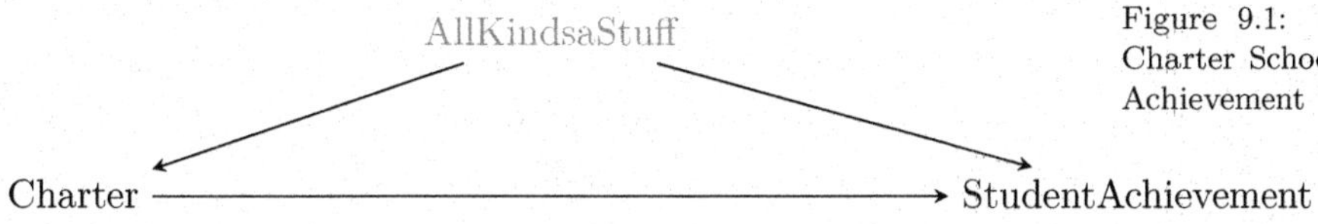

Figure 9.1: The Effect of Charter Schools on Student Achievement

However, our diagram doesn't stop there. A lot of charter schools have more interested students than they have slots, and many of them assign those slots by lottery, providing a convenient setting for lots of experimental analyses. So the diagram really looks like Figure 9.2.

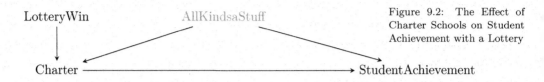

Figure 9.2: The Effect of Charter Schools on Student Achievement with a Lottery

Notice a few things: First, there are no back doors from Lottery to StudentAchievement. The effect of LotteryWin on StudentAchievement is identified in the data without any controls. Second, the only way that Lottery can affect StudentAchievement is through Charter. So if we calculate the effect of LotteryWin on StudentAchievement, that must really be telling us about the effect of Charter on StudentAchievement.

This all works because there are now two reasons why people go to Charters. The first is because of AllKindsaStuff—that part's fraught, let's avoid it. The second is because of the LotteryWin. If we can isolate just the part driven by Lottery, we've identified the answer to our research question.

WE CAN ANALYZE THE DATA FROM THIS EXPERIMENT IN A FEW WAYS:

1. Throw out all the data where Charter isn't driven by Lottery, and then look at the effect of Charter on StudentAchievement. In this context that means throwing out data from any schools without a lottery, or data from any students that got in through means other than the lottery, or any students who weren't even eligible for the lottery. Just take the students who were in the lottery and compare the ones who got into the Charter against the ones who didn't.[5]

    [5] This is what most people think of when they think of experiments—you only use the data *from the experiment.*

2. Use LotteryWin to *explain* or predict Charter. Then, take your prediction of whether someone goes to a Charter, which is based purely on their LotteryWin. Because this prediction is based only on LotteryWin and not AllKindsaStuff, the variation in the prediction contains none of the back doors of AllKindsaStuff. So then look at the relationship between the prediction and StudentAchievement to get the effect.[6]

    [6] This is the instrumental variable method, which will be covered more thoroughly in Chapter 19. Some other methods in the second part of this book apply variations on this idea.

In both approaches we isolate just the variation in Charter driven by LotteryWin and throw out the rest of the variation, either by tossing out data subject to the rest of the variation, or by focusing just on the variation we estimate to be due to LotteryWin.

This makes clear two things—first, that a randomized experiment very cleanly lets us identify the causal effect we're interested in, and second, that a randomized experiment requires us to focus on a very narrow slice of the data: the slice that is randomized. If that data doesn't represent the wider population, we won't get to our true effect no matter how big the sample is, or how clean the identification. This is one suspicion that some researchers have about charter school lotteries—that the schools that hold lotteries, and the students that enter them, aren't representative of the broader populations of charter schools and students, respectively, and so don't quite answer the question we want.[7]

[7] To account for all of this, we'd have to add some nodes onto our diagram like ChoosesToEnterLottery.

Putting that to the side, we definitely get a very clean identification. But to do this we need explicit randomization. This whole book is premised on the idea that this isn't always possible or feasible! So what else can we do?

9.3 What the World Can Do for Us

A "NATURAL EXPERIMENT" IS A REAL-WORLD SETTING IN WHICH SOME SORT OF RANDOMIZATION HAS BEEN DONE FOR US. In fact, you may have noticed a little cheat in the last section—that charter school study *isn't* a randomized controlled trial. It just so happens that the charter schools did some randomization. It wasn't researchers assigning things.[8] That already is an example of a natural experiment—the randomization occurred in the world; the researcher just came along to take advantage of it. The "wealth of lottery winners" example from earlier works like this too. Researchers don't decide who wins the lottery. But among lottery winners, there's randomization in who gets the big prize.

[8] At least not in most charter schools.

Natural experiment. When randomization of treatment occurs without a researcher controlling the randomization.

So the randomization doesn't need to be researcher-controlled. But can we go further? Does it even need to be as random as an explicit lottery? What does "randomization" mean, anyway?

We can think about natural experiments by considering what makes randomized experiments work in the first place. They work because they fix some of the variation in treatment to have no back doors. As long as we can make *that* magic happen, we have a working natural experiment.

SO WHAT WE NEED TO THINK ABOUT IS whether we can find a *source of variation in treatment* that has *no open back doors*. We can call this a "source of exogenous variation." Any path we can walk from the source of exogenous variation to the

Exogenous variation. Variation, or a variable, is exogenous, i.e., "coming from outside" if there is no other variable in the data generating process that causes it (perhaps after controlling for some things, making it "conditionally exogenous").

outcome must be (a) closed, or (b) contain our treatment.[9] An *ideal* source of exogenous variation is not caused by any other variable that belongs on the causal diagram.

This means that we can use plenty of things as sources of exogenous variation even if they're not purely random, as long as they're *as good as random in the context of our data generating process.*

What does the causal diagram for this look like? It looks... almost exactly like a randomized controlled trial! You can see for yourself in Figure 9.3.

[9] This is clearly true of actual randomization—there's no way for the randomization to affect the outcome except by causing you to be treated. So it's exogenous (not caused by anything else in the data generating process) and any link between the randomization and our outcome must be because of the treatment itself.

Figure 9.3: A Basic Natural Experiment Diagram

There are four real differences between randomized controlled experiments and natural experiments.

1. Sometimes there *will* be back doors from the NaturalRandomness to the Outcome, which doesn't happen with pure randomization. For example, let's say we're using the fact that Sesame Street became available at different times in different areas to examine the effects of kids watching Sesame Street. That's not purely random—it likely became available earlier in larger markets like urban areas, so there's a back door with SesameStreetTiming ← Urban → Outcome. But as long as we can control for something to shut the back door down, we're okay. The process is the same as when we're identifying the effect of our treatment by controlling for things—the idea here is that we're just picking a variable where the back doors are easier to control for.

2. Natural experiments are, well, more natural. People may not even realize they're a part of an experiment (in fact, nobody, including the researcher, may notice there's an experiment happening at all until long after it's over). So the observations you get may be more realistic. Sample sizes tend to be bigger, too. And people usually don't have to volunteer to be part of the experiment, so your sample isn't made up of a bunch of volunteer-types.

3. Because, just like in an experiment, we are isolating just the variation in treatment that is driven by the NaturalRandomness, we are tossing out any treatment that occurs for other reasons. So we are seeing the effect only among people who are sensitive to NaturalRandomness—if the effect would be different among another group of people, we won't see it for them. Some parents would never let their kids watch Sesame Street no matter whether it's available or not. Maybe Sesame Street would be even better for those kids than for the kids who *do* see it. But we'll never learn with a natural experiment![10]

4. People believe the exogeneity of pure randomization. But convincing people that your not-perfectly-random source of exogenous variation is exogenous in your data generating process, given that we're doing social science where everything is related to everything else, can be a tall order. Is there really only *one* back door from SesameStreetTiming to Outcome?

That last difference is a big one, and for some people it puts them off all but the purest and cleanest natural experiments. Does this whole process even work at all? Let's take a look at some example studies and see how far they can get.

9.4   Is it Too Good to be True?

FOR OUR FIRST STUDY, let's expand on what we already know: lotteries. Scott Hankins, Mark Hoekstra, and Paige Skiba published a study in 2011 on the effect of winning the lottery on declaring bankruptcy later.[11] Specifically, they looked at the Florida lottery, and only included people who had won *some amount* of money in the lottery.

They look only at people who had won some amount of money, rather than at everybody, to remove any variation based on the kind of people who play the lottery in the first place. If everyone in your sample plays the lottery, you've controlled for variables related to being the kind of person who plays the lottery.

Within that group, winning a really big prize should be completely random. They then compare people who win really big prizes (between $50,000 and $150,000) against people who won smaller prizes (less than $10,000). They find that winning a big prize reduces your chances of declaring bankruptcy *initially*, but after a while it doesn't seem to matter—the prize just pushed

[10] This issue will be discussed further in Chapter 10.

[11] Scott Hankins, Mark Hoekstra, and Paige Marta Skiba. The ticket to easy street? The financial consequences of winning the lottery. *Review of Economics and Statistics*, 93(3):961–969, 2011.

bankruptcy to a later date, rather than reducing the chances you go bankrupt. You can see their results in Figure 9.4. People who win small prizes (the dashed line) have basically the same bankruptcy rates whether it's before the prize or after, which makes sense. For people who win big prizes (the solid line), in the few years immediately following the prize, bankruptcies fall. However, by three years after the prize they spike back up, and then revert to matching those who won a small prize. In the aggregate, the effect on bankruptcy seems to be just pushing it from 1-2 years in the future to 3 years in the future.

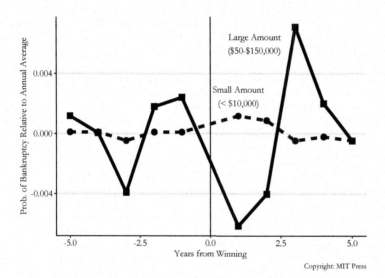

Figure 9.4: Winning the Lottery and Going Bankrupt, from Hankins, Hoekstra, and Skiba (2011)

Copyright: MIT Press

What needs to be true for this to work? It really needs to be the case that winning is random. So they do a number of things to make sure that's true. They check whether the chances of winning seem to be related to anything they observe, such as the characteristics of where the winner lives. Nope! No relationship, which is a relief. However, there is one thing they come up with—the rules of the Florida lottery changed over time, altering the number of small-prize winners. If bankruptcies also change over time, we have a Winner ← Time → Bankruptcy back door. So they control for the year in which the prize was given to account for this.

These results seem pretty solid![12] It's hard to imagine another reason why winning the lottery would be related to being bankrupt other than... because you won the lottery. Of course, these results only apply to the kind of people who play the lottery, but at least for that group in Florida, winning a big lottery

[12] At least in design. We might be slightly concerned that the big-prize winners in Figure 9.4 have bankruptcy rates that seem to jump up and down *before* they win, too. This might be an issue of noise introduced by not having *that* many big-prize winners.

prize doesn't seem to reduce your chances of going bankrupt in the long run.

SO WHEN THE WORLD PROVIDES US WITH ACTUAL LIT-
ERAL RANDOMIZATION, THAT GIVES US A CONVINCING DE-
SIGN. LET'S GO FURTHER and use a study where the source of exogenous variation is less clearly random: the wind. Pan He and Cheng Xu look at whether air pollution being worse causes people to drive more.[13] The research question makes sense—if it's smoggy and unpleasant outside, you're not going to want to walk, bike, or maybe even take the bus. But that's a problem. Cars cause pollution. So if pollution causes cars... well, that spiral isn't going anywhere good.

[13] Cheng Xu. *Essays on Urban and Environmental Economics.* PhD thesis, The George Washington University, 2019.

He and Xu get their data from Beijing, where pollution is quite bad. They look at whether people drive more on days when there is more pollution, and find that they do. However, pollution is related to all sorts of things that may also be related to driving, like whether the factories are running. Back doors are afoot! They find an exogenous source of pollution variation in the direction of the wind. In Beijing, a west-blowing wind blows pollution into the city. By isolating just the variation in pollution driven by wind direction, they find that an increase in daily pollution large enough to change the government's rating from "not polluted" to "polluted" increases driving by 3%.

Is the direction of the wind exogenous? It certainly seems unlikely that anything relevant to car-driving or pollution *caused* the wind.[14] However, it may well be that there are still back doors. For example, the direction of the wind might change with the season, and the season is certainly related to pollution and driving. The weather is likely to be related to all of these things, too. The causal diagram might look like Figure 9.5

[14] I'm actually not at all sure what causes the wind to blow in certain directions. Probably ask a meteorologist on that one.

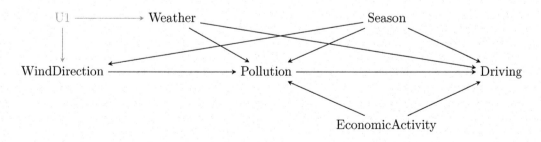

Figure 9.5: The Effect of Pollution on Driving in Beijing

So as long as we can control for season and weather—and they do—then they should be good to go. Of course, that's the thing! Are we certain that we've properly laid out all of the back doors? Are we missing anything? If we miss anything important, this won't work.

I find the use of wind direction here pretty compelling evidence. But it's certainly not as rock solid as the lottery. Remember, the difficulties of identifying the effect of X on Y by controlling to close back doors are the same here. We're just hoping to pick a variable that's easier to close back doors for than X.

HOW FAR FROM TRUE RANDOMIZATION CAN WE GO? As far as our assumptions are willing to take us—and as far as we're willing to carry those assumptions along.

Let's take a look at Camilleri and Diebold (2019),[15] who look at the effect of uncompensated care—medical care given by hospitals that they don't end up being paid for—on patient experience. You can imagine that if hospitals are spending a bunch of money giving care that they don't get paid for, they're going to have fewer resources available to give patients the best experience.[16]

The amount of uncompensated care a hospital gives is related to all sorts of back-door stuff. The kinds of patients they get and how likely those patients are to pay, the kinds of procedures they're known for, and so on. Camilleri and Diebold need a source of exogenous variation. They use the 2014 Medicaid expansion, in which some states but not others expanded access to the Medicaid program. Medicaid expansion considerably increased health insurance coverage in the states that accepted the additional aid. That additional coverage meant that there would be more compensation available for hospitals. Using this source of exogenous variation, they find that reductions in uncompensated care did improve patient experience, but only by a little bit.

Does this work as a source of exogenous variation? We can definitely imagine some back doors. After all, states didn't accept or reject Medicaid at random; the choice was highly politicized. States with different kinds of governments were more or less likely to expand Medicaid. The authors controlled for state and local characteristics to close these back doors.

Of course, the policy changed plenty of things about health care other than just improving hospital compensation. Medicaid expansion, and thus expanded access to insurance, should change lots of things about health care besides hospital compensation that might also be related to patient experience. We have to be willing to assume that the expansion really only

[15] Susan Camilleri and Jeffrey Diebold. Hospital uncompensated care and patient experience: An instrumental variable approach. *Health Services Research*, 54 (3):603–612, 2019.

[16] This takes place in the United States medical system where patients are expected to pay for care, and if they can't the hospital is sometimes on the hook for it.

affected hospital compensation in order to think of the variation in compensation driven by the Medicaid expansion as being exogenous.[17]

So does this study work? Sure! As long as the assumptions we've just laid out about Medicaid expansion only working though hospital compensation sound reasonable. If we aren't willing to swallow those assumptions, then no, it doesn't work.[18] There are lots of studies that use policy implementation as a source of exogenous variation like this. It's certainly a viable path, but we need to be very careful in thinking what assumptions we have to make about the data generating process, and whether those assumptions are true.

Isolating front paths is always feasible, just like identifying the effect of a treatment by closing back doors is always feasible, even if we don't have anything even *remotely* like purely-random variation as we would in a randomized experiment or even a lottery. However, the farther away we get from that pure randomization, the more things we need to control for, and the more assumptions we have to make, and perhaps the more *unbelievable* assumptions we have to make. This isn't a magic formula; we're just replacing the difficulty of finding and closing all back doors for a treatment variable with the difficulty of finding and closing all back doors for something else.

[17] Note that we're describing *front doors* from the Medicaid expansion here. If we wanted to know the effect of the Medicaid expansion itself, we'd want these included. But since we want to use Medicaid expansion to isolate just the part of uncompensated care that has no back doors, these are still bad paths that will mess up our research design.

[18] This is true for all studies, since all studies rely on assumptions. It's not that some studies have assumptions and others don't. Rather, we might just find one study's assumptions more believable than another's, given what we know about the real world.

9.5   Riding a Shooting Star

AH, BUT THERE IS ANOTHER WAY in which we can identify the causal effect of a treatment on an outcome by isolating front doors. It is called, appropriately, the "front door method," and it works very differently from the concepts covered so far in this chapter. It also sees a lot less usage in the real world and serves only as a tiny little coda to this chapter.

The reason the approach with the seems-pretty-important name of "front door method" gets tucked away in the last section of a chapter all about isolating front doors is that it only applies in very specialized scenarios. For a long time, applied researchers didn't really even think about it as a possible research design. And now that they do know about it in theory, they're scratching their heads thinking about when they could ever really use it in practice. So we'll talk about it here, briefly, and heck, maybe *you* can think about how it might be useful.

The front door method works when your causal diagram looks something like Figure 9.6, when there's a bad path that can't be closed, such as if W in that diagram can't be measured.

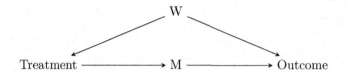

Figure 9.6: A Diagram the Front Door Method Works On

IN FIGURE 9.6, IF W CAN'T BE MEASURED, then we can't control for anything to identify the effect of Treatment on Outcome.

However, maybe we can identify something else. How about the effect of Treatment on M? We can identify that—the only back door is Treatment ← W → Outcome ← M. But Outcome is a collider on that path so it's already closed. We don't need to worry about it.

What else can we identify? How about the effect of M on Outcome? The only back door is M ← Treatment ← W → Outcome, and we can control for Treatment to close that back door path.

So we can identify both Treatment → M and M → Outcome. We just need to combine those two effects to get our effect Treatment → M → Outcome.

WE CAN USE THE CLASSIC EXAMPLE given when discussing the front door method:[19] smoking. It's difficult to figure out the effect of smoking on something like cancer rates because there are lots of things related to whether you smoke (background, income, health-mindedness, etc.) that can be hard to measure and would also be related to cancer rates. So we have a lot of back doors we can't close.

But what if we have something that sits between smoking and cancer—some measurable *reason why* smoking causes cancer? Let's say that thing is TarInLungs. In this simplified fantasy, the only reason Smoking causes Cancer is because it causes TarInLungs, and TarInLungs causes Cancer.[20] The diagram then looks like Figure 9.7.

Given this diagram, let's say that we look at the raw, unadjusted relationship between Smoking and TarInLungs and find that an additional cigarette per day adds an additional 15 grams of tar to your lungs over 10 years.

Then, we look at the relationship between TarInLungs and Cancer while controlling for Smoking and find that an additional

[19] Judea Pearl and Dana Mackenzie. *The Book of Why: The New Science of Cause and Effect.* Basic Books, New York City, New York, 2018.

[20] Although having more than one "thing in the middle" is acceptable, things can get complex when you do that, as Bellemare and Bloem (2019) show.

Marc F Bellemare and Jeffrey R Bloem. The paper of how: Estimating treatment effects using the front-door criterion. Technical report, Working Paper, 2019.

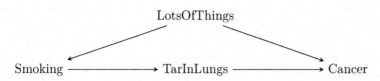

Figure 9.7: A Front-Door-Method Compatible Causal Diagram of Smoking

15 grams of tar in your lungs increases the chances of getting cancer by 2% over your lifetime.

So, then an additional cigarette per day increases the tar in your lungs by 15 grams, which in turn increases your probability of cancer by 2%. So an additional cigarette per day increases your probability of cancer by 2%.

THAT'S THE FRONT DOOR METHOD! So why is it not used very often? Largely because it requires that there be some variable like M or TarInLungs in those diagrams that exists entirely between Treatment and Outcome without being linked to anything else, and capturing a large portion of the reason why Treatment affects Outcome. That's a lot of conditions that need to be met before a method can be used. Worse, these conditions don't seem to pop up in the real world all that often. But hey, again, maybe you can figure it out.

10

Treatment Effects

10.1 For Whom the Effect Holds

FOR MOST OF THIS BOOK, AND INDEED IN THE TITLE, WE HAVE STUCK TO THE FICTION that there is such a thing as *the effect*. As though a treatment could possibly have a single effect—the same impact on literally everybody! That might be plausible in, say, physics. But in social science, everything affects everyone differently.[1]

To give a very simple example, consider a drug designed to reduce the rate of cervical cancer. This drug might be very effective. Perhaps it reduces the rate of cervical cancer by half... *for people with a cervix.* For people without a cervix, we can be pretty certain that the drug has absolutely no effect on the rate of cervical cancer.[2]

So at the very least, the drug has two effects—one for people with a cervix, and one for people without. But we don't need to stop there. Even if we just focus on people with a cervix, maybe the drug is highly effective for some people and not very effective for others. Something to do with body chemistry, or age, or dietary habits, who knows? The point is we might have

[1] If we'd just all become frictionless spheres, social science would be way easier. Downhill travel, too.

[2] Ugh, no fair.

DOI: 10.1201/9781003226055-10

a whole bunch of effects. Whenever we have a treatment effect that varies across a population (i.e., all the time), we can call that a *heterogeneous treatment effect*.

We can actually think of each individual has having their *own* treatment effect. Maybe the drug reduces the cancer rate by 1% for you and 0% for me and .343% for the woman who lives next door to me.

That's not even to say anything of the variation in the effect *outside* what you can see in the sample! The true effect of treatment is likely to vary from country to country, from year to year, and so on.[3] Just because, say, monetary stimulus improved the employment rate last time doesn't mean it will this time, at least not to the same degree. Everything varies, even effects.

We're all unique, with different circumstances, lives, physiologies, and responses to the world. Why would we start with an assumption that any two of us would be affected in exactly the same way? It's more of a convenience than anything.

SO WHAT CAN WE MAKE of the idea that we have heterogeneous treatment effects?

One thing we can try to do is to estimate those heterogeneous treatment effects. Instead of just estimating one effect, we can estimate a *distribution* of effects and try to predict, for a given person with a given set of attributes, what their effect might be.

This is a valid goal, and it is something that people try to do. This idea is behind concepts you might have heard of like "personalized medicine." It's also one thing that machine-learning types tend to focus on when they get into the area of causal inference.[4]

However, in addition to being a valid goal and the subject of some extremely cool work, it also gets highly technical very quickly. So in this chapter, we will instead focus on the other thing we can do with the concept of heterogeneous treatment effects: ask "if effects are so heterogeneous, then what exactly are we identifying anyway?"

After all, we've established in the rest of this book that we can identify causal effects if we perform the right set of adjustments. But whose causal effects are those? How can we tell?

It turns out that, if we've done our methods right, what we get is some sort of average of the individual treatment effects. However, it's often not the kind of average where everyone gets counted equally.

Heterogeneous treatment effect. An effect of a treatment on an outcome for which the effect itself varies across the population.

[3] The ability of a study with a given sample to produce an estimate that works in other settings is called *external validity*.

[4] If this kind of thing interests you, I recommend reading Chapter 21 and also go looking for anything and everything that the duo of Susan Athey and Guido Imbens have worked on together.

10.2   Different Averages

WHAT WE HAVE is the concept that each person has their own treatment effect. That means that we can think of there as being a *distribution* of treatment effects. This works just like any other distribution of a variable, like back in Chapter 3. The only difference is that we don't actually observe the treatment effects in our data.

And like any typical distribution, we can describe features of it, like the mean.

The mean of the treatment effect distribution is called, for reasons that should be pretty obvious, the *average treatment effect*. The average treatment effect, often referred to as the ATE, is in many cases what we'd *like* to estimate. It has an obvious interpretation—if you impose the treatment on everyone, then this is the change the average individual will see. If the average treatment effect of taking up gardening as a hobby is an increase of 100 calories eaten per day, then if everyone takes up gardening, some people will see an increase of less than 100 calories, some will see more, but on average it will be 100 calories extra per person.

However, estimating the average treatment effect is not always feasible, or in some cases even desirable.

Let's use the cervical cancer drug as an example. In truth, the drug will reduce Terry's chances of cervical cancer by 2 percentage points and Angela's by 1 percentage point, but Andrew and Mark don't have cervices so it will reduce their chances by 0.[5] The average treatment effect is $(.02 + .01 + 0 + 0) = .0075$, or .75 percentage points.

Now, despite your repeated pleas to the drug company, they refuse to test the drug on people without cervices, since they're pretty darn sure it won't do anything. They get a whole bunch of people like Terry and Angela and run a randomized experiment of the drug. They find that the drug reduces the chances of cervical cancer by, on average, $(.02 + .01) = .015$, or 1.5 percentage points.

That's not a *wrong answer*—in fact, their 1.5 is probably a better answer than your .75 for the research question we probably have in mind here—but it's definitely not the average treatment effect among the population.[6] So if it's not the population average treatment effect, what is it? We will want to keep in our back pocket some ideas of other kinds of treatment effect averages we might go for or might identify.

Treatment effect distribution. The distribution of the individual effect of treatment across the sample or population.

Average treatment effect. The mean of the treatment effect distribution.

[5] Aside on statistical terminology: when talking about changes in a *rate*, like the rate or probability of cervical cancer, a *percent* change is proportional, as percent changes always are, but a *percentage point* change is the change in the rate itself. So if a 2% chance rises to 3%, that's a $((.03/.02) - 1 =)$ 50% increase, or a $(.03 - .02 =)$ 1 percentage point increase. Beware percentage increases from low starting probabilities—they tend to be huge even for small actual changes.

[6] It *is* the average treatment effect among their sample, but we certainly wouldn't want to take that effect and assume it works for Andrew or Mark.

There are lots and lots and lots of different kinds of treatment effect averages,[7] but only a few important ones we really need to worry about. They fall into two main categories: (1) treatment effect averages where we only count the treatment effects of *some* people but not others, i.e., treatment effect averages conditional on something, and (2) treatment effect averages where we count everyone, but we count some individuals more than others.[8]

WHAT HAPPENS WHEN WE ISOLATE THE AVERAGE EFFECT for just a certain group of people? And how might we do it?

To answer this question, let's make some fake data. This will be handy because it will allow us to see what is usually invisible—what the treatment effect is for each person.

Once we have our fake data, we will be able to: (a) discuss how we can take an average of just some of the people, and (b) give an example of how we could design a study to *get* that average.

[7] I even have one of my own! It's called SLATE, and it's not very widely used but it's super duper cool and the way it works is hey where are you going?

[8] Technically, (1) is just a special case of (2) where some people count 100% and other people count 0%. But conceptually it's easier to keep them separate.

Name	Gender	Outcome Without Treatment	Outcome With Treatment	Treatment Effect
Alfred	Male	1	2	1
Brianna	Female	1	5	4
Chizue	Female	2	5	3
Diego	Male	2	4	2

Table 10.1: Fake Data for Four Individuals

We can see from Table 10.1 that these four individuals have different treatment effects. Keep in mind that this table is full of counterfactuals—we can't possibly see someone both treated and untreated. The table just describes what we *would see* under treatment or no treatment. If nobody were treated, then Alfred and Brianna would have an outcome of 1, and Chizue and Diego would have an outcome of 2. But with treatment, Alfred jumps by 1, Brianna by 4, Chizue by 3, and Diego by 2. The average treatment effect is $(1 + 4 + 3 + 2)/4 = 2.5$.

ONE COMMON WAY WE GET AN AVERAGE EFFECT FOR ONLY A CERTAIN GROUP is to literally pick a certain group. Notice in Table 10.1 that we have men and women. Let's say we run an experiment but only recruit men in our experiment for whatever reason.[9] So we get a bunch of guys like Alfred and a bunch of guys like Diego and we randomly assign them to get treatment or not. Our data ends up looking like Table 10.2.

Name	Treated	Outcome
Alfreds	Treated	2
Alfreds	Untreated	1
Diegos	Treated	4
Diegos	Untreated	2

Table 10.2: Men-Only Experiment

Then, using Table 10.2, we calculate the effect. We find that the treated people on average had an outcome of $(2+4)/2 = 3$, and the untreated had $(1+2)/2 = 1.5$ and conclude that the treatment has an effect of $3 - 1.5 = 1.5$. This is the exact same as the average of Alfred's and Diego's treatment effect, $(1+2)/2 = 1.5$. So we have an average treatment effect *among men*, or an average treatment effect *conditional* on being a man.

Again, this isn't a *wrong answer*. It just represents only a certain group and not the whole population. It's only a wrong answer if we think it applies to everyone.

ANOTHER COMMON WAY IN WHICH THE AVERAGE EFFECT IS TAKEN AMONG JUST ONE GROUP IS BASED ON WHO GETS TREATED. Based on the research design and estimation method, we might end up with the *average treatment on the treated* (ATT) or the *average treatment on the untreated* (ATUT), which averages the treatment effects among those who *actually* got treated (or not).

To see how this works, imagine that we can't randomize anything ourselves, but we happen to observe that Alfred and Chizue get treated, but Brianna and Diego did not. We do our due diligence of drawing out the diagram and notice that the outcome is completely unrelated to the probability that they're treated or not.[10] So we're identified! Great.

Conditional average treatment effect. An average treatment effect conditional on the value of a variable.

Average treatment on the treated. The average treatment effect among those who actually received treatment.

Average treatment on the untreated. The average treatment effect among those who did not actually receive treatment.

[10] Knowing the secret counterfactuals that we do, we can see that the average outcome *if treatment had never happened* is exactly $(1 + 2)/2 = 1.5$ for both the treated and untreated groups. In other words, there are no back doors between treatment and outcome. The differences arise only because of treatment.

Name	Treated	Outcome
Alfred	Treated	2
Brianna	Untreated	1
Chizue	Treated	5
Diego	Untreated	2

Table 10.3: Assigning Alfred and Chizue to Treatment

What do we get in our actual data? We can see in Table 10.3 that we get an average of $(2+5)/2 = 3.5$ among the treated people, and $(1+2)/2 = 1.5$ among the untreated people, giving us an effect of $3.5 - 1.5 = 2$. It's no coincidence that this is the average of Alfred's and Chizue's treatment effects, $(1+3)/2 = 2$.

In other words, we've taken the average treatment effect among just the people who actually got treated. ATT![11]

It's a bit harder to imagine how we might get the average treatment effect among the *untreated* (ATUT). And indeed this one doesn't show up as often. But one way it works is that you take what you know about how treatment varies, and what predicts who has big or small treatment effects, and then use that to predict what sort of effect the untreated group would see.

For example, say we get a sample of 1,000 Alfreds and 1,000 Briannas, where 400 Alfreds and 600 Briannas have been assigned to treatment on a basically random basis, leaving 600 Alfreds and 400 Briannas untreated.

The average outcome for treated people will be $(400 \times 2 + 600 \times 5)/1,000 = 3.8$, and for untreated people will be 1. However, we can run our analysis an extra two times, once just on Alfreds and once just on Briannas, and find that the average treatment effect conditional on being Alfred appears to be 1, and the average treatment effect conditional on being Brianna appears to be 4. Since we know that there are 600 untreated Alfreds and 400 untreated Briannas, we can work out that the average treatment on the untreated is $(600 \times 1 + 400 \times 4)/1,000 = 2.2$. ATUT!

The distinction between ATT and ATUT, and knowing which one we're getting, is an important one in nearly all social science contexts. This is because, in a lot of real-world cases, people are choosing for themselves whether to get treatment or not. This means that treated and untreated people are often different in quite a few ways (people who choose to do stuff are generally quite different from those who don't), and we might expect the treatment effect to be different for them too. Borrowing the example from the start of the chapter, if people could choose for themselves whether to take the cervical cancer drug, who would choose to take it? People with a cervix! The drug is more effective for them than for people without a cervix, so the ATT and ATUT aren't the same, and that's not something we can avoid—the drug being more effective for them is *why they chose to take it*.

One other way in which a treatment effect can focus on just a particular group is with the *marginal treatment effect*. The marginal treatment effect is the treatment effect of a person who is *just on the margin* of either being treated or not treated. This is a handy concept if the question you're trying to answer is "should we treat more people?" I won't go too much into the

[11] You can imagine how the ATT might crop up a lot. After all, we only see people getting treated if they're... actually treated. So you can see how we might pick up *their* treatment effects and get an ATT. It's almost hard to imagine how we could get anything else. How can we possibly ever get the average treatment effect, rather than the ATT, if we can't see what the untreated people are like when treated? Well, it comes down to setting up conditions where we can expect that the treatment effect is the same in treated and untreated groups. In this example, they clearly aren't. But if we truly randomized over a large group of people, there's no reason to believe the treated and untreated groups would have different effect distributions, so we'd have an ATE.

Marginal treatment effect. The treatment effect of the next person who would get treatment if treatment rates expanded.

marginal treatment effect here, as actually getting one can be a bit tricky. But it's good to know the idea is out there.

INSTEAD OF FOCUSING OUR AVERAGE JUST ON A GROUP OF PEOPLE, what if we include everyone, but perhaps weight some people more than others? We can generically think of these as being called "weighted average treatment effects."

In general, a weighted average is a lot like a mean. Let's go back to the average treatment effect—that was just a mean. The mean of 1, 2, 3, and 4 is $(1 + 2 + 3 + 4)/4 = 2.5$, as you'll recall from our fake data, reproduced below in Table 10.4. Now, let's not change that calculation, but just recognize that $1 = 1 \times 1$, $2 = 2 \times 1$, and so on.

Weighted average treatment effect. An average of individual treatment effects where different individuals count more than others. Each individual has a "weight."

Name	Gender	Outcome Without Treatment	Outcome With Treatment	Treatment Effect
Alfred	Male	1	2	1
Brianna	Female	1	5	4
Chizue	Female	2	5	3
Diego	Male	2	4	2

Table 10.4: Fake Data for Four Individuals

Substituting in 1×1 for 1, 2×1 for 2, and so on, our calculation for the mean is now $(1 \times 1 + 2 \times 1 + 3 \times 1 + 4 \times 1)/(1+1+1+1) = 2.5$. Here, everyone's number is getting multiplied by 1, and that's the same 1 for everybody. This is a weighted average where everyone gets the same weight (1).

BUT WHAT IF PEOPLE GOT *DIFFERENT* NUMBERS BESIDES 1? Continuing with the same fake-data example, let's say for some reason that we think Brianna should count twice as much as everyone else, and Diego should count half as much. Now our weighted average treatment effect is $(1 \times 1 + 4 \times 2 + 3 \times 1 + 2 \times .5)/(1 + 2 + 1 + .5) = 2.89$.

There are some applications where we get to pick what these weights are and apply them intentionally.[12] In the context of treatment effects, though, we rarely get to pick what the weights are. Instead, there's something about the *design* that weights some people more than others.

A common way this shows up is as *variance-weighted* average treatment effects. Statistics is all about variation. And the relationship between Y and X is a lot easier to see if X moves around a whole lot. If you don't see a lot of change in X, then it's hard to tell whether changes in Y are related to changes in

[12] Survey/sample weights, for example, as discussed in Chapter 13.

Variance-weighted average treatment effect. A treatment effect average where the kinds of people with lots of variation in treatment left after closing back doors are counted more heavily.

X because, well, *what* changes in X are we supposed to look for exactly? What's the relationship between living on Earth and your upper-body strength? Statistics can't help there, because pretty much everybody we can sample lives on Earth. We don't see a lot of people living elsewhere, so we can't observe how it makes them different to live elsewhere.

As a result, if some kinds of people have a lot of *variation in treatment* while others don't, our estimate may weight the treatment effect of of those with variation in treatment more heavily, simply because we can see them both with and without treatment a lot.

Let's say that we get a sample of 1,000 Briannas and 1,000 Diegos. For whatever reason, half of all Briannas have ended up getting treatment, but 90% of Diegos have. So our data looks like Table 10.5.

Name	N	Treated	Outcome
Brianna	500	Treated	5
Brianna	500	Untreated	1
Diego	900	Treated	4
Diego	100	Untreated	2

Table 10.5: Briannas and Diegos get Treatment at Different Rates

Now, we can't just compare the treated and untreated groups because we have a back door. "Being a Brianna/Being a Diego" is related both to whether you're treated, and to the outcome (notice that their outcomes would be different if nobody got treated). So we want to close that back door. One way we can do that is by subtracting out mean differences between Brianna and Diego, both for the outcome and the treatment.

When we do this, and reevaluate the treatment effect, we get an effect of 3.47.[13] This is closer to Brianna's treatment effect of 4 than to Diego's treatment effect of 2. We're weighting Brianna more heavily. Specifically, we are weighting them both by the variance in their treatment, and Briana has more variance. The variance in treatment among Briannas is $.5 \times .5 = .25$.[14] The variance in treatment among Diegos is $.9 \times .1 = .09$. The weighted average, then, is $(.25 \times 4 + .09 \times 2)/(.25 + .09) = 3.47$.

Our estimate of 3.47 is closer to Brianna's effect (4) than Deigo's (2) because we see a lot of her both treated and untreated, whereas Diego is mostly treated. Less variation in treatment means we can see the effect of that variation less. Note also that Diego counts less even though we see a lot of treated Diegos—this isn't the average treatment on the treated. We

[13] The math to get here gets a little sticky, although you can refer to the Conditional Conditional Means section of Chapter 4, or to Chapter 13. But basically, we subtract Brianna's outcome average of 3 from her outcomes, giving treated Briannas a 2 outcome and Untreated Briannas a -2 outcome, and her 50% treatment from her treatments, giving treated Briannas a ".5 treatment" and untreated Briannas a "$-.5$ treatment". Similarly, treated/untreated Diegos get $.2/-1.8$ for outcome and $.1/-.9$ for treatment. Fitting a straight line on what we have left tells us that a one-unit change in treatment gets a 3.47 change in outcome.

[14] The variance of a binary variable is always (probability it's 1)$\times$(probability it's 0)—that's worth remembering.

know we're getting a variance-weighted average treatment effect rather than the average treatment on the treated, because if we were getting ATT, we'd be closer to Diego and farther from Brianna.

Weighted average treatment effects pop up a lot whenever we start closing back doors. When we close back doors, we shut out certain forms of variation in the treatment. The people who really count are the ones who have a lot of variation left after we do that.

Variance-weighted treatment effects aren't the only kind of weighted average treatment effect. For example, if you close back doors by *selecting a sample* where the treated and untreated groups have similar values of variables on back door paths (i.e., picking untreated observations to *match* the treated observations), you end up with *distribution-weighted* average treatment effects, where individuals with really common values of the variables you're matching on are weighted more heavily.

ANOTHER FORM OF WEIGHTED TREATMENT EFFECTS that pops up often is based on how *responsive* treatment is.

In Chapter 9 we discussed the different ways that we can isolate just *part* of the variation in treatment. We either focus just on the part of the data in which treatment is determined exogenously (like running an experiment, and only including data from the experiment in your analysis) or use some source of exogenous variation to *predict* treatment, and then use those predictions instead of your actual data on treatment.

Of course, heterogeneous treatment effects don't *only* apply to the effect of treatment on an outcome. They can *also* apply to the effect of exogenous variation on treatment.

For example, suppose you're running a random experiment about diet where the treatment is having to eat 100 fewer calories per day than you normally would, and the outcome is your weight. Some people have pretty good willpower and control over their diet. If you tell them to eat less, they can do that. If you tell them to keep doing what they normally do, they can do that too.

Other people have less willpower (or less interest in satisfying a researcher).[15] They might only eat 90 fewer calories per day when told to eat 100 less. Or 50. Or 5. Or 0. Maybe a few people will be disappointed by being assigned to the "continue as normal" treatment and will cut their calories anyway.

So for some people, being assigned to treatment makes them eat 100 fewer calories. For some people it's 90, or 50, 0, or 10

[15] Or it's the middle of a pandemic and the Hot Cheetos are *right there in the pantry*.

more calories, or whatever. Heterogeneous treatment effects, but this time for the effect of treatment assignment on treatment, rather than the effect of treatment on outcome.

Naturally, if we limit our data to just the people in our experiment and look at the impact of the experiment, it's going to give us strange results.

When this happens—we have exogenous variation, but not everybody follows it, we limit our data to just the people in our experiment, and we look at the relationship between treatment *assignment* and the outcome— what we get is called the *intent-to-treat* estimate.[16] Intent-to-treat is the effect *of assigning treatment*, although not the effect of treatment itself, since not everybody follows the assignment.

Intent-to-treat gives us the average treatment effect of assignment, which is usually not what we want.[17] What does it give us for the effect of treatment? It's not *exactly* a weighted average treatment effect at that point. It does weight each person's treatment effect by *the proportion of their treatment effect they received*.[18] So if you got enough treatment to get 50% of its effects, you get a weight of .5. This weighting makes a lot of sense—if you get the full treatment, we see the full effect of your treatment when we start adding up differences. If you don't get the treatment you were assigned to, we still include you in our addition, but it couldn't have had an effect so you get a 0.[19]

The thing that makes it not exactly a weighted treatment effect is that instead of dividing by the sum of the weights, you divide by the number of individuals. In a weighted average treatment effect, a weight of 0 (you didn't respond to assignment at all) wouldn't affect the weighted average treatment effect. But in intent-to-treat, someone with a weight of 0 has no effect on the numerator, but they do affect the denominator, bringing the effect closer to 0.

Intent to treat. The average treatment effect *of assigning treatment*, which is not necessarily the same as the average treatment effect *of treatment*.

[16] More broadly, we get intent-to-treat when we have *exogenous variation* of some sort driving treatment, and we look directly at the relationship between that exogenous variation and the outcome.

[17] Unless we're going to use that same assignment in the real world. If I'm using "a policy that forces insurers to cover therapy" to understand the effect of therapy on depression, maybe I *do* want to know the effect of that policy, rather than the effect of therapy itself, since I have more control as a policymaker over that policy than I do over therapy.

[18] In most cases, this is just "actually got the treatment" or "didn't" so it's just 0 and 1.

[19] All of this applies even if treatment isn't 0/1! In those cases the weights are "how much more treatment you got."

Name	Gender	Outcome Without Treatment	Outcome With Treatment	Treatment Effect
Alfred	Male	1	2	1
Brianna	Female	1	5	4
Chizue	Female	2	5	3
Diego	Male	2	4	2

Table 10.6: Fake Data for Four Individuals

Returning to our fake data once more, if we recruited two Chizues and two Diegos and assigned one of each to treatment, but Chizue went along with assignment while Diego decided never to receive treatment, then in the treatment-assigned group we'd see Chizue's 5 and Diego's 2 (since Diego was never actually treated), and in the treatment-not-assigned group we'd see Chizue's 2 and Diego's 2. The calculated effect would be $(3.5 - 2) = 1.5$. This is also $(3 \times 1 + 3 \times 1 + 2 \times 0 + 2 \times 0)/(1 + 1 + 1 + 1) = 1.5$, or the effect of the two Chizues weighted by 1 (since they receive full treatment when assigned) plus the effect of the two Diegos weighted by 0 (since they never receive any treatment), divided by the number of people (4).

What if we take the other approach to finding front doors, where we use some source of exogenous variation to *predict* treatment, and then use those predictions instead of your actual data on treatment?

This turns out to do something very similar to the intent-to-treat. However, because this approach doesn't just say "were you assigned treatment or not?" but rather "how much more treatment do we think you got due to assignment?" we can now replace that "number of people" denominator with a "how much more treatment was there?" denominator.

Since "how much more treatment" was also our weight in the numerator, we're back to an actual weighted average treatment effect. Specifically, the weights are how much additional treatment each individual would get if assigned to treatment. We call this one the *local average treatment effect* (LATE).

For example, let's go back to Chizue and Diego, and Diego not going along with his treatment assignment. We look at assignment and at treatment, and notice that being assigned to treatment only seems to increase treatment rates by 50% (in the not-assigned group, nobody is treated; in the assigned group, 50% are treated). Based on that prediction, we expect to see only half of the treatment effect, and we can get back to the full treatment effect by dividing by .5.

This gives us an effect estimate of $(3.5 - 2)/.5 = 3$. We can also get this 3 from $(3 \times 1 + 3 \times 1 + 2 \times 0 + 2 \times 0)/(1 + 1 + 0 + 0) = 3$, which is the 3 effect of the two Chizues, each with a weight of 1 (since assignment increases their treatment from 0 to 1), and the 2 effect of the two Diegos, each with a weight of 0 (since assignment doesn't affect their treatment).

In other words, the LATE is a weighted average treatment effect where you get weighted more heavily the more strongly

Local average treatment effect. A weighted average treatment effect where the weights are *how much more treatment* that individual would get if assigned to treatment.

you respond to exogenous variation.[20] This is kind of a strange concept—why would we want to weight people who respond to irrelevant exogenous variation more strongly? Well, maybe we don't. But the LATE still looms large because it happens to be the weighted average treatment effect that pops up in a lot of research designs. Maybe not what you want, but what you get.

And along those lines, what *do* you get? How do we know, for a given research design, which of these treatment effect estimates we will end up with?

10.3   I Just Want an ATE, It Would Make Me Feel Great, What Do I Get?

BY THIS POINT WE KNOW that there are far more ways to get a single representative treatment effect than just averaging them (to get the average treatment effect). We can get the treatment effect just for certain groups, we can weight some individuals more heavily than others, we can weight people based on how the treatment was assigned.

Now, usually (not always), what we want is the average treatment effect—the effect we'd see on average if we took a single individual and applied the treatment to them.[21] The reason we bring up most of those *other* treatment effects at all is that we don't always get what we want!

The treatment effect you get isn't necessarily a choice you make. It's a consequence of the research design you have.[22] And since there aren't usually multiple available research designs that you can use to answer a given question, you're often stuck with the treatment effect average you get.

So for a given research design, which one do we get?

THE TREATMENT EFFECT WE GET IS ALMOST ENTIRELY DETERMINED BY THE SOURCE OF TREATMENT VARIATION WE USE. That's pretty much it. Ask where the variation in your treatment is coming from,[23] and you'll have a pretty good idea whose treatment effects you are averaging, and who is being weighted more heavily.

We've already discussed one example of this. If we perform a randomized experiment, then we will be ignoring everyone who isn't in our experiment. *The only treatment variation we are allowing is among the people in our sample—any variation outside our sample is ignored.* If our sample isn't representative of the broader population,[24] then we will be getting the average

[20] It is common in an econometrics class to hear that the LATE is "the average treatment effect among those who respond to assignment" and you might hear those who respond called "compliers." However, this is a simplification. If one person responds fully to assignment and another only has half a response, the LATE will not average them equally, even though both are compliers. It will weight the full-response person twice as much as the half-response person.

[21] Why might we not always want this? It depends what question we're trying to answer. If we want to know "what was the actual effect of this historical policy?" then we might want to know what effect treatment had on the people it actually treated (ATT). If we want to know "what would be the effect if we treated more people?" we might want the treatment on the untreated (ATUT) or the marginal treatment effect. If we want to know "is this more effective for men or women?" we would want some conditional treatment effects. And so on.

[22] And the estimator you use. While there are limits, for any given research design there are generally different ways of estimating the effect, which may give different treatment effect averages. Some of those estimators are specifically designed to give the ATE for research designs that don't normally produce it.

[23] After removing any variation you choose to remove by controlling for things, etc.

[24] And thus doesn't have the same average treatment effect as the broader population.

treatment effect *conditional on being in our sample*, a conditional average treatment effect.

Let's take another example. Let's say we're interested in the effect of being sent to traffic school on your future driving performance. Let's also say that we know there are only two reasons anyone goes to traffic school: making a terrible driving mistake, or having *someone else* make a terrible driving mistake that you are somehow punished for. This gives us the diagram in Figure 10.1.

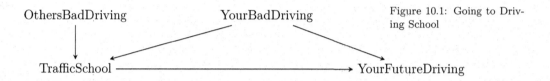

Figure 10.1: Going to Driving School

Recognizing the clear TrafficSchool ← YourBadDriving → YourFutureDriving back door, we decide to identify the effect by measuring and controlling for your own bad driving skills.

This will identify the effect, but it will also shut out any variation in TrafficSchool that's driven by YourBadDriving. So imagine two people, Rodney and Richard. Rodney has a 50% chance of not going to TrafficSchool, a 10% chance of going because of someone else's bad driving, and a 40% chance of going because of his own bad driving. Richard has a 50% chance of not going to TrafficSchool, a 30% chance of going because of someone else's bad driving, and a 20% chance of going because of his own bad driving.

We're tossing out that 40% for Rodney and 20% for Richard chances of going because of their own bad driving. There's only a 10% chance that Rodney goes to TrafficSchool *for the reason we still allow to count*, and similarly a 30% chance for Richard. That means there's *more remaining variation in treatment for Richard than for Rodney*, so Richard's treatment effect will be weighted more heavily than Rodney's will. A weighted average treatment effect!

FOLLOWING THIS LOGIC—WHICH TREATMENT VARIATION DO WE ALLOW TO COUNT—will tell us almost every time which treatment effect we're about to get.

We can go a little bit further and apply this logic ahead of time to develop some rules of thumb. These are just shortcuts to applying that same logic, but they're often easier to think about, and they work most of the time.

Rule of thumb 1: If you have true randomization in a representative sample and don't need to do any adjustment, you have an average treatment effect (ATE).

Rule of thumb 2: If you have true randomization only within a certain group, and you isolate that group so you can take advantage of that randomization, you have a conditional average treatment effect.

Rule of thumb 3: If you know that some variation in treatment is connected to back doors and so you close those back doors, using only the remaining variation, you have a weighted average treatment effect—variance-weighted if you're subtracting out explained variation, or weighted by how representative the observations are if you're picking a subsample of the data or picking control observations by matching them with treated observations.

Rule of thumb 4: If you are identifying your effect by assuming that some untreated group is what the treated group would look like if they hadn't been treated, then we have the average treatment on the treated (ATT).

Rule of thumb 5: If part of the variation in treatment is driven by an exogenous variable, and you isolate just the part driven by that exogenous variable, then you have a local average treatment effect (LATE).

These rules of thumb are, of course, rules of thumb and not true all the time. One important caveat here, already mentioned in a sidenote earlier in the chapter, is that *research design alone* isn't the only thing that determines which treatment effect average you get. The way you estimate the effect matters too. To give a basic example, take variance-weighted treatment effects that weight the effect by the variance of treatment. We can estimate the variance of treatment. What if we just... run our analysis while adding sample weights equal to the *inverse* of that variance? The variances will cancel out and we'll have an average treatment effect. The research design is a good place to start but the estimator matters. Part 2 of the book will talk about alternate ways of *estimating* particular research designs that get you different treatment effect averages.

10.4   Who Cares?

IT SEEMS ALMOST BEYOND THE POINT to worry too much about which kind of treatment effect average we get, doesn't it? After all, we've gone to all the work of identifying the effect in the first place. And each of these are averages of the actual treatment effects. Why should it matter?

We should care because we're interested in understanding causal relationships in the world!

The reason for paying attention to treatment effect averages (and which ones we are getting) is very clear if the reason we care about causal effects is that *we want to know what will happen if we intervene.*

Think way back to when we were defining causality back in Chapter 6—one way we talked about it was in the form of intervention. If we were to intervene to change the value of X, and Y changes as a result, then X causes Y.

This approach to causality is one reason why we care about getting causal effects in the first place. It's useful. If we know that X causes Y, then if we want to improve Y, we can change X. If aspirin reduces headaches, and you have a headache, then take an aspirin. We know what will happen because we've established the causal relationship.

Bringing in treatment effect averages changes considerably *what we can infer about what will happen* based on *the estimates we get in our analysis.*

For example, let's say we suspect that the presence of lead in drinking water has led to increased crime.[25] If we find evidence that lead in the drinking water *does* cause crime to rise, what would we use that information to do? Probably get the lead out of the drinking water, right?

[25] Which it might well do! See, for example, Reyes (2007).

However, what if it doesn't reduce crime for everyone? Let's say we found a number of localities that won government grants, awarded at random, to clean up the lead in their water. But among the localities that applied for the grants, there was no change in crime rates that followed. Perhaps their crime rates were already very low, or only localities with lead levels already too low to have an effect were the ones who applied for the grant.

In the case of this study, we got an average treatment effect conditional on being in the study. That conditional average treatment effect misrepresented the average treatment effect that we would get if we reduced lead levels in *everyone's* drinking water. If we don't pay attention to which treatment effect average we're getting, we might erroneously think that the effect is zero for everyone and decide not to bother removing lead from the water.

THIS CAN GO THE OTHER WAY TOO, where we estimate an average treatment effect but don't want that. For example, imagine you develop a new (and you think better) vaccine for the measles. You study your new vaccine with an experiment in the United States. And because you want to get a really representative average effect, you do a very careful job randomly recruiting everyone into your study, sampling people from all walks of life completely at random. For simplicity, let's assume nobody refuses being in your study.

This approach—selecting people completely at random and nobody opting out of the study—will give us an average treatment effect (at least among people in the United States).

Then you get the results back and you're shocked! The vaccine reduces the chances of measles, but only by a few tenths of a percent.

Well, that's probably because in the United States, north of 90% of people already have a measles vaccine, so your vaccine won't do much extra for them. What you wanted was the average treatment effect conditional on not already having had a measles vaccine.[26]

[26] Strangely, this does not count as an average treatment on the untreated.

IN GENERAL, WHAT YOU WANT is to think about *what intervention would look like*, whether it will be in the form of a policy that could be considered (changing how vaccinations occur, reducing lead for everyone, etc.) or in understanding how the world works (wages are going up for group X; what impact should we expect this will have on home ownership among group X?).

Once we know what intervention looks like, we want a treatment effect average that will match it. Planning to apply treatment to everyone, or at random? The average treatment effect is what you want. Just to a particular group? The conditional average treatment effect for that group. Wanting to expand an already-popular treatment to more people? Probably want the average treatment on the untreated or a marginal treatment effect. Planning to continue a policy that people opt into? Average treatment on the treated!

Understanding not just the overall effect, but who that effect is for, really fills in the gaps on making information from causal inference useful.

10.5   Treatment Effect Glossary

We've talked about a whole lot of different kinds of treatment effects. Let's remind ourselves what they are.

Average Treatment Effect. The average treatment effect across the population.

Average Treatment on the Treated. The average treatment effect among those who actually received the treatment in your study.

Average Treatment on the Untreated. The average treatment effect among those who did not actually receive the treatment in your study.

Conditional Average Treatment Effect. The average treatment effect among those with certain values of certain variables (for example, the average treatment effect among women).

Heterogeneous Treatment Effect. A treatment effect that differs from individual to individual.

Intent-to-Treat. The average treatment effect of assigning treatment, in a context where not everyone who is assigned to receive treatment receives it (and maybe some people not assigned to treatment get it anyway).

Local Average Treatment Effect. A weighted average treatment effect where the weights are based on how much more treatment an individual would get if assigned to treatment than if they weren't assigned to treatment.

Marginal Treatment Effect. The treatment effect of the next individual that would be treated if treatment were expanded.

Weighted Average Treatment Effect. A treatment effect average where each individual's treatment effect is weighted differently.

Variance-Weighted Average Treatment Effect. A treatment effect average where each individual's treatment effect is weighted based on how much variation there is in their treatment variable, after closing back doors.

11
Causality with Less Modeling

11.1 Confidence

EVERYTHING WE'VE DONE UP TO NOW when thinking about
identifying a causal effect has started with the same idea: draw
a causal diagram. Draw a causal diagram to map out our idea
of the data generating process, use that diagram to list out all
the paths from treatment to outcome, and then close pathways
so that we are left with just the Good Paths we want. Then
we've identified our effect!

So what if we can't get that first step? What if the causal
diagram itself is beyond our reach?

It's not just sheepishness at not being willing to put our assumptions down on paper. In complex situations—and most
social science is about highly complex situations—we may well
not know very much about what the causal diagram looks like.
Sure, we *could* draw a diagram, but it would have to be a massive
simplification, and we'd likely not even *think* of all the important
variables that should be on there. If we were being honest, we'd
map out what we know, and then have a big space labeled "I

DOI: 10.1201/9781003226055-11

DUNNO, SOME STUFF I GUESS" with arrows pointing from that to just about everything else on the diagram.

So what then? Do we just give up?

There are certainly some reasons why this very reasonable concern might turn us away from causal inference entirely. If we don't know what large portions of the causal diagram look like, then we will likely fail to recognize important back doors that need to be handled. Or perhaps we will try to close a back door, but by accident open up a path instead because there's a collider on it.

One approach we can take to this is to use the methods from Chapter 9, where instead of having to handle the mysterious mass that is "the entire set of back door paths," we can instead just try to focus on isolating a few front doors we know are there.

However, outside of that there are still some options. We'll never really, truly know what the actual causal diagram is. If we have a *pretty* good idea, we can just act as though that's the truth.[1] But if we know there's a lot of unknown spaces on our graph, we can ask ourselves what we can do to identify our causal effect the best we can, while reckoning with our own lacking knowledge. Ignorance is inevitable; let's make it not quite so painful.

[1] There's a good chance that we'll learn later that we made a fatal flaw, but sometimes it's hard to figure out how to do it better without making the mistake in the first place. Also, for now, this might be the best we can do.

11.2 Wide Open Spaces

WE'VE TALKED A LOT ABOUT MODELS. We've made complex models, we've drawn diagrams, we've carefully thought through the paths between treatment and outcome in a bunch of different directions. We've drawn messy models and clean models.

But for a lot of researchers, especially those who see the task of carefully modeling the data generating process as effectively impossible, the real model they're working with is a lot less complex. Rather, it looks like Figure 11.1.

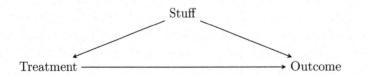

Figure 11.1: Completely Generic Causal Diagram for When you Don't Think A Full Model is Useful

Simple, straightforward. Treatment, outcome, and a back door through, uh, something. Some stuff.

Despite being simple, this diagram is actually far more daunting. By just writing "Stuff," we've given up even trying to figure out what all the back door paths are how to and close them. Sure, we can name a few things that might fall into the category of "Stuff," but we'll never name and measure them all.

This is a kind of principled ignorance. We know some things about what kinds of variables fit in "stuff," but we recognize that we don't know them all.[2] Let's take an econometric classic as an example—what's the effect of an additional year of education on your earnings? We can be pretty certain that demographic and socioeconomic background variables go into "stuff," as does intelligence, personality, the kind of school you're in, and so on. But even if we wrote out a list of fifty things and managed to control for them all, we would still say "I don't know everything, and the world is very complex. I must still be leaving something out. I don't believe I've identified a causal effect yet."

Unsurprisingly, people who think this way rarely believe that you can identify a causal effect just by controlling for variables.

So IF WE CAN'T SEE THE WHOLE DGP, WHAT CAN WE DO? Well, we can try to fill in whatever we can. That can get us a long way.

Instead of trying to map out the entire process, we will instead imagine that the data generating process looks more or less exactly like Figure 11.1, and spend all of our time asking what variables belong in "Stuff."

Surely the actual diagram isn't quite as simple as Figure 11.1. We know that. But we can probably think of most bad paths as simplifying in some way to this. In other words, we know it's a problem when some third variable causes both Treatment and Outcome. There are probably enough of those variables to use up all our energies anyway, so let's just focus on that.

That's the question, then—what variables out there likely cause both Treatment and Outcome? Don't worry about how those variables might cause each other, don't worry if there are potential collider effects going on, just worry about Stuff.

Let's take an example. We are interested in the concept of remittances, which are money that immigrants send back to their families in the country they came from. We want to know whether, among immigrants, the country that you emigrate to affects how much you send back in remittances.

What might cause both "destination country" and "remittance amount"? Anything that relates to the kind of work you do is likely to be a factor—training, education, intelligence, strength, and so on, since these will affect both which countries you're capable of emigrating to, and how much you earn once

[2] In economics, this question of approach—whether to try to map out the whole data generating process or to throw up your hands, call it impossible, and try to do things some other way—is the "structural vs. reduced form" debate. This book up to this chapter has been largely in the "structural" camp. Not to say that I think the reduced form side is wrong—much of my own work is reduced form. But I think most reduced-form types would agree they're doing as much structural work in their heads as they can before they face reality and switch gears for the actual analysis. That's why I think the structural side is a good place to start learning causality, regardless of where you end up going. This is as opposed to most econometrics textbooks, which are heavily reduced form (perhaps because structural econometrics, as opposed to structural theorizing, is much more difficult than reduced form, and economics as a field has decided to teach research design and statistical inference at the same time for some reason).

you get there (and thus can send back). The country you're coming from is a factor for similar reasons, although in this case it might affect what you send back because of how much your family needs it. Your culture, language, or religion might play a part—perhaps people prefer to emigrate to countries with similar cultures, languages, or religions where they might find it easier to fit in, and these same factors might influence what you can earn, or what expectations your family places upon you for remittances.

We could go on, but this is a decent list so far. We can think of these as "sources of bias"—if we don't control for them, we won't identify the effect. We might accidentally include some variables in here that are colliders on some path, or that are are actually irrelevant, but in general, asking "what variables cause both treatment and outcome" is on the easier end of the spectrum when it comes to laying out a causal diagram.

The key deviation at this point from the approach taken in the rest of the book is that we take this list of necessary controls and... *don't* use it to identify the effect.

Sure, we might control for some of them. But under this approach, we can take for granted that if we haven't already found a variable we *can't* control for, then that just means we didn't think hard enough about it. And if we think harder and do think of one, then either we can't control for it and we're stuck, or we can control for it and then we still have even more thinking to do.

Instead, this list of back-door fodder serves as a starting point. A set of considerations about the *kinds of things we need to control for*, but which we are going to control for *indirectly*, perhaps without controlling for anything at all.

HOW CAN WE CONTROL FOR STUFF WITHOUT CONTROLLING FOR STUFF? Well, we already know one way, and it might give you a hint for the other ways. In Chapter 9 we talked about how we can isolate just some causes of the treatment that are unrelated to the treatment.

If we can do this, then, in effect, we are controlling for Stuff without controlling for any Stuff at all. The back doors simply don't matter! And when you're thinking that there's no way to properly treat and close all the back doors, then doing something that makes all the back doors not matter at once sounds pretty good.

If we're doing this, why bother thinking about all the back-door Stuff in the first place? Because it encourages us to think about whether that exogenous source of treatment we need to find a front door is actually related to some of that Stuff—making it not so exogenous any more, or letting us know we do need to control for a little bit of Stuff too. In fact, we don't even need to *think* about whether our exogenous source is related to Stuff, we can actually test it ourselves—I'll talk about this in the next section.

The finding-front-doors method demonstrates that our goal here is to select methods that let us close a bunch of back doors at once, even if we can't measure the variables on those back doors. But it's not the only method for accomplishing this goal.

SOMETIMES, A CONTROL IN THE HAND IS WORTH AN INFI-NITE NUMBER IN THE BUSH. Some controls are both easy to measure and can close a lot of back doors.

Keeping in mind that the idea of controlling for a variable means *removing any variation explained by that variable*, then by controlling for a variable we will *also* be controlling for anything else that only varies *between* values of the first variable.

That was confusing. Let's try an example. Let's say we're doing a study on nature vs. nurture, and so to control for one aspect of nurture, we control for the house you grew up in—this would effectively result in comparing you against your siblings who grew up in the same house (or perhaps kids in other families in a multi-family household).

One other aspect of nurture we might want to control for is the geography of the region you grew up in—city vs. rural, the qualities of the neighborhood, and so on.

But once we've controlled for the house you grew up in, we don't need to control for that other stuff. The other kids who grew up in your house with you *also* grew up in your neighborhood, in your state, in your country, and so on. Something like neighborhood only varies between different houses. Once we've removed all variation related to your house, that will as a matter of course remove all variation related to your neighborhood.[3] So we don't need to control for neighborhood, or city vs. rural, or any number of other things that only differ *between* houses.

Most commonly, this comes up in the case of the method of "fixed effects," where individuals are observed multiple times, and a control for individual is added. This effectively controls for *everything* about that individual, whether it's easy to measure or not—their upbringing, their personality, and so on—as long as it's constant over time.[4] Plenty of controls knocked out in one fell swoop![5]

[3] For there to be any variation in neighborhood left after controlling for neighborhood, there would need to be houses that were in two different neighborhoods at once.

[4] So something about a person that changes over time, like their income, would not be accounted for by fixed effects.

[5] More about this method in Chapter 16.

SOMETIMES WE CAN JUST ASSUME THINGS ARE COMPARA-
BLE. Another approach that is often taken when there are far
more controls that need to be added than are feasible is to take
our group of treated people and compare them only to a group
that is, by assumption, on average comparable on all these controls in some way. A control group! If we assume that something
about these controls is identical between the treated and control groups, then we don't *need* to control for any of them, since
they're already the same.

This approach of selecting a control group is standard when
implementing an experiment. We enforce the comparability of
the groups in that case by randomly assigning people to treatment or control. On average, there shouldn't be any differences
on any of the variables on the back doors, since we randomly
assigned people without any regard for those variables. No need
to add them as controls then.

In cases where we haven't explicitly randomized, however,
this is a much heavier assumption to take on. In pretty much
any observational context, it would be completely unbelievable
to just claim that the treated and untreated groups are the
same for all variables on back doors, observable or unobservable.
You're basically just assuming you've identified the treatment
effect at that point, without doing anything, on no justification.[6]

So do we really do that? Just assume that treatment and
control are the same? No, of course not. Instead, we try to
find *particular comparisons* of treatment and control that are
the same.

This could be by finding a subset of the control group that
really does seem like it should be the same on average for all
of the Stuff, measured or not. For example, say you're interested in the effect of receiving community-support funds from
the government on the level of upward mobility in a community. Clearly, the communities that receive the funds are different from those that don't, so we wouldn't want to just compare
treatment and control.

But perhaps we could make a case that *among the communities that applied for funding*, the communities that got funding
are very similar to the ones that didn't, with funding awards
being semi-random. That's not quite as good as actual randomization, but it's certainly a start, and likely controls for a
lot of variables we couldn't measure, like "need for community
funding."[7]

[6] Of course nobody would
try to claim such a... ah, who
am I kidding.   Of course
there's a study that does
exactly this getting huge
press coverage and glow-
ing politically-based praise
as I write this, touting
a coronavirus cure.   Just
straight up assuming that
entire different countries
are the same on average
on all unmeasured back
doors, and having the gall
to call that a "country-
randomized controlled trial"
without even randomizing
anything. Hello from 2020.
If this book survives long
enough that you have no
idea what I'm talking about,
I can only hope that your
era-defining events are positive ones.

[7] For a standard method
that takes this sort of approach, see Chapter 20 on
regression discontinuity.

Another way we could make particular comparisons is by comparing treatment and control not *on average*, but rather *just within certain parts of the variation*. For example, rather than asking whether the treatment outcome average is different from the control outcome average (a comparison for which all of those variables on back doors are very much still active), we can ask whether *changes over time* in the treatment outcome average are different from *changes over time* in the control outcome average.

By looking at changes over time in the outcome, instead of the absolute level, we make the treatment/control comparison a lot more plausible, since we don't need the treatment and control groups to be the same on average for all the Stuff. We just need the *change over time* of that Stuff to be the same on average.[8] There are many other ways that we can knock out a lot of the variation in the Stuff, and rely on much narrower assumptions about what the treatment and control are comparable on.

SO THERE ARE SOME SAVING GRACES available to us that we can use to identify an effect, even if we're doubtful of our ability to truly map out the data generating process, or to *ever* figure out and measure the full set of necessary controls.

Each of these alternate approaches rely on their own sets of assumptions. And the whole idea here is that we didn't want to have to make a bunch of strong assumptions about the data generating process! It sort of seems like, instead of having the principled ignorance to admit what we don't know, we've just traded one set of assumptions for another.[9]

Thankfully, the researcher has yet another set of tools at their disposal that they can rely on to figure out whether their assumptions (which, if you remember from way way back at the beginning of the book, we *have to have assumptions* to identify anything) have indeed led them astray.

[8] For a standard method that takes this sort of approach, see Chapter 18 on difference-in-differences.

[9] This is unfair, but I *am* enjoying myself, so deal with it.

11.3   Yes, I'm Wrong, But Am I THAT Wrong?

HERE WE ARE, an estimate to our name, a set of controls, perhaps a control group or a way of isolating front doors, and still skeptical, ever skeptical, that we've managed to identify what we think we've identified. How can we make ourselves more certain?

Testing whether our assumptions are true is rarely possible— they often aren't about things we can observe in the first place. However, there are a few good avenues for checking whether our assumptions seem *false*. Distressingly, this is much easier to do (at least when it comes to checking assumptions using data rather than theory) than checking whether they seem true.

THIS BRINGS US TO THE WIDE WIDE WORLD OF "ROBUSTNESS TESTS." A robustness test is a way of either (1) checking whether we can disprove an assumption, or (2) redoing our analysis in a way that doesn't rely on that assumption and seeing if the result changes.

One way of doing robustness tests we already covered in Chapter 8. In that chapter, we looked at a causal diagram such as in Figure 11.2.

Robustness test. A test attempting to disprove an assumption the analysis makes, or seeing how much the results change when the assumption is relaxed.

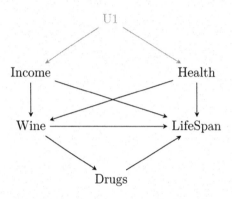

Figure 11.2: The Effect of Wine on Lifespan

This figure doesn't just imply what we need to control for in order to identify the effect of Wine on LifeSpan, it also exposes some assumptions we're making. For example, all paths between Drugs and Income on this diagram either contain a collider or contain Wine. So we are assuming that if we control for Wine, Drugs and Income should be unrelated. If we control for Wine and they *are* related, then that's evidence against one of the assumptions we made.

Most robustness tests work in a similar way to this. We detail an assumption we're making, which usually implies that a relationship between two things is not there, and then we test it. If that relationship *is* there, then that's evidence against our assumptions.

There are a whole bunch of available robustness tests for any given method you could be using. But let's take an example. Let's say we're relying on a control group, where we're comparing changes over time in the treatment group to changes over time in the control group.

Using this method assumes that the changes over time in all the back-door Stuff is the same in the treatment and control groups. So if we were to take some of those changes over time, we should find no relationship between changes over time and whether you're in the treatment group or the control group.

Now we have *the relationship that should not be,* and we're set up for our robustness test. We just need to take one of those back door control variables and see if its changes over time are related to being in the treatment or control group.

If that relationship is there, that's evidence against the assumptions we made for our research design. Sure, we could just add that variable as a control to fix the problem with the variable we checked, but even if we do that, we still have all the unmeasured variables in Stuff to worry about. We were operating under the assumption that we'd solved that problem. But checking the variable we *could* check showed we didn't fix it for that one, so why believe we did any better with the others?

Figuring out what kind of robustness test to do requires thinking carefully about the assumptions being made, and what sort of observable non-relationships those assumptions imply.[10] Then the tense moment while you hope really, really hard to not find a result... and, well... we'll see!

ONE FORM OF ROBUSTNESS TEST THAT CAN HELP FERRET OUT BAD ASSUMPTIONS, especially when you are using a method that compares a treated group to a comparable control group, is the *placebo test.* A placebo test is, as you might guess from the name, a test where you pretend that treatment is being assigned somewhere it isn't, and you check whether you estimate an effect. If you find an effect of "treatment," that tells you that there must be a bad assumption somewhere, since you're finding that the effect of *nothing* is *something!*[11] We discussed these before in Chapter 8 but they bear another look.

For example, say you're looking at the impact of an environmental conservation letter. A certain county tested a policy where, if you used above 1,200 kwh of electricity in a given month, you'd get a letter asking you nicely to use less. We think that untreated people who use 1,151—1,200 kwh are a pretty good control group for the treated people who use 1,201—1,250 kwh. Those usage rates are so close that we'd expect those groups to be similar on all back-door Stuff variables we can or can't measure. So we compare *next month's* usage among those who had 1,201—1,250 kwh last month to those who had 1,151—1,200 kwh last month. That's our design! Nice and simple.

[10] And, rarely, there are observable relationships that *should be* there that you could test and be concerned if they're *not* there. That would be a robustness test too; this kind is just less common.

Placebo test. An analysis performed after assigning a fake treatment to a group that did not receive treatment, in hopes that the estimated effect will be zero.

[11] This is different from the use of actual placebos in medicine, where the placebo *is* expected to have an effect because we can react to the belief that we are being treated (among other things; placebos are complex), and we just want to know if the real treatment has an effect beyond that placebo effect. In our usage of placebos, the placebo treatment occurs only inside our statistics software after the data is all collected—nobody actually received a placebo treatment. The effect really should be zero.

A standard robustness-check approach here would just be to see if the Stuff variables we can measure are different between the 1,151—1,200 kwh and 1,201—1,250 kwh groups.[12]

However, we can also do a placebo test. Let's imagine that the policy instead went into action at above 1,150 kwh. Now we can use the exact same design—this time comparing our fake-treatment group of 1,151-1,200 kwh against the new control group of 1,101-1,150 kwh. Now, there *shouldn't* be a big difference between these two groups in their next month's usage, because in actuality they both received the same treatment, which is no treatment at all. So if we *do* find a difference between these groups, that tells us that one of our assumptions is likely to be off. Perhaps groups whose usage differs by 50 kwh, in fact, are quite different on a lot of back-door-relevant variables!

To be honest, all of this is a little fanciful. The whole idea that we can test whether an assumption is true or not is silly. All assumptions are wrong at some level. We're not so much trying to prove these assumptions true or false as we are trying to prove them "not *so* wrong as to cause problems" or "too wrong to work with."

Do we have to do that? Surprisingly, no! There is another way. This is the *partial identification* approach.[13]

Under partial identification, we don't force ourselves to keep making assumptions until we've identified the effect. Instead, we make the assumptions that we're pretty certain about. Then, we use a *range of possibilities* about the remaining things we must assume. Finally, we figure out what our estimate is over that range, giving us a range of possibilities for the estimate itself.

Let's do an example. Say we're interested in the effect of owning a sports car on your tendency to drive over the speed limit. We add some controls—gender, age, income, parental income, and so on. We estimate the effect and find that owning a sports car increases your chances of speeding on a given drive by 5%. There are some unmeasured things remaining, though—tendency for risk-taking, for example.

We don't want to assume that getting a sports car is unrelated to risk-taking tendencies. That just doesn't sound plausible. But we can say "risk-taking is likely to be positively related to both sports car ownership and speeding," (even though we don't know how strongly) "so risk-taking being left out of the model makes the treatment effect look more positive. If we

[12] A table comparing a bunch of Stuff-type variables across the treated and control groups is often referred to as a "balance table." Balance tables will show up again in Chapter 14.

[13] Also known as "set identification." The term "sensitivity analysis" is related, and, in fact, the example here is sensitivity analysis. At the level of this book I think the ideas are closely related enough to be treated as one, but there are differences between them, as you'll find if you research further.
Partial identification. Relaxing some assumptions and so producing a range of possible treatment effect estimates rather than a single value, and perhaps adding stronger assumptions to narrow that range.

could control for risk-taking, the treatment effect would become more negative."

With that (much more plausible) assumption, we can't actually say what the effect of owning a sports car on speeding is precisely, but we can say that it's *no higher than 5%*. We have "bounded the effect from above" by 5%.

If we like, we can then go further. If we're willing to make assumptions a little stronger than just the sign, we can say something like "given all the other controls, the effect of risk-taking on buying a sports car is between 0 and X." This would let us say that the effect of sports car ownership on speeding is not just lower than 5%, but *between* 5% and some specific lower number, maybe 2%. We can adjust the strength of our assumptions as we like, getting more precise results with heavier assumptions, or less precise results with assumptions that don't say as much.

Partial identification is a wide-ranging area with plenty of methods.[14] Many of the details are unfortunately too technical for this book. But if you are interested, you can find a relatively gentle introduction in Gangl (2013).[15]

YOUR LAST LINE OF DEFENSE IS YOUR GUT. After all of that—checking the reasonableness of your assumptions and what they imply, checking how precise your results can get based on how strong your assumptions are—you're still left with a set of assumptions and a result.

And that result? Sometimes results just don't make sense.

That's an important thing to pay attention to. Is your result plausible at all? If it's not, then even if you can't figure out why or what you could do to fix it, then you have an assumption wrong somewhere. You just must.

For example, say you're studying the effect of drinking an extra glass of water once a week on your lifespan. You carefully design your research, control for all the necessary controls, do everything you can, and you find that the extra glass of water once a week makes you live twenty years longer.

Astounding! Alert the presses.

Well, probably not. There's just no way that's true. Completely unbelievable. You must have made an incorrect assumption, or made an error in your statistical code, or perhaps just got a huge fluke result in your data. What you didn't get is the real result. Not buying it.

There's a degree of subjectivity here. We want to be willing to accept surprising results—if we weren't, what's the point of doing research in the first place? But many results are not just

---

[14] Methods that go much deeper than just figuring out the direction that an uncontrolled variable biases the estimate.

[15] Markus Gangl. Partial identification and sensitivity analysis. In *Handbook of Causal Analysis for Social Research*, pages 377–402. Springer, 2013.

surprising but darn near impossible. And it doesn't even need to be as extreme as a twenty-year lifespan jump. If you found that a teacher training program increased student test scores by a third of a standard deviation, is that too good to believe?[16] If you found that a simple thirty-second audio message played once in a laboratory changed people's behavior *at all* five years later, is that plausible? Maybe not.

So that's the last line of defense you have. Your gut. Could this result ever possibly be? If it couldn't, then it isn't.

[16] Even well-regarded educational interventions considered quite successful generally have effects in the range of a tenth of a standard deviation.

Part II

The Toolbox

12

Opening the Toolbox

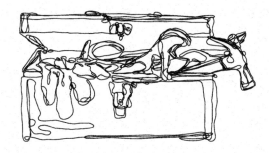

12.1   *Concept and Execution*

BASED ON THE CONTENT OF THIS BOOK THUS FAR, we have covered the concepts of data generating processes and causality. We've discussed how to isolate our paths of interest, and how to *identify* the paths we want by either shutting down the back door paths we don't want, or isolating the paths we want directly.

But that's all conceptual. How do we actually *do* those things?

While the first half of this book covered concepts and intuition, the second half covers execution. We'll be delving into the *toolbox* of methods that are commonly used by researchers.

Many of these methods, especially those past the "Regression," "Matching," and "Simulation" chapters, are based around the idea of *simplifying* a causal diagram. That is, in the real world, causal diagrams get so complex and intricate that it would be very difficult to measure and adjust for all the variables we need, like I talked about in Chapter 11.

But there are certain kinds of what we might call *template* causal diagrams that can be solved easily.[1] We can ask ourselves whether our context and research question of interest fits one of

[1] I attribute this concept of "template" diagrams to researcher Jason Abaluck, who has successfully used it to deeply annoy causal-diagram purists.

DOI: 10.1201/9781003226055-12

those templates. If it does, the associated method will give us a shortcut to identification that may be more plausible to people reading your work than trying to convince them you've really thought of and closed every back door.[2]

SO HOW TO USE THESE METHODS?

As always, first we want to model our data generating process, and draw a causal diagram! These methods are no replacement for understanding our data generating process, and indeed knowing which method to use relies on it.

Second, we want to ask ourselves *does our diagram look how it needs to look to use one of these methods?* For example, as you'll read in Chapter 19, to use the instrumental variables method there must be an "instrumental" variable that causes our treatment, and for which all paths from the instrumental variable to the outcome go through the treatment.

If it does, we can use the method. We've solved our research design problem. From that point, we can start concerning ourselves with statistical issues rather than design issues.

12.2　The Toolbox Chapters

The Toolbox chapters from Chapter 16 through Chapter 20 focus on "template" research designs in which the same sort of causal diagram, and thus design, applies in lots of different settings. These chapters will be structured the same, with three sub-chapters.

1. *How Does It Work?*: A conceptual overview of how the method identifies causal effects, a look at the kinds of diagrams that each of the methods works with, and a demonstration of how the method manipulates data to give you what you need.

2. *How Is It Performed?*: This sub-chapter gets a little more into the weeds, showing how the method is executed, usually using *regression* (which we'll talk more about in Chapter 13).

3. *How the Pros Do It*: If you want to actually use a given method in a real-world research project, the basic version is often not enough. This sub-chapter will discuss some of the additional considerations, adjustments, or methods that actual researchers often use when implementing these methods. It's impossible to cover everything that researchers actually do in these chapters. So these are designed not so much to

[2] Depending on who you talk to, these methods might be called "reduced form," or "quasiexperimental."

show you everything a researcher knows, but rather to make you aware of the kinds of things actual researchers are thinking about, and why.

BECAUSE THIS BOOK FOCUSES more on research design than econometrics proper, there is little in the way of statistical proofs. If you are interested in these, I recommend the excellent textbooks by Jeffrey Wooldridge.[3] Or, if you want the real advanced stuff, William Greene.[4]

It's also an undeniable fact that the *How the Pros Do It* sections do not tell you all the information you need to actually do it like the pros do it. This is because the way the pros *actually* do it is to read a voluminous and ever-changing literature on the newest approaches to these methods, or at least just read a bunch of other studies using the same general method and then largely follow their lead.[5] Trying to keep up in textbook form would be fruitless, and would require nearly a whole book on each method.

Rather, the How the Pros Do It sections focus on highlighting some of the most important caveats and extensions, and giving you what you need to go learn about the state of the art on your own. If you are hoping for additional up-to-date applications of these methods, or information on their history, I recommend Scott Cunningham's *Causal Inference: The Mixtape*.[6]

12.3 Code Examples

ALL OF THE CHAPTERS IN THE TOOLBOX will include code examples in R, Stata, and Python, showing you how methods can be executed in code.

These code chunks may rely on *packages* that you have to install. Anywhere you see `library(X)` or `X::` in R or `import X` or `from X import` in Python, that's a package **X** that will need to be installed if it isn't already installed. You can do this with `install.packages('X')` in R, or using a package manager like **pip** or **conda** in Python. In Stata, packages don't need to be loaded each time they're used, so I'll always specify in the code example if there's a package that might need to be installed. In all three languages, you only have to install each package once, and then you can load it as many times as you want.

One additional package you'll want to install to run these code examples is **causaldata**, which is a package of data sets I've made for this book (and several other books) and is available

[3] Jeffrey M Wooldridge. *Econometric Analysis of Cross Section and Panel Data*. MIT press, 2010.; and Jeffrey M Wooldridge. *Introductory Econometrics: A Modern Approach*. Nelson Education, 2016.

[4] William H Greene. *Econometric Analysis*. Pearson Education India, 2003.

[5] There are, of course, some pros who never move beyond what they learned in their textbooks. Depending on what they're doing, this is sometimes fine. Some research questions can be answered handily with tools that have been around long enough to make it into textbooks. Other times, well... there's plenty of work out there by pros that could be better.

[6] Scott Cunningham. *Causal Inference: The Mixtape*. Yale University Press, 2021.

for all three languages. Do `install.packages('causaldata')` in R, `ssc install causaldata` in Stata, or `pip install causaldata` (if using **pip**) in Python.

The datasets all come with documentation. Using the `mortgages` data as an example: in R, see the description with `help(mortgages, package = 'causaldata')` (or just `help(mortgages)` if you already loaded the package with `library(causaldata)`). You can also see the description of each variable as you work with `library(vtable)` and then `vtable(mortgages)` after loading the data. In Stata, variable labels can be seen in the Variable Explorer as normal, and you can get a description of the data set with the command `causaldata mortgages`. In Python, after loading the data with `from causaldata import mortgages`, you can see the data and variable descriptions with, respectively, `print(mortgages.DESCRLONG)` and `print(mortgages.NOTE)`.

These code examples have been run using R 4.1, Stata 15.1, and Python 3.8. If your version of the language is at that level or newer, you should be good to go! If it's older than that, you may want to upgrade, but while I haven't tested the code examples on all old versions, you're *probably* still fine as long as your R is version 3+ and your Stata is 14+ (except for the one example that relies on 16+, but I'll warn you about it). For Python, it's strongly recommended that you at least use 3.0+, as there are a number of major syntax changes from Python 2 to Python 3. For all three languages, you may still get somewhat different results based on updates to downloadable packages that occur after the publication of this book.[7] If you spot such a change, please feel free to contact me.

One final note on code, specifically in Stata: Stata doesn't naturally allow you to split one command onto two lines. However, some lines of code are going to be too long to fit on one line of the book and so must be split! This can be accomplished with the use of `///` at the end of a line, which means "the line isn't over yet, keep reading on to the next one." However, for some reason, Stata has decided that this only will only work if you are running code using the "Execute" button in the do-file editor. It doesn't work if you're just copy/pasting code into the Stata console. So if you see a `///` at the end of a line, either be sure to run that code using Execute from your do-file editor, or just erase the `///` and combine that line with the following line of code (and keep going until you hit a line that doesn't end in `///`).

[7] For example, the **model-summary** R package updated just before publication of this book and changed the significance star levels it displays—I had to change all the example code so it would keep producing the results I already had in the book! Who knows what package updates will occur *after* publication.

13

Regression

13.1 The Basics of Regression

REGRESSION IS THE MOST COMMON WAY IN WHICH WE FIT
A LINE TO EXPLAIN VARIATION. Using one variable to explain
variation in another is a key part of what we've been doing the
whole book so far!

When it comes to identifying causal effects, regression is the most common way of estimating the relationship between two variables while controlling for others, allowing you to close back doors with those controls.[1]
Naturally it's relevant to what we're doing. Not only will we be
discussing it further in this chapter, but many of the methods
described in other chapters of part 2 of the book are themselves
based on regression.

We've already covered the basics of regression, back in Chapter 4. So go back and take a look at that.

Some key points from that chapter:

- We can use the values of one variable (X) to predict the values of another (Y). We call this explaining Y using X. This

[1] Saying we "adjust" for those variables, rather than control for them, is probably more accurate, since we don't actually have control of anything here as we would in a controlled experiment. But "control" is the most common way of saying it.

DOI: 10.1201/9781003226055-13

is a purely statistical form of "explanation." It's not an explanation *why* Y happens (unless we've identified the effect of X on Y), just the ability to predict Y using X.

- While there are many ways to do this, one is to fit a *line* or *shape* that describes the relationship. For example, $Y = \beta_0 + \beta_1 X$. Estimating this line using ordinary least squares (standard, linear regression) will select the line that minimizes the sum of squared residuals, which is what you get if you take the prediction errors from the line, square them, and add them up. Linear regression, for example, gives us the *best linear approximation* of the relationship between X and Y. The quality of that approximation depends in part on how linear the true model is.

  - Pro: Uses variation efficiently
  - Pro: A shape is easy to explain
  - Con: We lose some interesting variation
  - Con: If we pick a shape that's wrong for the relationship, our results will be bad

- We can interpret the coefficient that multiples a variable (β_1) as a slope. So, a one-unit increase in X is associated with a β_1 increase in Y.

- With only one predictor, the estimate of the slope is the covariance of X and Y divided by the variance of X. With more than one, the result is sort of similar, but also accounts for the way the different predictors are correlated.

- If we plug an observation's predictor variables into an estimated regression, we get a prediction $\hat{Y}$. This is the part of Y that is explained by the regression. The difference between Y and $\hat{Y}$ is the unexplained part, which is also called the "residual."

- If we add another variable to the regression equation, such as $Y = \beta_0 + \beta_1 X + \beta_2 Z$, then the the coefficient on each variable will be estimated using the variation that remains after removing what is explained by the *other* variable. So our estimate of β_1 would not give the best-fit line between Y and X, but rather between *the part of Y not explained by Z* and *the part of X not explained by Z*. This "controls for Z."

- If we think the relationship between Y and X isn't well-explained by a straight line, we can use a curvy one instead.

OLS can handle things like $Y = \beta_0 + \beta_1 X + \beta_2 X^2$ that are "linear in parameters" (notice that the parameters β_1 and β_2 are just plain multiplied by a variable, then added up), or we can use nonlinear regression like probit, logit, or a zillion other options.

Those are the basics that more or less explain how regression works. And as long as that regression model looks like the population model (the true relationship is well-described by the shape we choose) it has a good chance of working pretty well. But there's still plenty left to cover. What else do we need to know?

We're going to add a few pieces of OLS that we haven't covered yet, or at least not thoroughly:

- The error term

- Sampling variation

- The statistical properties of OLS

- Interpreting regression results

- Interpreting coefficients on binary and transformed variables

That's a lot to get through! Let's get going.

13.1.1 Error Terms

A LOT OF THE DETAIL OF REGRESSION IS HIDDEN IN THE PLACES IT DOESN'T GO. I previously described fitting a straight line as figuring out the appropriate β_0 and β_1 values to give us the line $Y = \beta_0 + \beta_1 X$.

However, in our actual data, that line is clearly insufficient. It will rarely predict *any* observation perfectly, much less *all* of the observations. There's going to be a difference between the line that we fit and the observation we get. That's what I mean by "the places it doesn't go." We can add this difference to our actual equation as an "error term" which I'll mark as ε. So now our equation is:

$$Y = \beta_0 + \beta_1 X + \varepsilon \tag{13.1}$$

That difference goes by two names: *the residual*, which we've talked about, is the difference between the prediction we make with our fitted line and the actual value, and *the error* is the difference between *the true best fit-line* and the actual value.

Residual. The difference between a prediction and the actual outcome.

Error. The difference between the actual outcome and prediction we'd make if we had infinite observations to estimate our prediction.

Why the distinction? Well... sampling variation. Because we only have a finite sample, the best-fit line we *get* with the data we have won't quite be the same as the best-fit line we'd get if we had data on the whole population.[2] All we can really see will be the residual, but we need to keep that error in mind. As we know from Chapter 3, we want to describe the population. Those population-level errors will help us figure out what's going on.

[2] Remember Chapter 3?

You can see the difference between a residual and an error in Figure 13.1. The two lines on the graph represent the *true model* that I used to generate the data in the first place, and the OLS best-fit line I estimated using the randomly-generated sample. For each point, the vertical distance from that point to the OLS line is the residual (since the OLS line represents our prediction), and the vertical distance from that point to the true model is the error (since even if we knew the true model relating Y and X, we still wouldn't predict the point perfectly because of the other stuff that goes into Y).

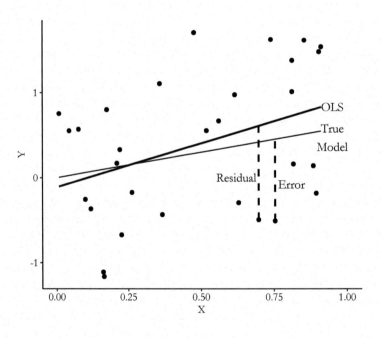

Figure 13.1: The Difference Between the Residual and the Error

So what's in an error, anyway? Where does it come from? Well, the error effectively contains *everything that causes Y* that is *not included in the model*. Y is on the left-hand-side of the $Y = \beta_0 + \beta_1 X + \beta_2 Z + \varepsilon$ equation, and it's determined by the right-hand-side. So if there's something that goes into Y

but isn't X or Z, then it has to come from somewhere. That somewhere is ε.

That ε includes both stuff we can see and stuff we can't. So, for example, if the true model is given by Figure 13.2, and we are using the OLS model $Y = \beta_0 + \beta_1 X + \beta_2 Z + \varepsilon$, then ε is made up of some combination of A, B, and C. We know that since we can determine from the graph that A, B, and C all cause Y, but aren't in the model.

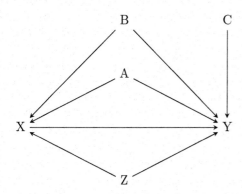

Figure 13.2: Depending on the Model Used, A, B, C, and Z Might be in the Error Term

13.1.2 Regression Assumptions and Sampling Variation

THINKING ABOUT WHAT'S IN THE ERROR TERM leads us to the first *assumption* we'll need to make about the error term. There are going to be a few of these assumptions, and I'll bring them up as they become necessary. People who use regression tend to be obsessed about those lil' εs and the assumptions we make about them. Go to any research seminar, and you'll hear endless questions about error terms and whether the assumptions we need to make about them are likely to be true.

The first assumption we'll talk about links the first and second parts of the book, and it's the *exogeneity* assumption. If we want to say that our OLS estimates of β_1 will, on average, give us the population β_1, then it must be the case that X is *uncorrelated* with ε.[3]

If that sounds familiar, it's because we're really just restating the conditions for identifying a causal effect! Every variable that causes Y in a causal diagram like Figure 13.2 ends up in the regression equation somehow, whether it's one of the variables in the model or just in ε. If a variable is *in the regression equation directly*, then that closes any causal paths that go through that variable. By adding a variable to the regression we "control

Exogeneity assumption. In a regression context, the assumption that the variables in the model (or perhaps just our treatment variable) is uncorrelated with the error term.

[3] Oddly, this does not require that X and ε are completely *unrelated*, just *uncorrelated*—there can be a relationship between the two, just not a *linear* relationship. For example, if the relationship between X and ε looks like a V or a U, then they're clearly related, but a straight line describing the relationship would be flat. The downward part would cancel out with the upward part. In reality, it's pretty rare that two variables are related but completely uncorrelated, so this doesn't usually let us get away with ignoring a relationship between X and ε.

for it" or "add it as a control." So with the regression $Y = \beta_0 + \beta_1 X + \beta_2 Z + \varepsilon$, the path $X \leftarrow Z \rightarrow Y$ is closed.

But if something is still in the error term and is correlated with X, we haven't closed the path. Since A is in ε, and we know from the diagram that $A \rightarrow X$, then we can say either that (a) we haven't closed the $X \leftarrow A \rightarrow Y$ back door path, and so the effect of X isn't identified, or, (b) X is correlated with ε and so is "endogenous" rather than exogenous, and the effect of X isn't identified.[4] These are saying, in effect, the same thing, just using lingo from different domains.[5]

Similarly, we can say in either set of lingo that we don't need to worry about C being in our model. It's in the error term—everything that predicts Y but isn't in the model is in the error term. But we can say either (a) there's no back door from X to Y that goes through C, or (b) X is unrelated to C and so C isn't leading X to be endogenous.

Speaking of lingo, if we *do* have that endogeneity problem, then on average our estimate of β_1 *won't* give us the population value. When that happens—when our estimate on average gives us the wrong answer, we call that a *bias* in our estimate. In particular, this form of bias, where we are biased because a variable correlated with X is in the error term, is known as *omitted variable bias,* since it happens because we omitted an important variable from the regression equation. Previously in the book we'd say we failed to close the back door that goes through A. In this lingo, we'd say that A gave us omitted variable bias, but C doesn't.

So that's our first assumption about the error term—the exogeneity assumption, that ε is uncorrelated with any variable we want to know the causal effect of. We know how to think about whether that assumption is true—the entire first half of this book is about figuring out what's necessary to identify a causal effect.

Of course, that's only the first thing we need to assume about the error term. Many of the others, though, relate to the sampling variation of the OLS estimates.

JUST LIKE THE MEANS WE DISCUSSED IN CHAPTER 3, RE-GRESSION COEFFICIENTS are estimates, and even though there's a true population model out there, the estimate we get varies from sample to sample due to sampling variation.

Conveniently, we have a good idea what that sampling variation looks like. We know that we can think of observations as being pulled from theoretical distributions. We also know that

[4] The terms endogeneity and exogeneity come from the biology terms "endogeny" and "exogeny," which roughly mean "from within the system" and "from outside the system." Endogenous variables are determined by other things in the system (other things that are part of determining the outcome).

[5] There are yet other ways to talk about this in other fields. You could also say that A is an "unobserved/unmeasured confounder" or that X and Y have a "spurious correlation."

Bias. An estimate that, on average, gives the wrong answer, no matter how big your sample size gets, is biased.

Omitted variable bias. Bias in a regression estimate that occurs because a variable correlated with the treatment is in the error term and otherwise omitted from the model.

statistics like the mean can also be thought of as being pulled from theoretical distributions. In the case of the mean, that's a normal distribution (see Chapter 3). If you drew a whole bunch of samples of the population and took the mean each time, the distribution of means across the sample would follow a normal distribution. Then we can use the estimated mean we get to try to figure out what the population mean is.

Regression coefficients also follow a normal distribution, and we know what the mean and standard deviation of that normal distribution is. Or, at least we do if we make a few more assumptions about the error term.

What is that normal distribution that the OLS coefficients follow?[6]

In a regression model with one predictor, like $Y = \beta_0 + \beta_1 X + \varepsilon$, an OLS estimate $\hat{\beta}_1$ of the true-model population β_1 follows a normal distribution with a mean of β_1 and a standard deviation of $\sqrt{\sigma^2/(var(X)n)}$, where n is the number of observations in the data, σ is the standard deviation of the error term ε, and $var(X)$ is the variance of X.[7]

If there's more than one variable in the model, the math starts to require matrix algebra, but it's the same idea. In the regression model $Y = \beta_0 + \beta_1 X + \beta_2 Z + \varepsilon$, the OLS estimates $\hat{\beta}_1$ and $\hat{\beta}_2$ follow a joint normal distribution, where their respective means are the population β_1 and β_2, and the standard deviations are the square roots of the diagonal of $\sigma^2 (A'A)^{-1}/n$, where A is a two-column matrix containing both X and Z. But you can think of $(A'A)^{-1}$ as saying "divide by the variances and covariances of X and Z," just like $\sqrt{\sigma^2/(var(X)n)}$ says "divide by the variance of X."

So how can we make an OLS estimate's sampling variation really small? There are only three terms in the standard deviation, so only three things to change. (1) We could shrink the standard deviation of the error term σ, i.e., make the model predict Y more accurately. (2) We could pick an X that varies a lot and has a big variation—an X that changes a lot makes it easier to check for whether Y is changing in the same way. Or (3) we could use a big sample so n gets big.

This standard deviation isn't generally *called* a standard deviation, though. The standard deviation of a sampling distribution is often referred to as a *standard error*. And that's what we'll call them here too.

How about those other assumptions we need to make about the error term? It turns out that those assumptions are

[6] The following calculations will assume that the exogeneity assumption mentioned earlier holds.

[7] Wait, how do we get the standard deviation of ε if we can't see it? Well, we don't know it exactly, so our idea of this sampling distribution will itself be an estimate based on our best guess of σ. We use the residual, which we *can* see, instead of the error, and base our estimate of σ on the standard deviation of the residual.

Standard error. The standard deviation of a sampling distribution.

necessary to assume that the stuff I've said up to now about the standard error is true. Let's tuck that away in our back pocket for now and ignore that looming problem. We'll come back to it in the "Your Standard Errors are Probably Wrong" section below. All of the things I'm going to say now about *using* standard errors will still hold up. There might just need to be some minor adjustments to the way we calculate them, which is what we'll discuss in that section.

13.1.3   Hypothesis Testing in OLS

OKAY, SO WHY DID WE WANT TO KNOW THE OLS COEFFI-CIENT DISTRIBUTION AGAIN? The same reason we want to think about *any* theoretical distribution—theoretical distributions let us use what we *observe* to come to the conclusion that certain theoretical distributions are unlikely.

And since the theoretical distribution of our OLS estimate $\hat{\beta}_1$ is centered around the population parameter β_1, that means we can use our estimate $\hat{\beta}_1$ to say that certain population parameters are very unlikely, for example "it's unlikely that the effect of X on Y is 13.2."

To recap what we're doing here, modifying only slightly what we have from Chapter 3:

1. We pick a theoretical distribution, specifically a normal distribution centered around a particular value of β_1

2. Estimate β_1 using OLS in our observed data, getting $\hat{\beta}_1$

3. Use that theoretical distribution to see how unlikely it would be to get $\hat{\beta}_1$ if indeed the true population value is what we picked in the first step

4. If it's super unlikely, that initial value we picked is probably wrong![8]

We can take this a step further towards the concept of *hypothesis testing*. Hypothesis testing is a way of formalizing the four steps above into a way of making a decision about whether we can "reject" the theoretical distribution, and thus the original β_1 we picked.[9]

Under hypothesis testing, we pick a "null hypothesis"—the initial β_1 value we're going to check against. In practice, this is almost always zero (although it certainly doesn't have to be), so let's simplify and say that our null hypothesis is $\beta_1 = 0$.

[8] Or, if you're Bayesian, that initial value needs to be revised a lot.

[9] There are plenty of reasons not to like hypothesis testing, but it's extremely common and so you have to know about it to figure out what's going on. What's wrong with it? It depends on the arbitrary and strangely sharp choice of rejection value ($p = .049$ rejects the null, but $p = .051$ doesn't?), and also relies on the idea that you've "found a result" when you've really just rejected something obviously false. In social science, very few relationships are truly exactly zero, and yet that's almost always our null-hypothesis value. So with enough sample size we *always* reject the null. My opinion: use significance testing; it is useful and it helps you converse with other people doing research. But don't get too serious about it. And certainly never choose your model based on significance! The first half of this book is about much better ways to choose models.

Then, we pick a "rejection value" α. If we calculate that the probability of getting our estimated result from the theoretical distribution based on $\beta_1 = 0$ is below α, then we say that's too unlikely, and conclude that $\beta_1 = 0$ is false, rejecting it. Commonly, this rejection value is $\alpha = .05$, so if the estimate we get (or something even stranger) has a 5% chance or less of occurring if the null value is true, then we reject the null.

Once we've rejected $\beta_1 = 0$, what conclusion can we come to? Not that our estimate is correct—we certainly don't know that, there are plenty of nonzero values we haven't rejected. All we can say is that we don't think it's likely to be 0. Knowing it's not zero can be handy—if it's not zero, that means there's *some* relationship there. And that's our conclusion.

To see this in action, I generated 200 random observations using the true model $Y = 3 + .2X + \varepsilon$, where ε is normally distributed with mean 0 and variance 1. Then, pretending that we don't know the truth is $\beta_1 = .2$, I estimated the OLS model using the regression $Y = \beta_0 + \beta_1 X + \varepsilon$.

The first estimate I get is $\hat{\beta}_1 = .142$. I can use the formula from the last section to also calculate that the standard error of β_1 is $se(\beta_1) = .077$. So the theoretical distribution I'm looking at is a normal distribution with mean 0 and standard deviation .077.

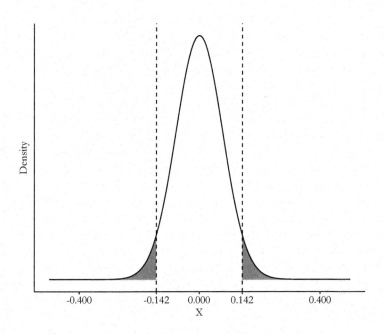

Figure 13.3: Comparison of Our .142 Estimate to a Theoretical Null Distribution Centered at 0

Under that distribution, the .142 estimate I got is at the 96.7th percentile of the theoretical distribution, as shown in Figure 13.3. That means that something as far from 0 as .142 is (or farther) happens $(100 - 96.7) \times 2 = 3.3 \times 2 = 6.6\%$ of the time.[10] If we started with $\alpha = .05$, then we would *not* reject the $\beta_1 = 0$ null hypothesis, since 6.6% is higher than 5%, even though we happen to know for a fact that the null is quite wrong.

We can see how this works from a different angle in Figure 13.4. This time I'm generating a bunch of samples of random data and showing the sampling distribution against a given null value. I'm magically giving myself the knowledge that $\beta_1 = .2$ is the true value to test against. I generate random data from that same true model $Y = 3 + .2X + \varepsilon$ a whole bunch of times, estimate $\hat{\beta}_1$ using OLS, and test it against the null of $\beta_1 = .2$. Sometimes I still reject the null, even though the null is literally true. But not often. In fact, I reject it exactly 5% of the time, because of the $\alpha = .05$ decision. Sampling variation will do that! This is a rejection of something that's true, and how often we do it is the "false positive" rate, or the terribly-named "type I error rate."[11]

[10] Note I've doubled the percentile here—that's because "as far away or farther" includes *both* sides of the distribution. This is a "two-tailed test," which is standard. If we did a "one-tailed test," caring only if it was as far away or farther *in the same direction*, we wouldn't do the doubling.

[11] Why is it a false "positive"? Because we sort of think about hypothesis testing as a "success" when we reject the null. This is silly, of course—we want to find the true value, not just go around rejecting nulls. A well-estimated non-rejection is much more valuable than a rejection of the null from a bad analysis. But there seems to be something in human psychology that makes us not act as though this is true.

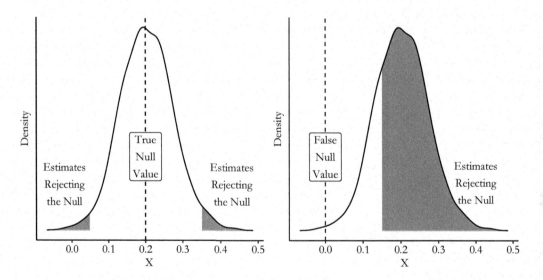

Figure 13.4: Testing 1,000 Randomly Generated Samples Against the True Null and a False Null

I then test those same estimates against the null of $\beta_1 = 0$. Sometimes I fail to reject the null, even though the null is false. Sampling variation does this too! Our failure to reject false nulls is the "false negative" rate or the "type II error rate."

What these figures demonstrate is that hypothesis testing really isn't about pinning down a true value, nor is it about hard-and-fast true-and-false results. It's about showing that certain theoretical distributions are unlikely. And we can then say that if they're *too* unlikely, we can forget about them.

In practice, when applied to regression in most contexts, we're talking about a normally-distributed regression coefficient and an $\alpha = .05$ false positive rate (with a two-tailed test), testing against a null that the coefficient is 0. So if the probability of our estimate (or something even farther from 0) occurring is less than 5%, we call that "statistically significant" and say the relationship is nonzero.

The calculation of the probability comes from computing the percentiles of the normal distribution with a mean of 0 and a standard deviation of our estimated $se(\hat{\beta}_1)$. If our estimate $\hat{\beta}_1$ is below the 2.5th percentile or above the 97.5th, that's statistically significant at $\alpha = .05$.

We can also look at the exact percentile we get and evaluate that. If, for example, we find that our estimate of $\hat{\beta}_1$ is at the 1.3rd percentile, then we can double that to 2.6, and say that we have a 2.6% probability of being that far away from the null or farther. This probability is called the "p-value." The lower that p-value is, the lower the chance is of finding a result that far from the null or farther by sampling variation alone. If you look back at Figure 13.3, the entire shaded area is the p-value, since it's the area as far away from the null as the actual estimate is, or farther. If the p-value is below α, that's statistical significance.

Using some known properties of the normal distribution, we can take a shortcut without having to go to the trouble of calculating percentiles. In this context, the *t-statistic* is the coefficient divided by its standard error $\hat{\beta}_1/se(\hat{\beta}_1)$. When we scale the estimate by its standard error, that puts us on the "standard normal" theoretical distribution where the standard deviation is 1. Then, if that t-statistic is below -1.96 (the 2.5th percentile) or above 1.96 (the 97.5th percentile), that's statistically significant at the 95% level, and a nonzero relationship.

p-value. The probability of getting an estimated result as far from the null (or farther), given the theoretical sampling distribution based on the chosen null-hypothesis value.

Using that sort of reading is what lets us read *regression tables*, which we'll get to in a second. But before we do, one aside on this whole concept of statistical significance. Rather, a plea.

Having taught plenty of students in statistical methods, and, further, talked with plenty of people who have received teaching in statistical methods, the *single greatest difference* between what students *are taught* and what students *learn* is about statistical significance. I think this is because significance is wily and tempting.[12] Powerful, tempting, seemingly simple, but so easy to misuse, and so easy to let it take you over and make you do bad things. So please keep the following things in mind and repeat them as a mantra at every possible moment until they live within you:

- An estimate *not* being statistically significant doesn't mean it's wrong. It just means it's not statistically significant.

- Never, ever, ever, ever, *ever*, ever, **ever** give in to the thought "oh no, my results aren't significant, I'd better change the analysis to get something significant." Bad.[13] I'm pretty sure professors always say to never do this, but students somehow remember the opposite.

- A lot of stuff goes into significance besides just "is the true relationship nonzero or not."—sampling variation, of course, and also the sample size, the way you're doing your analysis, your choice of α, and so on. A significance test isn't the last word on a result.[14]

- Statistical significance *only* provides information about whether a particular null value is unlikely. It doesn't say anything about whether the effect you've found *matters*. In other words, "statistical significance" isn't the same thing as "significant."[15] A result showing that your treatment improves IQ by .000000001 points is not a *meaningful* effect, whether it's statistically significant or not.

Keep in mind, generally: the point of significance testing is to think about not just the estimates themselves but also the precision of those estimates. If you've got a really cool result but the standard errors are huge, there's a good chance it's a fluke that could be consistent with lots of *true* relationships. That's the kind of thinking significance testing is meant to encourage. But there are other ways to keep precision in mind. You could just think about the standard errors themselves; ask if the estimate is precise regardless of whether it's far from a given null value. You could ask what the range of reasonable null values is (construct a confidence interval) instead of focus on one in particular. You could go full Bayesian and do whatever the heck it is those crazy cats get up to. Significance testing is just one way of doing it, and it has its pros and cons like everything else.

[12] If the Lord of the Rings were statistics, significance testing would be the One Ring. Not to mention all the people trying to throw it into a volcano. Of course if Lord of the Rings really were statistics they wouldn't have made those movies about it.

[13] Why? First off, maybe there truly just isn't much of an effect. Insignificant doesn't mean wrong! Also, changing the analysis to seek significance makes your future significance tests incorrect and meaningless—they all assume that you don't do this. Your results become garbage, even if they look nice and you can fool people into thinking they're good.

[14] That said, speculating on whether an insignificant result would become significant with, say, a bigger sample size generally isn't too useful.

[15] Even though researchers, myself included, will often just say "significant" as a shorthand for statistically significant.

13.1.4   Regression Tables and Model-Fit Statistics

WE HAVE A DECENT IDEA AT THIS POINT of how to think about
the line that an OLS estimation produces, as well as the statistical properties of its coefficients. But how can we interpret the
model as a whole? How can we make sense of a whole estimated
regression at once? For that we can turn to the most common
way that regression results are presented: the regression table.

We're going to run some regressions using data on restaurant and food inspections. We might be curious whether chain
restaurants get better health inspections than restaurants with
fewer (or only one) location.[16] We'll be using this data throughout this chapter. Some basic summary statistics for the data
are in Table 13.1. We have data on the inspection score (with a
maximum score of 100), the year the inspection was performed,
and the number of locations that restaurant chain has.

[16] This data comes from
Louis-Ashley   Camus   on
Kaggle.

Variable	N	Mean	Std. Dev.	Min	Pctl. 25	Pctl. 75	Max
Inspection Score	27178	93.643	6.257	66	90	100	100
Year of Inspection	27178	2010.337	5.949	2000	2006	2016	2019
Number of Locations	27178	64.766	84.267	1	27	71	646

Table 13.1: Summary Statistics for Restaurant Inspection Data

Now I'll run two regressions. The first just regresses inspection score on the number of locations the chain has:

$$InspectionScore = \beta_0 + \beta_1 Number of Locations + \varepsilon \quad (13.2)$$

and the second adds year of inspection as a control:

$$InspectionScore = \beta_0 + \beta_1 Number of Locations+$$
$$\beta_2 Year of Inspection + \varepsilon \quad (13.3)$$

I then show the estimated results for both in Table 13.2.

What can we see in Table 13.2? Each column represents a
different regression. The first column of results shows the results
from the Equation 13.2 regression, and the second column of
results show the results from the Equation 13.3 regression.

The first thing we notice is that each of the variables used
to predict Inspection Score gets their own set of two rows. The
"Intercept" also gets two rows—this is $\hat{\beta}_0$, and on some tables
is called the "Constant." The first row shows the coefficient
estimate. Our estimate $\hat{\beta}_1$, the coefficient on Number of Locations, is $\hat{\beta}_1 = -.0019$. It happens to be the same in both
regressions—the addition of Year as a control didn't change the
estimate enough to notice. There are also some asterisks—I'll
get to those in a second.

	Inspection Score	Inspection Score
(Intercept)	94.866***	225.333***
	(0.046)	(12.411)
Number of Locations	-0.019***	-0.019***
	(0.000)	(0.000)
Year of Inspection		-0.065***
		(0.006)
Num.Obs.	27178	27178
R2	0.065	0.068
R2 Adj.	0.065	0.068
F	1876.705	997.386

* p < 0.1, ** p < 0.05, *** p < 0.01

Table 13.2:   Regressions About Restaurants

Below the coefficient estimates, we have our measures of precision, in parentheses.[17] In particular, in this table these are the standard errors of the coefficients. $se(\hat{\beta}_1)$ in this regression is .0004—seems pretty precise to me. There are a few different ways you can measure the precision of a coefficient estimate. Putting standard errors here is the most common, but sometimes you'll see something like a confidence interval. In some fields, using a t-statistic (coefficient divided by the standard error) is more common than the standard error. Ideally a table will tell you which one they're doing, but not always.[18]

Finally, we have those asterisks, called "significance stars." These let you know at a glance whether the coefficient is statistically significantly different from a null-hypothesis value of 0. To be really precise, these aren't exactly significance tests. Instead, they're a representation of the p-value.[19] The lower the p-value is, the more stars we get.

Which p-value cutoff each number of stars corresponds to changes from field to field but should be described in the table note as it is here. A standard in many of the social sciences, and what you'll see in this book, is that * means that the p-value is below .1 (10%), ** means it's below .05 (5%), and *** means it's below .01 (1%).[20] So if we see ***, as we do on this table for the coefficient on Number of Locations, that would mean that *if we had decided that* $\alpha = .01$ (or higher), we would reject the null of $\beta_1 = 0$. If we see **, then we'd find statistical significance at $\alpha = .05$ or higher. Basically, it's a measure of *which α values you'd find significance with* for this coefficient.[21] These stars are a way of, at a glance, being able to tell which of the coefficients

[17] Some regression tables put the precision measures to the right of the coefficient rather than below, so you'd have two columns for each regression.

[18] If it doesn't say, it's usually a standard error, and if it's a single number it's almost always a standard error or a t-statistic. You can tell which of those two it is by looking at the significance stars—if precision values that are big relative to their coefficients give you significance, that's probably a t-statistic. If small values relative to the coefficients give you significance, that's probably a standard error.

[19] Remember, the p-value is the probability of being as far away from the null hypothesis value (or farther) as our estimate actually is. If the p-value is below α, that's statistical significance.

[20] In many fields, and in most R commands by default, everything is shifted over one step. * is below .05, ** is below .01, and *** is below .001. There's often another indicator in these cases like + for .1.

are statistically significantly different from 0, which all of these are.

Moving down the table, we have a bunch of things that aren't coefficients or precision measures. Which of these exact statistics are present will vary from table to table, but in general these are either descriptions of the analysis being run, or measures of the quality of the model. For the first, it's common to see any additional details about estimation listed down here (such as any standard error adjustments), and just about every table will tell you the number of Observations (or sometimes "N") included in estimating the regression.[22]

For the second, there are a billion different ways to measure the quality of the model. Probably two of the most common are included here—R^2 and Adjusted R^2. These are measures of the share of the dependent variable's variance that is predicted by the model. R^2 in the first model is .065, telling us that 6.5% of the variation in Inspection Score is predicted by the Number of Locations. If we were to predict Inspection Score with Number of Locations and then subtract out our prediction, we'd be left with a residual variable that has only $(100 - .065 =)$ 93.5% of the variance of the original. Adjusted R^2 is the same idea, except that it makes an adjustment for the number of variables you're using in the model, so it only counts the variance explained *above and beyond* what you'd get by just adding a random variable to the model.[23]

Finally, we have the "F-statistic." This is a statistic used to do a hypothesis test. Specifically, it uses a null that *all the coefficents in the model* (except the intercept/constant) *are all zero at once*, and tests how unlikely your results are given that null. It's pretty rare that this will be insignificant for any halfway-decent model, and so I for one mostly ignore this statistic.

Missing from this particular table, but present in a number of standard regression-table output styles, is the residual standard error, sometimes also called the root mean squared error, or RMSE. This is, simply, our estimate of the standard deviation of the error term based on what we see in the standard deviation of the residuals. We take our predicted Inspection Score values based on our OLS model and subtract them from the actual values to get a residual. Then, we calculate the standard deviation of that residual, and make a slight adjustment for the "degrees of freedom"—the number of observations in the data minus the number of coefficients in the model. The bigger

[21] This is a bit backwards—we generally want to choose a α value before we start. What's the point of knowing significance tests at higher α values? We already said that doesn't count. And if our α = .05, then *** doesn't really give us additional information—something is either significant or it's not. There's no such thing as "more significant." Of course, I'm describing an idealized version of significance testing that people don't actually do.

[22] Some other common appearances here might be an information criterion or two (like AIC or BIC, the Aikake and Bayes Information Criterion, respectively) or the sum of squares. I won't be covering these here.

[23] Yes, adding a random variable to the model does explain more of its variation. In fact, adding any variable to a model always makes the R^2 go up by some small amount, even if the variable doesn't make any sense.

this number is, the bigger the average errors in prediction for the model are.

What can we do with all of these model-quality measures? Take a quick look, but in general don't be too concerned about these. These are generally measures of how well your dependent variable is *predicted* by your OLS model. But if you're reading this book, you're probably not that concerned with prediction. You're interested in identifying causal effects, and in estimating *particular coefficients well*, rather than predicting the dependent variable overall.

If your R^2 or adjusted R^2 values are low, or your residual standard error is high, what this tells you is that there's *a lot going on with your dependent variable other than what you've modeled*. Is that a concern? Maybe. If you thought you wrote down a diagram that explained your dependent variable super well and covered all of the things that cause it, and you chose your model and which variables to control for based on that... and then you got a tiny R^2 anyway, then your initial assumption that you really understood the data generating process of your dependent variable might be wrong. But if you don't care about most of the causes of your dependent variable and are pretty sure you've included the variables in your model necessary to identify your treatment, then R^2 is of little importance.

Similar to statistical significance, I see R^2 values as another thing that students tend to fixate on. Gets their buns in a knot.[24] But while it can be a nice diagnostic, it's definitely not something to fixate on. Certainly don't design your model around maximizing R^2—build the right model for identifying the effect you want to identify and answering your research question, not the right model for predicting the dependent variable. Even if you are interested in prediction, R^2 has plenty of flaws in use for building predictive models too, although that's another book.

KNOWING WHERE ALL THE PIECES OF A REGRESSION ARE on a regression table, how can we interpret the results? Let's take a look at the regression table again in Table 13.3.

A key phrase to keep in mind when interpreting the results of an OLS regression is "a one-unit change in..." Regression coefficients are all about estimating a linear relationship between two variables, and reporting the results in terms of the slope, i.e., the relationship between a one-unit change in the predictor variable and the dependent variable.[25]

[24] I don't know what this means.

[25] Sometimes regression results are presented as "standardized coefficients" for which this becomes "a one-standard-deviation change in..." But you can also just think of it as a "one-unit change" for variables scaled to have a standard deviation of 1.

	Inspection Score	Inspection Score
(Intercept)	94.866***	225.333***
	(0.046)	(12.411)
'Number of Locations'	-0.019***	-0.019***
	(0.000)	(0.000)
'Year of Inspection'		-0.065***
		(0.006)
Num.Obs.	27178	27178
R2	0.065	0.068
R2 Adj.	0.065	0.068
F	1876.705	997.386

* $p < 0.1$, ** $p < 0.05$, *** $p < 0.01$

If we want to get real precise about it, the interpretation of an OLS coefficient β_1 on a variable X is "controlling for the other variables in the model, a one-unit change in X is linearly associated with a β_1-unit change in Y." If we want to get even more precise, we can say "If two observations have the same values of the other variables in the model, but one has a value of X that is one unit higher, the observation with the X one unit higher will on average have a Y that is β_1 units higher."

Let's start with the first column of results, with only one variable, a -0.019 on Number of Locations. Let's think briefly about our units here. Number of Locations is the number of locations in the restaurant chain, and the dependent variable, Inspection Score, is an inspector score on a scale that goes up to 100.

Since we have no other control variables in the model, that -0.019 means "a one-unit increase in the number of locations a chain restaurant has is linearly associated with a -0.019-point decrease in inspector score, on a scale of 0—100." Or, "comparing two restaurants, the one that's a part of a chain with one more location than the other will on average have an inspection score -0.019 lower."

Notice that however I'm wording it, I'm careful to avoid saying "a one-unit increase in number of locations decreases inspector score by -0.019." This would be implying that I've estimated a causal effect. I should only use this language if, based on what we learned in the first part of the book, we think we've identified the causal effect of number of locations on inspector score.

How about that constant term 94.866? This is our prediction for the dependent variable when all the predictor variables are zero. Think about it—our conditional mean of the dependent variable is whatever the OLS model spits out when we plug the appropriate values in. If we plug in 0 for everything, it all drops out and all we have left is our constant. So for a restaurant with zero locations, we'd predict an inspector score of 94.866. Of course, it's impossible to have a restaurant with zero locations, so this doesn't mean much.[26]

Let's move to the second column. We're introducing a control variable here, year of inspection. It doesn't seem to change the coefficient on number of locations, but it does change the interpretation.

Now we need to incorporate the fact that our interpretation of each coefficient is based on the idea that we are "controlling for" the other variables in the model.[27] We can say this a few ways.

We can say "controlling for year of inspection, a one-unit increase in the number of locations a chain restaurant has is linearly associated with a −0.019-point decrease in inspector score, on a scale of 0—100." Instead of "controlling for" we could instead say "adjusting for" or "adjusting linearly for" or even "conditioning on."[28]

We can say "comparing two inspections in the same year, the one for a restaurant that's a part of a chain with one more location than the other will on average have an inspection score −0.019 lower." After all, that's the idea of controlling for variables. We want to compare like-with-like and close back doors, and so we're trying to remove the part of the Number of Locations/Inspector Score relationship that is driven by Year of Inspection. The idea is that by including a control for Year, we are removing the part explained by Year, and can proceed as though the remaining estimates are comparing two inspections that effectively have the same year. There's no variation in year left—we have held year constant.[29] For this reason you will also often hear the interpretation "*holding year of inspection constant,* a one-unit increase in the number of locations is associated with a −0.019 decrease in inspector score."

This is a simple guide to interpreting a regression, and one that's focused pretty heavily on semantics. But training yourself to explain regression in these terms will truly help you interpret them better. The same intuition will carry over when you

[26] If you have predictor variables where the relevant range of the data is far away from zero, the constant term can sometimes give strange values. This isn't a problem—just don't try to make predictions far outside the range of your data. In general we don't need to worry ourselves too much about the constant term.

[27] i.e., closing the back doors that go through those variables.

[28] Remember the "conditional mean" stuff from Chapter 4? That's the kind of "conditioning" I mean.

[29] Or at least we've held its *linear* predictions constant. If Year of Inspection relates to Inspector Score or Number of Locations in ways that are best described by *curvy* lines, then including Year of Inspection as a linear predictor in the regression equation won't fully control for it.

start doing regression in more complex settings, as we'll discuss throughout the chapter.

13.1.5   Subscripts in Regression Equations

I'VE LEFT OUT ONE COMMON FEATURE OF EXPRESSING A RE-GRESSION.   When writing out the equation for a regression, people will commonly use *subscripts* on their variables (i.e., $X_{littletextdownhere}$).   We already have subscripts on our coefficients (the 1 in β_1, etc.), but they show up on the variables themselves, too. I've chosen mostly to omit these in this book,[30] but it's important to be familiar with the concept when you're reading papers or writing your own.

A regression might be expressed as

$$Y_i = \beta_0 + \beta_1 X_i + \varepsilon_i \qquad (13.4)$$

The i here tells us *what index the data varies across*. In other words, a single observation is a *what* exactly? Here we have i, which generally means "individual," i.e., an individual person or firm or country, depending on context. In this regression, Y and X differ across individuals.

Alternately, we might see

$$Y_t = \beta_0 + \beta_1 X_{t-1} + \varepsilon_t \qquad (13.5)$$

The t here would be shorthand for time period. This is describing a regression where each observation is a different time period. The X_{t-1} tells us that we are relating Y from a given period t to the X from the period before $(t-1)$.

The subscripts do their best work when there are multiple axes that things *could* vary along. Consider this example:

$$Y_{it} = \beta_g + \beta_t + \beta_1 X_{it} + \beta_2 W_i + \varepsilon_{it} \qquad (13.6)$$

This is describing a regression in which Y and X vary across a set of individuals i and across time t (a panel data set). There is a different intercept for each time period β_t, and also a different intercept for each value of g (and what's g? We'd have to look for where the researcher is describing their regression, but it might mean some *g*rouping that the individuals i are sorted into). We also see a control variable W_i. The i subscript here (and lack of a time subscript t) tells us that W only varies across individual and doesn't change over time.   Perhaps W is something like birthplace that is different for individuals but does not vary over time.

[30] I've got nothing against them!   But I think at the level we're working at they're more distracting than helpful.

13.1.6 *Turning a Causal Diagram into a Regression*

OKAY, SO WE'VE COVERED THE BASICS OF HOW REGRESSION WORKS. And we spent the first half of this whole book talking about how to identify causal effects using causal diagrams.

It's tempting to think that we automatically know how to link the two, but I've found that students often have difficulty making this jump. It's not a *super difficult* jump, mind you, but it's not one we can take for granted.

The chart in Figure 13.5 may help. On the diagram, we're thinking about what to do with a variable A. Is it the outcome variable, or is some other variable Y the outcome? Is it the treatment, or is some other variable X the treatment. If it's neither, where does it go on the diagram and also in a regression, if anywhere at all?

As shown in Figure 13.5, each variable may play one of several roles in a regression. It could be our outcome variable—we know all about those. Outcome variables become the dependent variables in regression, the $Y =$ part of $Y = \beta_0 + \beta_1 X$. We know all about treatment variables from our discussion of causal diagrams, too. Those become the $\beta_1 X$ part.

How about the rest? Our causal diagram tells us all about variables that need to be adjusted for in order to close back doors (and also those that *shouldn't* be controlled for to avoid closing front doors or opening back doors by controlling for colliders). Anything that should be included as a control should be, well, included as a control. That's the $+\beta_2 A$ part of $Y = \beta_0 + \beta_1 X + \beta_2 A$, where X is the treatment and Y is the outcome.

There are two things we haven't come to yet that show up in regression but not so much in causal diagrams. First, what if a variable A doesn't *need to be* controlled for, but also controlling for it won't break the identification? In that case we *can* include it as a control, and maybe we want to—as long as A is related to Y, then adding it will explain more of the variation in Y, reducing variance in the error term. And since the standard errors of the regression coefficients like $\hat{\beta}_1$ rely on the variance of the error term, the standard errors will shrink.

Finally, what about that part of the graph where it asks about whether the effect of treatment will differ across values of A? That's where *interaction terms* come in. We'll talk about those later in the chapter.

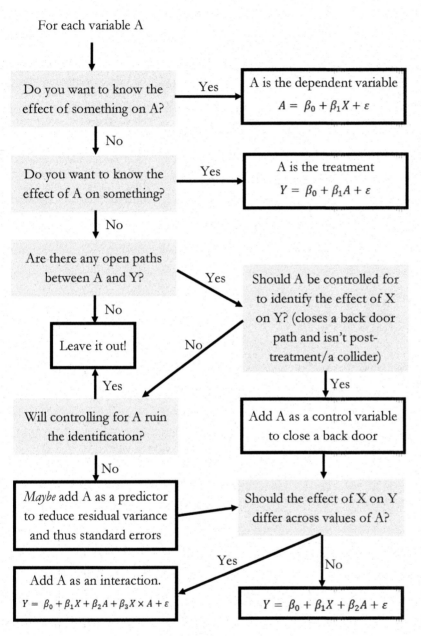

For each variable A

Do you want to know the effect of something on A?
→ Yes → A is the dependent variable
$$A = \beta_0 + \beta_1 X + \varepsilon$$

No

Do you want to know the effect of A on something?
→ Yes → A is the treatment
$$Y = \beta_0 + \beta_1 A + \varepsilon$$

No

Are there any open paths between A and Y?
→ Yes →

No

Leave it out!

Should A be controlled for to identify the effect of X on Y? (closes a back door path and isn't post-treatment/a collider)
← No

Yes

Will controlling for A ruin the identification?

Yes

Add A as a control variable to close a back door

No

Maybe add A as a predictor to reduce residual variance and thus standard errors

Should the effect of X on Y differ across values of A?

Yes

No

Add A as an interaction.
$$Y = \beta_0 + \beta_1 X + \beta_2 A + \beta_3 X \times A + \varepsilon$$

$$Y = \beta_0 + \beta_1 X + \beta_2 A + \varepsilon$$

Figure 13.5: Chart for Constructing a Regression Equation

13.1.7   Coding Up a Regression

BEFORE WE MOVE ON, WE SHOULD PROBABLY ACTUALLY DO A REGRESSION. Seems like a good idea.

The following code chunks will replicate Tables 13.2 and 13.3 in R, Stata, and Python. They will load in the restaurant inspection data, calculate the number of locations, regress inspection score on the number of locations, do another regression that also

includes year as a control, and then output a regression table to
file.

Note that while the code for performing a regression is fairly
standard in each language, the process for producing a regression table generally depends on downloadable packages. There
are plenty of options for these packages. I'll be using the **model-
summary** R package (`install.packages('modelsummary')`),
estout for Stata (`ssc install estout`), and **stargazer** for
Python (`pip install stargazer`). But some good alternatives
include `export_summs` from the **jtools** package, or **outreg2**,
regsave, or Stata 17 exclusive **collect** in Stata.[31] There don't
appear to be great Python alternatives for now.

[31] I recognize that **estout** is probably the better package for Stata regression tables and that's why I use it here, and I've heard good things about **regsave**, but I personally will use **outreg2** until the day that I die. You shall pry its arcane syntax, and the head-scratching fact that it is less up-to-date than **outreg**, from my cold dead fingers.

```
1  # R CODE
2  library(tidyverse); library(modelsummary)
3  res <- causaldata::restaurant_inspections
4
5  res <- res %>%
6      # Create NumberofLocations
7      group_by(business_name) %>%
8      mutate(NumberofLocations = n())
9
10 # Perform the first, one-predictor regression
11 # use the lm() function, with ~ telling us what
12 # the dependent variable varies over
13 m1 <- lm(inspection_score ~ NumberofLocations, data = res)
14
15 # Now add year as a control
16 # Just use + to add more terms to the regression
17 m2 <- lm(inspection_score ~ NumberofLocations + Year, data = res)
18
19 # Give msummary a list() of the models we want in our table
20 # and save to the file "regression_table.html"
21 # (see help(msummary) for other options)
22 msummary(list(m1, m2),
23     stars=TRUE,
24     output= 'regression_table.html')
25 # Default significance stars are +/*/**/*** .1/.05/.01/.001. Social science
26 # standard */**/*** .1/.05/.01 can be restored with
27 msummary(list(m1, m2),
28         stars=c('*' = .1, '**' = .05, '***' = .01),
29         output= 'regression_table.html')
```

```
1  * STATA CODE
2  * ssc install causaldata if you haven't yet!
3  causaldata restaurant_inspections.dta, use clear download
4
5  * Create NumberofLocations
6  by business_name, sort: g NumberofLocations = _N
7
8  * Perform the first, one-predictor regression
9  regress inspection_score NumberofLocations
10 * Store our results for estout
11 estimates store m1
12
13 * Now add year as a control
```

```
14 │ regress inspection_score NumberofLocations year
15 │ estimates store m2
16 │
17 │ * use esttab to create a regression table
18 │ * note esttab defaults to t-statistics, so we use
19 │ * the se option to put standard errors instead
20 │ * we'll save the result as regression_table.html
21 │ * (see help esttab for other options)
22 │ esttab m1 m2 using regression_table.html, se replace
```

```
 1 │ # PYTHON CODE
 2 │ import pandas as pd
 3 │ import statsmodels.formula.api as sm
 4 │ from stargazer.stargazer import Stargazer
 5 │ from causaldata import restaurant_inspections
 6 │
 7 │ res = restaurant_inspections.load_pandas().data
 8 │
 9 │ # Perform the first, one-predictor regression
10 │ # use the sm.ols() function, with ~ telling us what
11 │ # the dependent variable varies over
12 │ m1 = sm.ols(formula = 'inspection_score ~ NumberofLocations',
13 │ data = res).fit()
14 │
15 │ # Now add year as a control
16 │ # Just use + to add more terms to the regression
17 │ m2 = sm.ols(formula = 'inspection_score ~ NumberofLocations + year',
18 │ data = res).fit()
19 │
20 │ # Open a file to write to
21 │ f = open('regression_table.html', 'w')
22 │
23 │ # Give Stargazer a list of the models we want in our table
24 │ # and save to file
25 │ regtable = Stargazer([m1, m2])
26 │ f.write(regtable.render_html())
27 │ f.close()
```

13.2   Getting Fancier with Regression

REGRESSION IS A TOOL—a *very* flexible tool. And we've only really learned one way to use it! This section isn't even going to change how we use it, it's just going to change the variables that go into it. This in itself will open up a whole new world.

So far, we've talked about regression of the format $Y = \beta_0 + \beta_1 X + \beta_2 Z + \varepsilon$. We're still working with that. But we've also always been working with *continuous* variables included in the regression as their normal selves. Is there something else?

First, we might not have continuous variables. Instead, we might have discrete variables, which most often pop up as *binary* variables—true or false, rather than a particular number. How can we interpret β_2 if Z is "this person has blonde hair"? Or, how would we control for "hair color" if that's not a variable we've measured continuously and we instead have categories like "black," "brown," "blonde," and "red"?

Second, we might not use variables as they are but first *transform* them. We talked way back in Chapter 4 that some relationships might not be well-described by straight lines. Sometimes we want curvy lines. How can we do that, exactly?

Third, the *relationships themselves* might be affected by other variables. For example, back in Chapter 4 we talked about a study by Emily Oster that asked "is the relationship between taking vitamin E and health outcomes stronger during the period when vitamin E was recommended by doctors than during the periods when it wasn't?" We can model the relationship between taking vitamin E and health outcomes with $Outcomes = \beta_0 + \beta_1 VitaminE + \varepsilon$. But how can we model how *that relationship changes* over time? We need an *interaction*.

So those three things—all of which have to do with the kinds of variables we use as predictors in our OLS model—will be where we start. Once we've covered that, we'll move on to how we deal with standard errors, and when our *dependent variable* isn't continuous either.

13.2.1   Handling Discrete Variables

BINARY VARIABLES ARE ABSOLUTELY EVERYWHERE IN SOCIAL SCIENCE. Did you get the treatment or not? Are you left-handed or right-handed? Are you a man or a woman?[32] Are you Catholic or not? Are you married or not?

Binary variables are especially important for causal analysis, since a lot of the causes we tend to be interested in are binary in nature. Did you get the treatment or not?

Any time we have something that you *are* or *are not*, which happens any time we have some sort of qualitative description, we are dealing with a binary variable. These can be included in regression models just as normal. So we can still be working with $Y = \beta_0 + \beta_1 X + \beta_2 Z + \varepsilon$, but now maybe X or Z (or both) can only take two values: 0 ("are not"/false) or 1 ("are"/true).[33] If the binary variable is a control variable, we can think of it as we normally do—we're just shutting off back doors that go through the variable. But what if we are interested in the effect of that binary variable? How can we interpret the binary variable's coefficient?

Simply put, it's the *difference* in the dependent variable between the trues and the falses.[34] So for example, if we ran the

[32] Some people point out this should not be treated as a binary variable. I hope those people know how hard they're making regression analysis. Think of the researchers.

[33] Sometimes you'll see people use -1 and 1, or 1 and 2, instead of 0 and 1. This is usually just confusing though, and pretty rare in the social sciences.

[34] Why is this? OLS is trying to fit a line that minimizes squared residuals. But there are only two values the data can take on the x-axis: 0 and 1. Because of that, the OLS line can only produce two real predictions: one when it's 0 on the x-axis, and one when it's 1. The best prediction you can make in each case, which minimizes the squared residuals is just the mean of the outcome conditional on the binary variable being 0, or being 1. Then, the slope of the line is just how much the prediction increases when the x-axis variable increases by 1 (i.e., goes from 0 to 1).

regression $Sales = \beta_0 + \beta_1 Winter + \varepsilon$, where $Winter$ is a binary variable that's 1 whenever it's winter and 0 when it's not, then β_1 would be *how much higher* sales are on average in Winter than in Not Winter.[35]

[35] And if sales are lower on average in Winter, β_1 would be negative.

When we estimate the model, if we got $\hat{\beta}_1 = -5$, then we'd say that, on average, sales are 5 lower in Winter than they are in Not Winter. Take a look at the simulated data in Table 13.4. Notice that the coefficient on Winter (-5) is just the difference between the mean for non-winter (15) and the mean for winter (10). Also notice that the coefficient on the intercept (15) is the expected mean when all the variables are zero—when Winter is 0, we're in Not Winter, and average sales in Not Winter is 15.

	Mean	OLS Estimates
Winter	10.000	-5.000
		(1.103)
Not Winter	15.000	(Ref.)
(Intercept)		15.000
		(3.413)

Standard errors are in parentheses.
No significance stars shown.

Table 13.4: Average Sales by Season, or Regression of Sales on Winter

One important thing to note is that we *only include one side* of the yes/no question. The model is $Sales = \beta_0 + \beta_1 Winter + \varepsilon$, not $Sales = \beta_0 + \beta_1 Winter + \beta_2 NotWinter + \varepsilon$. Why is this? First off, imagine trying to interpret that. We want β_1 to compare Winter to Not Winter, but β_2 compares Not Winter to Winter. So... what's the difference between them? Interpretation would be pretty confusing. Second, this is to satisfy another OLS assumption we have to make, that there is no *perfect multicollinearity*. That is, you can't make a perfect linear prediction of any of the variables in the model with any of the other variables. Here, $Winter + NotWinter = 1$. We don't have a 1 in our model, though, so no problem, right? Wrong! There is an all-1s variable lurking in our model at all times—it's being multiplied by the constant!

Why can't we have that linear combination? Because OLS wouldn't be able to figure out how to estimate the parameters. Imagine the average sales in Winter is 10, and in Not Winter it's 15. If your regression is $Sales = \beta_0 + \beta_1 Winter + \beta_2 NotWinter + \varepsilon$, you could generate those exact same predictions with $\beta_0 = 15, \beta_1 = -5, \beta_2 = 0$. Or with $\beta_0 = 10, \beta_1 = 0, \beta_2 = 5$. Or with $\beta_0 = 3, \beta_1 = 7, \beta_2 = 12$. Or any infinite number of other ways! And while they'd all *work*, OLS has no way of picking one estimate out of those many, many options. It can't give you a best estimate any more. We need to limit its options by dropping NotWinter, in effect forcing it to choose the $\beta_0 = 15, \beta_1 = -5, \beta_2 = 0$ version.

WE CAN EXPAND OUR INTUITION ABOUT BINARY VARIABLES JUST A BIT and give ourselves the ability to include *categorical* variables in our model. Binary variables are just yes/no. But categorical variables can take any number of categories. These include variables like "what country do you live in?" or "what decade were you born in?" You can't answer these questions with yes and no, but they *are* discrete and mutually exclusive categories.[36]

We can handle categorical variables by just *giving each of the categories its own binary variable*. So instead of one variable for "which country do you live in?" it's one variable for "do you live in France?" and another for "do you live in Gambia?" and another for "do you live in New Zealand?" and so on. $Income = \beta_0 + \beta_1 France + \beta_2 Gambia + \beta_3 New Zealand + ...$ If we're including the categorical variable as a control, by including these binary variables for each category we can say that we've closed the back doors that go through, in this example, which country you live in.

And what if we want to interpret the coefficient on one of these binary categories? This is just a hair trickier than with a binary variable. Just like we couldn't include both sides of the binary variable in the model (we had to put just Winter in the model, not Winter and NotWinter), we also can't include every single category in the model. We need to drop one of them. The one we drop then becomes the "reference category." All the coefficients are *relative to that category*.

Before, with binary variables, the coefficient gave the difference between "yes" and "no." Now, with a categorical variable and a reference category, it's the difference between "this category" and "the reference category."

For example, if we drop France from our regression,[37] France becomes the reference category. If we then estimate $Income = \beta_0 + \beta_1 Gambia + \beta_2 New Zealand + ...$ and get $\hat{\beta}_1 = .5$ and $\hat{\beta}_2 = 3$, then that means *average income in Gambia is .5 higher than average income in France* and *average income in New Zealand is 3 higher than average income in France*. Both of the interpretations only take things relative to the reference category, not each other.

[36] Sometimes these categories might allow some overlap—some people do live in multiple countries, for example. But hopefully we can avoid these situations, or there are few enough of them we can just ignore them, as they make things much more difficult.

[37] This does not mean we're dropping all the observations from France. It just means we're removing the *France* binary variable from our regression.

We can see this in the simulated data in Table 13.5, which includes data only for France, Gambia, and New Zealand. Notice that average income for Gambia (30.5) is .5 higher than the average income for France (30), and the coefficient on Gambia when France is the reference category is .5. Similarly, the average income for New Zealand (33) is 3 higher than France, and the coefficient is 3. The intercept is the average income when all the predictors are 0—so it's not Gambia, and it's not New Zealand, meaning it must be France. The average income in France is 30, so the intercept is 30.

	Mean	OLS Estimates
France	30.000	(Ref.)
Gambia	30.500	.500
		(.103)
New Zealand	33.000	3.000
		(1.321)
(Intercept)		30.000
		(5.412)

Standard errors are in parentheses.
No significance stars shown.

Table 13.5: Average Income by Country among France, Gambia, and New Zealand, or Regression of Income on Country

This reference category stuff means that the coefficients on the categories (and their significance) have relatively little meaning on their own. They only have meaning relative to each other. The coefficients change completely if you change the reference category. If we made New Zealand the reference category instead, the coefficient on Gambia would change from .5 to -2.5.[38]

[38] Can you spot why?

If you want to know whether a categorical variable has a significant effect as a whole, you don't look at the individual coefficients. Instead, you look at *all* the category coefficients. This takes the form of a "joint F test," where you compare the predictive power of the model against a version where the categorical variable has been removed as a predictor.

In R, the `linearHypothesis` command in the **car** package can perform the joint F test. You give it the full list of names of coefficients for the categories (which you can make yourself using `paste` and `unique` with a little effort, or the `matchCoefs()` function). In Stata, `testparm i.cat` will do a joint significance test of all the categories of the `cat` variable. In Python, if you use `sm.OLS().fit()` to create a regression results object, let's say it's called `m1`, then you can do a joint F test with `m1.f_test()`.

You'll need to read the documentation to see how to create a matrix or string indicating the factor variables to be tested.

13.2.2   Polynomials

SOMETIMES, A STRAIGHT LINE ISN'T ENOUGH. OLS assumes that the relationship between the dependent variable and each of the predictor variables can be described as a straight line (linear). If that's not true, OLS will do a poor job. For example, look at Figure 13.6. The true relationship clearly follows the curvy line drawn on the right. If we try to estimate this relationship using a straight line, we get something that doesn't fit very well, shown by the line on the left.

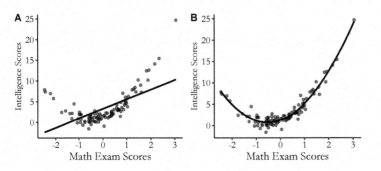

Figure 13.6: Math Scores and Intelligence: OLS and the True Model

We can still use OLS for these kinds of relationships though. We just need to be careful to model that curviness right in our regression equation.

There are two main ways we can do this: we can either *add polynomial terms*, or we can *transform the data*.

A POLYNOMIAL is when you have the same variable in an equation as itself, as well as powers of itself. So for example, $\beta_1 X + \beta_2 X^2 + \beta_3 X^3$ would be a "third-order polynomial" since it contains X to the first power (X), the second power (X^2), and the third power (X^3).[39],[40]

By adding these polynomial terms, it's possible to fit lines that aren't straight. In fact, add enough terms and you can mimic just about any shape.[41] This can be seen in Figure 13.7. Looking at the data points directly, this is clearly a nonlinear relationship. On top of the data points we have the best-fit line from a linear model $(Y = \beta_0 + \beta_1 X)$, which gives the straight line, a second-order polynomial model, which gives the curvy dashed line $(Y = \beta_0 + \beta_1 X + \beta_2 X^2)$, and a third-order

[39] These are also referred to as the linear, squared, and cubic terms, with quartic for the fourth power.

[40] Notice that this contains *all* the terms up to the third power. You pretty much *never* want to omit any, as in $\beta_1 X^2 + \beta_2 X^3$ which has the second- and third-order terms but not the first.

[41] Any functional relationship, that is—it still needs to make sense that we're trying to get a single conditional mean of Y for each value of X.

polynomial, which gives the curvy solid line ($Y = \beta_0 + \beta_1 X + \beta_2 X^2 + \beta_3 X^3$).

The curvy lines clearly do a better job fitting the relationship, and all we had to do was add those polynomial terms!

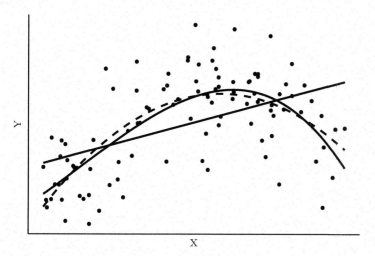

Figure 13.7: Regression Using a Linear, Square, and Cubic Model

That's the reason to add polynomial terms then. Add polynomial terms and you can better fit a line to a non-straight relationship. But this leaves us with two pressing questions: (1) how can we interpret the regression model once we have polynomial terms, and (2) why not just add a bunch of polynomial terms all the time?

HOW CAN WE INTERPRET A MODEL WITH POLYNOMIALS? For this one, we're going to need a little calculus I'm afraid. Let's take our cubic model from before:

$$Y = \beta_0 + \beta_1 X + \beta_2 X^2 + \beta_3 X^3 \qquad (13.7)$$

Our typical interpretation of a regression coefficient estimate like $\hat{\beta}_1$ would be "holding everything else constant, a one-unit increase in X is associated with a $\hat{\beta}_1$-unit increase in Y." However, that's a problem! We *can't* "hold everything else constant" because there's no way to change X without also changing X^2 and X^3. So an interpretation of the effect of X must take into account the coefficients on *each* of its polynomial terms. *When a regression model includes a polynomial for X, the individual coefficients on the X terms mean very little on their own, and must be interpreted together.*

How can we interpret them together? Let's try to get back our interpretation of "holding everything else constant, a one-unit increase in X is associated with a (???)-unit increase in Y." What is (???)? Well, what tool can we always reach for to figure out how one variable changes with another? The derivative! If we take the derivative of Y with respect to X, that will tell us how a one-unit change in X is related to Y.[42]

$$\frac{\partial Y}{\partial X} = \beta_1 + 2\beta_2 X + 3\beta_3 X^2 \qquad (13.8)$$

A one-unit change in X is associated with a $(\beta_1 + 2\beta_2 X + 3\beta_3 X^2)$-unit change in Y.[43] Notice that the relationship *varies with different values of X*. If $X = 1$, then the effect is $\beta_1 + 2\beta_2 + 3\beta_3$. But if $X = 2$ then it's $\beta_1 + 4\beta_2 + 12\beta_3$. The effect of X on Y depends on the value of X we already have. This is what we'd expect. Look back at Figure 13.7—as X increases, at first Y increases as well, but then Y stops increasing and starts declining. The effect of a one-unit change in X *should* vary as we move along the X-axis.

Let's use a more concrete example. Let's go back to our regression from Table 13.2 but add a squared term, which we can see in Table 13.6.

	Model 1
Constant	97.518***
	(0.059)
Number of Locations	-0.0802***
	(0.001)
Number of Locations^2	0.0001***
	(0.000)
Num.Obs.	27178
R2	0.194
R2 Adj.	0.194
F	3279.008

* p < 0.1, ** p < 0.05, *** p < 0.01

[42] This works in a linear model too, of course. In $Y = \beta_0 + \beta_1 X$, the derivative of Y with respect to X is just β_1. And thus we have our standard interpretation of β_1 from a linear model.

[43] Even if you don't know calculus, if all you've got is a polynomial, this is a derivative that's easy enough to do yourself. Just take each term, multiply it by its exponent, and then subtract one from the exponent (keeping in mind that $X = X^1$, and when you subtract one from that exponent you get $X^0 = 1$). So $\beta_3 X^3$ gets multiplied by its exponent of 3 to get $3\beta_3 X^3$, and then we subtract one from the exponent to get $3\beta_3 X^2$. Repeat with each term.

Table 13.6: A Regression with a Second-Order Polynomial using Restaurant Inspection Data

Here we see a coefficient of $-.0802$ on the linear term, and $.0001$ on the squared term. This means that a one-location increase in the number of locations associated with a restaurant chain is associated with a $\beta_1 + 2\beta_2 Number of Locations$, or $-.0802 + .0002 Number of Locations$, change in the health inspection score. So for a chain with only one location, adding a second one would be associated with a $-.0802+.0002(1) = -.08$ reduction in the health score. But for a chain with 1000 locations, adding its 1001st would be associated with a $-.0802 + .0002(1000) = .1198$ *increase* in its health score. There is no single effect; it depends on what part of the x-axis we're on. Leaving out the squared term would ignore that.

POLYNOMIALS SEEM PRETTY HANDY! So why not just add a whole bunch of them all the time? And while we're at it, why stop at just square terms and cubics? Why not add four-, five-, six-, or seven-order polynomials to all our models?

There are a few good reasons. For one, the model does get a little harder to interpret. Of course, if it's the right model we'd still want that anyway.

For another, often the higher-order polynomial terms don't really do anything. Take Figure 13.7 for example. The line fits from the second-order polynomial and the third-order polynomial models are almost exactly the same.[44] So in many cases, by adding more and more polynomial terms you make your model more complex without actually improving its fit.

There is also the issue that adding polynomial terms puts strain on the data and can lead to "overfitting," where a too-flexible model winds up bending in strange shapes to try to fit noise in the data, producing a worse model. Imagine, for example, fitting a 100-order polynomial on a data set with 100 observations in it. We'd predict every point perfectly, but clearly that line isn't really going to tell us much of anything.

Adding more polynomial terms can, in general, make the model very sensitive to little changes in the data, and can make its predictions kind of strange near the edges of the observed data. Take Figure 13.8, for example, which starts with Figure 13.7 and adds a regression fit using a ten-degree polynomial. The shape doesn't seem to follow the data all that much better than even the two-order polynomial. And we get strange up-and-down motion that seems to chase individual observations rather than the overall relationship. The worst offender is over on the far right, where at the last moment the line springs way up to try to fit the single point on the far right of the data. Those additional degrees in the polynomial aren't just unnecessary, they're actively making the model worse.

So we don't want "too many" polynomial terms. But how much is just right? Well, ideally, you want to have the number of polynomial terms necessary to model the actual true underlying

[44] And I can tell you, since I generated this data myself, that the true underlying model actually does have a cubic term. Despite this, the cubic term in the regression adds little to nothing.

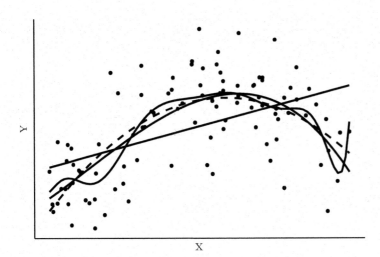

Figure 13.8: Regression Using a Linear, Square, Cubic, and Ten-degree Polynomial Model

shape of the relationship. But we don't actually know what that is, so that doesn't help much.

A good way to approach the question of how many polynomial terms to include (or whether to include them at all) is graphically. This works best when you don't have *too* many observations. Just draw yourself a scatterplot and see what the shape of the data looks like. Then, add your regression line and see whether it seems to explain the shape you see. If it doesn't, try adding another polynomial term until it does.

You can also graphically analyze the residuals. Try a regression model, calculate the residuals, and then plot *those* on the y-axis against your x-axis variable. If you have enough polynomial terms, there shouldn't be any sort of obvious relationship between X and the residuals. This can be seen in Figure 13.9. The same data from Figure 13.8 is estimated using a linear OLS and a second-order polynomial OLS, and the residuals from those regressions are plotted against X. After the linear regression, there's still a clear shape to the relationship between the residuals and X on the left. Not enough curviness to our line. But on the right, after our second-order polynomial, it's just a bunch of noise around 0. That's enough, we can stop!

Besides all that, just as a good rule of thumb, you almost never want to go beyond cubic unless you have a really good reason. At that point you have to start worrying that the the statistical issues with too many terms are going to be a problem. The cure of polynomials may be worse than the disease of nonlinearity.

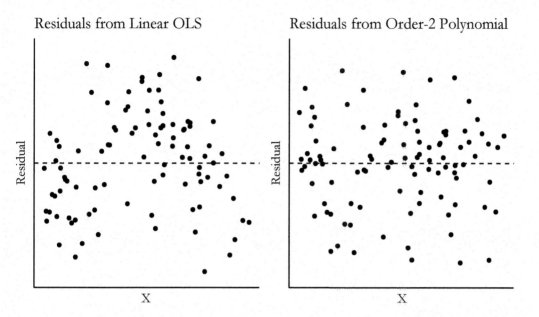

Residuals from Linear OLS   Residuals from Order-2 Polynomial

Figure 13.9: Residuals from
Linear OLS and Second-
Order Polynomial OLS

There are also more automatic approaches to selecting the
number of polynomials. You don't have to look far to find people
recommending that you try different amounts of polynomials,
and then omit higher-order terms that are insignificant. In other
words, try a quartic, and if the X^4 coefficient is insignificant,
drop it. Then try the cubic, and if the X^3 term is insignificant...
and so on. I don't recommend this approach, as you are at this
point making decisions about your model based on a statistical
significance test, and that's not what statistical significance tests
are meant to do.[45]

A somewhat better approach is to use a principled model-
selection algorithm like a Least Absolute Shrinkage and Selection Operator (LASSO). Information on the specifics of LASSO
will have to wait until the "Penalized Regression" section later
in the chapter. For now, you can just think of it as a variant of
OLS that tries to not just minimize the sum of squared errors,
but tries to do so while reducing the coefficients in the model.
Like dropping polynomial terms because they're insignificant, it
takes the approach that additional polynomial terms that don't
improve prediction much can be dropped. However, it does so
in a way that is actually designed for the purpose of figuring
out whether an additional term is "worth it," rather than using
significance testing, which only by happenstance sort of tries for
a similar goal.

[45] Remember, "insignificant" doesn't actually mean
"this coefficient is zero."
It means "we don't have
evidence to show that it's
not zero."   Omitting it
from the model forces it
to be zero, which may not
be right.

13.2.3   Variable Transformations

POLYNOMIAL TERMS ARE NOT THE ONLY WAY TO FIT A NON-
LINEAR SHAPE. In many cases it makes more sense to *transform
the variable itself* directly.

Generally this means running a variable through a function
before you use it. So for example, instead of running the regression

$$Y = \beta_0 + \beta_1 X + \varepsilon \qquad (13.9)$$

you might instead run

$$Y = \beta_0 + \beta_1 \ln(X) + \varepsilon \qquad (13.10)$$

where ln() is the base-e logarithm, or the "natural log."[46]

Why might we do this? There are two main reasons we might
want to transform a variable before using it, some of which we
already covered in Chapter 3.

THE FIRST REASON MIGHT BE to give the variable statistical
properties that play more nicely with regression. For example,
you may want to reduce *skew*. A variable is "right skewed" if
there are a whole lot of observations with low values and just
a few observations with *really high* values.[47] Income is a good
example of a skewed variable. Most people are somewhere near
the median income, but a few super-high earners like CEOs and
movie stars earn *way way way* more than the median.[48]

Skewed data are likely to produce *outlier* observations. An
outlier is an observation that has values that are very different
from the other observations. For example, if you have a hundred
people who all earn roughly $40,000 per year, and one person
who earns a million, that person is an outlier.[49]

There's nothing inherently wrong with outlier observations.
Sometimes people take big outliers as indications that the data
has not been recorded properly and that million-dollar-salary is
more likely a hundred-thousand with a typo, but often the data
is fine, it's just an outlier. However, even though the data may
be accurate, the presence of outliers can make regression models
statistically weak. Because the observations are way out there,
they tend to produce big errors, and so just a few observations
can have the effect of really tugging on the regression line and
having way too much influence.

Using a transformation that "scrunches down" the big observations to be closer to the others can reduce this skew, and
make the regression model behave a bit better. This can be seen
in Figure 13.10. On the left, we have raw data, with a single

[46] What makes it so "natu-ral"? Because whenever you have compounded growth (i.e., the interest you earn then starts earning its own interest, or the animal population reproduces, and then those babies grow up and re-produce as well), the number e will always be lurking around in your calculation somewhere, just like π will always be lurking around if you're doing calculations about circles and curves.

[47] Left skew would mean a whole lot of high values with just a few really low ones. This is less common than right skew in most social science research, though.

[48] For more on skewed vari-ables, refer back to Chapter 3.

[49] Outliers can also come from non-skewed data, too.

Outlier.   An outlier is an observation with a value that is much bigger or smaller than the other obser-vations.

big outlier way to the top-right of the rest of the data. I've estimated OLS twice on this data, once leaving out the outlier and once including it. The estimated OLS slope changes wildly!

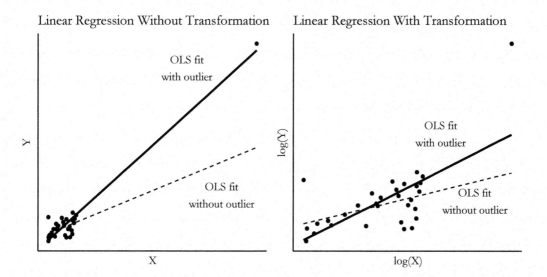

Linear Regression Without Transformation   Linear Regression With Transformation

Figure 13.10:   OLS Fit With and Without an Outlier, With and Without Log Transformations

On the right, I've taken the logarithm of both X and Y. You can immediately see that the outlier comes in closer to the rest of the data (this shows up on the graph as the other data spreading out, since we can zoom in more closely on it without losing the outlier). The distance between observations is shrunk at the high end. Also, including or excluding the outlier has a much more muted effect on what the regression line looks like.

And keep in mind, it's fine for the presence of the outlier to affect the slope of the OLS line. In fact, it should.[50] *All* observations should have some sort of effect on the regression line. The problem with the graph on the left is that it has *too much* influence for just being a single observation, and we can ease that a little bit with the logarithms.

THE SECOND REASON TO TRANSFORM A VARIABLE is to try to get a linear relationship between the variables. Remember, OLS naturally fits a straight line. It can only fit curvy lines if you transform the variables, either by using a polynomial or some other transformation.

This comes down to what kind of relationship you think you're modeling. Should a one-unit increase in X relate to a β_1-unit increase in Y? Then you're talking about a linear relationship and you don't need a transformation. But what if a one-unit increase in X should relate to some *percentage* increase

[50] If it didn't, we could save ourselves a lot of time with this transformation business and just drop all the outliers.

in Y? Or if a certain *percentage* increase in X should relate to a certain increase in Y? Or something like that?

For example, if we were regressing your current wealth on the interest rate of your investments as of ten years ago, the true relationship is $Wealth = InitialWealth \times e^{10 \times InterestRate}$. The effect of an increase in $InterestRate$ should be *multiplicative*, not linear. So we need a transformation to "linearize" the relationship. In this case, a logarithm comes in handy again. If we take the logarithm of both sides, we get

$$\ln(Wealth) = \ln(InitialWealth) + 10 \times InterestRate \quad (13.11)$$

and that's a linear relationship.[51] OLS will have an easy time with that one. The logarithm is handy for turning multiplicative relationships like this into additive linear ones, since $\log(XY) = \log(X) + \log(Y)$.

So if you have a theoretical relationship, you want to think about *how those two variables relate. Should* the relationship be linear? Or will you need some sort of transformation to get a linear relationship out of it?

WHAT KINDS OF TRANSFORMATIONS ARE THERE, THEN? We have our reasons for transforming now, but what kinds of transformations can we do?

There are a lot of options here and I don't plan to go through all of them. But a few tend to pop up regularly in social science research.

The first is, unsurprisingly, the logarithm. We've talked about this one.

The second, third, and fourth come from the problems we have with the logarithm. In particular, the logarithm can't handle values of zero. $\log(0)$ is undefined. So what if we want the nice skewness-handling properties of the logarithm but also have data with zeroes in it?

There are three popular options. Our second transformation, after the logarithm, is to simply add one to the variable before taking the logarithm, $\log(x + 1)$. This will do the job but also is very much a kludge that tends to lose some of the nice things we liked about the logarithm in the first place. Without going too deep in the weeds, I don't recommend it.

The third transformation is the *square root*. Like the logarithm, taking the square root of a variable "squishes down" its big positive outliers. But it can handle a zero, with $\sqrt{0} = 0$. However, while using a square root does tamp down on big

[51] The ln() means "natural log," i.e., a logarithm with base e. It's pretty rare that you'll see any other kind of logarithm in the empirical social sciences.

observations, it doesn't do so nearly as *much* as a logarithm does. For example, going from 4 to 10,000, the square root is multiplied by 50 ($\sqrt{10000}/\sqrt{4} = 100/2 = 50$), whereas the logarithm is only multiplied by about 6.6 ($\log(10000)/\log(4) \approx 9.2/1.4 \approx 6.6$). Also, you lose the nice "percent increase" interpretation that logarithms have, which we'll talk about in a bit.

The fourth is the *inverse hyperbolic sine* transformation, or "asinh," which is equal to $\ln(x + \sqrt{x^2 + 1})$. This transformation has the nice property of accepting 0 ($asinh(0) = 0$) as well as negative numbers.[52] Plus, for reasonably large values (say, above 10) it provides results that are *very* similar to a plain-ol' logarithm. So the nice stuff we like about logarithms we also get with the inverse hyperbolic sine. For skewed data with zeroes where your only goal is to reduce the effect of the skew, the inverse hyperbolic sine is usually what I recommend.

However, if the goal is to get some sort of theoretical percentage-increase interpretation out of the variable, then any sort of ad-hoc method to deal with zeroes (which by nature don't really belong with percentage increases) is going to get some strange results. There are some ways to lessen the problem, like ensuring the mean of the variable is large by scaling it up (using dollars rather than thousands of dollars, for example).[53] You may also just be better off with a method that naturally deals with zeroes rather than either $asinh(x)$ or $\ln(1 + x)$, like Poisson regression.

THE FIFTH TRANSFORMATION WE'LL DISCUSS doesn't really deal with skew or linearizing a relationship, but it does deal with outliers. *Winsorizing*, especially popular in finance, is the process of taking some data and very forcefully squishing in its ends. You simply take any values that are far enough away from the center and reduce them towards the center. To Winsorize the top X%, you take every observation above the Xth percentile, and replace it with the Xth percentile.

For example, if we had 100 observations from 1 to 100, and we Winsorized the top and bottom 5%, that would mean leaving the 6—95 observations as they were, and then taking the 1, 2, 3, 4, and 5 and replacing them with 6, and also taking the 96, 97, 98, 99, and 100 observations and replacing them with 95.

This is a very brute-force way of dealing with outliers, but it tends to work. It leaves most of the data untouched, which is nice if you think the true relationship really is linear, but you don't want your model to be too influenced by outliers.

[52] Although using it for variables that have both positive and negative values probably won't do what you want.

[53] Marc F Bellemare and Casey J Wichman. Elasticities and the inverse hyperbolic sine transformation. *Oxford Bulletin of Economics and Statistics*, 82(1): 50–61, 2020.

THE SIXTH AND FINAL TRANSFORMATION we'll consider doesn't really improve the statistical properties of the model at all. It doesn't account for skew, outliers, or linearization. This is the transformation of *standardizing* a variable by subtracting its mean and dividing by its standard deviation, or $(X - mean(X))/sd(X)$.[54]

Why do this if it doesn't improve the statistical properties of the model at all? Because it can make the model easier to interpret. We know that in the model $Y = \beta_0 + \beta_1 X + \varepsilon$, the interpretation of β_1 relies on a "one-unit increase in X." But that means we need to keep in mind the scale of all of our variables so we know how big a one-unit increase is. But if we standardize first, then β_1 instead has an interpretation based on a "one-standard-deviation increase in X," which can be easier to think about in some cases.

Same goes for standardizing Y. Imagine Y is student test scores, and X is time spent watching TV per day. If $\hat{\beta}_1 = -5$, then a one-unit increase in TV hours per day reduces test scores by 5. Is that a big effect? A small effect? We'd need to know way more about the test itself. But if we standardize Y first, and $\hat{\beta}_1 = -.5$, then an extra hour of TV a day reduces test scores by half a standard deviation—pretty big! We can more immediately interpret the result now. Naturally, this only works for variables where we have a sense of the standard deviation more than we have a sense of the units.

AND THAT'S IT—those are the six transformations we'll cover. There are many, many more transformations in the world, but in my experience these are the most common in social science research. Honestly, I'm not sure I've ever had occasion to use a transformation other than one of these six, or a polynomial.

That said, just having listed these transformations doesn't mean we're quite done with them yet. In particular, logarithms—which I just can't seem to stop going on about—can make our results a bit trickier to interpret. So how can we do that?

INTERPRETING THE RESULTS OF A REGRESSION WITH A LOGARITHM TRANSFORMATION appears at first to be no problem. Then you think a little about it, and it seems like a huge problem. Then you learn even more about it, and it once again becomes no problem at all. Such is life.

Let's look first at a regression with a log transformation on X (we'll get to Y later).

$$Y = \beta_0 + \beta_1 \ln(X) + \varepsilon \qquad (13.12)$$

[54] Some people call this "normalizing" instead of "standardizing," but this is misleading, as it sounds like you're forcing the data to have a normal distribution. Standardizing has nothing to do with producing a normal distribution, contrary to the popular misconception.

So how does it appear at first? No problem! We can interpret the $\hat{\beta}_1$ as "a one-unit change in $\ln(X)$ is associated with a $\hat{\beta}_1$-unit change in Y." Great.

Now let's think a little about it—what exactly *is* a one-unit change in $\ln(X)$? What does that even mean? It's not the same thing as a one-unit change in X.

Let's learn even more about it, at which point it will become easy again. Linear changes in $\ln(X)$ are related to *proportional* changes in X. Think about it. We know that $\ln(AB) = \ln(A) + \ln(B)$. So if we start with $\ln(X)$ and we *add 1 to it*, we get $1 + \ln(X) = \ln(e) + \ln(X) = \ln(eX)$.[55] *Adding* something to $\ln(X)$ is equivalent to *multiplying* something by X.

In particular, whatever we add to $\ln(X)$, let's call the thing we add c, we are in effect multiplying X by e^c. Now we can use a convenient little fact: If that c is close to 0, then e^c is close to $1 + c$. Why is that convenient? Because it means that a .05 increase in $\ln(X)$ is approximately equivalent to multiplying X by 1.05, or in other words a .05 increase in $\ln(X)$ is approximately equivalent to a *5% increase in X*.[56] Now you know why I was talking earlier about relationships that were proportional in nature.[57]

Let's take this intuition and apply it so we can interpret a log no matter where it shows up in our equation.

First, back to our original equation with a log transformation for X: $Y = \beta_0 + \beta_1 \ln(X) + \varepsilon$. Let's use a 10% change rather than a 100% change here for X, so we're more in the realm of reality (for most social science studies, 10% changes show up a lot more often then 100% changes) and so the $1 + c \approx e^c$ approximation works better. A 10% increase in X translates to a $10/100 = .1$-unit increase in $\ln(X)$, which is associated with a $.1 \times \beta_1$-unit change in Y. So we can interpret β_1 by saying that a 10% increase in X is associated with a $.1 \times \beta_1$-unit change in Y.

Next, let's try putting the log on Y instead: $\ln(Y) = \beta_0 + \beta_1 X + \varepsilon$. Now, a 1-unit change in X is associated with a β_1-unit change in $\ln(Y)$, which then means a $\beta_1 \times 100\%$ change in Y.

Finally, we can put the log on both. $\ln(Y) = \beta_0 + \beta_1 \ln(X) + \varepsilon$. Now, a 10% increase in X is associated with a $10/100 = .1$-unit increase in $\ln(X)$, which is associated with a $.1 \times \beta_1$-unit change in $\ln(Y)$, which translates into a $.1 \times \beta_1 \times 100\%$, or $10 \times \beta_1\%$ change in Y. We can actually skip the part about picking an initial percentage increase in X, since as you can see, whatever we pick just shows up in the percentage change for Y as well. So

[55] Remember, $\ln()$ is a base-e logarithm, so $\ln(e) = 1$.

[56] It's *exactly* equivalent to a 5.271% increase in X, but generally researchers will simplify and just go with saying that the c increase in $\ln(X)$ is just a $c \times 100\%$ change in X. You may want to get more precise for big percentage changes though—a 1-unit change in $\ln(X)$ is an $e^1 = e \approx 2.71$ change in X, i.e., a 171% increase, not 100%. Pretty far away! But for small percentage changes, which is mostly what's relevant in social science anyway, the approximation works fine. Once you get above .1/10% it starts to get a bit dicier.

[57] You could fix this by using a different base—for example, $\log_{1.1}(X)$ (the base-1.1 logarithm of X), which is equivalent to $\ln(X)/\ln(1.1)$, will increase by exactly one unit for a 10% increase in X. No approximation required; you just need to pick the size of your percentage increase ahead of time. This approach isn't widespread, but if I have anything to say about it, it soon will be. If the second edition of this book takes this out of the footnote and into the main text, you'll know I won.

with logs on both X and Y we can just say that any percentage change in X is associated with β_1 times that percentage change in Y.

And that's it! You can memorize those three interpretations if you want, but it's probably a better idea to work through the logic each time instead so you know what's going on. Plus, as a bonus, this kind of thinking works for *any* transformation. Not all transformations have percentage interpretations, but you can always think through the logic of "what does a one-unit change in this transformed variable *mean* in terms of the original variable?" and work out your interpretation from there.

13.2.4   Interaction Terms

POLYNOMIALS AND TRANSFORMATIONS ARE GREAT FOR LET-TING the relationship between Y and X be flexible. We can fit any sort of curvy line we want, so that the relationship between Y and X can change depending on the value of X. But we're still missing something. What if the relationship between Y and X differs not based on the value of X, *but based on the value of a different variable Z?*

For example, what's the relationship between the price of gas and how much an individual chooses to drive? For people who own a car, that relationship might be quite strong and negative. For people who don't own a car, that relationship is probably near zero. For people who own a car but mostly get around by bike, that relationship might be quite weak. The relationship between gas prices and miles driven *differs depending on car ownership*. We might want to allow the relationship to vary so we can model it properly. We might also be interested in whether the relationship *is* different between these groups.[58]

However, we have no way of representing this change in relationship with either a polynomial or a transformation. Instead we will need to use *interaction terms*.

An interaction term is when you multiply together two different variables and include their product in the regression, like $Y = \beta_0 + \beta_1 X + \beta_2 Z + \beta_3 XZ + \varepsilon$.

In terms of putting the model together, there's not a lot more to it than that. If you think that the relationship between Y and X is different for different values of Z, then simply include $X \times Z$ in your model, as well as Z by itself.[59]

The real difficulty with interaction terms is *interpreting* them. Interpretation can be a bit of a brainbender, which is

[58] And plenty of studies are designed around looking at whether a relationship changes with the value of some other variable. The Oster study we looked at in Chapter 4 is one such paper.

Interaction term. The product of two variables multiplied together and included in a regression, usually alongside the un-multiplied versions.

[59] Why do we need to include the interacting variable by itself, too? Because otherwise, the coefficient on $X \times Z$ accounts for not only the interaction between the two, but also the direct effect of Z itself. It confuses how the relationship between Y and X changes with Z with the effect between Y and Z. No good! So we include Z by itself to move that direct relationship to a different term and out of the interaction.

unfortunate; you can't walk two steps in a standard social science research design without running into a study where the interaction term is the most important part.

So how can we interpret interaction terms? The question of interpreting a model with interaction terms is actually two questions. The first is *what is the effect of a variable X when there's an interaction between X and something else in the model?* The second is *how can I interpret the interaction term?*

The intuition for both comes back to good ol' calculus. If you don't know calculus you should still be able to follow along with the intuition. But if you do know calculus, hey, calculus is pretty neat right? Let's be in a calculus club.[60]

Let's tackle the first question: what is the effect of X when there's an interaction between X and something else in the model? To solve this, we can write out the regression equation, and then just take the derivative with respect to X. Done! After all, the derivative of Y with respect to X is just how Y changes as X changes, which is also the interpretation of a regression coefficient. If you don't know calculus, this is a very easy derivative. As long as there aren't any polynomial or transformation terms for X, just gather all the terms that have an X in them, and then remove the X. Done. If there *are* polynomial or transformation terms for X, well... there are calculators online that can do derivatives for you.

So if our regression is

$$Y = \beta_0 + \beta_1 X + \beta_2 Z + \beta_3 XZ + \varepsilon \qquad (13.13)$$

then we have

$$\frac{\partial Y}{\partial X} = \beta_1 + \beta_3 Z \qquad (13.14)$$

and that's the effect of X on Y.[61]

Of course, we're not entirely done yet, as the effect of X that we've calculated has a Z in it. This means that there is no single effect of X.[62] Rather, we need to ask what the effect of X is at *a given value of Z*. We can say what the effect of X is at $Z = 0$ (it's just β_1), or at $Z = 16$ (it's $\beta_1 + 16\beta_3$), or any other value.

This also means that the individual coefficients themselves don't tell us much, just like with polynomials. β_1 no longer gives us "the effect of X." So if you're looking at effect sizes, you need to consider the entire effect-of-X expression. Just looking at β_1

[60] As someone who values their own explaining-concepts-clearly skill enough to bother writing a textbook many hundreds of pages long, I realized recently that I did not remember what it was like to not be able to take a derivative, and had forgotten how to work through a fairly basic math problem without using the calculus tools that make it a snap, as opposed to the more laborious tools usually used to teach it at a more basic level. If you ever see a calculus textbook with my name on it, you can be assured that it is either bad, or ghostwritten, or ghostwritten and bad.

[61] Again, if you're not a calculus person, notice how we just took the terms with X in them—$\beta_1 X$ and $\beta_3 XZ$—, and took the X out, leaving us with β_1 and $\beta_3 Z$.

[62] As there shouldn't be. The whole point of this was that the effect of X would change with Z.

alone tells you very little (and, further, the *significance* of β_1 tells you very little).

So that's how we can get the effect of X when it's involved in an interaction term. How about our other question? What is the interpretation of the interaction term itself? Formally, the same approach of taking the derivative works too—in the regression equation $Y = \beta_0 + \beta_1 X + \beta_2 Z + \beta_3 XZ$, the interaction term β_3 is the cross derivative of Y with respect to both X and Z, i.e., $\frac{\partial^2 Y}{\partial X \partial Z}$. That's a little harder to wrap our heads around, though, so we may want to back out and think of it in a more intuitive, less mathematical way.

One good way to think about interpreting interaction terms is this: in the regression $Y = \beta_0 + \beta_1 Z$, the coefficient on Z tells us *how our prediction of Y changes when Z increases by one unit*, right? But as we've already established, we can get *the effect of X* in the equation $Y = \beta_0 + \beta_1 X + \beta_2 Z + \beta_3 XZ$ by taking the derivative (or gathering the X terms and dropping X) to get $\frac{\partial Y}{\partial X} = \beta_1 + \beta_3 Z$. This looks familiar, doesn't it? Now, the coefficient on Z in *this* equation, β_3, tells us how our prediction of $\frac{\partial Y}{\partial X}$ changes when Z increases by one unit. In other words, β_3 is *how much stronger the effect of X on Y gets* when Z increases by one unit.[63]

Often, the variable being interacted with is binary. This doesn't change the interpretation, but it makes it a bit easier to work through. Say we have the regression equation $Y = \beta_0 + \beta_1 X + \beta_2 Child + \beta_3 X \times Child$, where $Child$ is a binary variable equal to 1 for children and 0 otherwise. What do we have from this equation? We can say that the effect of X on Y is $\beta_1 + \beta_3 Child$, just as before, and that β_3 is how much stronger the effect of X on Y gets from a one-unit increase in $Child$.[64]

But $Child$ can only increase from 0 to 1—those are the only values it takes. So if the effect of X on Y is $\beta_1 + \beta_3 Child$, then that means that β_1 is the effect of X on Y *when Child = 0*, i.e., the effect of X on Y for non-children is β_1. We get that by just taking $Child = 0$ and plugging it in to get $\beta_1 + \beta_3 0 = \beta_1$. To get the effect of X on Y for children, we plug in $Child = 1$ and get $\beta_1 + \beta_3 1 = \beta_1 + \beta_3$.

The difference between those two effects—$\beta_1 + \beta_3$ for children and β_1 for non-children, is β_3. This is the difference in the effect of X on Y between children and non-children. Or, put another way, it's how much stronger the effect of X on Y gets when you increase $Child$ by 1, going from $Child = 0$ to $Child = 1$, i.e., non-child to child. That's the same interpretation we had

[63] It goes the other way too—it's also how much stronger the effect of Z on Y gets when X increases by one unit. If it feels strange that we're forcing those two things to be the same number, that is indeed a bit strange. But it's just a consequence of how derivatives work.

[64] If you want to really flex those interpretation-muscles, walk through the interpretation of a *three-way interaction*, i.e., $Y = \beta_0 + \beta_1 W + \beta_2 X + \beta_3 Z + \beta_4 WX + \beta_5 WZ + \beta_6 XZ + \beta_7 WXZ$. The same logic works; there's just another step. β_7 is how much a one-unit increase in Z increases the extent to which a one-unit increase in W increases the effect of a one-unit increase in X on Y.

before. If we find a meaningful (or, if you prefer, statistically significant) coefficient on β_3, then we can say that "the effect of X on Y differs between children and non-children."

Table 13.7: Regressions with Interaction Terms

	Inspection Score	Inspection Score	Inspection Score
Number of Locations	-0.019***	-1.466***	-0.019***
	(0.000)	(0.154)	(0.000)
Weekend	1.432***	1.459***	1.759***
	(0.419)	(0.418)	(0.488)
Year of Inspection	-0.065***	-0.108***	-0.065***
	(0.006)	(0.008)	(0.006)
Number of Locations x Year of Inspection		0.001***	
		(0.000)	
Number of Locations x Weekend			-0.010
			(0.008)
Num.Obs.	27178	27178	27178
R2	0.069	0.072	0.069
R2 Adj.	0.069	0.072	0.069
F	669.086	525.621	502.255

* p < 0.1, ** p < 0.05, *** p < 0.01

In Table 13.7 we can see a continuation from Table 13.2 earlier in the chapter, but this time with some interaction terms included. In the second column we have an interaction between $Number of Locations$ and $Year of Inspection$, while in the third column we have an interaction between $Number of Locations$ and $Weekend$, a binary variable indicating whether the inspection was done on a weekend or not.

In the second column, we have an effect of $Number of Locations$ of -1.466 (which is the coefficient on $Number of Locations$) $+.001 Year of Inspection$ (the coefficient on the interaction term). How can we interpret this? The -1.466 doesn't tell us much— that's the relationship between a one-unit increase in $Number of Locations$ and $Inspection Score$ if we plug in $Year of Inspection = 0$, i.e., in the year 0! That's way outside our data and doesn't reflect a realistic prediction.[65] But we could plug in a value like $Year of Inspection = 2008$, which is in the data, and say that the relationship between a one-unit increase in $Number of Locations$ and $Inspection Score$ in 2008 is $-1.466+.001\times2008 = .542$. Positive! In fact, the effect is positive for the entire range of years actually in the data, despite that big negative -1.466.

[65] Told ya the individual coefficients don't mean as much when you have an interaction. You need to take it all in at once.

Don't be tricked by the coefficients on their own. Always interpret them with the relevant interactions. We can also say that every year, the relationship between a one-unit increase in *NumberofLocations* and *InspectionScore* gets more positive by .001.

In the third column, we have an effect of *NumberofLocations* of $-.019 - .010Weekend$. From this, we can say that on a non-weekend, the association between a one-unit increase in *NumberofLocations* and *InspectionScore* is $-.019$, and on the weekend, it's $-019 - .010 = -.029$.[66] We can also say that the relationship is .010 more negative on weekends than it is on weekdays.[67]

THERE ARE A FEW THINGS TO KEEP IN MIND when it comes to using interaction terms. The first is to think very carefully about *why* you are including a given interaction term, because they are prone to fluke results if you go fishing. Think way back to Chapter 1—if we just let the data do whatever, we'll find results that don't mean much. We need to approach research and data with a specific design. Have a strong theoretical reason *why* you'd expect the effect of X to vary across different values of Z before adding the $X \times Z$ interaction term. "I dunno, maybe the effect is different" isn't a fantastic reason.

It's always a good idea to approach data with a design and an intention rather than just fishing around, interaction terms or not. Why do I emphasize this so strongly when it comes to interaction terms? First, because there are a *lot* of different interactions you *could try*, and a lot of them seem tantalizing. Does the effect of job training differ between genders? Between races? Between different ages? Sure, it might. We can tell ourselves a good story about why each of those might be great things to interact with job training. But if we try *all* of them, we're likely to find a significant interaction pretty quickly even if there's really nothing there. So trying interactions all willy-nilly tends to lead to false positives.

The temptation to try a bunch of stuff is extra-hard to resist when you have a null effect. Sure, maybe we didn't find any effect of job training. But maybe the effect is only there for women! Try a gender interaction and look for $\beta_1 + \beta_3$ to be big/significant even when β_1 isn't. Or maybe it's only there for Pacific Islanders! Try a race interaction. Or maybe only in Iowa. Try a state interaction. However, because there are nearly infinite different subgroups we *could* look at (especially once you consider going deeper—maybe the effect is only there

[66] Why is the effect always positive in the second column but negative in the third? Isn't that odd? It is! This is the kind of strange finding that would make us explore what's going on in our model and data a little more deeply to see what's happening.

[67] Although this difference doesn't seem very meaningfully large, and is not statistically significant.

for women Pacific Islanders in Iowa!), we're almost certain to find some effect if we check every single one. In general, you should be highly skeptical of an effect that *isn't there at all* for the whole sample, but *is* there for some subgroup. Surely there are plenty of effects that are, in truth, only there for a certain subgroup. But fishing around with a bunch of interactions is going to give you way too many false positives. So stick to cases where there's a strong theoretical reason *why* you'd expect the effect to only be there for a subgroup, or differ across groups, and make sure you believe your story before you check the data.

Second, even if we do have a strong idea of a difference in effect that might be there, interaction terms are *noisy*. They're basically looking for the *difference between* two noisy things— the effect for one group and the effect for the other group, both of which have sampling variation.[68] The difference between two noisy things will be noisier still.[69] You need a *lot* more observations to precisely estimate an interaction term than to precisely estimate a full-sample estimate. Even if the difference is as large as one group having an effect twice the size of the other group, you may need sixteen times as many observations to have adequate statistical power for the interaction term as you'd need for the main term.[70]

Why is noisiness a problem? Because estimates that are noisy will more often get surprisingly big effects. The sampling variation is very wide with noisy estimates. So there's a better chance that when you *do* find an interaction term big enough to call it interesting, it's just sampling variation, even if the result is statistically significant!

That said, interaction terms are still very much worthwhile if you have a solid reason to include them. And they're super important to learn for causal inference. In fact, many of the research designs in this second half of the book are based around interaction terms. Turns out you can get a lot of identification mileage out of looking at how the effect of your treatment variable differs between, say, conditions where it should have a causal effect and conditions where any effect it has is just backdoor paths. So give yourself some time to practice interpreting and using them. It will definitely pay off.

[68] This explanation uses a binary interaction but the same logic applies to continuous interactions.

[69] In most cases. If they're correlated in a particular way, then differencing them may cancel out some of their noise.

[70] Andrew Gelman. You need 16 times the sample size to estimate an interaction than to estimate a main effect. Statistical Modeling, Causal Inference, and Social Science, 2018.

13.2.5 *Coding Up Polynomials and Interactions*

In these two coding blocks we're going to be continuing to work with the restaurant-inspection data we've been covering in this chapter.

First, we'll do a regression with a polynomial in it. Then, we'll calculate the effect of *Number of Locations* for a given value of the variable in such a way that also gives us a standard error on the effect at that value.

```
1   # R CODE
2   # Load packages and data
3   library(tidyverse); library(modelsummary); library(margins)
4   df <- causaldata::restaurant_inspections
5
6   # Use I() to add calculations of variables
7   # Including squares
8   m1 <- lm(inspection_score ~ NumberofLocations +
9                            I(NumberofLocations^2) +
10                           Year, data = df)
11
12  # Print the results to a table
13  msummary(m1, stars = c('*' = .1, '**' = .05, '***' = .01))
14
15  # Use margins to calculate the effect of locations
16  # at 100 locations.
17  # variables is the variable we want the effect of
18  # and at is a list of values to set before looking for the effect
19  m1 %>%
20      margins(variables = 'NumberofLocations',
21      at = list(NumberofLocations = 100)) %>%
22      summary()
23
24  # The AME (average marginal effect) is what we want here
25  # And the standard error (SE) is reported, too
```

```
1   * STATA CODE
2   causaldata restaurant_inspections.dta, use clear download
3
4   * ## interacts variables
5   * Interact a variable with itself to get a square!
6   * use c. to let Stata know it's continuous
7   reg inspection_score c.numberoflocations##c.numberoflocations year
8
9   * look at the results in a nicer table
10  esttab
11
12  * Use margins with the dydx() option to get the effect of numberoflocations
13  * and use at() to specify that you want it when there are 100 locations
14  margins, dydx(numberoflocations) at(numberoflocations = 100)
15  * The first term is the effect, second is the SE
```

```
1   # PYTHON CODE
2   import statsmodels.formula.api as sm
3   from stargazer.stargazer import Stargazer
4   from causaldata import restaurant_inspections
5   df = restaurant_inspections.load_pandas().data
```

```
 6
 7   # Use I() to insert calculations of your variables
 8   # and ** to square
 9   m1 = sm.ols(formula = '''inspection_score ~
10       NumberofLocations +
11       I(NumberofLocations**2) +
12       Year''', data = df).fit()
13
14   # m1.summary() would do here, but if we wanted
15   # to write to file we could extend Stargazer...
16   Stargazer([m1])
17
18   # Use t_test to test linear combinations of coefficients
19   # be sure to test them against 0 to get the appropriate result
20   # coef is the estimate here, std err the SE.
21   # We know this is the right equation to use because we know
22   # the derivative - we need to figure that out first
23   m1.t_test('NumberofLocations + 2*I(NumberofLocations ** 2)*100 = 0')
```

Next, we will include an interaction term. Similarly, we will calculate the effect of locations at a specific value of the interacting variable (remember, the interaction term itself already shows the *difference* between effects, we don't need that from a separate command).

```
 1   # R CODE
 2   # Load packages
 3   library(tidyverse); library(modelsummary); library(margins)
 4   df <- causaldata::restaurant_inspections
 5
 6   # Use * to include two variables independently
 7   # plus their interaction
 8   # (: is interaction-only, we rarely use it)
 9   m1 <- lm(inspection_score ~ NumberofLocations*Weekend +
10                           Year, data = df)
11
12   # Print the results to a table
13   msummary(m1, stars = c('*' = .1, '**' = .05, '***' = .01))
14
15   # Use margins to calculate the effect of locations
16   # on the weekend
17   # variables is the variable we want the effect of
18   # and at is a list of values to set before looking for the effect
19   m1 %>%
20       margins(variables = 'NumberofLocations',
21               at = list(Weekend = TRUE)) %>%
22       summary()
23
24   # The AME (average marginal effect) is what we want here
25   # And the standard error (SE) is reported, too
```

```
 1   * STATA CODE
 2   causaldata restaurant_inspections.dta, use clear download
 3
 4   * ## interacts variables
 5   * use c. to let Stata know locations is continuous
 6   * and i. to let it know weekend is binary
 7   reg inspection_score c.numberoflocations##i.weekend year
 8
```

```
 9 | * look at the results in a nicer table
10 | esttab
11 |
12 | * Use margins with the dydx() option to get the effect of numberoflocations
13 | * and use at() to specify that you want it when it's the weekend
14 | margins, dydx(numberoflocations) at(weekend = 1)
15 | * The first term is the effect, second is the SE
```

```
 1 | # PYTHON CODE
 2 | import statsmodels.formula.api as sm
 3 | from stargazer.stargazer import Stargazer
 4 | from causaldata import restaurant_inspections
 5 | df = restaurant_inspections.load_pandas().data
 6 |
 7 | # Use * to include two variables independently
 8 | # plus their interaction
 9 | # (: is interaction-only, we rarely use it)
10 | m1 = sm.ols(formula = '''inspection_score ~
11 |            NumberofLocations*Weekend +
12 |            Year''', data = df).fit()
13 |
14 | # m1.summary() would do here, but if we wanted
15 | # to write to file we could extend Stargazer...
16 | Stargazer([m1])
17 |
18 | # Use t_test to test linear combinations of coefficients
19 | # be sure to test them against 0 to get the appropriate result
20 | # coef is the estimate here, std err the SE.
21 | # We got the proper coefficient names from reading the reg table
22 | m1.t_test('NumberofLocations + NumberofLocations:Weekend[T.True] = 0')
```

13.2.6   Nonlinear Regression

SO FAR, WHILE WE'VE ALLOWED THE PREDICTOR VARIABLES
TO BE ALL KINDS OF THINGS—binary, skewed, related in nonlinear ways to the dependent variable—we've kept a tighter leash
on the outcome variable. We've always assumed that we can
predict the conditional mean of the dependent variable using a
linear function of the predictors. In other words, we can predict
the dependent variable by taking our predictors, multiplying
them each by their coefficients, and adding everything up.[71]

However, that won't always work. To be able to predict something with a linear function, it needs to behave like, well, a
line. It needs to change smoothly as the predictors change (so
it can't, for example, be a discrete value that jumps up rather
than climbs up). Also, it needs to continue off forever in either
direction, i.e., take any value—if your estimated model is, for
example $NumberofChildren = 4.5 - .1IncomeinThousands$,
then your prediction for someone with $50,000 in $Income$ is $-.5$
children. That doesn't make much sense.

[71] Even if that linear
function sometimes in-
cludes transformations or
polynomials of individual
predictors, they still show
up as linear in the equation.
$Y = \beta_0 + \beta_1 X + \beta_2 X^2$ is a
linear function of X and X^2.

That brings us to *nonlinear regression*. This is a big wide world of approaches to performing regression that don't rely on a linear relationship between the predictors and the outcome variable. These approaches are usually tailored to the exact kind of dependent variable you're dealing with. So in this chapter (unfortunately), all we're going to be able to cover are the probit and logit models for dealing with binary dependent variables. But there's also Poisson regression for count variables, multinomial logit and probit for categorical data, tobit for censored data, and so on and so on.[72]

Even though using linear regression for these strange dependent variables doesn't make sense, some people do it anyway. There are actually some good reasons to do OLS anyway even when the dependent variable isn't quite right. All of these nonlinear regression models add their own assumptions, and generally don't have closed-form solutions,[73] making their estimates a little less stable. They can have a hard time with things like controls for categorical variable with lots of categories. Plus, while using OLS with a binary outcome guarantees your error terms are heteroskedastic (we'll talk about this in a few sections), the presence of heteroskedasticity actually makes the *coefficients wrong* in logit and probit models, and a specialized heteroskedasticity-robust version of the estimators is required. In the context of binary dependent variables, using OLS anyway is called the "linear probability model," and it does see a fair bit of use.

That said, while in some cases the nonlinear-regression cure is worse than the incorrect-linear-model disease, generally it is not, and you want to use the appropriate model. Why exactly is that? As I mentioned earlier, I'll be going through why we do this, and how it works, specifically in application to binary dependent variables.

GENERALIZED LINEAR MODELS ARE A WAY OF EXTENDING WHAT WE ALREADY KNOW ABOUT REGRESSION to a broader context where we need some nonlinearity. Generalized linear models, or GLM, aren't the only way that nonlinear regression is done, but it's how probit and logit are often done, and it applies in some other contexts too. And it's pretty easy to think about once we've covered regression.

GLM says this: take that regression model, and pass it through a function in order to make your prediction.[74] We call this function $F()$ a "link function," and we call the input to that function the "index." That's it! So our regression equation is

[72] There are also methods like nonlinear least squares that deal with nonlinearity in the *predictors*, for example if you think the model is $Y = \beta_0 \beta_1^X$.

[73] "Closed-form solutions" means that you can derive a function that tells you how to find the right coefficients. In OLS it's the equation you get from using calculus to minimize the sum of squared errors. Plug in your data and it tells you the exact correct answer. A lot of these nonlinear regression methods can't do that, so instead of plugging in to get the coefficient they have to basically try a bunch of different coefficients until they get coefficients that work pretty well. While they do this in a way that's a lot smarter than I'm making it sound here, it doesn't always work well.

[74] In case you've forgotten, a *function* is a mathematical object that takes some number of inputs and spits out an output. So if our function were $F(x) = x + 2$, then $F(5)$ would be $5 + 2 = 7$.

Generalized linear model. A linear model that runs its linear index through a function in order to produce a prediction.

$$Y = F(\beta_0 + \beta_1 X) \tag{13.15}$$

Notice that the $\beta_0 + \beta_1 X$ index inside the function $F()$ is the exact same regression equation we were working with before.[75] Also notice that if we make $F(x)$ be just $F(x) = x$, then we're back to $Y = \beta_0 + \beta_1 X$ again and are basically doing OLS.

How does GLM let us run nonlinear regression? As long as we pick a function that mimics the properties that Y has, then we can ensure that we never make an unrealistic prediction.

In the case of binary variables, we are predicting the probability that Y is equal to 1. In other words, $F(\beta_0 + \beta_1 X)$ should give us a probability. Probabilities should be between 0 and 1. So what are we looking for in a link function that works with a binary Y?

- It should be able to take any value from $-\infty$ to ∞ (since $\beta_0 + \beta_1 X$ can take any value from $-\infty$ to ∞)

- It should only produce values between 0 and 1

- Whenever the input $(\beta_0 + \beta_1 X)$ increases, the output should increase

There are infinitely many functions that satisfy these criteria, but two are most popular: the *logit link* function and the *probit link* function.

The logit link function is

$$F(x) = \frac{e^x}{1 + e^x} \tag{13.16}$$

and the probit link function is $\Phi(x)$, where Φ is the "cumulative distribution function" of the standard normal distribution, i.e., $\Phi(x)$ tells you the percentile of the standard normal distribution (mean 0, standard deviation 1) that x is.

These two link functions—probit and logit—produce nearly identical regression predictions. You don't generally need to worry too hard about which of the two you use.[76] The important thing, though, is that neither of them will make predictions outside of the range of 0 and 1, and their coefficients will be estimated in a context that is aware of those boundaries and so better-suited for those non-continuous dependent variables.

WHAT'S SO BAD ABOUT PEACE, LOVE, AND LINEARITY? Why not just stick with OLS? What we're really interested in is how increases in X cause changes in Y. Surely getting identification sorted is the real important step, and we don't necessarily

[75] And it would continue to be just like the models we're used to when we add other controls, etc.

[76] If you're planning to do some algebra in order to prove some property of the model, or derive its marginal effects, the logit is a lot easier to work with. It can also be computationally easier to estimate. But generally don't worry about it, at least not until you start estimating multinomial models.

care what values of Y are *predicted* as long as we still get those effects, right?

Two problems with this argument. First, *any* time you're estimating a model that's not the right model, strange results can creep in even in unexpected ways, and in ways that can be hard to predict ahead of time.[77] Second, not properly accounting for nonlinearity *can* mess up your estimates of how X affects Y.

We can see this in action in Figure 13.11. Here we have a bunch of observations that are either $Y = 0$ or $Y = 1$ at different values of X. On top of that graph we have the fitted OLS model as well as a fitted logit model (the probit would look very similar).

[77] That applies to any kind of model incorrectness. Run a nonlinear regression but don't have the right polynomial terms or controls? That will give you strange results too.

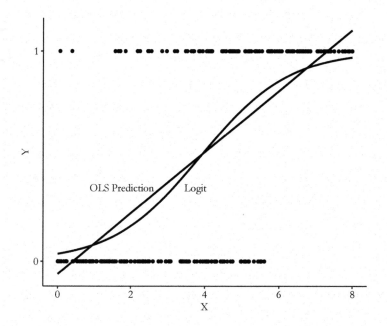

Figure 13.11: OLS and Logit Fit on Binary Dependent Variable

What do we see? As previously mentioned, we see the OLS prediction going outside the range of 0 and 1. But beyond that, *the slopes of the OLS and logit models are different.* This is especially true near the right or left sides of the graph. What does this mean? It means that the reason we were thinking about using OLS anyway—getting the effect of X on Y rather than caring about realistic predictions—falls apart anyway.

But why is this? Well, think about it—if X has a positive effect on Y, but you already predict that Y has a .999 chance of occurring, then what should a one-unit increase in X do? Almost nothing, right? There's simply not that much more that Y can increase by. But if your current prediction of Y is

.5, then there's a lot more room to grow, so the effect can be bigger. It *should* be bigger! OLS is unable to account for this.[78]

OLS is unable to account for how the effect of X on Y should differ. This works in a different way to how the effect differed when we talked about polynomials. The curviness shouldn't be based on *the value of X*, as it is with a polynomial, it should be based on *the current predicted value of Y*. When the sample average of Y is in the middle—near .5 or so—using OLS isn't a huge problem, although you still might want to use probit or logit anyway. But if the average of Y is anywhere near 0 or 1, you have a big problem and absolutely need a nonlinear model.

INTERPRETING A GENERALIZED LINEAR MODEL can be a little trickier than interpreting an OLS model. In one, very narrow and pretty useless sense, the interpretations are exactly the same. That is to say, each coefficient still means "a one-unit change in this variable is associated with a (coefficient-sized) change in the index." That's true whether we're talking about linear or nonlinear regression. The difference, however, is that with OLS, the index function is just a regression equation giving us the conditional mean of the dependent variable, and we know how to interpret a change in the dependent variable.[79] In GLM, the index function is the thing that gets put through the link function. So it's in terms of "units of change in the thing that gets put through the link function," which is way harder to think about.

It's common to present the results of probit and logit models, and nonlinear regression more generally, in terms of their *marginal effects*. Marginal effects basically translate things back into linear-model terms. While the coefficient on X in a logit model gives the effect of a change in X on the index, the marginal effect of X gives the effect of a change in X on the probability that the dependent variable is 1.

Mathematically, the marginal effect is straightforward, although we do have to dip back into the calculus well once again. The marginal effect of X on the probability that $Y = 1$ is just the derivative of $Pr(Y = 1)$ with respect to X, which has a conveniently simple solution for these binary-dependent variable models:

$$\frac{\partial Pr(Y = 1)}{\partial X} = \beta_1 Pr(Y = 1)(1 - Pr(Y = 1)) \qquad (13.17)$$

So the marginal effect of X is equal to the coefficient on X, β_1, multiplied by the predicted probability that $Pr(Y = 1)$ (which comes from the model, and is conditional on all the predictor

[78] Even using a polynomial in OLS, while it would introduce a curvy line that might even look a bit like the logit line, can't quite get this job done in a consistent way.

[79] Technically, OLS doesn't have a "link function," although linear GLM does. But I think this is still a clear way to explain it so... let's stick with it.

Marginal effect. In a probit or logit model, the effect/association of a one-unit change in a variable X on the probability that the binary dependent variable Y is 1/"true."

values) and one minus that same probability. Note that $Pr(Y = 1)(1 - Pr(Y = 1))$ is tiny when $Pr(Y = 1)$ is near 0 or 1, and at its biggest when $Pr(Y = 1) = .5$, so we retain the result from Figure 13.11 that the slope is steepest in the "middle" of the graph and flattest on the sides.

There is, however, a catch. That catch is that there is no one marginal effect for a given variable. I just said to look back at Figure 13.11. Well... keep looking at it! As X increases, the logit-predicted probability that $Y = 1$ might go up by a tiny amount (on the left or right side of the graph) or by quite a bit (in the middle). In fact, each observation in the data has its own marginal effect, based on the $Pr(Y = 1)$ for that observation, as we'd expect based on the $\beta_1 Pr(Y = 1)(1 - Pr(Y = 1))$ equation for the marginal effect. If the model predicts you're 90% likely to have $Y = 1$, your marginal effect is $\beta_1(.9)(.1)$. If it predicts you're 40% likely, your marginal effect is $\beta_1(.4)(.6)$. So what value do we give as "the" marginal effect?[80]

There are two common approaches to this issue, one of which I will heavily recommend over the other. The first, which I will not recommend, is the "marginal effect at the mean." The marginal effect at the mean first takes the mean of all the predictor variables. If we're just working with X, then it takes the mean of X. Then, it uses that mean to predict $Pr(Y = 1)$. And finally it uses that predicted $Pr(Y = 1)$ to calculate the marginal effect, $\beta_1 Pr(Y = 1)(1 - Pr(Y = 1))$. This gives the marginal effect of an observation with completely average predictors. Why don't I recommend this? Because it's calculating the marginal effect for no person in particular. Who is this marginal effect for, anyway?[81] Plus, marginal effects at the mean take the mean of each variable independently, when really these variables are likely to be correlated. It's strange to take the mean of everything, for example, if someone with a near-mean value of X almost always has a near-minimum value of Z.

What I recommend instead is the *average marginal effect*. The average marginal effect calculates each individual observation's marginal effect, using that observation's predictors to get $Pr(Y = 1)$ to get $\beta_1 Pr(Y = 1)(1 - Pr(Y = 1))$. Then, with each individual observation's marginal effect in hand, it takes the average to get the average marginal effect. That's it! Then you have the mean of the marginal effect across the sample, which gives you an idea of the representative marginal effect.[82]

[80] Crucially, this variation in effect across the sample is *not the same* as treatment effect heterogeneity as we explored in Chapter 10 or when talking about interaction terms. The marginal effect varying with $P(Y = 1)$ only captures one narrow aspect of variation in the effect—only the part that has to do with "not having much room to grow" near the maximum or minimum Y predictions. Not to do with, say, a treatment being more effective for men than for women. All that treatment effect heterogeneity stuff applies in probit/logit models too, but you have to deal with it explicitly like you would in OLS.

[81] Statistics has no shortage of creating representative values that don't actually represent any particular individual—that happens any time we use a mean. But in this case, there are other issues, too, and we have better options. The nobody-really-has-2.3-children problem isn't enough of a problem to make us give up on means *entirely*, as means are invaluable for other reasons, but it can be enough to give up on marginal effect at the mean, which has superior alternatives.

Average marginal effect. The mean of the calculated marginal effect across every observation in the sample.

[82] Of course, nothing requires you to get the mean—you can do whatever you want with the effect distribution. Take the median, standard deviation, whatever. Or look at the whole distribution. But in general people focus on the mean here.

To take a quick example, let's say we estimated the logit model $Pr(Y = 1) = logit(.1 + .05X - .03Z)$. And for simplicity's sake let's say we did this with only two observations in the sample. The first observation has $X = 1$ and $Z = 2$, and the second has $X = 0$ and $Z = 20$.

The predicted value for the first observation is $Pr(Y = 1) = logit(.1 + .05(1) - .03(2)) = logit(.99) = \frac{e^{.99}}{1+e^{.99}} \approx .729$. Similarly, the predicted value for the second observation is $Pr(Y = 1) = logit(.1 + 0 - .03(20)) = logit(-.5) = \frac{e^{-.5}}{1+e^{-.5}} \approx .622$. Then, the marginal effect of X, since the coefficient on X is .05, would be $.05(.729)(1 - .729) = .010$ for the first observation, and $.05(.622)(1 - .622) = .012$ for the second. Average those out to get the average marginal effect of $(.010 + .012)/2 = .011$.

LET'S CODE UP SOME PROBIT AND LOGIT MODELS! In the following code, I'll show how to estimate a logit model, and then how to estimate average marginal effects for those models. What about probit? Conveniently, all of the following code samples can be used to estimate probit models, simply by replacing the word "logit" with the word "probit" wherever you see it.

In each case, we'll be seeing whether weekend inspections have become more or less prevalent over time, once again with our restaurant inspection data.

In the case of the R code, you will often see guides online telling you to use the `logitmfx` function from the **mfx** package, instead of margins. This works fine, but do be aware that it defaults to the marginal effect at the mean instead of the average marginal effect.

```
1   # R CODE
2   # Load packages and data
3   library(tidyverse); library(modelsummary); library(margins)
4   df <- causaldata::restaurant_inspections
5
6   # Use the glm() function with
7   # the family = binomial(link = 'logit') option
8   m1 <- glm(Weekend ~ Year, data = df,
9            family = binomial(link = 'logit'))
10
11  # See the results
12  msummary(m1, stars = c('*' = .1, '**' = .05, '***' = .01))
13
14  # Get the average marginal effect of year
15  m1 %>%
16      margins(variables = 'Year') %>%
17      summary()
```

```
1   * STATA CODE
2   causaldata restaurant_inspections.dta, use clear download
3
4   * Use the logit command to regress
```

```
5 │ logit weekend year
6 │
7 │ * and the margins command
8 │ margins , dydx(year)
```

```
 1 │ # PYTHON CODE
 2 │ from stargazer.stargazer import Stargazer
 3 │ import statsmodels.formula.api as sm
 4 │ from causaldata import restaurant_inspections
 5 │ df = restaurant_inspections.load_pandas().data
 6 │
 7 │ # sm.logit wants the dependent variable to be numeric
 8 │ df['Weekend'] = 1*df['Weekend']
 9 │
10 │ # Use sm.logit to run logit
11 │ m1 = sm.logit(formula = "Weekend ~ Year", data = df).fit()
12 │
13 │ # See the result
14 │ # m1.summary() would also work
15 │ Stargazer([m1])
16 │
17 │ # And get marginal effects
18 │ m1.get_margeff().summary()
```

13.3   Your Standard Errors are Probably Wrong

13.3.1   Why Your Standard Errors are Probably Wrong

ALL OF THE THINGS WE'VE DONE WITH REGRESSION UP TO THIS POINT have been focused on getting our models right, setting up any nonlinearities properly, interpreting the coefficients, and, of course, identifying our effects. But the estimated model isn't going to mean much for you if you don't have some estimate of the precision of your estimates, i.e., a standard error.[83]

As you'll recall from earlier in the chapter, from estimating our model $Y = \beta_0 + \beta_1 X + \varepsilon$ we have our OLS coefficient estimate $\hat{\beta}_1$ and our estimate of the standard error $se(\hat{\beta}_1)$ (which we base on $var(X), n$, and our estimate of σ). Then, our best guess for the sampling distribution of $\hat{\beta}_1$ is a normal distribution with a mean of $\hat{\beta}_1$ and a standard deviation of $se(\hat{\beta}_1)$. *This is the purpose of the standard error—it tells us about that sampling distribution, which lets us figure out things like the range of plausible estimates, confidence intervals, statistical significance, and eliminating certain theoretical/population distributions as unlikely.*

That brings us to those error term assumptions I mentioned and said we'd come back to later. We want very badly to know that we understand what the sampling distribution is. We bother calculating the standard error because we want to

[83] I want to emphasize that standard errors really aren't just for the purpose of calculating statistical significance! They give information about the sampling distribution of your estimate, which is good for *lots* of different approaches to using observed data to learn about theoretical/population distributions and relations.

know the standard deviation of that sampling distribution. But if certain assumptions are violated, then the standard error will *not* describe the standard deviation of the sampling distribution of the coefficient.

The first assumption we must return to is that we have to assume that the *error term ε itself* must be normally distributed.[84] That assumption about the normality of the error term is what lets us prove mathematically that the OLS coefficients are normally distributed.

Conveniently, this assumption isn't *too* important. OLS estimates tend to have sampling distributions pretty close to normal even if ε is pretty far from normal. Not perfect, but you don't generally have to get too up in arms about this one. That said, there are some weird error-term distributions out there that really do mess things up. But that's for another time.

THE SECOND ASSUMPTION IS that the error term is *independent and identically distributed*. That is, we need to assume that the theoretical distribution of the error term is *unrelated to the error terms of other observations and the other variables for the same observation* (independent) as well as *the same for each observation* (identically distributed). It doesn't matter *which* data point you look at, the error term might be different from observation to observation but it will always have been drawn from the same distribution.

[84] A common misconception is that the *variables in the model* must be normally distributed, but this is incorrect; we don't need that.

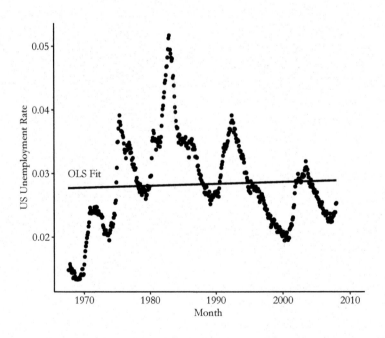

Figure 13.12: US Unemployment Rate Over Time, with OLS Fit

What does this mean exactly? It's easier to think about when it fails. One way it could fail is *autocorrelation*, where error terms are *correlated with each other* in some way. This commonly pops up when you're working with data over multiple time periods (temporal autocorrelation) or data that's geographically clustered (spatial autocorrelation).

Autocorrelation. When the error terms are correlated with each other.

Taking temporal autocorrelation as an example, say you're regressing the US unemployment rate growth on year. The economy tends to go through up and down spells that last a few years, as in Figure 13.12. Think about the errors here—there tend to be a few positive errors in a row, then a few negative errors. So the distribution can't be independent since the error from last period predicts the error this period.

ANOTHER COMMON WAY THAT BEING INDEPENDENT AND IDENTICALLY DISTRIBUTED COULD FAIL is in the presence of *heteroskedasticity*. A big long word! Now this is a real textbook.

Heteroskedasticity. When the variance of the error term is related to the variables in a regression model.

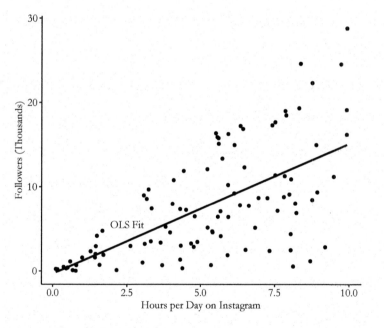

Figure 13.13: Simulated Instagram Follower Data

Heteroskedasticity is when the *variance* of the error term's distribution is related to the variables in the model. For example, say you were regressing how many Instagram followers someone has on the amount of time they spend posting daily. You might find that, as shown in Figure 13.13, among people who spend almost no time, there's very little variation in follower count. But among people who spend a lot of time, there's huge amounts of variation. Small amounts of time means little variation in follower count and, accordingly, little variation in the error term. But there's lots of variation in the error term for large amounts of time spent. The error distribution isn't identical because the variance of the distribution changes with values of the variables in the model.

When it comes to failures of the "independent and identically distributed" assumption, we again have a convenient result. These failed assumptions don't actually make the OLS coefficients non-normally distributed. However, they *do* make it so that our $\sqrt{\sigma^2/(var(X)n)}$ estimate of the standard error is wrong. This means that we just have to figure out what the standard deviation of that sampling distribution *is*. This calculation will require some way of accounting for that autocorrelation or that heteroskedasticity.

13.3.2   *Fixing Your Standard Errors*

IN THIS SECTION I'M GOING TO DISCUSS a few common fixes to standard errors that we can use in cases where we think the independent and identically distributed assumption fails.[85]

First we'll consider the issue of heteroskedasticity, which is when the variance of the error term ε changes depending on the value of X (or any of the predictors in the model, for that matter). The problem with heteroskedasticity, when it comes to estimating standard errors, is that in the areas where variance is higher, the conditional mean of Y in that area will move around more from sample to sample. Depending on whether this occurs for higher, lower, or middling values of X, this might make $\hat{\beta}_1$ change more from sample to sample (increasing the standard deviation of the sampling distribution) so as to meet those changing values. But since regular OLS standard errors simply estimate the same variance for the whole sample, it will understate how much things are changing.

To provide one highly simplified example of this in action, see Figure 13.14, which contains two data sets with the same level of *overall* error term variance across the sample. On the left, we have heteroskedasticity. Imagine picking one observation from the cluster on the left and one from the cluster on the right and drawing a line between them. That's our OLS estimate from a sample size of 2. Now imagine doing it over and over.[86] With the Y value fairly pinned down on the right, but moving a lot on the left, the slope moves a lot from sample to sample—it's that high-variance cluster driving this, which won't be accounted for

[85] Before getting too far into it, I want to point out that there are *far more ways than these* that standard errors can be wrong, and also more ways than these that they can be fixed. The Delta method, Krinsky-Robb... so many fixes to learn. Here I will show you only the *beginning* of the ways your standard errors will be wrong. If you decide to continue doing research, you can prepare for a lifetime of being wrong about your standard errors in fantastic ways that you can only dream of today.

[86] If you're more of a physical learner, hold a pencil in your hands—the height of each hand is the conditional mean of Y in your sample, and the angle of the pencil as you hold it is $\hat{\beta}_1$. Now move your hands up and down, independent of each other. Sometimes this changes the angle a lot—if the left hand goes up but righty goes down—but sometimes it doesn't—if both go up. That's regular sampling variation. Now try moving *just one hand* up and down. The angle changes much more rapidly! That's one form of heteroskedasticity.

in the sample-wide estimate of the residual variance. Now do the same thing on the right. The slope is still going to vary from sample to sample, but the extent of this variation is driven by the variance in each cluster, which is the same, and so when we estimate that variation using the sample-wide residual variation, we'll estimate it properly.

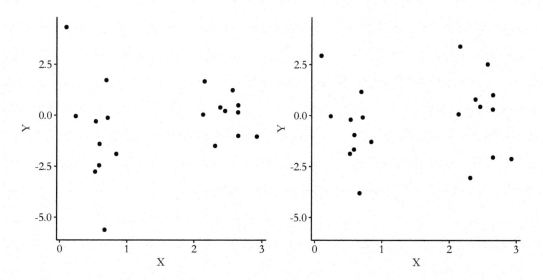

Given this problem, what can we do? Well, we can simply have our estimate of the standard error personalize what we think about the error variance. Instead of $\sqrt{\sigma^2/(var(X)n)}$ we can use a "sandwich estimator," which, in short, allows the individual X values that go into that $var(X)$ calculation to be scaled up, or down, or even be multiplied by other observations' X values.[87]

One of the more common heteroskedasticity-robust sandwich estimator methods is Huber-White. In Huber-White standard errors, the scaling of the individual X values works by taking each observation's residual, squaring it, and applying that squared residual as a weight to do the scaling. This, in effect, weights observations with big residuals more when calculating the variance.

There are actually several different estimators that, in very similar ways but with minor tweaks, adjust standard errors for the presence of heteroskedasticity. In general, you're looking here for "heteroskedasticity-robust standard errors" or just "robust standard errors," most of which are sandwich estimators, and with Huber-White standard errors as a standard application.[88]

Figure 13.14: A Very Simplified Demonstration of Why Heteroskedasticity Messes Up Standard Errors

[87] It's called a "sandwich estimator" because in the matrix-algebra form of the estimator, which is $(X'X)^{-1}(X'\Sigma X)(X'X)^{-1}$, the matrix Σ that does the scaling (and, later, lets the observations be correlated) gets "sandwiched inside" a bunch of Xs.

[88] Some users of regression, especially economists, tend to assume that heteroskedasticity is always a problem and so use heteroskedasticity-robust standard errors by default. I can't say they're wrong.

WE CAN THEN MOVE ALONG TO ERRORS THAT FAIL TO BE INDEPENDENT. In other words, they're correlated with each other. Why would correlated errors require us to change how we calculate standard errors? Correlated errors will change the sampling distribution of the estimator in the same way that correlated *data* would change a sampling distribution of a mean. Say you were gathering samples of people and calculating their height. If you randomly sampled people from all over the globe each time, that would give you a certain standard deviation for the sampling distribution of the average height. But if instead you randomly sampled *families* each time and calculated average height, the values would be correlated—your height is likely more similar to your parents than to some stranger. We'd be more likely than in the randomly-sample-everyone approach to get all tall people and thus a very tall average, or all short people and thus a very short average. So the mean is swingier, and the sampling distribution has a larger standard deviation. Some adjustment must be made.

When figuring out what adjustment to make, we're in a bit of a pickle—we have to figure out *how* the errors correlated with each other. After all, it doesn't make much sense to say "my errors are correlated"—they can't *all* be correlated. Instead, we'd say they're correlated *according to some feature*. Each of the different ways they can be correlated calls for a different kind of fix.

One common way in which errors can be correlated is *across time*—this is the time-based autocorrelation we discussed earlier. In the presence of autocorrelation we can use heteroskedasticity- and autocorrelation-consistent (HAC) standard errors.[89]

A common approach to HAC standard errors is the Newey-West estimator, which is another form of sandwich estimator. This estimator starts with the heteroskedasticity-robust standard error estimate, and then makes an adjustment.[90] That adjustment comes from first picking a number of lags—the number of time periods over which you expect errors to be correlated. Then, we calculate how much autocorrelation we observe in each of those lags, add them together—with shorter-length lags being featured more heavily in the sum—and that's used to make the adjustment factor.[91]

Autocorrelation across time is not the only form of autocorrelation, with correlation across *geography* also being highly consequential. While I won't go deeply into the subject here,

[89] Autocorrelation-consistent standard errors are only one tool out of like a billion in the back pocket of anyone working with time series data. These tools are almost entirely absent from this book. While time series does pop up in a few places here, like in Chapter 17 on event studies, for a serious treatment of time series you're better off looking at a more traditional econometrics textbook.

[90] Why start with robust SEs? Why not just a fix for autocorrelation and not heteroskedasticity? Because autocorrelation by its nature tends to introduce heteroskedasticity anyway. If errors are correlated, then at times where one observation has a small error, other nearby observations will also have small errors, producing low error variance. But at times where one observation has a large error, other nearby observations will also have large errors, producing high variance. Low variance at some times and high variance at other times. Heteroskedasticity!

[91] In a bit more detail, we multiply the de-meaned time series by the error terms, and calculate the correlation between that and its lag. We sum up those correlations, weighting the one-period lag by just below 1, then with declining weights for further lags. Take that sum, multiply it by two, and add one. That's the adjustment factor.

Conley spatial standard errors are another form of sandwich estimator, which calculates standard errors while accounting for the presence of autocorrelation among geographic neighbors.

ANOTHER COMMON WAY FOR ERRORS TO BE CORRELATED is in a *hierarchical structure*. For example, say you're looking at the effect of providing laptops to students on their test scores. Well, a given classroom of students isn't just given laptops or not, they are also in the same classroom with the same teacher. Whatever we predict about their test scores based on having laptops, they're likely to all be similarly above-prediction or below-prediction together because of the similar environments they face—their errors will be correlated with each other. We'd say these errors are *clustered* within classroom.

Conveniently, in these cases, we can apply *clustered standard errors*. The most common form of clustered standard errors are Liang-Zeger standard errors, which are again a form of sandwich estimator. With clustered SEs, you explicitly specify a grouping, such as classrooms. Then, instead of using the sandwich estimator to scale up or down individual X values, it instead lets X values *interact with other X values in the same group*.[92],[93]

The use of clustered standard errors can account for any sort of correlation between errors *within* each grouping.[94] This sort of flexibility is a nice relief. No need to make those restrictive assumptions about error terms having correlations of zero. Of course, statistical precision comes from assumptions. If you won't assume anything, the data won't tell you anything. So clustering at too broad a level (say, if your laptop-test score analysis has two schools and you cluster at the school level, or just clustering among the whole sample—everyone's error is correlated with everyone else's) is the equivalent of saying "uh... I dunno what's happening here, could be just about anything I guess," and you'll get very large standard errors that might understate what your data can actually tell you. So at some point you've got to put your foot down on just how far that correlation is allowed to spread.

What's the right level to cluster at, then? First off, if you have a strong theoretical idea of when errors *should* be clustered together, such as in a classroom, go for it. Beyond this, another common approach is to *cluster at the level of treatment*. So, in the laptop example, if laptops were assigned to some classrooms but not others, you'd cluster at the classroom level. But if they were assigned to some *schools* and not others, you'd cluster at the school level. These are not hard-and-fast rules, however, and

[92] For the matrix algebra fans: remember that Σ matrix that's "sandwiched between" all the Xs from a few notes ago? Well, for clustered standard errors, this matrix is block-diagonal, where each grouping has unrestricted nonzero values within its block, but the rest of the matrix is 0.

[93] What's the difference between including a control for a group indicator in the model, and clustering at that group level? First off, you can do both, and people often do. Second, including the indicator in the model suggests that group membership is an important *predictor of the outcome* and possibly *on an important back door*, while clustering suggests that group membership *is related to the ability of the model to predict well*.

[94] They can even be used to account for autocorrelation of any sort in panel data where you have multiple time series all in one data set—just cluster by the time series.

clustering decisions require a lot of knowledge about the specific domain you're working in. See if you can find other people working on your topic and figure out how they're clustering.

Something important to remember about the Liang-Zeger clustered standard errors, though, is that they work really well for *large numbers of clusters*. Say, more than 50. Fewer clusters than that and they're still an improvement on regular standard errors, but they're nowhere near as accurate as you want them to be.[95] For smaller numbers of clusters, you may instead want to use wild cluster bootstrap standard errors, which are sort of a mix between clustered standard errors, which we're discussing now, and bootstrap standard errors, which I'll discuss in the next section. See more about wild cluster bootstrap standard errors in Chapter 15.

SO IS IT JUST SANDWICHES ALL THE WAY DOWN? No! There's another way of estimating standard errors that has *very* broad applications, and that's *bootstrap standard errors*. The bootstrap is about as close to magic as statistics gets. What are we *trying to do* with standard errors? What's the point of them? We want to provide a measure of sampling distribution, right? We want to get an idea of what would happen if we estimated the same statistic in a bunch of different samples.

So the bootstrap says: "let's estimate the same statistic in a bunch of different samples and see what the distribution is across those samples. Ta-da! Sampling distribution." Oh. Huh, that makes sense. Why didn't we think of that at the start?

But hold on, how is it possible to estimate the statistic in a bunch of different samples? We only have the one sample! That's the genius of the bootstrap. It *uses the one sample we have, but re-samples it.* The process for a bootstrap is:[96]

1. Start with a data set with N observations.

2. Randomly sample N observations from that data set, allowing the same observation to be drawn more than once.[97] This is a "bootstrap sample."

3. Estimate our statistic of interest using the bootstrap sample.

4. Repeat steps 1—3 a whole bunch of times.

5. Look at the distribution of estimates from across all the times we repeated steps 1—3. The standard deviation of that distribution is the standard error of the estimate. The 2.5th and 97.5th percentiles show the 95% confidence interval, and so on.

[95] Or even if you have a lot of clusters, but treatment is assigned along cluster lines and you only have a few *treated* clusters.

[96] I'm describing bootstrap standard errors here, but the concept of a bootstrap is broader than just standard errors, and you may see it pop up elsewhere. It's certainly popular in machine learning.

[97] Depending on the structure of the data—for example in the context of panel data where the same individual is measured multiple times and so has multiple rows—it may make sense to not sample totally at random. The "cluster bootstrap," for example, randomly samples *chunks* of data, perhaps individuals or groups rather than observations.

Does this actually work? Let's take a very simple example. Four observations in our data: 1, 2, 3, 4, and we're estimating the mean. The sample mean, of course, is $(1+2+3+4)/4 = 2.5$. The sample variance of the data is $((-1.5)^2 + (-.5)^2 + (.5)^2 + (1.5)^2)/(4-1) = 1.67$. So the standard error that we'd calculate by our typical process is $\sqrt{1.67/4} = .645$.

All right, now let's try the bootstrap. We get our first bootstrap sample and it's 1, 1, 4, 4—remember, the same observation can be resampled, which means that some observations from the original data show up multiple times and others don't show up at all. The mean of this bootstrap sample is $(1+1+4+4)/4 = 2.5$. We resample again and get 1, 3, 4, 4, with a mean of 3. Do it again! The next one is 1, 2, 4, 4, mean of 2.75. Next! 1, 2, 2, 3, mean of 2. So far the average of the bootstrap samples we've taken is $(2.5+3+2.75+2)/4 = 2.56$, and the standard deviation is .427.

.427 is not exactly the .645 that we know is correct, but we're not anywhere near done. When you do a bootstrap you generally want to do it hundreds if not thousands of times. Remember, we're trying to mimic a theoretical distribution here by using real-world resampling. We're going to want to get a bit closer to the *theoretical infinite* number of observations that the theoretical distribution is based on.

Let's go a bit bigger. Instead of our observations being 1 to 4 let's do 1 to 1,000, and we'll calculate 2,000 bootstrap samples instead of four bootstrap samples. Now, our sample mean is 500.5, and the standard error of that mean calculated using the standard method is 9.13. The mean of our bootstrap distribution is 500.5 as well, and the standard deviation of our 2,000 bootstrap samples turns out to be 9.06.[98] A bit smaller than 9.13, but we're basically getting what we thought we'd get. This is showing us that maybe those theoretical-distribution assumptions were actually a little *pessimistic* about the precision in this particular instance.[99]

This is the basic process by which bootstrap standard errors can be calculated. A bit computationally intensive, definitely. But on the flip side, they can sometimes point out where the traditional estimates are too conservative (or not conservative enough), they let you explore *any part* of the sampling distribution you're curious about—median, different percentiles, skew, etc.—not just the standard deviation, and they *are very useful in estimating standard errors for unusual stuff.* What if we have an estimator that we're not sure follows a normal

[98] This exact number will be a bit different each time you do bootstrap since there is randomness in the process. If you do bootstrap and want your results to be replicable, be sure to set a random seed in your software before doing it.

[99] Remember, those theoretical distributions are only approximations as far as the actual data goes.

distribution, or any distribution we know about? No problem! Just bootstrap it.

Bootstrap does come with its own set of assumptions and weaknesses, of course—it may be magic, but while there may be such a thing as magic there's no such thing as a free lunch. The bootstrap doesn't perform particularly well with small sample sizes,[100] it doesn't perform well for "extreme value" distributions (super highly skewed data) and it has some difficulties that at the very least need to be addressed if the observations in the data are related to each other, for example in panel data where multiple observations represent the same individual over time. For another example, the standard bootstrap doesn't do well under autocorrelation. But for an estimate with a decent sample size and some well-behaved independently-distributed data, the bootstrap can be a great option, especially if the regular standard errors are difficult to calculate in that context. Also, there are some alternate versions of the bootstrap discussed in Chapter 15 that address some of these problems.

LET'S PUT SOME OF THOSE STANDARD ERROR ADJUSTMENTS INTO ACTION. In each of the following code samples, I will show first how to implement heteroskedasticity-robust standard errors, heteroskedasticity- and autocorrelation-consistent standard errors, and then clustered standard errors.

First, robust standard errors. We will continue to use our restaurant inspection data here.

One important thing to know about robust standard errors, especially when it comes to coding them up, is that there are actually many different ways of calculating them. Each method uses different minor adjustments on the same idea. Some versions work better on small samples; others work better on certain kinds of heteroskedasticity. All of the methods below provide access to multiple different kinds of robust estimators (often given names like HC0, HC1, HC2, HC3, and so on). But different languages select different ones as defaults.[101] For example, Stata uses HC1, while most R methods default to HC3. Always be sure to check the documentation to see which one you're getting, and think about which one you should use.

[100] If you're wondering why I switched from a sample of 4 to a sample of 1,000 for my example above, that's why.

[101] This leads to much confusion when trying to replicate your results from one language in another language.

```
1   # R CODE
2   # Load packages
3   library(tidyverse)
4   # For approach 1
5   library(AER); library(sandwich)
6   # For approach 2
7   library(modelsummary)
8   # For approach 3
```

```r
 9  library(fixest)
10
11  df <- causaldata::restaurant_inspections
12
13  # In R there are many ways to get at robust SEs
14  # For the first two methods we'll look at, we need to estimate
15  # the regression on its own
16  # (most types of regressions will work, not just lm()!)
17  m1 <- lm(inspection_score ~ Year + Weekend, data = df)
18
19  # First, the classic way, sending our regression
20  # object to AER::coeftest(), specifying the kind
21  # of library(sandwich) estimator we want
22  # vcovHC for heteroskedasticity-consistent
23  coeftest(m1, vcov = vcovHC(m1))
24
25  # Second, we can take that same regression object
26  # and, do the robust SEs inside of our msummary()
27  msummary(m1, vcov = 'robust',
28                  stars = c('*' = .1, '**' = .05, '***' = .01))
29
30  # Third, we can skip all that and use a regression
31  # function with robust SEs built in, like
32  # fixest::feols()
33  feols(inspection_score ~ Year + Weekend, data = df, se = 'hetero')
34  # (this object can be sent to modelsummary
35  # or summary as normal)
```

```stata
1  * STATA CODE
2  causaldata restaurant_inspections.dta, use clear download
3
4  * robust SEs are built into Stata's very DNA!
5  * for basically any regression command, just add ", robust"
6  reg inspection_score weekend year, robust
```

```python
 1  # PYTHON CODE
 2  import statsmodels.formula.api as sm
 3  from causaldata import restaurant_inspections
 4  df = restaurant_inspections.load_pandas().data
 5
 6  # We can simply add cov_type = 'HC3'
 7  # to our fit!
 8  m1 = sm.ols(formula = 'inspection_score ~ Year + Weekend',
 9              data = df).fit(cov_type = 'HC3')
10
11  m1.summary()
```

Next, we will address autocorrelation using Newey-West standard errors. Keep in mind that in each case you must select a maximum number of lags (or have one be selected for you). This book doesn't focus much on time series so I won't go that deeply into it. But basically, closely examine the autocorrelation structure of your data to figure out how far to let the lags go.

Because these are intended for use with time series data, we will first need to adjust our data to be a time series.

```
 1 │ # R CODE
 2 │ # Load packages
 3 │ library(tidyverse)
 4 │ # For approach 1
 5 │ library(AER); library(sandwich)
 6 │
 7 │ df <- causaldata::restaurant_inspections
 8 │
 9 │ # Turn into a time series
10 │ df <- df %>%
11 │     group_by(Year) %>%
12 │     summarize(Avg_Score = mean(inspection_score),
13 │     Pct_Weekend = mean(Weekend)) %>%
14 │     # Only the years without gaps
15 │     filter(Year <= 2009)
16 │
17 │ # Perform our time series regression
18 │ m1 <- lm(Avg_Score ~ Pct_Weekend, data = df)
19 │
20 │ # Use the same coeftest method from robust SEs, but use the
21 │ # NeweyWest matrix. We can set maximum lags ourselves,
22 │ # or it will pick one using an automatic procedure (read the docs!)
23 │ coeftest(m1, vcov = NeweyWest)
```

```
 1 │ * STATA CODE
 2 │ causaldata restaurant_inspections.dta, use clear download
 3 │
 4 │ * Turn our data into a single time series
 5 │ collapse (mean) weekend inspection_score, by(year)
 6 │
 7 │ * Only use the run of years without gaps
 8 │ keep if year <= 2009
 9 │
10 │ * Tell Stata we have a time series dataset
11 │ tsset year
12 │
13 │ * Use "newey" instead of "regress"
14 │ * We must set maximum lags with lag()
15 │ newey inspection_score weekend, lag(3)
```

```
 1 │ # PYTHON CODE
 2 │ import statsmodels.formula.api as sm
 3 │ from causaldata import restaurant_inspections
 4 │ df = restaurant_inspections.load_pandas().data
 5 │
 6 │ # Get our data into a single time series!
 7 │ df = df.groupby('Year').agg([('mean')])
 8 │
 9 │ # Only use the years without a gap
10 │ df = df.query('Year <= 2009')
11 │
12 │ # We can simply add cov_type = 'HAC' to our fit, with maxlags specified
13 │ m1 = sm.ols(formula = 'inspection_score ~ Weekend',
14 │             data = df).fit(cov_type = 'HAC',
15 │             cov_kwds={'maxlags':1})
16 │
17 │ m1.summary()
18 │ # Note that Python uses a "classic" form of HAC that does not apply
19 │ # "pre-whiting", so results are often different from other languages
```

For the last of our sandwich estimators, we can use clustered standard errors. We don't really have any grouping that makes a whole lot of sense to cluster by in the data. But since this is only a technical demonstration, we can do whatever. We'll be clustering by whether the inspection was on a weekend. There are only two values of this, which is not a lot of clusters. We could get some strange results.

```r
# R CODE
# Load packages
library(tidyverse)
# For approach 1
library(AER); library(sandwich)
# For approach 2
library(modelsummary)
# For approach 3
library(fixest)

df <- causaldata::restaurant_inspections

# In R there are many ways to get at clustered SEs. For the first
# two methods we'll look at, we need to estimate the regression on
# its own. (most commands will work, not just lm()!)
m1 <- lm(inspection_score ~ Year + Weekend, data = df)

# First, the classic way, sending our regression object to
# AER::coeftest(), specifying the kind of library(sandwich) estimator.
# We want vcovCL for clustering.
# Note the ~ before Weekend; this is technically a formula
coeftest(m1, vcov = vcovCL(m1, ~Weekend))

# Second, we can pass it to msummary which will cluster if we send
# a formula with the cluster variable to vcov
msummary(m1, vcov = ~Weekend,
             stars = c('*' = .1, '**' = .05, '***' = .01))

# Third, we can use a regression function with clustered SEs built in,
# like fixest::feols(). Don't forget the ~ before Weekend.
feols(inspection_score ~ Year + Weekend,
    cluster = ~Weekend,
    data = df)
```

```stata
* STATA CODE
causaldata restaurant_inspections.dta, use clear download

* for basically any regression command,
* just add ", vce(cluster)"
reg inspection_score weekend year, vce(cluster weekend)
```

```python
# PYTHON CODE
import statsmodels.formula.api as sm
from causaldata import restaurant_inspections
df = restaurant_inspections.load_pandas().data

# We can simply add cov_type = 'cluster' to our fit, and specify the groups
m1 = sm.ols(formula = 'inspection_score ~ Year',
            data = df).fit(cov_type = 'cluster',
            cov_kwds={'groups': df['Weekend']})
```

```
10
11 || m1.summary()
```

Finally, we come to bootstrap standard errors. There's a
necessary choice here of *how many bootstrap samples to make.*
So... how many should you make? There's not really a hard-
and-fast rule. For most applications, a few thousand should be
fine. So let's do that. The code below only shows a straightforward each-observation-is-independent bootstrap. But in general
something like a cluster boostrap, where certain observations
are all included or excluded together, isn't far off if you read the
documentation, or Chapter 15.

```
1  || # R CODE
2  || # Load packages and data
3  || library(tidyverse); library(modelsummary); library(sandwich)
4  || df <- causaldata::restaurant_inspections
5  ||
6  || # Let's run our standard model from before
7  || m <- lm(inspection_score ~ Year + Weekend, data = df)
8  ||
9  || # And use the vcov argument of msummary with vcovBS from the
10 || # sandwich package to get boostrap SEs with 2000 samples
11 || msummary(m, vcov = function(x) vcovBS(x, R = 2000),
12 ||              stars = c('*' = .1, '**' = .05, '***' = .01))
```

```
1  || * STATA CODE
2  || causaldata restaurant_inspections.dta, use clear download
3  ||
4  || * for most commands, just add ", vce(boot, reps(number-of-samples))"
5  || reg inspection_score weekend year, vce(boot, reps(2000))
```

```
1  || # PYTHON CODE
2  || import pandas as pd
3  || import numpy as np
4  || import statsmodels.formula.api as sm
5  || from causaldata import restaurant_inspections
6  || df = restaurant_inspections.load_pandas().data
7  ||
8  || # There are bootstrap estimators in Python but generally not designed for OLS
9  || # So we'll make our own. Also see the Simulation chapter.
10 || def our_reg(DF):
11 ||     # Resample with replacement
12 ||     resampler = np.random.randint(0,len(df),len(df))
13 ||     DF = DF.iloc[resampler]
14 ||
15 ||     # Run our estimate
16 ||     m = sm.ols(formula = 'inspection_score ~ Year',
17 ||             data = DF).fit()
18 ||     # Get all coefficients
19 ||     return(dict(m.params))
20 ||
21 ||     # Run the function 2000 times and store results
22 ||     results = [our_reg(df) for i in range(0,2000)]
23 ||
24 ||     # Turn results into a data frame
25 ||     results = pd.DataFrame(results)
```

```
26
27        # Mean and standard deviation are estimate and SE
28        results.describe()
29
30   our_reg(df)
```

13.4   Additional Regression Concerns

ALL THAT AND WE STILL AREN'T DONE WITH REGRESSION?
How long is this chapter?[102] Alas, as with any statistical method,
there are infinite little crevices and crannies to look into and
wonder whether the method is truly appropriate for your data.
And for something as widely-used as regression, it's worth our
time to explore what we can.

In this final section we'll look at four additional tricks or concerns we'll want to keep in mind when working with regression.

13.4.1   Sample Weights

EVERYTHING WE'VE SAID ABOUT USING A SAMPLE RELATION-
SHIP TO LEARN ABOUT A POPULATION RELATIONSHIP makes
the assumption that the sample is drawn randomly from that
population. So if we have a sample of, say, Brazilians, then
each person in Brazil is equally likely to end up in our sample,
whether they live in Rio de Janeiro or the countryside, whether
they're rich or poor, whether they're educated or not, and so on.

This is rarely true for a number of reasons. Some people are
easier to survey or gather data on than others. Some sampling
methods don't even try to randomly sample everyone, instead
sampling people based on convenience, or location, or any number of other characteristics.

When this occurs, obviously, the results are better-
representative of the people who were more likely to be sampled. We can solve this issue with the application of *sample
weights*, where each observation in the sample is given some
measure of importance, with some observations being treated
as more important (and thus influential in the estimation) than
others.

In application to a bivariate regression, our estimate of the
slope is $Cov(X, Y)/Var(X)$. Without a sample weight, we estimate the covariance by taking $X - \bar{X}$ and multiplying it by
$Y - \bar{Y}$ for each observation, then taking the average of that
product over all the observations. With a sample weight w, we
do the exact same thing, but we take our $(X - \bar{X})(Y - \bar{Y})$ for

[102] And if you're tired of reading it, think about how I feel *writing* it. This chapter is topping 30,000 words— novella length! My *fingers* are tired. But I must press on. You WILL understand linear modeling. Oops, wait, hold on. My ring finger fell off.

Sample weight. A scaling applied to each observation when calculating sample statistics.

each observation and multiply *that* by the observation's weight w before taking the average of that whole thing.[103] The higher your w, the more your values will count when calculating the covariance, variance, and so on, and so you'll be more represented in the final estimates.

This whole process is known, unsurprisingly, as *weighted least squares*.

THERE'S MORE THAN ONE REASON TO WEIGHT. The problem I just mentioned—where certain people are more likely to be sampled than others—is a very common problem in the social sciences. It's so common, in fact, that many surveys, and certainly most of the big and well-performed surveys like the census in most countries, will contain sample weights already prepared for you. The survey planners have a good idea about how their sampling was done, and so have a decent idea of how likely different individuals, households, firms, or whatever, were to be included in the sample.

These weights are sometimes created by comparing the proportion of people with a given attribute in the data against a population value. If the population has 500 men and 500 women (50% of each), and our sample of 100 people has 70 men and 30 women, then we might say that each man had a $70/500 = 14\%$ chance of being included, and each woman had a $30/500 = 6\%$ chance. Those probabilities would inform our sample weights.

Then, we'd weight each individual by the *inverse* of those probabilities. So men would get a $1/.14 \approx 7.143$ weight, and women would get a $1/.06 \approx 16.667$ weight. This weights people who were less likely to be sampled more heavily, giving us back the demographic proportions in the population, making our estimates better represent the population.

For example, if we estimated the proportion of men using our sample alone, we'd estimate that the population has 70% men. Clearly wrong! But if we weight it, we'd get $(7.143 \times 70 + 16.667 \times 0)/(7.143 \times 70 + 16.667 \times 30) = 50\%$.[104] Perfect! Same idea goes when estimating regression coefficients. Without the weights, they're skewed by the sampling process. With weights, we better represent the relationship in the population.

Sample weights work. If your data has sample weights (and they're estimated properly, which in a professional survey they likely would be), read carefully about what they mean, and use them. Be sure to use the ones that fit the analysis you're doing—survey data will often have different weights for different observation levels. If you want to generalize to individuals,

[103] Often the weight will be standardized so the average of all weights over the whole sample is 1. This prevents the estimate from, say, calculating an *overall bigger* variance rather than just a variance that better-represents some observations. Also, *formally* what's happening here is that every value of every variable is being multiplied by $\sqrt{w}$. Then when you multiply two variables together, as in XY for the covariance, or XX for the variance, that becomes the weight $\sqrt{w}^2 = w$.

Weighted least squares. Ordinary least squares performed with sample weights.

Inverse sample probability weights. A sample weight where each observation is weighted by the inverse of the probability that it was included in the sample.

[104] That denominator (the $7.143 \times 70 + 16.667 \times 30$) is the sum of all the weights in the sample, and is there to scale the weights so they average out to 1.

use the individual-person weights. If you want to generalize to households use the household weights, and so on.

THERE ARE OTHER REASONS TO WEIGHT, THOUGH. A big one comes if you're using *aggregated data.* For example, maybe you're doing an analysis of the impact of distributing laptops on test scores. But instead of using data on the individual students and their individual test scores, maybe you use classroom averages.[105]

Now, analyzing the classroom-average data directly would have some problems. Some classrooms are bigger than others. Bigger classrooms represent more students, so if you want a result of the form "a laptop increases a student's scores by $\hat{\beta}_1$," then your aggregated analysis will be hard to figure out, since your results would really be on the classroom-average level. Second, whatever average data you have will be less noisy if it came from a bigger classroom, since the sampling distribution of the mean has a smaller standard deviation if it comes from a bigger sample. So some of the means in your data will be better-estimated than others, but OLS will treat them as the same.

These are two different problems with two different (but similar) solutions.

The first problem, and solution, is very simple. One classroom represents 30 people while another represents 40? Easy! Just count the first classroom 30 times and the second classroom 40 times. These are *frequency weights*, where the weight is simply the number of observations the variable represents. This returns your results to being interpretable on the individual, non-aggregated level. It's not quite as good as actually having the non-aggregated data,[106] since you lose out on that between-individual variation. But it does give you easier interpretation, and gives you a result that accounts for differences in size.

The second problem, relating to differences in the variance of the mean, is a bit trickier, but not much. For this problem we have inverse variance weighting. We can calculate the variance of each aggregated observation's mean. Then, weight each observation by the inverse of that variance.

The variance of a sample mean is σ^2/N, where N is the number of observations in the sample, i.e., the number of observations aggregated into the observation. The inverse of that is N/σ^2. Under the assumption that σ^2 is constant across the sample, that part drops out of the weighting (since it applies to everyone, and the weights are all relative to each other), leaving

[105] In many real-world cases, you might only have access to the data on classroom averages. If you do have the individual data, often that's what you want to use.

Frequency weights. A sample weight equal to the number of observations represented in aggregated data, used so that the interpretation is on the non-aggregated level.

[106] Unless each observation in the group is, truly, exactly the same.

us with weights of N. Each aggregate observation is weighted by the number of observations that got aggregated into it.

Hold on... frequency weights weight by the number of observations used to aggregate, too. So what's the difference between frequency weights and inverse variance weights? There actually isn't a difference when it comes to estimates—frequency weights and inverse variance weights will produce the exact same coefficients. However, there is a difference when it comes to standard errors. Frequency weights act as though each aggregated observation is truly a collection of separate, independent, completely identical observations. So when it calculates $\sigma/var(X)$ for the standard errors, it uses *the unaggregated sample size* to calculate σ. Inverse variance weights, however, recognize its data as being a set of means, and so uses *the aggregated sample size* to calculate σ. Frequency weight standard errors will then always be smaller than inverse variance weight standard errors.[107]

Inverse variance weighting has plenty of other applications too, besides aggregated data. There are lots of other situations in which we know that some observations represent a lot more variance than others. For example, treatment itself may be a lot more variable for some observations than others, and those observations will be weighted more heavily in the estimate, which we can correct by inversing that variance to make the weight. Problem solved. One application of this to the fixed effects method can be seen in Chapter 16.

Another place where inverse variance weighting pops up that we *won't* have room to cover is in meta-analysis. In meta-analysis, you look at a bunch of different studies of the same topic and try to estimate the overall effect, aggregating together their results. Some studies have estimates with big standard errors, and others have estimates with small standard errors. Inverse variance weighting helps weight the more precisely-estimated effects more strongly.

LET'S ESTIMATE SOME REGRESSIONS WITH SAMPLE WEIGHTS. Most commands will let you just give them a weighting variable. The task of figuring out which kind of weighting to do is on you, and you just pass off the final weight. So if you want inverse sampling probability weights, better calculate that inverse yourself!

This also means that if a certain approach to weighting requires you to adjust the standard errors, that won't happen. If you want to do frequency weights, for example, you'll either need to adjust the standard errors yourself, or skip the

[107] So be really sure that "a collection of independent, completely identical observations" describes your data before using frequency weights. An example of this would be a randomized experiment with a binary outcome, so your four aggregated observations are "no treatment, $Y = 0$," "no treatment, $Y = 1$," "treatment, $Y = 0$," and "treatment, $Y = 1$," which would truly represent collections of independent observations.

use of weights altogether and just copy each observation a number of times equal to its weight and run the regression without weights. You can do this in R by passing the data `df` to `slice(rep(1:nrow(df), weight))` after loading the **tidyverse**.

Python is a bit trickier—you can do it with `expanded_df = df.loc[numcopies]`, but this requires you to calculate the variable `numcopies`, the number of times you want to copy each row, yourself. The **itertools** package may help with that.

Stata is a an exception here, since it has several weight types (`aweight`, `fweight`, `iweight`, `pweight`) that it treats differently, and not all commands accept all kinds of weights. You'd generally use `pweight` for inverse-sampling-probability weights, `aweight` for variance weighting, and `fweight` for frequency weighting, which will properly handle the standard error adjustment for you.[108] Look at Stata's `help weight` before using weights in Stata.

For these examples, we'll be doing some inverse variance weighting with an aggregated version of our restaurant inspection data. First, we'll aggregate the data to the different restaurant chains, and count how many inspections we're averaging over. Specifically, we'll see if average inspection score is related to the first year that the chain shows up in the sample. Then, since the variance of the sample mean is proportional to the inverse of the number of observations in the sample, we'll weight by the inverse of that inverse, i.e., the number of observations in the sample.

[108] Although you could also do the copy-each-observation approach like with R and Python by using `expand weight`.

```r
# R CODE
# Load packages and data
library(tidyverse); library(modelsummary)
df <- causaldata::restaurant_inspections

# Turn into one observation per chain
df <- df %>%
    group_by(business_name) %>%
    summarize(Num_Inspections = n(),
    Avg_Score = mean(inspection_score),
    First_Year = min(Year))

# Add the weights argument to lm
m1 <- lm(Avg_Score ~ First_Year,
        data = df,
        weights = Num_Inspections)

msummary(m1, stars = c('*' = .1, '**' = .05, '***' = .01))
```

```stata
* STATA CODE
causaldata restaurant_inspections.dta, use clear download

* Aggregate the data
```

```
5  g Num_Inspections = 1
6  collapse (mean) inspection_score (min) year ///
7      (sum) Num_Inspections, by(business_name)
8
9  * Since we want inverse-variance weights,
10 * we use aweights, or aw for short
11 reg inspection_score year [aw = Num_Inspections]
```

```
1  # PYTHON CODE
2  import statsmodels.formula.api as sm
3  from causaldata import restaurant_inspections
4  df = restaurant_inspections.load_pandas().data
5
6  # Aggregate the data
7  df['Num_Inspections'] = 1
8  df = df.groupby('business_name').agg(
9      {'inspection_score': 'mean',
10         'Year': 'min',
11         'Num_Inspections': 'sum'})
12
13 # Here we call a special WLS function
14 m1 = sm.wls(formula = 'inspection_score ~ Year',
15             weights = df['Num_Inspections'],
16             data = df).fit()
17
18 m1.summary()
```

These code examples, of course, just show how to run a single regression with weights. If you are working with survey data that has complex weighting structures that you'll need to reuse over and over again, you might want to have the language explicitly acknowledge that and help you. In R you can do this with the **survey** package. In Stata you can look at the range of svy commands and prefixes with `help svy` and `help svyset`.

13.4.2   Collinearity

WHEN I DISCUSSED CATEGORICAL PREDICTOR VARIABLES, I talked about *perfect multicollinearity*. Perfect multicollinearity is when a linear combination of some of the variables in the model perfectly predicts another variable in the model. When this happens, the model can't be estimated. If the model is $Sales = \beta_0 + \beta_1 Winter + \beta_2 NotWinter$, then $\beta_0 = 15, \beta_1 = -5, \beta_2 = 0$ produces the exact same predictions (and residuals) as $\beta_0 = 10, \beta_1 = 0, \beta_2 = 5$, as well as an infinite number of other sets of estimates. OLS can't pick between them. It can't estimate the model.[109]

But what happens if we don't have *perfect* multicollinearity, but we do have predictor variables that are *super highly correlated*? After all, one way of thinking about $Winter$ and $NotWinter$ is they're just two variables with a correlation of

[109] In practice, regression software will estimate the model, but only by dropping one of the variables so the perfect multicollinearity goes away.

−1. What if we had two variables with a correlation of .99 or −.99 instead of 1 or −1? Or even .9 or −.9? We'd call that a high degree of *collinearity* (but not perfect).

Outside of the categorical-variable example, high collinearity often comes up when you have multiple variables that measure effectively the same thing. For example, say you wanted to use an "Emotional Intelligence" score to predict whether someone would get a promotion. You give people an emotional intelligence test with two halves with tiny variations on the same questions. Then, you use the first half of the test and the second half as separate predictors in the model, hoping to get an overall idea of how emotional intelligence as a whole predicts promotions. The scores from those two halves are likely highly correlated.[110]

High levels of collinearity tend to make the estimates very sensitive and noisy, leading the sampling distributions to be very wide and driving the standard errors upwards.

WHY DOES COLLINEARITY INCREASE STANDARD ERRORS? Let's stick with our Winter/Sales example, Let's replace the Winter variable with WinterExceptJan1. So now our predictors are WinterExceptJan1, which is 1 for all non-January 1 days in Winter, and NotWinter, which is 1 for all non-Winter days. January 1 is zero for both of them. The correlation between the two variables would be very nearly −1. $WinterExceptJan1 + NotWinter = 1$ is true for all observations except for January 1. So we are close to a perfect linear combination but not quite.

In that setting, with the model $Sales = \beta_0 + \beta_1 WinterExcept Jan1 + \beta_2 NotWinter$, $\beta_0 = 15, \beta_1 = -5, \beta_2 = 0$ will not produce the exact same predictions (and residuals) as $\beta_0 = 10, \beta_1 = 0, \beta_2 = 5$, but it will be very close. Every day but January 1 will be predicted exactly the same. Now, unlike with perfect multicollinearity, we *can* pick a "best" set of estimates. Simply set β_0 equal to *Sales* on January 1, β_1 equal to the average difference between *Sales* on the other Winter days and on January 1, and β_2 equal to the average difference between *Sales* on non-Winter days and on January 1.

However, those estimates we picked were *entirely* dependent on the January 1 value. If Sales on January 1 are 5, then we get $\beta_0 = 5, \beta_1 = 5, \beta_2 = 10$. If *Sales* on January 1 just happens to be 15 instead, then everything changes wildly to $\beta_0 = 15, \beta_1 = -5, \beta_2 = 0$. A single observation changing value shifts every coefficient by about the amount of the shift! β_1 then moves around just as much as a single observation of *Sales* does.

Collinearity. When a predictor variable in the model is *very strongly* predicted by a linear combination of the other variables in the model.

[110] And if they're not, that test probably isn't very good.

The standard error of β_1 basically *is* the standard deviation of *Sales*, rather than shrinking quickly with the sample size as it normally would.[111]

WHAT CAN WE DO ABOUT COLLINEARITY THEN? The first step is, as always, to *think through our model carefully.* Even before thinking about whether a calculated correlation is really high, ask yourself "why am I including these predictors in the model?" Remember, adding another predictor to the model means you're *controlling for it* when estimating the effect of the other variables. Do you want to control for it?

A tempting thought when you have multiple measures of the same concept is that including them all in the model will in some way let them complement each other, or add up to an effect. But in reality it just forces each of them to show the effect of the variation that each measure has that's unrelated to the other measures.

Going back to the example with the two halves of the emotional intelligence test, why would we want to know the effect of the first half *while controlling for the second half?* "Controlling for the second half of the exam, an additional point on the first half increases..." what would that even mean? The coefficient on the first half would represent the relationship between score on the first half and promotion probability after removing the part of that relationship that's explained by the second half, but that explained part should basically be all of the relationship. There won't be much variance left, making $var(X)$ tiny and $\sigma^2/var(X)$ big, inflating those standard errors. In any case, where the correlation really is that strong, you probably aren't closing any back doors, you're just controlling away your treatment.

In cases like this, where we have multiple measures of the same thing, one good approach to including all the predictors while avoiding collinearity is the use of *dimension reduction* methods, which take multiple measures of the same thing and distill them down into their *shared* parts, capturing the underlying concept you think they all represent (rather than their non-shared parts, which is what including them all in a regression would do).

Dimension reduction is a large field, largely growing out of psychometrics, and I won't even try to do it justice here. But methods like latent factor analysis and principal components analysis take multiple variables and produce a small set of factors that represent different shared aspects of those variables,

[111] This particular result is based on this very specific example of two categories with only one observation left out, but the same general idea applies to any kind of high-collinearity situation.

and which are unrelated to each other (and so produce no collinearity at all).

HOW CAN WE ADDRESS COLLINEARITY IN OTHER CONTEXTS, where the issue isn't multiple measures of the same thing, but multiple measures of different things that happen to be strongly correlated?

For example, maybe you want to look at the impact of an increase in wages on a firm's profitability, controlling for stock market indices in both Paris and Brussels, where it trades a lot. Paris and Brussels stock prices are likely correlated strongly, but they're not the same thing. We can imagine how we might want to control for both, but maybe the correlation is too high and we'd have a collinearity problem. Should we remove one, or use dimension reduction? How much correlation is "too much"?

You could just check the correlation between all the predictors and seek out high correlations, but this wouldn't account for cases where it takes more than one variable to form a very close linear prediction of a variable.

A common tool in checking whether there's too much collinearity,[112] which takes into account the entire set of variables at once, is the variance inflation factor. The variance inflation factor is different for each variable in the model. Specifically, for variable j, it's $VIF_j = 1/(1 - R_j^2)$, where R_j^2 is the R^2 from a regression of variable j on all the other variables in the model.

As a very rough rule of thumb, if a variable has a VIF above 10 (i.e., if regressing that variable on the other predictors produces an R^2 above .9), then you might want to consider removing it from the model, or performing dimension reduction with the other variables it's strongly related to. Other people use a VIF above 5 as a cutoff. In general, using hard cutoffs as decision rules rarely makes sense in statistics. Use high VIFs as an opportunity to think very carefully about your model and think about whether removing things makes sense.

You can calculate the VIF for each variable in the model in R using vif(m1) from the **car** package, where m1 is your model. In Stata, run your regression, and then follow that with estat vif.

Python has the variance_inflation_factor function which you can import from statsmodels.stats.outliers_influence. However, it works a bit differently from the R and Stata versions because it operates on a data frame of the predictors instead of on the model, and only calculates the VIF for one of them at a time, so you have to loop through them if you

[112] In a linear regression, anyway.

want everything. If X is a data frame consisting only of your predictors, you can calculate the VIF for each of them with `[variance_inflation_factor(X.values, i) for i in range(X.shape[1])]`.

13.4.3   Measurement Error

IT SEEMS LIKE A PRETTY REASONABLE ASSUMPTION THAT THE VARIABLES IN THE DATA ARE THE VARIABLES IN THE DATA. But what if... they're not?

What I mean by this is that it's pretty common for the values you've recorded in the data to not be perfectly precise. It's also common for the variable you have in the data to only be a *proxy* for the variable you want, rather than the real thing.

Imprecision is a natural part of life. No measurement is perfectly precise. If you have a data set of heights, the "5 foot 6 inch" and "6 foot 1 inch" in the data probably are hiding more accurate values like "5.5132161 feet" and "6.0781133 feet" that you don't have access to. If the only problem is that you've rounded your values to a certain reasonable level of precision, that's probably not a huge issue. But if the mismeasurement is more severe—if you're rounding both of those very-different heights to 6 feet, or if that "5 foot 6 inch" is really "5 foot 8 inch" but someone wrote it down wrong, or used the ruler incorrectly, then the values in the data don't match the real values, so there's no way for our estimates to pick up on actually-existing values in the data.

The use of proxies is common as well. A proxy is a variable used in place of another, because you don't have the variable you actually want. If you want to know the effect of giving students laptops on future test scores, you may think that mathematical ability exists on a back door and so want to control for a student's mathematical ability. You can't actually see their mathematical ability, but you have access to their scores on a math test, which isn't exactly the same thing but should be closely related. The test score is a proxy for mathematical ability, but it's not the same thing. If two students have the same ability but different test scores, the model will treat them as different when really they're the same, and thus will get it wrong.

Imprecision and proxies are both examples of *measurement error*, also known as *errors-in-variables*. The idea is that the variables in your model represent the true *latent value* you want, with some error.

Proxy. A variable that should be strongly related to the variable you actually want in your model, used in its place because you don't have measurements of the variable you actually want.

Measurement error. When a variable you're using is an inexact representation of the variable you actually want, due to mismeasurement, imprecision, or using a proxy.

Latent value/latent variable. A value or variable that is a part of your model, but not a part of your data. Generally the term is only used when you have other variables you're trying to use to represent the latent variable.

$$X = X^* + \varepsilon \qquad\qquad (13.18)$$

where X is the variable you have in your data, X^* is the latent unobserved variable, and ε is an error term.[113]

WHY DO WE CARE ABOUT MEASUREMENT ERROR? It's pretty easy to imagine how measurement error can lead our estimates astray. In the extreme, imagine using some completely unrelated variable as a proxy. If I want to know the effect of laptops on test scores, but instead of using the variable "has a laptop" I use "has brown hair," we have no reason to believe that would give us the effect of having a laptop.

But what about in less extreme cases? There are two main kinds of measurement error, and they have different implications.

The easiest kind of measurement error to handle is *classical measurement error,* which is measurement error where the error ε is unrelated to the latent variable X^*. When this happens, the measured variable X is just "the true latent variable, plus some noise."

Classical measurement error has clear implications: in the regression $Y = \beta_0 + \beta_1 X$, if X suffers from classical measurement error, then the estimate $\hat{\beta}_1$ will be *closer to* 0 than the true β_1 you'd get if you could run $Y = \beta_0 + \beta_1 X$.[114] It will "shrink towards zero," also called "attenutation."

Why does it do this? Well, X^* has an effect of β_1 on Y, right? Whenever X^* moves, we expect Y to move as well. But if we see X, the noisy version of X^*, then we won't see all those X^* moves. We'll see Y dancing around, responding to changes in X^* that we can't see in X. And the more changes in Y there are that *aren't* related to X, the less related X and Y appear to be. So the estimate shrinks towards zero.

Classical measurement error also has some clear fixes, or, rather, it offers the opportunity not to fix it. Most commonly, people will just run regular OLS without actually fixing anything, and then treat the estimate as a "lower bound." They know the effect is *at least* as far away from zero as the estimate they get, although they don't know how much larger.

You may have noticed that I've only discussed measurement error in X. What about Y? Turns out classical measurement error in Y just isn't much of a problem. Sure, it will reduce the model's R^2 (since there's now some more noise in Y that we can't predict), but it won't affect the coefficients, at least not

[113] This equation shows the error term as being added to the true value, but it could easily work in some other way, for example being multiplied by the true value.

[114] This is also true in a model with controls.

on average. Those coefficients are already used to there being noise in Y it can't predict. There's just a bit more of it now. That noise will go into the regression model's error term.

IF THAT'S CLASSICAL MEASUREMENT ERROR, WHAT'S *NON-CLASSICAL* MEASUREMENT ERROR? This is when the error *is* related to the true value, and it makes things much trickier. Why might measurement error be non-classical? One example is if X^* is binary, as is X, so both can only be 0 or 1. If $X^* = 1$, then the error can only possibly be 0 or -1. If $X^* = 0$, then the error can only possibly be 0 or 1. The value of X^* determines what the error can be, making them related.

Non-classical measurement error occurs with continuous variables, too. Say you have a bunch of tax data on business owners and are measuring income. You know that some people are going to misrepresent their income on their taxes, so as to pay less in tax. Misrepresenting income is a lot easier with cash-based businesses than with businesses that mostly get paid with credit cards or checks. Cash-based businesses also on average tend to have lower incomes than others. So we'd expect more negative measurement errors (underreported income) among low-income businesses than high-income businesses. The error is related to the true value.

For another example, imagine a study that relies on a variable where people report how much they exercise each day. Someone who exercises a lot may feel comfortable reporting the truth to the survey-taker, while someone who exercises very little may bump their numbers up to look better.

Non-classical measurement error is more unpredictable than classical measurement error, because without knowing more about the exact way the error is related to X^*, we don't know whether β_1 will be shrunk towards zero, inflated away from zero, biased upwards, biased downwards, or not biased at all! It could be anything. Also, unlike classical measurement error, non-classical measurement error is a problem when it happens to Y as well as when it happens to X.

So if you think you have measurement error, put some serious thought into whether you think there's any way for that error to be related to the latent variable. If it is, you have a bigger problem on your hands than you thought.

IF WE HAVE MEASUREMENT ERROR AND WANT TO FIX IT, HOW CAN WE DO THAT? If you have some information about the measurement error itself, you might be able to do some adjustments to get closer to the true β_1. I won't go deeply into these methods—you'll have to look elsewhere. Measurement error is one of those topics where diving in takes you really deep. Every textbook I can find focusing on the subject is much more advanced than the one you're reading now, but it still may be worth your time to look at a book like *Measurement Error Models* by Wayne Fuller.[115]

Without going too deep, though (or, rather, rattling off the keywords you can Google for later), what can we do? If you have an idea of the ratio of the variance of the regression error to the variance of the measurement error, you can perform a Deming regression or Total Least Squares, which adjusts for the shrinkage towards zero.

There are other options out there that you can look for, including the general method of moments (GMM). Another common approach is the use of instrumental variables, although unlike with the use of instrumental variables in Chapter 19, the goal here is not to identify a causal effect but rather adjust for measurement error. If you have two measurements of the same latent variable, and both measurements are noisy but have unrelated errors (for example you have two imperfect thermometers both recording temperatures a mile apart, and you want the temperature in that area), you can use one measurement as an instrumental variable to predict the other, and then use the prediction in the model. This isolates the shared part of the two measurements, which will be closer to the true latent value than either alone.

That's not even getting into the whole class of methods specially designed for use with measurement error in binary predictors! Many of these try to estimate how often a 0 turns into a 1, and a 1 turns into a 0, and they incorporate that into their estimates of the regression model.

Of course, these methods I've listed all relate to measurement error when you have a *linear* model. What if you're working with a nonlinear model like logit, probit, or something else? Well... you'll have to look for variants of these measurement error corrections specifically designed for those methods. They're out there. Usually they are variants on the models I've already talked about, such as GMM or instrumental variables. Happy hunting.

[115] Wayne A Fuller. *Measurement Error Models*, volume 305. John Wiley & Sons, 2009.

13.4.4   Penalized Regression

THE ENTIRE FIRST HALF OF THIS BOOK BEAT A VERY REPET-
ITIVE DRUM: construct your model from theory. Use what you
know to decide very carefully which controls are necessary and
which should be excluded. And so on.

However, it's not uncommon for this process to leave us with
far too many candidates for inclusion as controls. What do we
do if we have thirty, fifty, a thousand potential control variables
that *might* help us close a back door? Even if we have a good
reason for including all of them, actually doing so would be a
statistical mess that would be very difficult to interpret, and
would give us a collinearity nightmare. Plus, the data would
likely be *overfit*—with that many predictors, the model will fit
the sample very closely, but will be responsive to noise (sort of
for the same reasons as collinearity), and so will not match the
true model well, or predict well if applied to a new sample.

We probably want to drop some of those controls. But which
ones? Ideally we'd have some sort of principled *model selection
procedure* that would do the choosing for us.

Enter *penalized regression*. Penalized regression is a tool coming from the world of machine learning. Machine learning methods are generally much more concerned with out-of-sample prediction than with causal inference,[116] but we're going to be able
to put one of their tricks to use here. Remember how OLS picks
its coefficients by minimizing the sum of squared residuals? We
can represent that goal as

$$argmin_\beta \left\{ \sum (Y - \hat{Y})^2 \right\} \qquad (13.19)$$

In other words pick the β (arg)uments that (min)imize $\sum (Y - \hat{Y})^2$, which is the sum ($\sum$) of squared (^{2}) residuals ($Y - \hat{Y}$). Note
that β shows up inside that $\hat{Y}$, which is $\hat{\beta}_0 + \hat{\beta}_1 X$ (plus whatever
other coefficients and predictors you have).

Penalized regression says "sure, minimizing the sum of squared
residuals is great. But I want you to try to achieve two goals at
once. Make the sum of squared residuals as small as you can,
sure, but *also* make this function of the β coefficients small."
The goal now becomes

$$argmin_\beta \left\{ \sum (Y - \hat{Y})^2 + \lambda F(\beta) \right\} \qquad (13.20)$$

where $F()$ is some function that takes all the coefficients and
aggregates them in some way, usually a way that gets bigger as

[116] But not always. I recommended many, many chap-
ters ago to look at the work
of Susan Athey and Guido
Imbens, and that recom-
mendation still holds. Also
see some of the sections in
Chapter 21.

[117] And this is a function
only of the β coefficients—
X doesn't show up here, nor
does Y.

the βs grow in absolute value,[117] and λ is a penalty parameter—a tiny λ means "don't worry too much about the penalty, just minimize the sum of squared residuals, please," and a big λ means "it's really important to keep the βs small, worry about that more than minimizing the residuals."[118]

There are plenty of functions you could choose for $F()$, but the most common are "norm" functions, specifically the L1-norm, which gives you "LASSO regression," and the L2-norm which gives you "ridge regression." The L1-norm function just adds up the absolute values of the βs ($\sum |\beta|$, where $||$ is the absolute value function), and the L2-norm function adds up the squares of the βs ($\sum \beta^2$).[119]

Having this penalty encourages the estimator to pick β values closer to 0, so as to avoid the penalty. Why is this desirable? Because less-bold predictions will back off a bit on fitting the *current* data, but as a tradeoff, you'll probably get better predictions of *future* data, since your estimate won't be distracted as much by the noise in the sample.

But these methods don't just shrink coefficients uniformly; it sort of works with a coefficient "budget," and it spends wisely. It only makes coefficients bigger if they really help in reducing that sum of squared residuals. The L2-norm (ridge regression) ends up giving you shrunken coefficients. But the L1-norm (LASSO) version, also known as the "least absolute shrinkage and selection operator," doesn't just shrink coefficients, it also tends to send a lot of coefficients to exactly zero, effectively tossing those variables out of the model entirely—that's why it's a "selection operator".

And that's the way we can apply LASSO to our problem of having too many covariates. It figures out which variables it can drop while doing the least damage to the sum of squared residuals. It then ends up shrinking the coefficients that remain, to boot.

BUT IF WHAT WE WANT IS TO IDENTIFY A CAUSAL EFFECT, WON'T LASSO GIVE US A BAD MODEL? There are two ways in which LASSO could get us in trouble. The first is that some covariates need to be included to close back doors, even if they don't help prediction that much. That's an easy fix—just tell LASSO not to drop specific necessary covariates. Or, after it tells you to drop them, just... don't do it. Add them back in.

The second way LASSO can get us in trouble is that our goal here is very much *not* to get the best out-of-sample prediction as the machine learning types aim for (and designed LASSO

[118] I'm presenting penalized regression in its linear/OLS form here, but you can just as easily penalize any other regression method, like logit or probit.

[119] You can also mix and match the two by weighting the L1 and L2 norms and adding them. This is called elastic net regression.

for), but rather to get the best estimate of a causal effect that we can. And here comes LASSO, explicitly *not even trying to get the best estimate of* β—that part where it shrinks the coefficients doesn't make them more accurate estimates of the coefficients, it just makes the out-of-sample prediction better. In fact, *LASSO will in general give you biased estimates of coefficients.* But that's an easy fix, too. If you only want to use LASSO for variable selection, do that. Then, take the variables it says to keep in (as well as any you need to close back doors, as previously mentioned), and rerun good ol' non-penalized OLS using those variables.

How can we actually perform LASSO? There are a few practical things to keep in mind when running LASSO.

One important thing to point out here is that before estimating a LASSO model, it's important to standardize all your variables. That is, we want to take each variable, subtract out its mean, and divide by its standard deviation.

Why do we want to do this? Because the penalty term, $\sum |\beta|$ cares only about the size of the coefficient. And the size of the coefficient is sensitive to the scale of the variable! For example, comparing the models $Y = \beta_0 + \beta_1 X$ and $Y = \beta_0 + \beta_1 HalfofX$, where $HalfofX = X/2$, if we estimate $\hat{\beta}_1 = 3$ in the first model, we'll get the exact same prediction if $\hat{\beta}_1 = 6$ in the second model. The /2 from $HalfofX$ is exactly balanced by doubling the β_1. This isn't a problem with regular OLS, since our interpretation just changes to match. A one-unit change in $HalfofX$ corresponds to a two-unit change in X. A one-unit change in $HalfofX$ is thus twice the change of a one-unit change in X, so it makes sense it has twice the effect (6 compared to 3).

The fact that smaller scale means bigger coefficients means that simply dividing X by 2, which shouldn't make any meaningful difference in the model, doubles the penalty that the coefficient on X receives, and gives LASSO more reason to shrink it or drop it.

We don't want that, so we make sure all variables are on the same scale by standardizing first. Some LASSO commands will do this standardization for you (or have an option to do so). For others you'll need to do it yourself.

The second thing to keep in mind is that LASSO lets you *try stuff* and see what needs to be kept. So it's relatively common to run LASSO *while including a bunch of polynomial terms and interactions for basically everything.* If the polynomial's not important, it will drop out.

You can only go so wild with this, of course. If you truly interact everything with everything else and LASSO tells you to keep some strange interactions, will you be able to make sense of the model? And if you include a polynomial and it tells you only to keep the cube term and not the linear or square, you'll have to add those lower terms back in anyway.

Third, we've left out an important step of LASSO—picking the λ value that tells us how much to penalize the coefficients.[120] You can set this as you like, choosing a higher value to really toss out a lot of variables, or a low value to only toss out a few. Often, people will use *cross-validation* to select the value.

Cross-validation is a method where you split the data up into random chunks. Then you pick a value of λ. Then you drop each of those chunks one at a time, and use the remaining chunks to predict the values of the dropped chunk, letting you in effect predict out-of-sample. Then bring that dropped chunk back and drop the next chunk, estimating the model again and predicting *those* dropped values. Repeat until all the chunks have been dropped once, and evaluate how good the out-of-sample prediction is for the λ you picked. Then, pick a different λ value and do it all over again. Repeat over and over, and finally choose the λ with the best out-of-sample prediction.

That said, let's run some LASSO models.[121] If you run this code, you might get slightly different results as to what to drop. Why? Because there's some randomness in the LASSO estimation procedure. To get perfectly consistent results you'd need to set the random seed.[122]

[120] Sometimes, instead of λ (lambda) it's called α (alpha).

[121] This code can also be found on the LASSO page on https://lost-stats.github.io/.

[122] So why don't I do that here? To demonstrate how not setting a seed can make your code irreproducible. Set those seeds!

```
1  # R CODE
2  # Load packages and data
3  library(tidyverse); library(modelsummary); library(glmnet)
4  df <- causaldata::restaurant_inspections
5
6  # We don't want business_name as a predictor
7  df <- df %>%
8      select(-business_name)
9
10 # the LASSO command we'll use takes a matrix of predictors
11 # rather than a regression formula. So! Create a matrix with
12 # all variables and interactions (other than the dependent variable)
13 # -1 to omit the intercept
14 X <- model.matrix(lm(inspection_score ~ (.)^2 - 1,
15                      data = df))
16 # Add squared terms of numeric variables
17 numeric.var.names <- names(df)[2:3]
18 X <- cbind(X, as.matrix(df[, numeric.var.names]^2))
19 colnames(X)[8:9] <- paste(numeric.var.names, 'squared')
```

```
20
21   # Create a matrix for our dependent variable too
22   Y <- as.matrix(df$inspection_score)
23
24   # Standardize all variables
25   X <- scale(X); Y <- scale(Y)
26
27   # We use cross-validation to select our lambda
28   # family = 'gaussian' does OLS (as opposed to, say, logit)
29   # nfolds = 20 does cross-validation with 20 "chunks"
30   # alpha = 1 does LASSO (2 would be ridge)
31   cv.lasso <- cv.glmnet(X, Y,
32                         family = "gaussian",
33                         nfolds = 20, alpha = 1)
34
35   # I'll pick the lambda that maximizes out-of-sample prediction
36   # which is cv.lasso$lambda.min and then run the actual LASSO model
37   lasso.model <- glmnet(X, Y,
38                         family = "gaussian",
39                         alpha = 1,
40                         lambda = cv.lasso$lambda.min)
41   # coefficients are shown in the beta element - . means LASSO dropped it
42   lasso.model$beta
43
44   # The only one it dropped is Year:Weekend
45   # Now we can run OLS without that dropped Year:Weekend
46   m1 <- lm(inspection_score ~
47                         (.)^2
48                         - Year:Weekend
49                         + I(Year)^2
50                         + I(NumberofLocations)^2,
51                         data = df)
52
53   msummary(m1, stars = c('*' = .1, '**' = .05, '***' = .01))
```

```
1    * STATA CODE
2    causaldata restaurant_inspections.dta, use clear download
3
4    * NOTE: this code relies on functions introduced in Stata 16.
5    * If your Stata is older, instead do ssc install lars and use
6    * the lars function (the syntax will be a bit different)
7
8    * Standardize all variables
9    foreach var of varlist inspection_score numberoflocations ///
10                   year weekend {
11        summ `var'
12        replace `var' = (`var' - r(mean))/r(sd)
13   }
14
15   * Use the lasso command to run LASSO while including all interactions
16   * and squares, using sel(cv) to select lambda using cross-validation
17   * "linear" gives us OLS, but logit or probit would work
18   lasso linear inspection_score ///
19       c.numberoflocations##c.year##c.weekend ///
20       c.numberoflocations#c.numberoflocations c.year#c.year, sel(cv)
21   * get list of included coefficients
22   lassocoef
23
24   * Looks like the only one not included is numberoflocations*weekend,
25   * so we can rerun OLS without that.
```

```
26 | * first, reload the non-standardized values
27 | import delimited "restaurant_data.csv", clear
28 | g weekend = weekend == "TRUE" if !missing(weekend)
29 |
30 | reg inspection_score c.numberoflocations##c.year  ///
31 |     c.year##c.weekend c.numberoflocations#c.year#c.weekend ///
32 |     c.numberoflocations#c.numberoflocations c.year#c.year
```

```
1  | # PYTHON CODE
2  | import pandas as pd
3  | import numpy as np
4  | import statsmodels.formula.api as sm
5  | from sklearn.linear_model import LassoCV
6  | from sklearn.preprocessing import PolynomialFeatures
7  | from causaldata import restaurant_inspections
8  | df = restaurant_inspections.load_pandas().data
9  |
10 | # Create a matrix with predictors, interactions, and squares
11 | X = df[['NumberofLocations','Year','Weekend']]
12 | X_all = PolynomialFeatures(2, interaction_only = False,
13 |                           include_bias = False).fit_transform(X)
14 |
15 | # Use LassoCV to pick a lambda with 20 chunks. normalize = True
16 | # standardizes the variables. This particular model has trouble
17 | # converging so let's give it more iterations to try with max_iter
18 | reg = LassoCV(cv = 20, normalize = True,
19 |               max_iter = 10000,
20 |               ).fit(X_all,df['inspection_score'])
21 |
22 | reg.coef_
23 | # Looks like Weekend, squared Year, Year * NumberofLocations,
24 | # and Year * Weekend all get dropped. So we can redo OLS without them
25 |
26 | m1 = sm.ols(formula = '''inspection_score ~
27 | NumberofLocations + I(NumberofLocations**2) +
28 | NumberofLocations:Weekend + Year''', data = df).fit()
29 | m1.summary()
```

Something you might notice if you read the code for all three languages is that the results are different across languages. Each of the LASSO estimates was different in selecting what to drop, despite being given the exact same data (except for the Stata version, which was also given the interaction between all three variables). This is not as uncommon as you might expect in statistical software—results can be different across different languages all the time,[123] especially for methods like LASSO that aren't "solved" and so you have to search over a bunch of different coefficients to find ones that work. Those search methods can differ a lot over software packages.

Also, the *default options* may differ a lot. Even if the underlying estimation code were the same, these different languages may have still produced different results based on issues like where to start the estimation search, how to perform cross-validation, and which λ values to check in cross-validation.

[123] Bruce D McCullough and Hrishikesh D Vinod. Verifying the solution from a nonlinear solver: A case study. *American Economic Review*, 93(3):873–892, 2003.

Whatever language you use, be sure to read the documentation, be sure you know what you're asking it to estimate, and ideally try your estimate in more than one language to see if it's consistent.

14
Matching

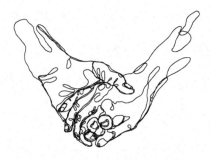

14.1 Another Way to Close Back Doors

REGRESSION GETS A LOT OF ATTENTION, but it's not the only way to close back doors. Through the first half of this book I was pretty insistent about it, in fact: *choosing a sample in which there is no variation in a variable W closes any open back doors that W sits on.*[1]

That sounds simple enough, and indeed it's an intuitive idea. "We picked two groups that look pretty comparable, and compared them," is a lot easier to explain to your boss than "we made some linearity assumptions and minimized the sum of squared errors so as to adjust for these back-door variables."[2] But once you actually start trying to *do* matching you quickly realize that there are a bunch of questions about *how exactly you do such a thing* as picking comparable groups. It's not obvious! And the methods for doing it can lead to very different results. Surely some smart people have been thinking about this for a while? Yes, they have.

Matching is the process of closing back doors between a treatment and an outcome by *constructing comparison groups that*

[1] As opposed to regression's "finding all the variation related to variation in W and removing it closes any open back doors that W sits on."

[2] Although in reality most bosses (and audiences) are either stats-savvy enough to understand regression anyway, or stats *un*-savvy enough that you can just say "adjusting for differences in X," and they won't really care how you did it.

DOI: 10.1201/9781003226055-14

are similar according to a set of matching variables. Usually this is applied to binary treated/untreated treatment variables, so you are picking a "control" group that is very similar to the group that happened to get "treated."[3,4]

To suggest a very basic example of how matching might work, imagine that we are interested in getting the effect of a job-training program on your chances of getting a good job. We notice that, while the pool of unemployed people *eligible* for the job-training program was about 50% male/50% female, the program just happened to be advertised heavily to men. So the people actually in the program were 80% male/20% female. Since labor market outcomes also differ by gender, we have a clear back door *Outcomes ← Gender → JobTrainingProgram.*

The matching approach would look at all the untreated people and would construct a "control group" that was also 80% male/20% female, to compare to the already 80—20 treated group. Now, comparing the treated and untreated groups, there's no gender difference between the groups. The *Gender → Job TrainingProgram* arrow disappears, the back door closes, and we've identified the *JobTrainingProgram → Outcomes* effect we're interested in.[5]

Matching and regression are two different approaches to closing back doors, and while identifying an effect using a set of control/matching variables assumes in both cases that our set of control/matching variables is enough to close all back doors, the other assumptions we rely on are different (but not better or worse) using the two approaches. To give one example, it's fairly easy to use matching without relying on the assumption of *linearity* that we had to rely on or laboriously work around in the regression chapter. To give another, matching is a lot more flexble in its ability to give you the kind of treatment effect average that you want (as in Chapter 10). Neither method is necessarily better than the other, and in fact they can be used together, with their different assumptions used to complement each other (as will be discussed later in the chapter).

Ugh. If matching were purely worse, we could simply ignore it. Or if it were better, we could have skipped that whole regression chapter. But it's neither, so we should probably learn about both. Naturally, there are a lot of details to nail down when we decide to do matching. That's exactly what this chapter is about.[6]

[3] You could apply matching to non-binary treatments, and there are some methods out there for doing so, for example by picking certain values, or ranges of values, of the treatment variable and matching on those. But matching is largely applied in binary treatment cases, and the rest of this chapter will focus on that case.

[4] It's also common to match by picking a set of *treated* observations that are most similar to the *control* observations. For simplicity of explanation I'm going to completely ignore this for the first half or so of the chapter. But we'll get there.

[5] Assuming there are no other back doors left open.

[6] The question of "which is more important to learn" comes down, oddly, to which field you're in. In some fields, like economics, regression is standard and matching is very rare. In others, like sociology, you're likely to see both. In still others, matching is very common and regression rare. You can just go ahead and learn whatever's important in your field, and skip the other if it's rare. Or... you can be a *cool rebel* and learn it all, and use *whatever you think is best for the question at hand.* So cool. Take that, *squares.* You can try to keep us down, but you'll never kill rock n' roll! Uh, I mean matching.

14.2   *Weighted Averages*

FIRST OFF, WHAT ARE WE EVEN TRYING TO DO with matching? As I said, we want to make our treatment and control groups comparable. But what does that mean, exactly?

Matching methods create a set of *weights* for each observation, perhaps calling that weight w. Those weights are designed to make the treatment and control groups comparable.

Then, when we want to estimate the effect of treatment, we would calculate a *weighted mean* of the outcomes for the the treatment and control groups, and compare those.[7]

AS YOU MAY RECALL FROM EARLIER CHAPTERS, a weighted mean multiplies each observation's value by its weight, adds them up, and then divides by the sum of the weights. Take the values 1, 2, 3, and 4 for example. The regular ol' non-weighted mean actually *is* a weighted mean, it's just one where everyone gets the same weight of 1. So we multiply each value by its weight: $1 \times 1 = 1, 2 \times 1 = 2, 3 \times 1 = 3, 4 \times 1 = 4$. Then add them up: $1 + 2 + 3 + 4 = 10$. Finally, divide the total by the sum of the weights, $1 + 1 + 1 + 1 = 4$—to get $10/4 = 2.5$.[8]

Now let's do it again with unequal weights. Let's make the weights $.5, 2, 1.5, 1$. Same process. First, multiply each value by its weight: $1 \times .5 = .5, 2 \times 2 = 4, 3 \times 1.5 = 4.5, 4 \times 1 = 4$. Then, add them up: $.5 + 4 + 4.5 + 4 = 13$. Finally, divide by the sum of the weights: $.5 + 2 + 1.5 + 1 = 5$, for a weighted mean of $13/5 = 2.6$.

We can write the weighted mean of Y out as an equation as:

$$\frac{\sum wY}{\sum w} \tag{14.1}$$

or in other words, multiply each value Y by its weight w, add them all up (Σ), and then divide by the sum of all the weights Σw. You might notice that if all the weights are $w = 1$, this is the same as taking a regular ol' mean.

BUT WHERE DO THE WEIGHTS COME FROM? There are many different matching processes, each of which takes a different route to generating weights. But they all do so using a set of "matching variables," and using those matching variables to construct a set of weights so as to close any back doors that those matching variables are on. The idea is to create a set of weights such that there's no longer any variation between the treated and control groups in the matching variables.

[7] There are other ways you can use the weights to estimate an effect, and I'll get to those later in the chapter.

[8] Some approaches to weighted means skip the "divide by the sum of the weights" step, and instead require that the weights themselves add up to 1.

In the last section I gave the 80/20 men/women example. Let's walk through that a bit more precisely.

Let's say we have a treated group, each of whom has received job training, consisting of 80 men and 20 women. Of the 80 men, 60 end up with a job and 20 without. Of the women, 12 end up with a job and 8 without.

Now let's look at the control group, which consists of 500 men and 500 women. Of the men, 350 end up with a job and 150 without. of the women, 275 end up with a job and 225 without.

If we look at the raw comparison, we get that $(60+12)/100 = 72\%$ of those with job training end up with jobs, while in the control group $(350+275)/1000 = 62.5\%$ end up with jobs. That's a treatment effect of 9.5 percentage points. Not shabby! But we have a back door through gender. The average job-finding rates are different by gender, and gender is also related to whether someone got job training.

There are many different matching methods, each of which might create different weights, but one method might create weights like this:

- Give a weight of 1 to everyone who is treated

- Give a weight of $80/500 = .16$ to all untreated men

- Give a weight of $20/500 = .04$ to all untreated women

With these weights, let's see if we've eliminated the variation in gender between the groups. The treated group will still be 80% men—giving all the treated people equal weights won't change anything on that side. How about the untreated people? If we calculate the proportion male, using a value of 1 for men and 0 for women, the weighted mean for them is $(.16 \times 500 + .04 \times 0)/(.16 \times 500 + .04 \times 500) = 80\%$. Perfect![9]

Now that we've balanced gender between the treated and control groups, what's the treatment effect? It's still 72% of people who end up with jobs in the treated group—again, nothing changes there. But in the untreated group it's $(.16 \times 350 + .04 \times 275)/(.16 \times 500 + .04 \times 500) = 67\%$,[10] for a treatment effect of $.72 - .67 = 5$ percentage points. Still not bad, although not as good as the 9.5 percentage points we had before. Some of that gain was due to the back door through gender.

[9] Where do these numbers come from? $.16 \times 500$ is each man, who has a "man" value of 1, getting a weight of .16. So each man is $.16 \times 1$. Add up all 500 men to get $.16 \times 500$. Women have a "man" value of 0 and a weight of .04—add up 500 of those and you get $.04 \times 0$. Then in the denominator, it doesn't matter whether you're a man or a woman, your weight counts regardless. So we add up the 500 .16 weights for men and get $.16 \times 500$, and add that to the 500 .04 weights for women, $.04 \times 500$.

[10] That's 350 employed men with a weight of .16 each, plus 275 employed women with a weight of .04 each, divided by the sum of all weights—500 men with a weight of .16 each plus 500 women with a weight of .04 each.

14.3   Matching in Concept: A Single Matching Variable

I'VE SAID THERE ARE MANY WAYS TO DO MATCHING. SO WHAT ARE THEY? For a demonstration, we'll start by matching on a single variable and see what we can do with it. It's relatively uncommon to match on only one variable, but it's certainly a lot easier to think about, and lets us separate out a few key concepts from some important details I'll cover in the multivariate section.

This will be easier still if we have an example. Let's look at defaulting on credit card debt. Sounds fun. I have data on 30,000 credit card holders in Taiwan in 2005, their monthly bills, and their repayment status (pay on time, delay payment, etc.) in April through September of that year.[11] We want to know about the "stickiness" of credit problems, so we're going to look at the effect of being late on your payment in April (*LateApril*) on being late on your payment in September (*LateSept*). There are a bunch of back doors I'm not even going to try to address here, but one back door is the size of your bill in April, which should affect both your chances of being behind in April and September. So we'll be matching on the April bill (*BillApril*).

In performing our matching on a single variable, there will be some choices we'll make along the way:

1. What will our matching criteria be?

2. Are we *selecting matches* or *constructing a matched weighted sample*?

3. If we're *selecting matches*, how many?

4. If we're *constructing a matched weighted sample*, how will weights decay with distance?

5. What is the worst acceptable match?

(1) WHAT WILL OUR MATCHING CRITERIA BE? When performing matching, we are trying to select observations in the control group that are similar to those in the treated group (usually). But what does "similar to" even mean? We have to pick some sort of definition of similarity if we want to know what we're aiming for.[12]

There are two main approaches to picking a matching criterion: distance matching and propensity score matching.

[11] Moshe Lichman. UCI machine learning repository, 2013.

[12] Or, more often, if we want to know what the computer should be looking for.

Distance matching says "observations are similar if they have similar values of the matching variables." That's it. You want to minimize the *distance* (in terms of of how far the covariates are from each other) between the treatment observations and the control observations.

There's a strong intuitive sense here of why this would work— we're forcefully ensuring that the treatment and control groups have very little variation in the matching variables between them. This closes back doors. The diagram in Figure 14.1 looks familiar, and it's exactly the kind of thing that distance matching applies to. We have some back doors. We can measure at least one variable on each of the back doors. We match on those variables. Identification!

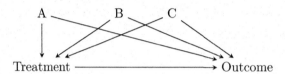

Figure 14.1: Standard Backdoor Causal Diagram for Distance Matching

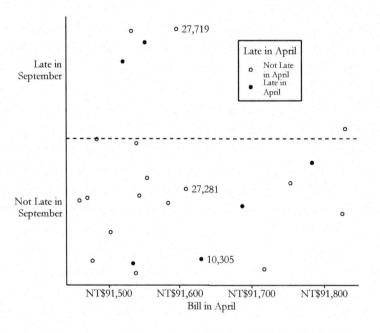

Figure 14.2: Matching on April 2005 Credit Card Bill Amount in Taiwanese Data (Data Range Limited)

Using our credit card debt example, let's pick one of the treated (late payment in April) observations: lucky row number 10,305. This person had a *BillApril* of 91,630 New Taiwan dollars (NT$), their payment was late in April (*LateApril* is true), but their payment was not late in September (*LateSept* is false).

Since our matching variable is *BillApril*, we'd be looking for untreated matching observations with *BillApril* values very close to 91,630. A control with a *BillApril* of NT\$113,023 (distance of $|91,630 - 113,023| = 21,393$) or 0 (distance of $|91,630 - 0| = 91,630$) wouldn't be ideal matches for this treated observation. Someone with a *BillApril* value of 91,613 (distance $|91,630 - 91,613| = 17$) might be a very good match, on the other hand. There's very little distance between them.

If we were to pick a single matching control observation for row number 10,305,[13] we'd pick row 27,281, which was *not* late in April (and so was untreated), and has a *BillApril* of NT\$91,609 (distance $|91,630 - 91,609| = 21$), which is the closest in the data to NT\$91,630 among the untreated observations. We can see this match graphically in Figure 14.2. We have our treated observation from row 10,305, which you can see near the center of the graph. The matches are the untreated observations that are closest to it on the x-axis (our matching variable). The observation from row 27,281 is just to the left—pretty close. Only a bit further away, also on the left, is the observation from row 27,719—we'll get to that one soon.

The other dominant approach to matching is propensity score matching. Propensity score matching says "observations are similar if they were equally likely to be treated," in other words have equal treatment propensity.[14,15]

Like with distance matching, there's a logic here too. We're not *really* interested in removing all variation in the matching variables, right? We're interested in identifying the effect of the treatment on the outcome, and we're concerned about the matching variables being on back doors.[16,17] Propensity score matching takes this idea seriously and figures that if you match on treatment propensity, you're good to go.[18]

The causal diagram in mind here looks like Figure 14.3. The matching variables A, B, and C are all on back doors, but those back doors go through *TreatmentPropensity*, the probability of treatment. The probability of treatment is unobservable, but we can estimate it in a pretty straightforward way using regression of treatment on the matching variables.[19]

Returning to our credit card example, if I use a logit regression of treatment on *BillApril* in thousands of New Taiwan dollars, I get an intercept of .091 and a coefficient on *BillApril* of .0003. If we are again looking for a match for our treated

[13] And picking a single match is not the only way to match, of course—we'll get there.

[14] I recommend Caliendo and Kopeinig (2008) as a very-readable guide for propensity score matching. I have taken guidance from them a few times in writing this chapter.

[15] Marco Caliendo and Sabine Kopeinig. Some practical guidance for the implementation of propensity score matching. *Journal of Economic Surveys*, 22(1): 31–72, 2008.

[16] And actually, matching on variables that don't close back doors, even if they predict treatment, can actually harm the estimate, as in Bhattacharya and Vogt (2007).

[17] As I'll explore more in the "Checking Match Quality" section, one implication of this approach is that propensity score matching doesn't really try to *balance* the matching variables. If a one-unit increase in A increases the probability of treatment by .1, while a one-unit increase in B increases the probability of treatment by .05, then propensity score matching says that someone with $A = 2$ and $B = 1$ is a great match for someone with $A = 1$ and $B = 3$. Distance matching would disagree.

[18] One implication of the propensity score approach is that it doesn't work quite as well if we can only close *some* of the back doors. If variables on the open back doors are related to variables you've matched on, then the propensity score can actually *worsen* the match quality along those still-open paths See Brooks and Ohsfeldt (2012) for a broader explanation.

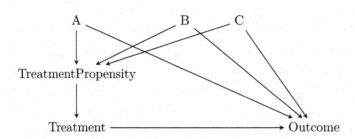

Figure 14.3: Causal Diagram Amenable to Propensity Score Matching

friend in row 10,305 with a *BillApril* of NT$91,630, we would first predict the probability of treatment for that person. Plugging in 91.630 for *BillApril* in thousands, we get a predicted probability of treatment of .116.[20]

Now we calculate the predicted probability of treatment for all the control observations, too. We want to find matches with a very similar predicted probability of treatment. Once again we find a great match in row 27,281, also with a predicted probability of .116. Hard to get closer than that![21]

(2) ARE WE *SELECTING MATCHES* OR *CONSTRUCTING A MATCHED WEIGHTED SAMPLE*? A few times I've referred to the matching process in the context of "finding matches," and other times I've referred to the construction of weights. Both of these represent different ways of matching the treatment and control groups. Neither is the default approach—they both have their pros and cons, and doing matching necessarily means some choices between different options, neither of which dominate the other.

The process of selecting matches means that we're picking control observations to either be *in* or *out* of the matched control sample. If you're a good enough match, you're in. If you're not, you're out. Everyone *in* the matched sample receives an equal weight.[22]

In the previous section, we selected a match. We looked at the treated observation on row 10,305, and then noticed that the control observation on row 27,281 was the closest in terms of both distance between matching variables and the value of the propensity score. If we were picking only a single match to be "in," we'd select row 27,281, and that row would get a weight of 1. Everyone else would be "out" and get a weight of 0 (at least unless they're the best match for a different treated observation).

That's just one method though. How else might we determine which controls are in and which are out? For that you'll need to look at step (3).

[19] This innocuous line hides one of the difficulties with using propensity score as opposed to distance—we need to choose a regression model. We just covered a whole chapter on regression, so you know what kind of issues we're inviting back.

[20] The use of logit or probit to estimate propensity scores is extremely standard. However, propensity scores can also be estimated without the parametric restrictions of logit and probit, and in large samples this may actually make the final estimate more precise. See Hahn (1998).

[21] With only one matching variable, the differences between distance and propensity score matching are trivial, so it's not surprising that we got the same best match both times. If I showed Figure 14.2 again it would look the exact same but with a relabeled *x*-axis. This will change when we expand to more than one matching variable.

How about the alternate approach of constructing a matched weighted sample? This process, like the process of selecting matches, entails looking at all the control observations and seeing how close they are to the treated observations. However, instead of simply being in or out, each control will receive a *different* weight depending on how close it is to a treated observation, or how handy it will be in making the matched control group look like the treated group.

Often, although not always, the matched-weighted-sample approach entails weighting observations more heavily the closer they are to a given treated observation, and less heavily the farther away they are. So for example we found that the difference in the matching variable between rows 10,305 and 27,281 was $|91,630 - 91,609| = 21$. But the *second*-best control match is in row 27,719, with a difference of $|91,630 - 91,596| = 34$. We might want to include both, but weight the observation with the smaller difference (27,281) more than the observation with the bigger difference (27,719). Maybe we give row 27,281 a weight of 1/21 and give row 27,719 a weight of 1/34.[23]

How exactly would we construct these weights? That's a question for step (4).

We can contrast these two approaches in Figure 14.4. On the left, we look for matches that are very near our treated observation. In this case, we're doing one-to-one matching and so only pick the observation in row 27,281, the closest match, as being "in." Everything else gets grayed out and doesn't count at all. On the right, we construct our comparison estimate for 10,305 using *all* the untreated data (the other treated observations get dropped here, but we'd want to come back later and do matching for them too). But they don't all count equally. Observations near our treated observation on the x-axis get counted a lot (big bubbles). As you move farther away they count less and less. Get far enough away and the weights fall to zero.

So which should we do? Should we select matches or should we give everyone different weights?[24] Naturally, there's not a single right answer—if there was I could skip telling you about both approaches.

The selecting-matches approach has some real intuitive appeal. You're really just constructing a comparable control group by picking the right people. It's also, often, a little easier to implement than something with more fine-tuned weights, and it's easier to see what your match-picking method is doing. You also avoid scenarios where you accidentally give one control

[22] Or at least they usually receive an equal weight, but in some methods this might not be the case. For example, if treatment observation 1 has a single matched control, but treatment observation 2 has three matched controls, some methods might give an equal weight to all four controls, but some other methods might count the control matched to treatment observation 1 three times as much as the three matched to treatment observation 2. Or, perhaps the same control is matched to multiple treatment observations and so gets counted multiple times.

[23] This exact weighting scheme wouldn't be *standard* but it does help you get the idea.

[24] Technically, selecting matches *is* a form of constructing a weighted matched sample—everyone getting weights of 0 or 1 is one way of making a weighted matched sample. But they still seem pretty different, so I'll treat them as different here.

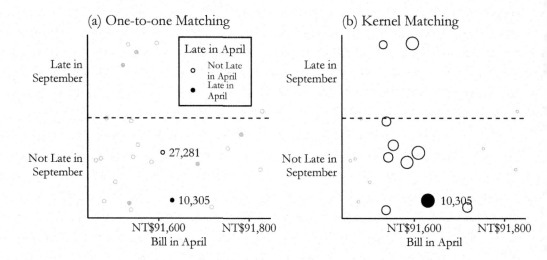

Figure 14.4: Selecting Matches or Weighting by Kernel

observation an astronomical weight—what if the difference between row 10,305 and row 27,281 was only .000001 instead of 21? It would receive an enormous 1/.000001 weight, and the weighted control-group mean would be almost entirely determined by that one observation! There are, of course, ways to handle that, but it's a concern you have to keep in mind.

On the other hand, the weighted-group approach has some nice statistical properties. It's a little less sensitive—"in" or "out" is a pretty stark distinction, and so tiny changes in things like caliper (we'll get to it) or measurement error can change results significantly, making the selecting-matches approach a little noisier. Varying weights also offer nice ways of accounting for the fact that different observations will simply have better or worse matches available in the data than others.

One place you should *definitely* use the weighted-group approach is if you're using propensity scores. Despite the practice being very popular, propensity scores probably shouldn't be used to select matches on an "in"/"out" basis.[25] The combination of these approaches can actually serve to *increase* the amount of matching-variable variation between treatment and control, due to the way that different variables can both affect propensity, so you might match a high value of one with a high value of the other.

(3) IF WE'RE *SELECTING MATCHES*, HOW MANY? If we do decide to go with the selecting-matches approach, we need to figure out *how many* control matches we want to pick for each treatment observation.

[25] Gary King and Richard Nielsen. Why propensity scores should not be used for matching. *Political Analysis*, 27(4):435–454, 2019.

The three main approaches here are to pick the *one* best match (one-to-one matching), to pick *the top* k of best matches (k-nearest-neighbor matching), or to pick *all* the acceptable matches, also known as radius matching since you accept every match in a radius of acceptability.

These procedures all pretty much work as you'd expect. With one-to-one matching, you pick the single best control match for each treated observation. With k-nearest-neighbor matching, you pick the best control match... and also the second-best, and the third best... until you get to k matches (or run out of potential matches). And picking all the acceptable matches just means deciding what's an acceptable match or not (see step 5), and then matching with all the acceptable matches.

With each of these procedures, we also need to decide if we're matching *with replacement* or *without replacement*. What do we do if a certain control observation turns out to be the best match (or one of the k-best matches, as appropriate) for two different treated observations? If we're matching without replacement, that control can only be a match for one of them, and the other treated observation needs to find someone different. If we're matching with replacement, we can use the same control multiple times, giving it a weight equal to the number of times it has matched.

How can we choose between these different options? It largely comes down to a tradeoff between bias and variance.

The fewer matches you have for each treated observation, the *better* those matches can be. By definition, the best match is a better match than the second-best match. So comparing, for example, one-to-one matching with 2-nearest-neighbors matching, the 2-nearest-neighbors match will include some observations in the control group that aren't *quite* as closely comparable to the treated group. This introduces bias because you can't claim quite as confidently that you've really closed the back doors.

On the other hand, the *more* matches you have for each treated observation, the *less noisy* your estimate of the control group mean can be, and so the more precise your treatment effect estimate can be. If you have 100 treated observations, the mean of 100 matched control observations will have a wider sampling distribution than the mean of *200* matched control observations. Simple as that! More matches means less sampling variation and so lower standard errors.

So the choice of how many matches to do will be based on how important bias and precision are in your estimate, and *how bad*

One-to-one matching. A matching procedure in which only one control observation is matched to each treatment observation, as opposed to "one-to-many" in which each treated observation might be matched to multiple control observations.

k-nearest-neighbor matching. A matching procedure in which the k best control matches are used as matches for each treatment observation.

the matches will get if you try to do more matches. If you have zillions of control observations and will be able to pick a whole bunch of super good matches for each treated observation, then you're likely not introducing much bias by allowing a bunch of matches, and you can reduce variance by doing so. So do it! But if your control group is tiny, your third-best match might be a pretty bad match and would introduce a lot of bias. Not worth it.

How about the with-replacement/without-replacement choice? A bias/variance tradeoff comes into play here, too. Matching with replacement ensures that each treated observation gets to use its best (or k best, or all acceptable) matches. This reduces bias because, again, this approach lets us pick the best matches. However, this approach means that we're using the same control observations over and over—each control observation has more influence on the mean, and so sampling variation will be larger (what happens if one *really good match* is matched to 30 treatments in one sample, but isn't there in a different sample?). In the extreme, imagine only having one control observation and matching it to all the controls—the sample mean for the controls would just be that one observation's outcome value, and would have a standard error that's just equal to the standard deviation of the outcome. Dividing by $\sqrt{N}$ to get the standard error doesn't do much if $N = 1$.

Matching with replacement does have something else to recommend it, though, in addition to having lower bias—it's not *order-dependent*. Say you're matching without replacement. Treated observations 1 and 2 both see control observation A as their best match. But the second-best match is B for observation 1, and C for observation 2. Who gets to match with A? If observation 1 does, then C becomes part of the control group. But if observation 2 does, then B becomes part of the control group. The makeup of the control group, and thus the estimate, depends on who we decide to "award" the good control A to. There are some principled methods for doing this (like giving the good control to the treated observation with the worse backup), but in general this is a bit arbitrary.

(4) IF WE'RE *CONSTRUCTING A MATCHED WEIGHTED SAMPLE*, HOW WILL WEIGHTS DECAY WITH DISTANCE? Both of the main approaches to matching—selecting a sample or constructing weights—have some important choices to make in terms of how they're done. In a matched weighted sample approach, we will be taking our measure of distance, or the propensity score,

and using it to weight each observation separately. But how can we take the distance or propensity score and turn it into a weight?

Once again, we have a few options. The two main approaches to using weights are *kernel matching*, or more broadly *kernel-based matching estimators* on one hand, and *inverse probability weighting* on the other hand.

Kernel-based matching estimators use a *kernel function* to produce weights. In the context of matching, kernel functions are functions that you give a *difference* to, and it returns a weight. The highest value is at 0 (no difference), and then the value smoothly declines as you move away from 0 in either direction.[26] Eventually you get to 0 and then the weight stays at 0. This approach weights better matches highly, less-good matches less, and bad matches not at all.

There are an infinite number of potential kernels we could use. A very popular one is the Epanechnikov kernel.[27] The Epanechnikov kernel has the benefit of being very simple to calculate:[28]

$$K(x) = \frac{3}{4}(1 - x^2) \qquad (14.2)$$

where $K(x)$ means "kernel function" and this function only applies in the bounds of x being between -1 and 1. The kernel is 0 outside of that. Figure 14.5 shows the Epanechnikov kernel graphically.

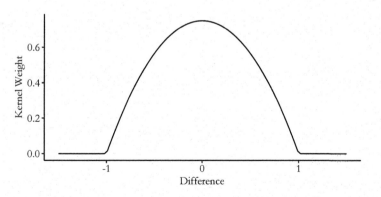

Note the hard-encoded range—this kernel only works from a difference of -1 to a difference of 1. It is standard, then, to standardize the distance before calculating differences to give to the kernel.[29] This ensures that you don't get different matching results just because you had a differently-scaled variable.

[26] An exception here is the uniform kernel, which uses the same weight for anyone in a certain range, and then suddenly drops to 0.

[27] The Epanechnikov kernel is popular not only for kernel-based matching but also other uses of kernels like estimating density functions.

[28] In addition to being simple to calculate (plenty of kernels are simple to calculate), the Epanechnikov kernel is popular because it minimizes asymptotic mean integrated squared error, which is a measure of how different a variable's estimated and actual densities are. You probably don't need to know this.

Figure 14.5: The Epanechnikov Kernel

[29] Standardizing means we subtract its mean and divide by its standard deviation. In this case the part about subtracting the mean doesn't actually matter since we're about to subtract one value from another, so that mean would have gotten subtracted out anyway.

Once the kernel gives you back your weights,[30] you can use them to calculate a weighted mean. Or... I did say *kernel-based* matching estimators. There are some other methods that start with the kernel but then use that kernel to estimate the mean in more complex ways. I'll cover those in the Estimation with Matched Data section below.

How does this work in our credit card debt example?[31] We already calculated two differences. The difference between the *BillApril* matching variable in rows 10,305 and 27,281 was $|91,630 - 91,609| = 21$, and the difference between row 10,305 and row 27,719 was $|91,630 - 91,596| = 34$. Next we standardize— the standard deviation of *BillApril* is 59.994, making those two differences into $21/59.994 = .35$ and $34/59.994 = .56$ standard deviations, respectively. Next we pass them through the Epanechnikov kernel, giving weights of $\frac{3}{4}(1 - .35^2) = .658$ and $\frac{3}{4}(1 - .56^2) = .515$, respectively. Now, we'd want to repeat this process for *all* the controls, but if we imagine these are the only two close enough to matter, then we compare the treated mean outcome *LateSept* value of 0 (false) for row 10,305 against the values of 0 for row 27,281 and 1 for 27,719, which we average together with our weights to get $(.658 \times 0 + .515 \times 1)/(.658 + .515) = .561$. Finally we get a treatment effect of $0 - .561 = -.561$.

How about *inverse probability weighting* then? Inverse probability weighting, which descends from work by Horvitz and Thompson (1952) via Hirano, Imbens, and Ridder (2003),[32] is specifically designed for use with a propensity score. Then, it weights each observation by the *inverse of the probability that it had the treatment value it had.* So if you were actually treated, and your estimated propensity score is .6, then your weight would be 1/.6. If you *weren't* actually treated and your estimated propensity score was .6, then your weight would be one divided by the chance that you *weren't* treated, or $1/(1 - .6) = 1/.4$.

We weight by the *inverse* of the probability of what you *actually are* to make the treated and control groups more similar. The treated-group observations that get the biggest weights are the ones that are *most like* the untreated group—the ones with small propensities (and thus big *inverse* propensities) least likely to have gotten treated who got treated anyway. Similarly, the control-group observations with the biggest weights are the ones most like the treated group, who were most likely to have gotten treated but didn't for some reason.

[30] And if a control observation is matched to multiple treatments, you then add all its weights up.

[31] We've finally come far enough along to estimate a treatment effect!

[32] Daniel G Horvitz and Donovan J Thompson. A generalization of sampling without replacement from a finite universe. *Journal of the American Statistical Association*, 47(260):663–685, 1952.; and Keisuke Hirano, Guido W Imbens, and Geert Ridder. Efficient estimation of average treatment effects using the estimated propensity score. *Econometrica*, 71 (4):1161–1189, 2003.

Inverse probability weighting has a few nice benefits to it. You don't have a lot of choices to make outside of specifying the propensity score estimation regression, so that's nice. Also, you can do inverse probability weighting without doing any sort of matching at all—no need to check each treated observation against each control, just estimate the propensity score and you already know the weight. Plus, as Hirano, Imbens, and Ridder (2003) show, weighting is the most precise way to estimate causal effects as long as you've got a big enough sample and a flexible way to estimate the propensity score.[33]

There are downsides as well, big surprise. In particular, sometimes you get a really unexpected treatment or non-treatment, and then the weights get huge. Imagine someone with a .999 propensity score who ends up not being treated—a one in a thousand chance (but with a big sample, likely to happen sometime). That person's weight would be $1/(1 - .999) = 1,000$. That's really big! This sort of thing can make inverse probability weighting unstable whenever propensity scores can be near 0 or 1.

There are fixes for this. The most common is simply to "trim" the data of any propensity scores too near 0 or 1. But this can make the standard errors wrong in ways that are a little tricky to fix. Other fixes include first turning the propensity score into an odds ratio ($p/(1 - p)$ instead of just p), and then, *within the treated and control groups separately*, scaling the weights so they all sum to 1.[34]

Let's bring inverse probability weighting to our credit card debt example. We previously calculated using a logit model that the treated observation in row 10,305 had a .116 probability of treatment, as did the control observation in 27,281. Onto that we can add row 27,719, which has a propensity of .116 as well (we did pick them for being similar, after all. With such similar *BillApril* values, differences in predicted probability of treatment don't show up until the fifth decimal place).

What weights does everyone get then? Row 10,305 is actually treated, so we give that observation a weight of 1 divided by the probability of treatment, or $1/.116 = 8.621$. It gets counted a lot because it was unlikely to be treated, and so looks like the untreated groups, but was treated.

The control rows will both get weights of 1 divided by the probability of *non*-treatment, or $1/(1 - .116) = 1.131$. They're likely to be untreated, and so aren't like the treated group, and

[33] Without making it seem like I've discovered the *one true path* or anything like that, when I do matching, or when people ask me how they should do matching, I usually suggest inverse probability weighting of the propensity score, or entropy balancing (described in the Entropy Balancing section in this chapter). As I've shown throughout the chapter, there are plenty of situations in which other methods would do better, but for me these are what I reach for first.

[34] Matias Busso, John Di-Nardo, and Justin McCrary. New evidence on the finite sample properties of propensity score reweighting and matching estimators. *Review of Economics and Statistics*, 96(5):885–897, 2014.

don't get weighted too heavily. From here we can get weighted means on both the treated and control sides.

(5) WHAT IS THE WORST ACCEPTABLE MATCH? The purpose of matching is to try to make the treated and control groups comparable by picking the control observations that are most similar to the treated observations. But what if there's a treated observation for which there *aren't* any control observations that are *at all* like it? We can't very well use matching to find a match for that treated observation then, can we?

Most approaches to matching use some sort of *caliper* or *bandwidth* to determine how far off a match can be before tossing it out.

The basic idea is pretty straightforward—pick a number. This is your caliper/bandwidth. If the distance, or the difference in propensity scores, is bigger in absolute value than that number, then you don't count that as a match. If that leaves you without *any* matches for a given treated observation, tough cookies. That observation gets dropped.

Usually, the caliper/bandwidth is defined in terms of *standard deviations* of the value you're matching on (or the standard deviation of the propensity score, if you're matching on propensity score), rather than the value itself, to avoid scaling issues. As mentioned in the previous section, in our credit card debt example, the treated observation 10,305 had a distance of .35 standard deviations of *BillApril* with control observation 27,281, and .56 standard deviations of difference with observation 27,719. If we had defined the caliper as being .5 of a standard deviation, then only one of those matches would count. The other one would be dropped.

Some matching approaches end up using calipers/bandwidths naturally. Any kernel-based matching approach will place a weight of 0 on any match that is far enough away that the kernel function you've chosen sets it to 0. For the Epanechnikov kernel that's a distance of 1 or greater.

Another matching approach that does this naturally is *exact matching*. In exact matching, the caliper/bandwidth is set to 0! No ifs, ands, or buts. If your matching variables differ *at all*, that's not a match. Usually, exact matching comes in the form of *coarsened exact matching*,[35] which is exact matching except that whenever you have a continuous variable as a matching variable you "coarsen" it into bins first. Otherwise you wouldn't be able to use continuous variables as matching variables, since it's pretty unlikely that someone else has, for

[35] Stefano M Iacus, Gary King, and Giuseppe Porro. Causal inference without balance checking: Coarsened exact matching. *Political Analysis*, 20(1):1–24, 2012.

example, the exact same *BillApril* as you, down to the hundredth of a penny. Coarsened exact matching is actually a pretty popular approach, for reasons I'll discuss in the Matching on Multiple Variables section.

How can we select a bandwidth if our method doesn't do it for us by helpfully setting it to zero? Uh, good question. There is, unfortunately, a tradeoff. And yes, you guessed it, it's a tradeoff between bias and variance.

The wider we make the bandwidth, the more potential matches we get. If we're allowing multiple matches, we'll get to calculate our treated and control means over more observations, making them more precise. If we're only allowing one match, then fewer observations will be dropped for being matchless, again letting us include more observations.

However, when we make the bandwidth wider we're allowing in more bad matches, which makes the quality of match worse and the idea that we're closing back doors less plausible. This brings bias back into the estimation.

There is a whole wide literature on "optimal bandwidth selection,"[36] which provides a whole lot of methods for picking bandwidths that best satisfy some criterion. Often this is something like "the bandwidth selection rule that minimizes the squared error comparing the sample control mean and the actual control mean." Software will often implement some optimal bandwidth selection procedure for you, but also it's a good idea to look into what those procedures are in the software documentation so you know what's going on.

In general, when making any sort of decision about matching, including this one, the choice is often between fewer, but better, matches that produce estimates with less bias but less precision, or more, but worse, matches that produce estimates with more bias but more precision.

AND THAT'S JUST ABOUT IT. While there are still some details left to go, those are the core conceptual questions you have to answer when deciding how to go about doing an estimation with matching. And maybe a few more detail-oriented questions too. This should at least give you an idea of what you're trying to do and why. But, naturally, there are more details to come.

[36] I was gonna cite some representative paper here, but honestly there isn't one. There's a lot of work on this. Head to your favorite journal article search service and look for "matching optimal bandwidth selection." Go wild, you crazy kids.

14.4   Multiple Matching Variables

PHEW, THAT SURE WAS A LOT IN THE PREVIOUS SECTION! Distance vs. propensity, numbers of matches, bias vs. variance tradeoff... no end of it. And that was all just about matching on *one* variable. Now that we're expanding to matching on *multiple* variables, as would normally be done, surely this is the point where we'll get way more complex.

Well, not really. In fact, going from matching on one variable to matching on multiple variables, there's really only one *core* question to answer: how do we, uh, turn those multiple variables into one variable so we can just do all the stuff from the previous section?

There are, of course, details to that question. But that's what it boils down to. We have many variables. Matching relies on picking something "as close as possible" or weighting by how close they are, which is a lot easier to do if "close" only has one definition. So we want to boil those many variables down into a single measure of difference. How can we do that? We could do it by distance, or by propensity score, or by exact matching. We could mix it up, requiring exact matching on some variables but using distance or propensity score for the others.[37] In each case, there are choices to make. That's what this section is about.

Throughout this section and the rest of the chapter, I'll be adding example code using data from Broockman (2013).[38] In this study, Broockman wants to look at the *intrinsic movitations* of American politicians. That is, what is the stuff they'll do even if there's no obvious reward to it? One example of such a motivation is that Black politicians in America may be especially interested in supporting the Black American community. This is part of a long line of work looking at whether politicians, in general, offer additional support to people like them.

To study this, Broockman conducts an experiment. In 2010 he sent a whole bunch of emails to state legislators (politicians), simply asking for information on unemployment benefits.[39] Each email was sent by the fictional "Tyrone Washington," which in the United States is a name that strongly suggests it belongs to a Black man.

The question is whether the legislator responds. How does this answer our question about intrinsic motivation? Because Broockman varies whether the letter-writer claims to live *in* the legislator's district, or in a city far away from the district. There's not much direct benefit to a legislator of answering an

[37] For example, maybe you're looking at the impact of some policy change on businesses, and want to match businesses that are more similar along a number of dimensions, but *require* that businesses are matched exactly to others in the same industry. That way you're not comparing a movie theater to a motorcycle repair shop, no matter how similar they look otherwise.

[38] David E Broockman. Black politicians are more intrinsically motivated to advance blacks' interests: A field experiment manipulating political incentives. *American Journal of Political Science*, 57(3):521–536, 2013.

[39] This is a pretty good choice of topic for the email, if you ask me. Answering the email could make a material improvement in the person's life, not just advance a policy issue.

out-of-district email. That person can't vote for you! But you still might do so out of a sense of duty or just being a nice person.

Broockman then asks: do Black legislators respond less often to out-of-district emails from Black people than in-district emails from Black people? Yes! Do non-Black legislators respond less to out-of-district emails from Black people than in-district emails from Black people? Also yes! Then the kicker: is the in-district/out-of-district difference *smaller* for Black legislators than non-Black ones? If so, that implies that Black legislators have additional intrinsic motivation to help the Black emailer, evidence in favor of the intrinsic-motivation hypothesis and the legislators-help-those-like-themselves hypothesis.[40]

Where does the matching come in? It's in that last step where we compare the in/out-of-district gap between Black and non-Black legislators. Those groups tend to be elected in very different kinds of places. Back doors abound. Matching can be used to make for a better comparison group. In the original study, Broockman used median household income in the district, the percentage of the district's population that is Black, and whether the legislator is a Democrat as matching variables.[41]

14.4.1   Distance   Matching   with   Multiple   Matching Variables

LET'S START WITH DISTANCE MEASURES. How can we take a lot of different matching variables and figure out which two observations are "closest?" We have to boil them down into a single distance measure.

Finally, we can wipe our brow: there's a single, largely agreed-upon approach to doing this: the *Mahalanobis distance*.[42]

The Mahalanobis distance is fairly straightforward. We'll start with a slightly simplified version of it. First, take each matching variable and divide its value by its standard deviation. Now, each matching variable has a standard deviation of 1. This makes sure that no variable ends up being weighted more heavily just because it's on a bigger scale.

After we've divided by the standard deviation, we can calculate distance. For a given treated observation A and a given control observation B, the Mahalanobis distance is the sum of the squares of all the differences between A and B. Then, after you've taken the sum, you take the square root. In other words, it's the sum of squared residuals we'd get if trying to predict the

[40] Even stronger evidence would repeat the same experiment but with a White letter-writer. We *should* see the same effect but in the opposite direction. If we don't, then the proper interpretation might instead be that Black legislators are just more helpful people on average.

[41] Specifically, he used coarsened exact matching followed by a regression. We'll talk about both of those steps in this chapter.

[42] I certainly won't say it's the *only* way to do it. And data scientists are scratching their heads right now because Mahalanobis isn't anywhere near the most common pick for them, Euclidian (which skips the step of dividing out shared variation—we'll get there) is. But come on, give a textbook author a victory where they can find one.

matching variables of A using the values of B, if the standard deviation of each matching variable was 1. That's it! That's the Mahalanobis distance.

Or it's a simplified version of it, anyway. The one piece I left out is that we're not just dividing each variable by its standard deviation. Instead, we're dividing out *the whole covariance matrix* from the squared values of the variables. If all the matching variables are unrelated to each other, this simplifies to just dividing each variable by its standard deviation. But if they *are* related to each other, this removes the relationships between the matching variables, making them independent before you match on them.

This requires some matrix algebra to express, but even if you don't know matrix algebra, all you really need to know is that $x'x$ can be read as "square all the xs, then add them up." Then you can figure out what's going on here if you squint and remember that taking the inverse ($^{-1}$) means "divide by." The Mahalanobis distance between two observations x_1 and x_2 is:

$$d(x_1, x_2) = \sqrt{(x_1 - x_2)'S^{-1}(x_1 - x_2)} \qquad (14.3)$$

where x_1 is a vector of all the matching variables for observation 1, similarly for x_2, and S is the covariance matrix for all the matching variables.

Why is it good for the relationships between different variables to be taken out? Because this keeps us from our matching relying *really strongly* on some latent characteristic that happens to show up a bunch of times. For example, say you were matching on beard length, level of arm hair, and score on a "masculinity index." All three of those things are, to one degree or another, related to "being a male." Without dividing out the covariance, you'd basically be matching on "being a male" three times. Redundant, and the matching process will probably refuse to match any males to any non-males, even if they're good matches on other characteristics. If we divide out the covariance, we're still matching on male-ness, but it doesn't count multiple times.

Returning once again to our credit card debt example, let's expand to two matching variables: *BillApril* and also the *Age*, in years, of the individuals. Comparing good ol' treated observation 10,305 and untreated observation 27,281, we already know that the distance between them in their *BillApril* is $|91,630 - 91,609| = 21$. The difference in ages is $|23 - 37| = -14$.

Let's calculate Mahalanobis distance first using the simplified approach where we ignore the relationship between the matching variables. The standard deviation of *BillApril* is NT\$59,554, and so after we divide all the values of *BillApril* by the standard deviation, we get a new difference of $|1.5386 - 1.5382| = .0004$ standard deviations. Not much! Then, the standard deviation of *Age* is 9.22, and so we end up with a distance of $|2.49 - 4.01| = 1.52$ standard deviations.

We have our differences in standard-deviation terms. Square them up and add them together to get $.0004^2 + 1.52^2 = 2.310$, and then take the square root—$\sqrt{2.310} = 1.520$—to get the Mahalanobis distance of 1.520. We would then compare this Mahalanobis distance to the distance between row 10,305 and all the other untreated rows, ending up in a match with whichever untreated observation leads to the lowest distance.

Let's repeat that, but properly, taking into account the step where we divide by the covariance matrix. We have a difference of 21 in *BillApril* and a difference of -14 in age. We put that next to its covariance matrix and get a Mahalanobis distance of

$$d(x_1, x_2) = \left(\begin{bmatrix} 21 \\ -14 \end{bmatrix}' \begin{bmatrix} 59,544^2 & 26,138 \\ 26,138 & 85 \end{bmatrix}^{-1} \begin{bmatrix} 21 \\ -14 \end{bmatrix} \right)^{1/2} =$$

$$\sqrt{.0000257 + 2.312} = 1.521 \quad (14.4)$$

Okay, not a huge difference from the simplified version in this particular case (1.521 vs. 1.520). But still definitely worth doing. It's not like we *knew* it wouldn't matter before we calculated it both ways.

The Mahalanobis distance has a convenient graphical interpretation, too. We're squaring stuff, adding it up, and then taking the square root. Ring any bells? Well, for two matching variables that's just the distance between those two points, as we know from Pythagoras' theorem. The Mahalanobis distance is the length of the hypotenuse of a triangle with the difference of the first matching variable for the base and the difference of the second matching variable for the height.

You can see the distance calculation in Figure 14.6. It's just how long a straight line between the two points is, after you standardize the axes. That's all! This generalizes to multiple dimensions, too. Add a third matching variable and you're still drawing a line between two points; it's just in three dimensions this time. Look around you at any two objects and imagine a

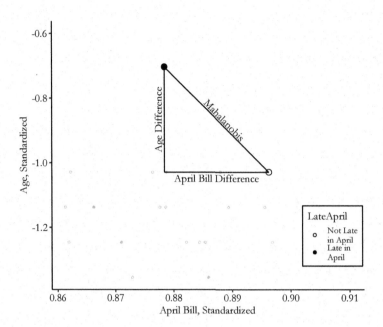

line connecting them. That's the Mahalanobis distance between them if you're matching on the three spatial dimensions: width, length, and height.

And so on and so forth. Four dimensions is a bit harder to visualize since we don't see in four dimensions, but it's the same idea. Add as many dimensions as you like.

This does lead us to one downside of Mahalanobis matching, and distance matching generally—the *curse of dimensionality*. The more matching variables you add, the "further away" things seem. Think of it this way—imagine a top-down map of an office building. This top-down view has only two dimensions: north/south and east/west. Wanda on the southwest corner of the building is right next to Taylor who's also on the southwest. A great match! Now move the camera to the side so we can see that same building in three dimensions: north/south, east/west, and up/down. Turns out Wanda's in the southwest of the 1st floor, but Taylor's in the southwest of the 22nd floor. No longer a great match.

The curse of dimensionality means that the more matching variables you add, the less likely you are to find an acceptable match for any given treated observation. There are a few ways to deal with this. One approach is to try to limit the number of matching variables, although that means we're not going to be closing those back doors—no bueno. Another is to just have a whole bunch of observations. Matches being hard to find is okay if you have eight zillion observations to look through to find one. That sounds good, but also, we only have so many observations available.

A third approach is to simply be a little more forgiving about match quality, or at least the value of the caliper/bandwidth, as the number of dimensions goes up. Sometimes people will just try different bandwidths until the amount of matches looks right. One more disciplined approach is to use your normal bandwidth, but divide the Mahalanobis score by the square root of the number of matching variables.[43]

Let's take Broockman to meet Mahalanobis![44] We'll be using one-to-one nearest neighbor matching with a Mahalanobis distance measure, calculated using our three matching variables (*medianhhincom*, *blackpercent*, *leg_democrat*). We'll be matching districts with Black legislators to those without (*leg_black*). There are two parts to the code here: the calculation of matching weights, and the estimation of the treatment effect on the outcome variable. I'll show how to do both. In the case of the Broockman example, the outcome variable is *responded*, whether the legislator responded to the email. Of course, the actual design was a bit more complex than just looking at the effect of *leg_black* on *responded*, but we'll get to that later. For now we're just showing the relationship between *leg_black* and *responded* after matching on *medianhhincom*, *blackpercent*, and *leg_democrat*.

[43] Colin Mahony. Effects of dimensionality on distance and probability density in climate space. The Seasons Alter, 2014.

[44] An important note about the code examples in this chapter: I'll be showing descriptions of how to use these commands, but when it comes to matching, the *default options* on some important things vary quite a bit from package to package. What standard errors to use? How to handle ties in nearest-neighbor matching, where multiple observations are equally good matches? Should it try to get an average treatment effect or an average treatment on the treated? Don't expect the different languages to produce the same results. If you want to see what details lead to the different results, and how you could make them consistent, you'll need to dive into the documentation of these commands.

```
1  # R CODE
2  library(Matching); library(tidyverse)
3  br <- causaldata::black_politicians
4
5  # Outcome
6  Y <- br %>%
7      pull(responded)
8  # Treatment
9  D <- br %>%
10     pull(leg_black)
11 # Matching variables
12 X <- br %>%
13     select(medianhhincom, blackpercent, leg_democrat) %>%
14     as.matrix()
15
16 # Weight = 2, oddly, denotes Mahalanobis distance
17 M <- Match(Y, D, X, Weight = 2, caliper = 1)
18
19 # See treatment effect estimate
20 summary(M)
21
22 # Get matched data for use elsewhere. Note that this approach will
23 # duplicate each observation for each time it was matched
24 matched_treated <- tibble(id = M$index.treated,
25                           weight = M$weights)
26 matched_control <- tibble(id = M$index.control,
```

```
27                              weight = M$weights)
28  matched_sets <- bind_rows(matched_treated,
29                              matched_control)
30  # Simplify to one row per observation
31  matched_sets <- matched_sets %>%
32                      group_by(id) %>%
33                      summarize(weight = sum(weight))
34  # And bring back to data
35  matched_br <- br %>%
36      mutate(id = row_number()) %>%
37      left_join(matched_sets, by = 'id')
38
39  # To be used like this! The standard errors here are wrong
40  lm(responded~leg_black, data = matched_br, weights = weight)
```

```
1  * STATA CODE
2  causaldata black_politicians.dta, use clear download
3
4  * Create an id variable based on row number,
5  * which will be used to locate match IDs
6  g id = _n
7
8  * teffects nnmatch does nearest-neighbor matching using first the outcome
9  * variable and the matching variables then the treatment variable.
10 * The generate option creates variables with the match identities
11 teffects nnmatch (responded medianhhincom blackpercent leg_democrat) ///
12     (leg_black), nneighbor(1) metric(mahalanobis) gen(match)
13 * Note this will produce a treatment effect by itself
14 * Doing anything else with the match in Stata is pretty tricky
15 * (or using a caliper). If you want either of those you may want to
16 * look into the mahapick or psmatch2 packages
17 * Although these won't give the most accurate standard errors
```

```
1  # PYTHON CODE
2  import pandas as pd
3  import numpy as np
4  # The more-popular matching tools in sklearn
5  # are more geared towards machine learning than statistical inference
6  from causalinference.causal import CausalModel
7  from causaldata import black_politicians
8  br = black_politicians.load_pandas().data
9
10 # Get our outcome, treatment, and matching variables
11 # We need these as numpy arrays
12 Y = br['responded'].to_numpy()
13 D = br['leg_black'].to_numpy()
14 X = br[['medianhhincom', 'blackpercent', 'leg_democrat']].to_numpy()
15
16 # Set up our model
17 M = CausalModel(Y, D, X)
18
19 # Fit, using Mahalanobis distance
20 M.est_via_matching(weights = 'maha', matches = 1)
21
22 print(M.estimates)
23 # Note it automatically calcultes average treatments on
24 # average, on treated, and on untreated/control (ATC)
```

PERHAPS MAHALANOBIS DOESN'T GO FAR ENOUGH. We can forget about trying to boil down a bunch of distances into one, and really any concerns about one variable being more important than the others, by making the case that *all the variables are infinitely important*. No mismatch at all. We're back to exact matching, and specifically coarsened exact matching, which I first mentioned in the section on matching concepts.[45]

In coarsened exact matching, something only counts as a match if it *exactly* matches on each matching variable. The "coarsened" part comes in because, if you have any continuous variables to match on, you need to "coarsen" them first by putting them into bins, rather than matching on exact values. Otherwise you'd never get any matches. What does this coarsening look like? Let's take *BillApril* from the credit card debt example. We might cut the variable up into ten deciles. So for our standby treated observation on row 10,305, instead of matching in that row's exact value of NT$91,630 (which no other observation shares), we'd recognize that the value of NT$91,630 is between the 80th and 90th percentiles of *BillApril*, and so just put it in the category of "between NT$63,153 (80th percentile) and NT$112,110 (90th percentile)" (which about 10% of all observations share).[46]

Once you've determined how to bin each continuous variable, you look for exact matches. Each treated observation is kept only if it finds at least one exact match, and is dropped otherwise. Each control observation is kept only if it finds at least one exact match as well, and given a weight corresponding to the number of treated-observation matches it has, divided by the number of control observations matched to that treated observation. Then all of that is further multiplied by the total number of matched control observations divided by the total number of matched treatment observations. In other words, $(MyTreatedMatches/MyControlMatches) \times (TotalControl$ $Matches/TotalTreatedMatches)$. This process ensures, without a doubt, that there are absolutely no differences in the matching variables (after binning) between the treated and control groups.[47]

Coarsened exact matching does require some serious firepower to work, especially once the curse of dimensionality comes into play. Zhao (2004) looks at an example of some fairly standard data,[48] and an attempt to match on twelve variables. This creates about six million cells after interacting all twelve

[45] Stefano M Iacus, Gary King, and Giuseppe Porro. Causal inference without balance checking: Coarsened exact matching. *Political Analysis*, 20(1):1–24, 2012.

[46] That's a pretty big bin. We'd want to think carefully about whether we think it's reasonable to say we've closed the back door through *BillApril* if the way we do it is comparing someone with a bill of NT$64,000 to someone with a bill of NT$112,000. Maybe we need narrower bins for this variable. Construct those bins carefully!

[47] This doesn't mean there are truly no differences. As mentioned above, any coarsened continuous variable matches exactly *on bins*, but not on the actual value, so there are differences in the value within bins. There could also be differences in the variables you *don't* use for matching. And if the curse of dimensionality encourages you to leave matching variables out when using coarsened exact matching, uh oh!

[48] Zhong Zhao. Using matching to estimate treatment effects: Data requirements, matching metrics, and Monte Carlo evidence. *Review of Economics and Statistics*, 86(1):91–107, 2004.

variables. The data set had far fewer than six million observations. Not everyone is going to find a match.

The need for a really big sample size is no joke or something you can skim over, either. Coarsened exact matching, if applied to moderately-sized samples (or any size sample with too many matching variables), can lead to lots of treated observations being dropped. Black, Lalkiya, and Lerner (2020)[49] replicated five coarsened exact match studies and found that lots of treatment observations ended up getting dropped, which made the treatment effect estimates much noisier, and can also lead the result to be a poor representation of the average treatment effect if certain kinds of treated observations are more likely to find matches than others. One of those studies was Broockman (2013)—uh oh! If you've been wondering why the list of matching variables in the Broockman (2013) code examples is so short, it's because even this list already leads quite a few observations to be dropped with coarsened exact matching, and any more would worsen the problem. Black, Lalkiya, and Lerner (2020) recommend forgetting the method entirely, although I wouldn't go that far. I would say that you should probably limit its use to huge datasets, and you *must* check how many treated observations you lose due to not finding any matches. If that's a lot, you should switch to another method.

The need for a lot of observations, and the relative ease and low computational requirements of calculation,[50] makes coarsened exact matching fairly popular in the world of big-data data science. There you might hear the exact matches referred to as "doppelgangers."[51] It's also fairly popular in the world of enormous administrative data sets, where a researcher has managed to get their hands on, say, governmental records on the entire population of a country.[52]

Let's look at Klemick, Mason, and Sullivan (2020) as one example of coarsened exact matching at work in neither of those contexts. They are interested in the effect of the Environmental Protection Agency's (EPA) Superfund Cleanup project on the health of children. This was a project where the EPA located areas subject to heavy environmental contamination and then cleaned them up. Did the cleanup process improve childrens' health?[53]

Their basic research design was to compare the levels of lead in the blood of children in a neighborhood before the cleanup to after, expecting not only that the lead level will drop after cleanup, but that it will improve more in areas very near the

[49] Bernard S Black, Parth Lalkiya, and Joshua Y Lerner. The trouble with coarsened exact matching. *Northwestern Law & Econ Research Paper Forthcoming*, 2020.

[50] No square roots or logit models to calculate—it matches or it doesn't. Plus, you don't need to really match each observation. You can just count the number of treated and untreated observations in each "bin" and create weights on that basis.

[51] One other place that coarsened exact matching pops up is in *missing-data imputation*. Is your data set incomplete, with some data points gone missing? You could look for some exact matches on all your *non-missing* variables and see what value of the missing variable *they* have. I discuss missing-data imputation more in Chapter 22.

[52] Yes, these data sets exist. You often see them for Nordic countries, where you're really tracked cradle-to-grave. They also exist in more limited forms in other countries. For example there are a few studies that use the entire IRS database of United States tax information. And before you ask, yes of *course* every researcher is deeply jealous of the people who have access to this data. Would I give a finger for access to this data? Hey, I've got nine others.

[53] Heather Klemick, Henry Mason, and Karen Sullivan. Superfund cleanups and children's lead exposure. *Journal of Environmental Economics and Management*, 100:1022–1089, 2020.

site (less than 2 kilometers) than it will in areas a little farther away (2—5 kilometers), which likely weren't as affected by the pollution when it was there.[54]

[54] This design is a preview of what you'll see in Chapter 18 on difference-in-differences.

As always, there are back doors between distance to a Superfund site and general health indicators, including blood lead levels. The authors close some of these back doors using coarsened exact matching at the neighborhood level, matching neighborhoods on the basis of which Superfund site they're near, the age of the housing in the area (what percentage of it was built before 1940), the percentage of the population receiving welfare assistance, the percentage who are African American, and the percentage of the population that receives the blood screenings used to measure the blood lead levels used as the dependent variable in the study.

They start with data on about 380,000 pepole living within 2 kilometers of a Superfund site, and 900,000 living 2—5 kilometers from a Superfund site. Then, the matching leads to a lot of observations being dropped. The sample after matching is about 201,000 from those living within 2 kilometers of a Superfund site, and 353,000 living 2—5 kilometers from a site.

The large drop in sample may be a concern, but we do see considerably reduced differences between those living close to the site or a little farther away. The matching variables look much more similar, of course, but so do other variables that potentially sit on back doors. Percentage Hispanic was not a matching variable, but half of the pre-matching difference disappears after matching. There are similar reductions in difference for traffic density and education levels.

What do they find? Superfund looks like it worked to reduce blood lead levels in children. Depending on specification, they find *without* doing any matching that lead levels were reduced by 5.1%—5.5%. *With* matching the results are a bit bigger, from 7.1%—8.3%.

How can we perform coarsened exact matching ourselves? Let's head back to Broockman (2013), who performed coarsened exact matching in the original paper.

```
1   # R CODE
2   library(cem); library(tidyverse)
3   br <- causaldata::black_politicians
4
5   # Limit to just the relevant variables and omit missings
6   brcem <- br %>%
7       select(responded, leg_black, medianhhincom,
8       blackpercent, leg_democrat) %>%
9       na.omit() %>%
```

```
10      as.data.frame() # Must be a data.frame, not a tibble
11
12 # Create breaks. Use quantiles to create quantile cuts or manually for
13 # evenly spaced (You can also skip this and let it do it automatically,
14 # although you MUST do it yourself for binary variables). Be sure
15 # to include the "edges" (max and min values). So! Six bins each:
16 inc_bins <- quantile(brcem$medianhhincom, (0:6)/6)
17
18 create_even_breaks <- function(x, n) {
19         minx <- min(x)
20         maxx <- max(x)
21
22         return(minx + ((0:n)/n)*(maxx-minx))
23 }
24
25 bp_bins <- create_even_breaks(brcem$blackpercent, 6)
26
27 # For binary, we specifically need two even bins
28 ld_bins <- create_even_breaks(brcem$leg_democrat,2)
29
30 # Make a list of bins
31 allbreaks <- list('medianhhincom' = inc_bins,
32                   'blackpercent' = bp_bins,
33                   'leg_democrat' = ld_bins)
34
35 # Match, being sure not to match on the outcome
36 # Note the baseline.group is the *treated* group
37 c <- cem(treatment = 'leg_black', data = brcem,
38         baseline.group = '1',
39         drop = 'responded',
40         cutpoints = allbreaks,
41         keep.all = TRUE)
42
43 # Get weights for other purposes
44 brcem <- brcem %>%
45     mutate(cem_weight = c$w)
46 lm(responded~leg_black, data = brcem, weights = cem_weight)
47
48 # Or use their estimation function att. Note there are many options
49 # for these functions including logit or machine-learing treatment
50 # estimation. Read the docs!
51 att(c, responded ~ leg_black, data = brcem)
```

```
1 * STATA CODE
2 * If necessary: ssc install cem
3 causaldata black_politicians.dta, use clear download
4
5 * Specify our matching variables and treatment. We can also specify
6 * how many bins for our variables - (#2) means 2 bins.
7 * We MUST DO this for binary variables - otherwise
8 * it will try to split this binary variable into more bins!
9 cem medianhhincom blackpercent leg_democrat(#2), tr(leg_black)
10
11 * Then we can use the cem_weights variable as iweights
12 reg responded leg_black [iweight = cem_weights]
```

```
1 # PYTHON CODE
2 import pandas as pd
3 import statsmodels.formula.api as sm
4 # There is a cem package but it doesn't seem to work that well
```

```
 5   # So we will do this by hand
 6   from causaldata import black_politicians
 7   br = black_politicians.load_pandas().data
 8
 9   # Create bins for our continuous matching variables
10   # cut creates evenly spaced bins
11   # while qcut cuts based on quantiles
12   br['inc_bins'] = pd.qcut(br['medianhhincom'], 6)
13   br['bp_bins'] = pd.qcut(br['blackpercent'], 6)
14
15   # Count how many treated and control observations
16   # are in each bin
17   treated = br.loc[br['leg_black'] == 1
18   ].groupby(['inc_bins','bp_bins','leg_democrat']
19   ).size().to_frame('treated')
20   control = br.loc[br['leg_black'] == 0
21   ].groupby(['inc_bins','bp_bins','leg_democrat']
22   ).size().to_frame('control')
23
24   # Merge those counts back in
25   br = br.join(treated, on = ['inc_bins','bp_bins','leg_democrat'])
26   br = br.join(control, on = ['inc_bins','bp_bins','leg_democrat'])
27
28   # For treated obs, weight is 1 if there are any control matches
29   br['weight'] = 0
30   br.loc[(br['leg_black'] == 1) & (br['control'] > 0), 'weight'] = 1
31   # For control obs, weight depends on total number of treated and control
32   # obs that found matches
33   totalcontrols = sum(br.loc[br['leg_black']==0]['treated'] > 0)
34   totaltreated = sum(br.loc[br['leg_black']==1]['control'] > 0)
35   # Then, control weights are treated/control in the bin,
36   # times control/treated overall
37   br['controlweights'] = (br['treated']/br['control']
38   )*(totalcontrols/totaltreated)
39   br.loc[(br['leg_black'] == 0), 'weight'] = br['controlweights']
40
41   # Now, use the weights to estimate the effect
42   m = sm.wls(formula = 'responded ~ leg_black',
43   weights = br['weight'],
44   data = br).fit()
45
46   m.summary()
```

14.4.2   Entropy Balancing

THERE IS A NEWCOMER in the world of distance matching that, while it hasn't been quite as popular in application relative to other options, seems to have some really nice properties. The newcomer is *entropy balancing*, which grows out of Hainmueller (2012).[55]

In other matching methods, you aggregate together the matching variables in some way and match on *that*. Then you hope that the matching you did removed any differences between treatment and control. And as I'll discuss in the Checking

[55] Jens Hainmueller. Entropy balancing for causal effects: A multivariate reweighting method to produce balanced samples in observational studies. *Political Analysis*, 20(1): 25–46, 2012.

Match Quality section, you look at whether any differences remain and cross your fingers that they're gone.

The basic concept of entropy balancing is this: instead of aggregating, matching, and hoping, we instead *enforce restrictions* on the distance between treatment and control. We say "no difference between treatment and control for these moment conditions for these matching variables," where moment conditions are things like means, variances, skew, and so on. For example "no difference between treatment and control for the mean of X_1." You can add a lot of those restrictions. Add another restriction that the mean of X_2 must be the same, and X_3, and so on. The descriptive statistics don't have to be means, either. Entropy balancing lets you require that things like the variance or skewness are the same, too, if you like.[56]

Once you have your set of restrictions, entropy balancing gets to work searching for a set of weights that satisfies those restrictions. As long as the restrictions are *possible* to satisfy, you'll get them, too.[57] That's just about it! Entropy balancing gives the assured-no-difference result that coarsened exact matching does, but without having to limit the number of matching variables or drop a bunch of treated observations. Why didn't we do this before?

This section is short for two reasons: first, the intuitive explanation of entropy balancing is quite simple, but to go into any sort of detail would require going a *lot* more complex. And second, because entropy balancing is not as well-known as other approaches at the time of this writing. Perhaps this is because it requires a fair bit more technical detail than other methods.[58]

But it's still worth learning about. Perhaps it will become more popular and expected in the future. Perhaps you'll use it and *that will make it* more popular and expected in the future. Or at the very least, it can act as a stand-in for the many other matching approaches that didn't make this book. And there are a few. What, the eight million combinations of options for the matching procedures we have aren't enough? No! There's genetic matching, subclassification, optimal matching, etc. etc..

Let's code up an entropy balance estimation. The examples will use R and Stata. There is a working Python library **empirical_calibration** that includes entropy balancing, but you're in for a ride, as even installing it is not as straightforward as other packages. I'll leave that one out for now, but you can pursue it on your own.[59]

[56] The way I think of entropy balancing myself is "sorta like the method of moments but for matching." That's a line that will be complete nonsense to most people reading this. But if it's not nonsense to you, then it took me all of nine words to explain entropy balancing pretty accurately. People do actually use method of moments to generate matching weights in a very similar manner to entropy balancing, too.

[57] If you can recall your multivariate calculus, it's really just applying a Lagrange multiplier.

[58] Also, more of that detail requires trust in your software programmer—what, are *you* going to code up the weight-searching method? Okay, fine, maybe you are. But for most people it's a bit more out of their reach than the other methods in this chapter. "I enforced moment conditions" also sounds a bit more like magic than "I found people with similar values of the matching variables" if you're trying to explain what you're doing to someone less well-versed in statistics.

[59] See   https://github.com/google/empirical_calibration.

```
 1 || # R CODE
 2 || library(ebal); library(tidyverse); library(modelsummary)
 3 || br <- causaldata::black_politicians
 4 ||
 5 || # Outcome
 6 || Y <- br %>%
 7 ||     pull(responded)
 8 || # Treatment
 9 || D <- br %>%
10 ||     pull(leg_black)
11 || # Matching variables
12 || X <- br %>%
13 ||     select(medianhhincom, blackpercent, leg_democrat) %>%
14 ||     # Add square terms to match variances if we like
15 ||     mutate(incsq = medianhhincom^2,
16 ||     bpsq = blackpercent^2) %>%
17 ||     as.matrix()
18 ||
19 || eb <- ebalance(D, X)
20 ||
21 || # Get weights for usage elsewhere
22 || # Noting that this contains only control weights
23 || br_treat <- br %>%
24 ||     filter(leg_black == 1) %>%
25 ||     mutate(weights = 1)
26 || br_con <- br %>%
27 ||     filter(leg_black == 0) %>%
28 ||     mutate(weights = eb$w)
29 || br <- bind_rows(br_treat, br_con)
30 ||
31 || m <- lm(responded ~ leg_black, data = br, weights = weights)
32 || msummary(m, stars = c('*' = .1, '**' = .05, '***' = .01))
```

```
 1 || * STATA CODE
 2 || * If necessary: ssc install ebalance
 3 || causaldata black_politicians.dta, use clear download
 4 ||
 5 || * Specify the treatment and matching variables
 6 || * And then in targets() specify which moments to match
 7 || * 1 for means, 2 for variances, 3 for skew
 8 || * Let's do means and variances for our continuous variables
 9 || * and just means for our binary matching variable (leg_democrat)
10 || * and store the resulting weights in wt
11 || ebalance leg_black medianhhincom blackpercent leg_democrat, ///
12 ||     targets(2 2 1) g(wt)
13 ||
14 || * Use pweight = wt to adjust estimates
15 || reg responded leg_black [pw = wt]
```

14.5   Propensity Score Weighting with Multiple Matching Variables

THE PROPENSITY SCORE IS PROBABLY THE MOST COMMON WAY of aggregating multiple matching variables into a single value that can be matched on. Remember, the propensity score is the estimated probability that a given observation would have gotten treated. Propensity score matching often means selecting a set of matched control observations with similar values of the propensity score. However, as previously mentioned, modern advice tends to be towards using the propensity score to conduct inverse probability weighting.[60]

In this section I'll focus on the estimation of the propensity score itself. Then, as suggested by the section title, you'll go on to use these propensity scores to construct matching weights, rather than select a set of matched observations. The specific details on how those weights are constructed from the propensity score will be shown in the code examples, and also detailed further in the "Matching and Treatment Effects" section at the end of the chapter, since the calculation of weights depends on the kind of treatment effect you want.

PROPENSITY SCORES ARE USUALLY ESTIMATED BY REGRESSION. This is highly convenient for me because I just wrote a whole dang chapter about regression, and you can just go read that.[61]

Specifically, propensity scores are usually estimated by logit or probit regression.[62] The main work of doing propensity score estimation by regression is in constructing the regression model.

What choices are there to make in constructing the regression model? You've got your standard model-construction problems— anything that's a determinant of treatment should definitely be in there. But there are also important functional form concerns. Which matching variables should be included with polynomials? Or interactions?

You're trying to predict something here—the propensity score— but you do want to keep your thinking in the realm of causal diagrams. After all, you're trying to close back doors here. So including things as predictors if they're on back doors, and excluding them if they're on front doors, is still key when constructing your model.

You're not totally in the dark (or reliant entirely on theory) for the selection of your model, though. After all, a good propensity score should serve the purpose of closing back doors, right? We can check this. Back doors in Figure 14.3 are of the form $Treatment \leftarrow TreatmentPropensity \leftarrow A \rightarrow Outcome$. Controlling for $TreatmentPropensity$ should close all doors between $Treatment$ and the matching variable A. So, as Dehejia and Wahba (2002) suggest,[63] you can split the propensity score up into bins. Then, within each bin (i.e., controlling for $TreatmentPropensity$), you can check whether each matching

[60] Gary King and Richard Nielsen. Why propensity scores should not be used for matching. *Political Analysis*, 27(4):435–454, 2019.

Inverse probability weights. Matching weights based on the propensity score, often one divided by the propensity of receiving the level of treatment that was actually received.

[61] And if you figured you'd just read the matching chapter and skip regression because of how long the regression chapter is, *gotcha*.

[62] The negatives of the linear probability model are very evident here, since you're using the predicted values, so anything outside the 0 or 1 range is unusable. So the linear probability model is out.

[63] Rajeev H Dehejia and Sadek Wahba. Propensity score-matching methods for nonexperimental causal studies. *Review of Economics and Statistics*, 84(1): 151–161, 2002.

variable is related to treatment. If it is, you might want to try adding some more polynomial or interaction terms for the offending matching variables. This is called a "stratification test," and is a form of balance checking, which I'll cover more in the Balance Checking section below.

Once you've estimated the logit or probit model, you get the predicted probabilities,[64] and that's it! That's your propensity score.

[64] Careful! Using a standard predict function with a logit or probit won't default to predicting the probabilities in some languages, but will instead give you the index. Check the documentation of your predict function, and also check the values of what you get to make sure it's between 0 and 1.

```r
1   # R CODE
2   library(causalweight); library(tidyverse)
3   br <- causaldata::black_politicians
4
5   # We can estimate our own propensity score
6   m <- glm(leg_black ~ medianhhincom + blackpercent + leg_democrat,
7            data = br, family = binomial(link = 'logit'))
8   # Get predicted values
9   br <- br %>%
10       mutate(ps = predict(m, type = 'response'))
11  # "Trim" control observations outside of
12  # treated propensity score range
13  # (we'll discuss this later in Common Support)
14  minps <- br %>%
15       filter(leg_black == 1) %>%
16       pull(ps) %>%
17       min(na.rm = TRUE)
18  maxps <- br %>%
19       filter(leg_black == 1) %>%
20       pull(ps) %>%
21       max(na.rm = TRUE)
22  br <- br %>%
23       filter(ps >= minps & ps <= maxps)
24
25  # Create IPW weights
26  br <- br %>%
27       mutate(ipw = case_when(
28       leg_black == 1 ~ 1/ps,
29       leg_black == 0 ~ 1/(1-ps)))
30
31  # And use to weight regressions (The standard errors will be wrong
32  # here unless we bootstrap the whole process - See the code examples
33  # from the doubly robust estimation section or the simulation chapter)
34  lm(responded ~ leg_black, data = br, weights = ipw)
35
36  # Or we can use the causalweight package!
37  # First, pull out our variables
38  # Outcome
39  Y <- br %>%
40       pull(responded)
41  # Treatment
42  D <- br %>%
43       pull(leg_black)
44  # Matching variables
45  X <- br %>%
46       select(medianhhincom, blackpercent, leg_democrat) %>%
47       as.matrix()
48
```

```
49   # Note by default this produces average treatment effect,
50   # not average treatment on the treated, and trims propensity
51   # scores based on extreme values rather than matching treated range
52   IPW <- treatweight(Y, D, X, trim = .001, logit = TRUE)
53
54   # Estimate and SE
55   IPW$effect
56   IPW$se
```

```
1    * STATA CODE
2    causaldata black_politicians.dta, use clear download
3
4    * We can estimate our own propensity score with probit or logit
5    logit leg_black medianhhincom blackpercent leg_democrat
6    * Get predicted values for our propensity score
7    predict ps, pr
8
9    * "Trim" control observations outside of
10   * treated propensity score range
11   * (we'll discuss this later in Common Support)
12   summ ps if leg_black
13   replace ps = . if ps < r(min) | ps > r(max)
14
15   * Get inverse probability weights
16   g ipw = 1/ps if leg_black
17   replace ipw = 1/(1-ps) if !leg_black
18
19   * Use inverse weights in regression
20   reg responded leg_black [pw = ipw]
21   * (or see the bootstrap standard error example in the section on doubly
22   * robust estimation, or the simulation chapter) Note that simple vce(bootstrap)
23   * won't work since it won't bootstrap the matching process
24
25   * Or do everything in teffects
26   * (This will use improved standard errors,
27   * unless you've done bootstrap,
28   * but won't let us do the trim)
29   teffects ipw (responded) (leg_black medianhhincom ///
30       blackpercent leg_democrat, logit)
```

```
1    # PYTHON CODE
2    import pandas as pd
3    import numpy as np
4    import statsmodels.formula.api as sm
5    # The more-popular matching tools in sklearn
6    # are more geared towards machine learning than statistical inference
7    from causalinference.causal import CausalModel
8    from causaldata import black_politicians
9    br = black_politicians.load_pandas().data
10
11   # Get our outcome, treatment, and matching variables as numpy arrays
12   Y = br['responded'].to_numpy()
13   D = br['leg_black'].to_numpy()
14   X = br[['medianhhincom', 'blackpercent', 'leg_democrat']].to_numpy()
15
16   # Set up our model
17   M = CausalModel(Y, D, X)
18
19   # Estimate the propensity score using logit
20   M.est_propensity()
```

```
21
22   # Trim the score with improved algorithm trim_s to improve balance
23   M.trim_s()
24
25   # If we want to use the scores elsewhere, export them
26   # (we could have also done this with sm.Logit)
27   br['ps'] = M.propensity['fitted']
28
29   # We can estimate the effect directly (note this uses "doubly robust" methods
30   # as will be later described, which is why it doesn't match the sm.wls result)
31   M.est_via_weighting()
32
33   print(M.estimates)
34
35   # Or we can do our own weighting
36   br['ipw'] = br['leg_black']*(1/br['ps']
37   ) + (1-br['leg_black'])*(1/(1-br['ps']))
38
39   # Now, use the weights to estimate the effect (this will produce
40   # incorrect standard errors unless we bootstrap the whole process,
41   # as in the doubly robust section later, or the Simulation chapter)
42   m = sm.wls(formula = 'responded ~ leg_black',
43   weights = br['ipw'],data = br).fit()
44
45   m.summary()
```

THERE ARE OTHER WAYS TO ESTIMATE THE PROPENSITY SCORE. Really, any method you could use to run a regression of a binary variable on multiple predictors and get a predicted value between 0 and 1 *could* work. That doesn't mean that all of those methods would be good ideas.

Other approaches have turned out to be particularly useful among the bevy of alternatives. In general, the methods aside from logit and probit that are useful manage to do something that logit and probit can't.[65] Two things that logit and probit have trouble with are *high degrees of nonlinearity* and *high dimensions.*

[65] Makes sense.

High degrees of nonlinearity is a problem? Aren't logit and probit *nonlinear* regression models? Well, yes, but they do require linearity *in the index function,* if you look back to the section on logit and probit in Chapter 13. You can add polynomial terms, of course. But how many? Is it enough? Too many and things will start to get weird.

High dimensions occur when you have *lots* of matching variables. Or, alternately, when you have *lots of interactions* between all your matching variables in your regression. Again, this is stuff that it's theoretically possible for regression to handle, but as the number of terms creeps higher and higher, regression ceases to be the best tool for the job.

So what is the best tool for the job? There are lots of options. One that's relatively well-known in health fields is high-dimensional propensity score estimation,[66] which just picks a set of "best" covariates from a long list based on theoretical importance, and which seem to predict really well, before running logit.

A bit further afield, but seeing good results, are machine learning methods. Machine learning is great at predicting stuff given data that has a lot of dimensions. And predicting treatment using a lot of dimensions is exactly what we want to do.

Two popular approaches to propensity score estimation using machine learning methods are regularized regression, as discussed in Chapter 13, where you toss everything in and let a penalty parameter decide what to down-weight or throw out, and boosted regression.

Boosted regression is an iterative method that starts with a binary regression model (like a logit), checks which observations are particularly poorly predicted, and then runs itself again, weighting the poorly-estimated observations more heavily so the prediction model pays more attention to them, reducing their prediction error. Then it continues doing this a bunch of times. In the end, you combine all the different models you ran to produce some sort of aggregated score. In the context of propensity scores, this leads to pretty good estimates of the propensity score.[67]

14.6   Assumptions for Matching

14.6.1   Conditional Independence Assumption

LIKE WITH ANY STATISTICAL METHOD, however well-thought out our methods are, they always rely on assumptions. Using matching can relax some assumptions, like linearity (unless we're using a regression to estimate a propensity score). But that certainly doesn't make it assumption-free.

Some assumptions carry over from our other work. The main one of those is the *conditional independence assumption*. Remember in the Regression chapter where I talked about omitted variable bias being the regression lingo for "open back doors"? Well, the conditional independence assumption is matching lingo for "all the back doors are closed."

In other words, the conditional independence assumption says that the set of matching variables you've chosen is enough to

[66] Sebastian Schneeweiss, Jeremy A Rassen, Robert J Glynn, Jerry Avorn, Helen Mogun, and M Alan Brookhart. High-dimensional propensity score adjustment in studies of treatment effects using health care claims data. *Epidemiology*, 20(4):512, 2009.

[67] Beth Ann Griffin, Daniel F McCaffrey, Daniel Almirall, Lane F Burgette, and Claude Messan Setodji. Chasing balance and other recommendations for improving nonparametric propensity score models. *Journal of Causal Inference*, 5(2), 2017.

close all back doors. The entirety of the relationship between treatment and outcome is either one of the causal front-door paths you want, or is due to a variable that you've measured and included as a matching variable.

That's just new lingo for an assumption we've already covered. What else do we need to think about?

14.6.2   Common Support

THE MATCHING PROCESS, NO MATTER HOW YOU DO IT, ASSUMES THAT THERE ARE APPROPRIATE CONTROL OBSERVATIONS TO MATCH WITH. But what if there aren't? Chaos ensues! This is called a failure of *common support*.

Why is this a problem? Let's take an extreme example. You're interested in the effect of early-release programs on recidivism. These are programs that allow people who have been sentenced to prison time to leave before their sentence is up, often under certain terms and conditions.[68] You've got a group of treated ex-prisoners who were released early, and a group of untreated ex-prisoners who served their full sentence, and information on whether they committed any additional crimes in the ten years after they got out.

[68] Parole is an example of an early-release program.

You figure that the prisoner's behavior while in prison might be related both to getting an early release and to their chances of recidivism. So you match on a "behavior score" given by the prison review board that decides whether someone gets early release.

But uh-oh! Turns out that the review board score, which goes from 1 to 10, translates too directly to early release. Everyone with an 8—10 gets early release, and everyone with a 1—7 doesn't. So when you go looking for matches for your treated observations, you end up trying to find ex-prisoners who got a score of 8—10 but didn't get early release. And there aren't any of those. The analysis lacks *common support*—there simply aren't any comparable control observations.

Or imagine that instead of trying to select matches, you're constructing weights. What weight do you give to the 1—7s so that the control group is comparable to the treatment group? Uh, zero I guess. So everything gets a weight of zero and... you again have nothing to compare.

The problem is still there even if it isn't quite as stark. Maybe all 9—10s get early release while some 8s do and other 8s don't. We still don't have any good matches for the 9—10s. Or maybe there *are* a handful of 9s and 10s that don't get early release. Now we have matches, but how heavily do we want to rely on that tiny number of matches?

The assumption of common support says that there must be *substantial* overlap in the distributions of the matching variables comparing the treated and control observations. Or in the context of propensity scores, there must be substantial overlap in the distribution of the propensity score.

COMMON SUPPORT IS FAIRLY EASY TO CHECK FOR, and there are a few ways to do it.

The first approach to checking for common support is simply to look at the distribution of a variable for the treated group against the distribution of that same variable for the untreated group, as in Figure 14.7 (a), which shows the distribution of the percentage Black in district (*blackpercent*) matching variable in our Broockman (2013) study.

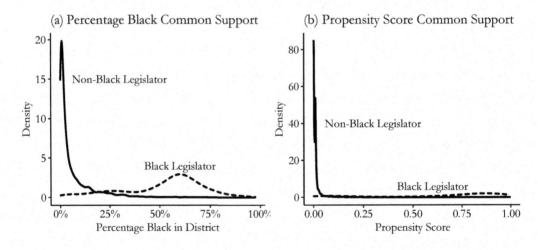

Figure 14.7: Common Support of the Percentage Black Variable and the Propensity Score in Broockman (2013) Data

Is there good common support for this variable? It's a bit iffy. It's not a problem that the distributions are *different*—we can use matching or re-weighting to scale up the parts of the control distribution until they're the same as the treated distribution... *assuming that those parts of the control distribution exist to be scaled up*. If there's nobody home at that part of the distribution, not much we can do. And there are certainly some regions here where there's not much to compare to. There aren't that many districts with a more than 50% Black population but a non-Black legislator. That's a problem for common support, as many of the Black legislators we do have represent areas with a 50%—80% Black population.

If there *are* areas with a 50%—80% Black population but a non-Black legislator, then those areas will receive huge weights and increase noise in the estimate. Or if there *aren't* any, then those areas with 50%—80% Black population and a Black legislator will have to be dropped. And that's the bulk of the treated data! The comparison won't be all that representative of the actual groups with Black legislators.

This same approach of contrasting the distribution works for propensity scores. In Figure 14.7 (b) you can see the distribution of the propensity score for the treated and control groups. As you would expect, the distribution is further to the right for the treated group than the control group.[69] That's not a problem. The problem arises if there are big chunks of the distribution for one group (particularly the treated group) where there is *no or very very little* weight in the control group. Again, we do have some issues here, where there simply aren't that many districts with non-Black legislators but high propensity scores. In fact, the propensity is so bunched down near 0 for that group that the density in that region is enormous, squashing down the rest of the graph and making it hard to see.

In the context of propensity scores, we may as well take this opportunity to check something else, too—in propensity score estimation we need to assume that the true propensity score is never exactly 0 or 1.[70] If we see big parts of the distribution right around 0 or 1, this should raise a concern.

Another way to check for common support when you're selecting matches is simply to see how successful you are at finding matches. If 1% of your treated observations fail to find an acceptable match, that's pretty good. If 90% of your treated observations fail to find an acceptable match, then you lack common support.

WHAT TO DO ABOUT SUPPORT? The first step we can take in dealing with support is to avoid trying to match where we don't have any support. That is, we want to drop observations that don't match well.

Some matching methods do this for us. If we're picking matches with a caliper, then bad matches won't be selected as matches. Observations without common support won't make it to the matched sample. Done!

Other methods require a little extra work after the fact. Propensity score approaches, especially inverse probability weighting, won't automatically drop observations outside the range of common support. For this reason it's common to "trim" the

[69] If it weren't that would be weird, as it would imply that the untreated group was more likely to be treated than the treated group.

[70] Why? It's easy to see why when we're weighting—if you're a treated observation with a 100% probability of treatment, then your weight would be 1/0. Oops! Divide by 0. Similarly, an untreated observation with 0% probability of treatment would get 1/0.

propensity score. For the treated and control groups separately, look for the range of values of the propensity score where the density distribution is zero (or close to zero, below some cut-off value). Then, drop all observations in the *other* group that have propensity scores in the range that doesn't exist in *your* group.[71] In other words, limit the data to the range of common support.

Figure 14.8 repeats Figure 14.7 (b), but then trims to the region of common support, requiring at least ten control observations in a given propensity score bin .02 wide.[72] In the data we have left, every observation has at least a few observations that have comparable propensity scores. We now know that when we weight to make each group more like the other, we're not trying to jam in truly incomparable observations. Only the ones we *can* use to build a comparable group.

[71] Marco Caliendo and Sabine Kopeinig. Some practical guidance for the implementation of propensity score matching. *Journal of Economic Surveys*, 22(1): 31–72, 2008.

[72] This "minimum to count as common support" approach I have here is arbitrary. I made it up. There's not really a rule of thumb on how much is enough support if it's nonzero.

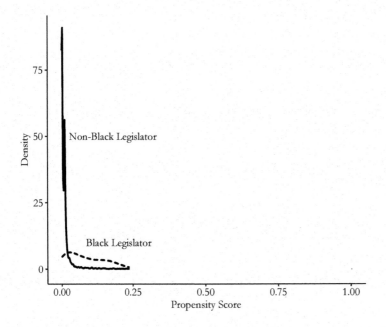

Figure 14.8: Propensity Scores After Trimming to Common Support

This is not the only way to trim for common support. A common approach is to trim outliers—any observation with a propensity score in the top or bottom X% gets trimmed. Or instead of working with percentiles, simply trim anything near 0 or 1, maybe anything from 0—.01 and .99—1—remember, we are assuming we don't have any true 0 or 1 propensities anyway. Another is to just look for *any* match, rather than a sufficient number of matches in the region. So for example, look at the minimum and maximum propensity score for the treated observations, and get rid of any control observations outside that range.

All of these approaches to enforcing common support entail dropping data. This brings us to the stinging welt of the whole thing—if this process leads to us dropping a *lot* of observations, especially treated observations, then at best we can say that we'll be able to generate an estimate of the treatment effect *among the individuals that match well,* whatever that means, and at worst we can say the matching has failed. Matching relies on the existence of comparable observations, and those might not be around. Maybe a different matching method would work, but more likely matching isn't going to be able to close your back doors.

14.6.3   Balance

THE GOAL OF MATCHING IS TO CLOSE BACK DOORS THROUGH THE MATCHING VARIABLES. So, uh, did it? *Balance* is the assumption that our approach to selecting a matched group has closed back doors for the variables we're interested in. In theory (and in a perfect world), the weights we select should lead to treated and control groups with the exact same values of the back-door variables. No variation between treatment and control for that back-door variable means there's no back door any more. It has been closed.

"The exact same values" isn't really going to happen though. So how can we check if we have *acceptable* levels of balance? And what happens if we don't?[73] Like anything else that allows for worse matches, bad balance leads to bias. If the balance is bad, the back doors aren't being closed, so the treatment effect gets contaminated with all those nasty back doors.

Thankfully we don't have to do much theorizing about whether there's a problem. We can check in the matched data we have whether there are meaningful differences between the treated and control groups in variables for which there should be no differences—variables that should be balanced.

And if we find that we do have a balance problem, by any of these methods? Then it's back to the matching procedure. It's not uncommon for matching-based estimation to be iterative. Match the data, check balance, and if it's bad, change something about the matching procedure (more variables, different

[73] Most of this section also applies to other contexts in which there *shouldn't be* any differences between two groups in a set of variables. You could apply these tools to, for example, a randomized experiment, where we'd expect the randomization to lead to balance. Or something like Regression Discontinuity in Chapter 20, where there should be balance for a certain subset of the data.

functional form for a propensity score model, a tighter caliper, etc.) and repeat the process.[74]

A COMMON WAY OF CHECKING FOR BALANCE IS A BAL-ANCE TABLE. Balance tables are pretty straightforward. Take a bunch of variables that should have balance. Then, show some of their summary statistics (usually, mean and standard deviation, maybe some other stuff) separately for the treated and control groups. Finally, do a difference-of-means test for each of the variables to see if the treated and control groups differ.[75] You might add some other bells and whistles like a standardized difference in means, where the difference between treated and control groups is divided by their shared standard deviation—fancy! There's also a variant on standardized difference-in-means called "standardized bias," a measure you might see on some balance tables, which is standardized difference-in-means with a slightly different standardization.[76]

Often, the balance table is created twice: once with the raw, un-matched data to show the extent of the balance problem that matching is to solve, and then again with the matched data to ideally show that the balance problem is gone. What does good balance look like? It looks like no meaningfully large differences in means. And if you're doing a test, you should find no statistically significant differences in means, or at least no more than you'd expect by chance.[77]

Balance tables are often fairly easy to code. In R, there are many good options in `MatchBalance` in the **Matching** package and `sumtable` in the **vtable** package, both of which can either be run both without and with matching weights, or with the output from `Match`, to compare balance before and after matching. In Stata you can follow a `teffects` analysis with `tebalance summarize` (although this won't do significance tests). For a more standard balance table in Stata you can use the **balanc-etable** package, adding weights to check the post-matching balance. In Python, if you've created a `CausalModel()` called M using the **causalinference** package, you can get a balance table using `print(M.summary_stats)`.

We can see an example of a balance table for Broockman (2013), which is a subset of the output produced by `MatchBalance` in R following the Mahalanobis matching from earlier in this chapter. Importantly, there's no one standard way to present a balance table, so the format that your software gives you may not match this one. For each variable, before and after matching, it shows the mean of the variable for the treated group,

[74] This is one of the selling points for methods like coarsened exact matching and entropy balancing, which by construction avoid balance problems, to the point where the seminal coarsened exact matching paper is titled "Causal Inference Without Balance Checking." No iterative matching/balance checking!

[75] If you managed to make your way this far in the book without having taken basic statistics, well, I'm not sure how I feel about that. But in any case, if you did, you might not have done a difference-in-means test. It's just a test of whether the mean of a variable is different in two groups. You can even do it using regression by regressing the variable on a treatment indicator. Done.

[76] Specifically this is the difference in means divided by a shared variance term, which you get by adding up the variance in the treatment group with the variance in the control group, multiplying the result by .5, and then taking the squared root.

Paul R Rosenbaum and Donald B Rubin. Constructing a control group using multivariate matched sampling methods that incorporate the propensity score. *The American Statistician*, 39(1):33–38, 1985.

[77] If you've got a huge long list of variables on your balance table, then it's likely not a problem if a couple have significant differences, especially if the difference-of-means itself is meaningfully small.

the mean of the variable for the control group, the standardized difference (a.k.a. standardized bias from earlier in this section), and a p-value from a *t*-test for the mean being different in the treated and control groups.

	Before Matching	After Matching
Median Household Income		
Mean Treatment	3.33	3.333
Mean Control	4.435	3.316
Std. Mean Diff	-97.057	1.455
t-test p-value	<.0001	0.164
Black Percent		
Mean Treatment	0.517	0.515
Mean Control	0.063	0.513
Std. Mean Diff	224.74	1.288
t-test p-value	<.0001	0.034
Legislator is a Democrat		
Mean Treatment	0.978	0.978
Mean Control	0.501	0.978
Std. Mean Diff	325.14	0
t-test p-value	<.0001	1

Table 14.1: Broockman (2013) Balance Table Before and After Mahalanobis Matching

What do we see here? First, before matching takes place, there's a serious amount of imbalance. The mean values are both meaningfully and statistically significantly different between the treatment and control groups. How about after matching? Balance looks pretty good. There's no meaningfully large difference in the means between treatment and control for any of the three variables we're looking at. For "Legislator is a Democrat" (*leg_democrat* in the code), there's literally zero difference in the mean.

We do see a statistically significant (at the 95% level) difference in the percentage of the district that is Black after matching, with a *t*-statistic p-value below .05. However, this goes along with a meaningfully very close mean value (.515 vs .513). Additionally, to see if we should be concerned, we might want to look at the full distributions and see whether *those* are different.[78]

BALANCE TABLES TEND TO FOCUS ON THE MEAN. Is the mean the same in the treated and control groups? Sure, you

[78] The full table also contains a Kolmogorov-Smirnov test in addition to a *t*-test. This test checks whether two distributions are the same. The p-value for this test for Black percentage is very high.

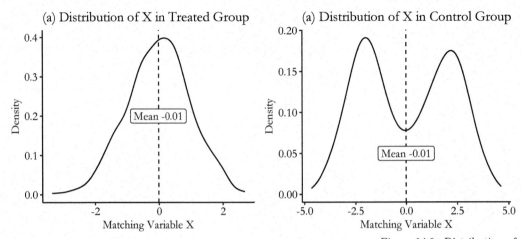

Figure 14.9: Distribution of a Matching Variable After Matching in the Treated and Control Groups

could use a balance table to check if other things are the same, but you don't see that a lot.

And that's a problem! Would you really believe that you'd closed a back door if the treated and control groups had distributions of a matching variable that looked like Figure 14.9? Sure the means are the same. But there's clearly a difference there.

Figure 14.10: Overlaid Distribution of a Matching Variable After Matching in the Treated and Control Groups

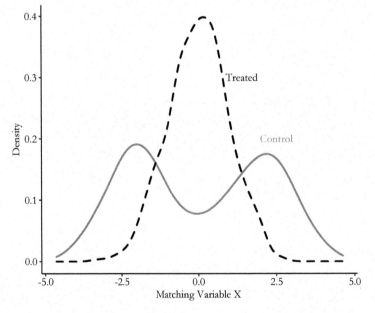

How can we check for differences between the treated and control groups in the distribution of a variable? Well... I already sorta answered that with Figure 14.9. Just plot the distribution of the variable for both the (matched) treated and control

groups, and see how badly they differ. The only difference from Figure 14.9 is that you'd typically overlay the distributions on top of each other so you can more easily see any differences. Overlaying the distributions from Figure 14.9 in Figure 14.10 makes it pretty easy to spot the problem.

We wouldn't necessarily throw out matching for *any* mismatch between the distributions, but egregious mismatches like the one in Figure 14.10 might be a cause to go back to the drawing board. You can make these graphs for any continuous matching variable, and you can use the tools for discrete distributions from Chapter 3 to do the same for categorical or binary matching variables.

One place that this overlaid density graph is almost mandatory is when you're working propensity scores. You've only got one real matching variable that matters—the propensity score—so you'd better hope the density lines up pretty well.

You can see this in action in Figure 14.11, which shows overlaid density plots for the Broockman study for the matching variable "percentage Black in district," and for the propensity score, both after applying inverse probability weights (and trimming any propensity scores from $0 - .02$ or from $.98-1$).

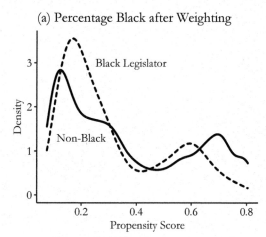

(a) Percentage Black after Weighting

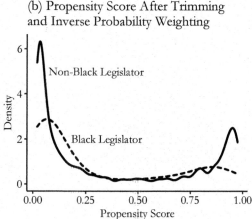

(b) Propensity Score After Trimming and Inverse Probability Weighting

Figure 14.11: Percentage Black in District and Propensity Score Distributions Before and After Inverse Probability Weighting

How do these look now? They're... not perfect. With a really good propensity score model, the propensity score distributions for treated and control observations should really be nearly identical. These are a lot better than they were before the matching, but they're still pretty dissimilar. The same thing goes for the percentage Black in district. This might be a good indication to go back to the propensity score model and try adding some things to it to improve the balance we see.

Overlaid density plots are fairly easy to create by hand, but in some cases there are prepackaged versions we can go with.

```r
# R CODE
library(tidyverse)
br <- causaldata::black_politicians

# We can estimate our own propensity score
m <- glm(leg_black ~ medianhhincom + blackpercent + leg_democrat,
    data = br, family = binomial(link = 'logit'))
# Get predicted values
br <- br %>%
    mutate(ps = predict(m, type = 'response'))

# Create IPW weights
br <- br %>%
    mutate(ipw = case_when(
    leg_black == 1 ~ 1/ps,
    leg_black == 0 ~ 1/(1-ps)))

# Density plots for raw data
ggplot(br, aes(x = medianhhincom, color = factor(leg_black))) +
    geom_density()

# And using our matching weights
ggplot(br, aes(x = medianhhincom, color = factor(leg_black),
    weight = ipw)) + geom_density()
```

```stata
* STATA CODE
causaldata black_politicians.dta, use clear download

* If we start with teffects...
teffects ipw (responded) (leg_black medianhhincom ///
    blackpercent leg_democrat, logit)

* We can check the balance of a given matching variable
* using tebalance, density which shows both raw and weighted balance
tebalance density medianhhincom
```

```python
# PYTHON CODE
import pandas as pd
import numpy as np
import seaborn as sns
from causaldata import black_politicians
br = black_politicians.load_pandas().data

# Overlaid treatment/control density in raw data
fig1 = sns.kdeplot(data = br,
x = 'medianhhincom',
hue = 'leg_black',
common_norm = False)
fig1.plot()

# Start the new plot
fig1.get_figure().clf()

# Add weights from any weighting method to check post-matching density
```

```
19 │ # Here we have the ipw variable from our propensity score matching
20 │ # in previous code blocks (make sure you run those first!)
21 │ fig2 = sns.kdeplot(data = br, x = 'medianhhincom',
22 │ hue = 'leg_black', common_norm = False, weights = 'ipw')
23 │ fig2.plot()
```

THE USE OF A PROPENSITY SCORE ALLOWS US TO TEST
ANOTHER ASPECT OF BALANCE. In particular, if our sample is
properly balanced and the matching has worked to close back
doors, then *controlling for the propensity score*, we should also
see no relationship between the matching variables and treatment status, i.e. balance. This is evident from our propensity
score causal diagram, reproduced in Figure 14.12. Shut down
TreatmentPropensity and there should be no path from A, B,
or C to *Treatment*.

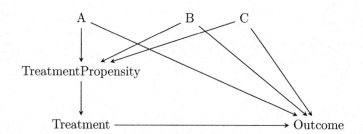

Figure 14.12: Causal Dia-
gram Amenable to Propen-
sity Score Matching

Dehejia and Wabha (2002) offer a way of testing this implication of our diagram using a *stratification test*.[79] In a stratification test we repeatedly look *within limited ranges of the
propensity score*. By limiting ourselves to a narrow chunk of
the propensity score, we are in effect controlling for it. So split
the whole range of propensity scores up into chunks and check
whether the matching variables and treatment are related within
each chunk.[80]

14.7   Estimation with Matched Data

14.7.1   Estimating Mean Differences

WE'RE NEARLY TO THE END OF THE CHAPTER, AND WE'RE
JUST NOW GETTING TO ESTIMATING A TREATMENT EFFECT?
Well, yeah. When it comes to matching, most of the work—
picking a matching method, doing the matching and making
the weights, checking common support and balance, and maybe
going back to tweak the matching method if the balance is bad—
is already done before we need to estimate anything. But now
we've done that stuff, and it's time.[81]

[79] Rajeev H Dehejia and
Sadek Wahba. Propen-
sity score-matching methods
for nonexperimental causal
studies. *Review of Eco-
nomics and Statistics*, 84(1):
151–161, 2002.

[80] How many chunks? They
should be narrow enough
that the *propensity score
itself* should be balanced
within each chunk. If it's
not, go narrower. But not
so narrow that there aren't
enough observations within
each strata to do a test.

[81] Sound the alarm!

With a set of matching weights in hand, how can we estimate the effect of treatment? Get the weighted mean of the outcome for the treated and control groups and compare them. Okay, that's it, bye.

All right, well, that's not *really* it. That is quite a lot of it, though. If all you're interested in doing is getting the effect of treatment on the mean of the outcome, comparing the weighted means will do it.

ONE IMPORTANT DETAIL TO KEEP IN MIND is that while the treatment effect is that simple to estimate, the *standard errors* on that treatment effect won't be quite as simple. Calculating the standard error of a weighted mean is easy. But the standard error you calculate *won't remember that the weights had to be estimated* and perhaps that some observations were dropped in the matching process. The step of preprocessing the data to match observations and create weights introduces some uncertainty, and so increases the standard errors. Incorporating this uncertainty into your standard errors is important.

There are a few different ways to calculate the appropriate standard error adjustments, most of them having to do with figuring out the variance of the outcome variable conditional on treatment and on the matching variables (or on the weights that come from the matching variables). However, they're all quite complex, and not really the aim of this book. Until you're ready to delve into the econometric papers that describe the standard error adjustments you'd want to do,[82] you're probably going to be using software that does it for you. So the key takeaway here is "unless you're willing to put in the time to do it right, don't calculate your own standard errors for treatment effects under matching, let matching-based estimation software do it for you."

ONE EXCEPTION IS THE USE OF BOOTSTRAP STANDARD ERRORS, as discussed in Chapter 13 on regression. Bootstrap standard errors can be applied to matching estimates. In application of a bootstrap to matching, you first randomly resample the data with replacement (so you get the same sample size you started with, but some observations are replaced with multiple copies of others), then re-perform the matching from scratch, and finally estimate the treatment effect in your bootstrap sample. Repeat this process many, many times, and the standard deviation of the treatment effect across all the bootstrap samples is the standard error of your treatment effect. You can see how this process naturally incorporates any uncertainty we get

[82] Once again, Caliendo and Kopeinig (2008) pops up as a good place to start.

from the matching process—it allows that uncertainty to feed right into our treatment effect distribution.

Both intuitively (because you are actively incorporating how the matching process affects the sampling distribution) and mechanically (because bootstrap standard errors are fairly simple to do yourself by hand, it takes only a little coding if it's not already in the command you're using), bootstrap standard errors can be a great option for matching estimates. That said, bootstrap standard errors can only be used with the "constructing a weighted matched sample" approach to matching. They simply don't work properly for the "selecting matches" approach, because the sharp in/out decisions that process makes don't translate into the bootstrap having the proper sampling distribution.[83] So if you're using a weighting approach, go for it. If not, either go with an appropriate standard error adjustment in your software, or if you're feeling real spicy you can go find one of the bootstrap variants designed to work properly with selecting-matches approaches.

Bootstrapping a matching process can be a little tricky to code up if you're not using a command that offers it already, since you aren't just bootstrapping a single command (or asking for the "bootstrap standard error" option in a regression). You need to bootstrap *the whole matching process from start to finish*. I offer some coding examples later in the chapter in application to doubly robust estimation, and also in Chapter 15. Slight tweaks on this code should help you provide bootstrap standard errors for all kinds of matching processes.

14.7.2   Combining Matching and Regression

IN MANY CASES WE WANT TO DO SOMETHING A LITTLE MORE COMPLEX than comparing means between the treated and control groups. For example, maybe our research design depends on a specific functional form of the relationship between treatment and outcome, which would be difficult to capture with matching but easy with regression. But we still want to close back doors by matching.[84] What then? We can combine our matching weights with a regression model to do *regression adjustment*.

The approach here is pretty straightforward. We already have a set of selected matched observations, or a set of matching weights, from our matching procedure. We can apply that subsample, or those weights, to our regression model, just as we did with sample weights in Chapter 13.

[83] Alberto Abadie and Guido W Imbens. On the failure of the bootstrap for matching estimators. *Econometrica*, 76(6): 1537–1557, 2008.

[84] This will come back in Chapter 18 on difference-in-differences, which relies on comparing a treated and control group (which might be matched), and pinpointing a particular discontinuity in the relationship between outcome and time.

In fact, we've already done this procedure—the basic difference in weighted means we talked about before can be achieved by applying matching weights to a regression with only treatment as a predictor, i.e. $Outcome = \beta_0 + \beta_1 Treatment + \varepsilon$. The estimate of the effect would be $\hat{\beta}_1$. This is the method I've been using in the code examples. It's not a big jump to start adding other things to that regression.

Using weights isn't the only way to go (and may not be ideal if you already have other plans for those weights, like survey sample weights). Another approach to using matching with regression is to think carefully about what the propensity score is—it's a variable that, if controlled for, should block all back doors from *Treatment* to *Outcome*. So... control for it! The propensity score (or the inverse probability weight based on the propensity score) can be added to regression as a control.

In fact, now that we're at this section we can finally come back around and run the analysis that Broockman (2013) actually ran. He didn't just do matching and estimate a treatment effect from that. No! He took those matching weights and used them in a regression that let him not just see the difference in response rates between Black legislators and others, but see whether the *difference between in-district vs. out-of-district* response rates for these letters from Black people was different between Black legislators and others. That calls for an interaction term in a regression. Not to mention we can add plenty of other controls.

```
1    # R CODE
2    library(cem); library(tidyverse); library(modelsummary)
3    br <- causaldata::black_politicians
4
5    # This copies the CEM code from the CEM section
6    # See that section's code for comments and notes
7
8    # Limit to just the relevant variables and omit missings
9    # (of which there are none in this data)
10   brcem <- br %>%
11       select(responded, leg_black, medianhhincom,
12       blackpercent, leg_democrat) %>%
13       na.omit() %>%
14       as.data.frame()
15
16   # Create breaks
17   inc_bins <- quantile(brcem$medianhhincom, (0:6)/6)
18
19   create_even_breaks <- function(x, n) {
20           minx <- min(x)
21           maxx <- max(x)
22           return(minx + ((0:n)/n)*(maxx-minx))
23   }
```

```
24
25   bp_bins <- create_even_breaks(brcem$blackpercent, 6)
26   ld_bins <- create_even_breaks(brcem$leg_democrat,2)
27
28   allbreaks <- list('medianhhincom' = inc_bins,
29       'blackpercent' = bp_bins,
30       'leg_democrat' = ld_bins)
31
32   c <- cem(treatment = 'leg_black', data = brcem,
33       baseline.group =  '1', drop = 'responded',
34       cutpoints = allbreaks, keep.all = TRUE)
35
36   # Get weights for other purposes.  Note this exact code only
37   # works because we didn't have to drop any NAs. If we did,
38   # lining things up would be trickier
39   br <- br %>%
40       mutate(cem_weight = c$w)
41   m1 <- lm(responded~leg_black*treat_out + nonblacknonwhite +
42       black_medianhh + white_medianhh + statessquireindex +
43       totalpop + urbanpercent, data = br, weights = cem_weight)
44   msummary(m1, stars = c('*' = .1, '**' = .05, '***' = .01))
```

```
1   * STATA CODE
2   causaldata black_politicians.dta, use clear download
3
4   * Use any matching method to get weights
5   * We'll use coarsened exact matching like in the original
6   cem medianhhincom blackpercent leg_democrat(#2), tr(leg_black)
7
8   reg responded i.treat_out##i.leg_black nonblacknonwhite leg_democrat ///
9       leg_senator south blackpercent black_medianhh white_medianhh ///
10      statessquire totalpop urbanpercent [iweight=cem_weights]
```

```
1   # PYTHON CODE
2   import pandas as pd
3   import statsmodels.formula.api as sm
4   from causaldata import black_politicians
5   br = black_politicians.load_pandas().data
6
7   # This copies the CEM code from the CEM section
8   # See that section's code for comments and notes
9   br['inc_bins'] = pd.qcut(br['medianhhincom'], 6)
10  br['bp_bins'] = pd.qcut(br['blackpercent'], 6)
11  treated = br.loc[br['leg_black'] == 1
12  ].groupby(['inc_bins','bp_bins','leg_democrat']
13  ).size().to_frame('treated')
14  control = br.loc[br['leg_black'] == 0
15  ].groupby(['inc_bins','bp_bins','leg_democrat']
16  ).size().to_frame('control')
17
18  # Merge the counts back in
19  br = br.join(treated, on = ['inc_bins','bp_bins','leg_democrat'])
20  br = br.join(control, on = ['inc_bins','bp_bins','leg_democrat'])
21
22  # Create weights
23  br['weight'] = 0
24  br.loc[(br['leg_black'] == 1) & (br['control'] > 0), 'weight'] = 1
25  totalcontrols = sum(br.loc[br['leg_black']==0]['treated'] > 0)
26  totaltreated = sum(br.loc[br['leg_black']==1]['control'] > 0)
27  br['controlweights'] = (br['treated']/br['control'])
```

```
28 )*(totalcontrols/totaltreated)
29 br.loc[(br['leg_black'] == 0), 'weight'] = br['controlweights']
30
31 # Now, use the weights to estimate the effect
32 m = sm.wls(formula = '''responded ~ leg_black*treat_out +
33 nonblacknonwhite + black_medianhh + white_medianhh +
34 statessquireindex + totalpop + urbanpercent''',
35 weights = br['weight'],
36 data = br).fit()
37 m.summary()
```

From these regressions we get a coefficient of -.142 on *leg_black*, -.351 on *treat_out*, and .229 on *leg_black* × *treat_out*. Recalling what we learned about interpreting a model with interaction terms from Chapter 13, we know that this means that Black legislators were 14.2 percentage points less likely than White legislators to respond to in-district emails, in-district emails were 35.1 percentage points less likely to get a response than out-of-district emails from White legislators, but the in-district vs. out-of-district gap is 22.9 percentage points *smaller* for Black legislators than White ones, which is evidence in favor of the idea that Black legislators have an intrinsic motivation to help Black people.

IF WE PLAY OUR CARDS RIGHT, REGRESSION AND MATCHING CAN COMBINE TO BE MORE THAN THE SUM OF THEIR PARTS. *Doubly robust estimation* is a way of combining regression and matching that works even if there's something wrong with the matching *or* something wrong with the regression—as long as it's not both.

What do I mean by "something wrong?" It doesn't cover everything—you still need to, you know, actually measure for and include the variables that allow you to close back doors.[85] Doubly robust estimation won't let you get around that. But regression and matching both rely on assumptions. For example, perhaps your regression model is *misspecified*—you've got all the right variables in there, but maybe the functional form isn't quite right. The same thing can happen when specifying the model you use to estimate your propensity score.

Doubly robust estimation is any method that still works as long as *one of those two models* is properly specified. The other one can be wrong.

Double-robustness is *a thing that some estimators have* rather than being a specific estimator. So there are actually lots of methods that are doubly robust.[86] I could list a dozen right here without even thinking too hard. But we don't have all day. So we'll look at three, the first one in detail.

Doubly robust estimation. An estimation method that identifies the effect even if one of the regression or matching models is misspecified.

[85] Failing to do this would make *neither* method work—remember, at least one needs to work.

[86] How do we know if a particular estimator is doubly robust? Someone writes a paper with a lot of equations proving that it is. There are a lot of these papers.

The first estimator we'll look at is the one described in Wooldridge (2010, Chapter 21.3.4). This one is appealing because the process of doing it is *almost* like you'd guess doubly robust estimation works on your own if you'd just heard about it and had to come up with it yourself. And it actually does work!

The steps are fairly straightforward:

1. Estimate the propensity score p for each observation using your matching variables

2. Use the propensity score to produce the inverse probability weights: $1/p$ for treated observations and $1/(1-p)$ for untreated observations

3. Using *only treated observations*, estimate your regression model, using the matching variables as predictors (and perhaps some other predictors too, depending on what you're doing), and inverse probability weights

4. Repeat step (3) but using only untreated observations

5. Use the models from steps (3) and (4) to produce "treated" and "untreated" predicted observations for the whole sample

6. Compare the "treated" means to the "untreated" means to get the estimate of the causal effect

7. To get standard errors, use bootstrap standard errors, or the heteroskedasticity-robust errors described in Wooldridge (2010) (not just the same as asking for regular heteroskedasticity-robust standard errors)

While there are a lot of steps here, none of them are particularly complex, and all are fairly easy to do by hand. In fact, let's work through the code for this approach.

```
1   # R CODE
2   library(boot); library(tidyverse)
3   br <- causaldata::black_politicians
4
5   # Function to do IPW estimation with regression adjustment
6   ipwra <- function(br, index = 1:nrow(br)) {
7           # Apply bootstrap index
8           br <- br %>% slice(index)
9
10          # estimate and predict propensity score
11          m <- glm(leg_black ~ medianhhincom + blackpercent + leg_democrat,
12                  data = br, family = binomial(link = 'logit'))
13          br <- br %>%
14              mutate(ps = predict(m, type = 'response'))
15
```

```
16          # Trim control observations outside of treated PS range
17          minps <- br %>%
18              filter(leg_black == 1) %>%
19              pull(ps) %>%
20              min(na.rm = TRUE)
21          maxps <- br %>%
22              filter(leg_black == 1) %>%
23              pull(ps) %>%
24              max(na.rm = TRUE)
25          br <- br %>%
26              filter(ps >= minps & ps <= maxps)
27
28          # Create IPW weights
29          br <- br %>%
30              mutate(ipw = case_when(
31              leg_black == 1 ~ 1/ps,
32              leg_black == 0 ~ 1/(1-ps)))
33
34          # Estimate difference
35          w_means <- br %>%
36              group_by(leg_black) %>%
37              summarize(m = weighted.mean(responded, w = ipw)) %>%
38              arrange(leg_black)
39
40          return(w_means$m[2] - w_means$m[1])
41  }
42
43
44  b <- boot(br, ipwra, R = 200)
45  # See estimate and standard error
46  b
```

```
1   * STATA CODE
2   causaldata black_politicians.dta, use clear download
3
4   * Wrap this in a program so we can bootstrap
5   capture program drop ipwra
6   program def ipwra, rclass
7   quietly{
8           * Estimate propensity
9           logit leg_black medianhhincom blackpercent leg_democrat
10          * Get predicted values for our propensity score
11          predict ps, pr
12
13          * "Trim" control observations outside of
14          * treated propensity score range
15          summ ps if leg_black
16          replace ps = . if ps < r(min) | ps > r(max)
17
18          * Get inverse probability weights
19          g ipw = 1/ps if leg_black
20          replace ipw = 1/(1-ps) if !leg_black
21
22          * Regress treated and nontreated separately,
23          * then predict for whole sample
24          regress responded medianhhincom blackpercent ///
25                  leg_democrat [pw = ipw] if !leg_black
26          predict untreated_outcome
27          regress responded medianhhincom blackpercent ///
28                  leg_democrat [pw = ipw] if leg_black
```

```
29            predict treated_outcome
30
31            * Compare means
32            summarize untreated_outcome
33            local unt_mean = r(mean)
34            summarize treated_outcome
35            local tr_mean = r(mean)
36
37            local diff = `tr_mean' - `unt_mean'
38    }
39
40    drop ps ipw untreated_outcome treated_outcome
41    return scalar diff = `diff'
42    end
43
44    * And a version for bootstrapping
45    capture program drop ipwra_b
46    program def ipwra_b, rclass
47        preserve
48        * Create bootstrap sample
49        bsample
50        * Estimate
51        ipwra
52        * Bring back original data
53        restore
54        return scalar diff = r(diff)
55    end
56
57    * Run once with original data
58    ipwra
59    local effect = r(diff)
60    * Then boostrap
61    simulate diff = r(diff), reps(2000): ipwra_b
62    local N = _N
63    bstat, stat(`effect') n(`N')
64
65    * Or skip all that and just use teffects ipwra (no trim here though)
66    causaldata black_politicians.dta, use clear download
67    teffects ipwra (responded medianhhincom blackpercent leg_democrat) ///
68         (leg_black medianhhincom blackpercent leg_democrat, logit)
```

```python
1    # PYTHON CODE
2    import pandas as pd
3    import numpy as np
4    import statsmodels.formula.api as sm
5    from causaldata import black_politicians
6    br = black_politicians.load_pandas().data
7
8    # As mentioned, est_via_weighting from causalinfernece already does
9    # doubly robust estimation but it's a different kind!
10   # Let's do Wooldridge here.
11   def ipwra(br):
12       # Estimate propensity
13       m = sm.logit('leg_black ~ medianhhincom + blackpercent + leg_democrat',
14       data = br)
15       # Get fitted values and turn them into probabilities
16       m = m.fit().fittedvalues
17       br['ps'] = np.exp(m)/(1+np.exp(m))
18
19       # Trim observations outside of treated range
```

```
20 ║    minrange = np.min(br.loc[br['leg_black']==1, 'ps'])
21 ║    maxrange = np.max(br.loc[br['leg_black']==1, 'ps'])
22 ║    br = br.loc[(br['ps'] >= minrange) & (br['ps'] <= maxrange)]
23 ║
24 ║    # Get inverse probability score
25 ║    br['ipw'] = br['leg_black']*(1/br['ps']
26 ║    ) + (1-br['leg_black'])*(1/(1-br['ps']))
27 ║
28 ║    # Regress treated and nontreated separately,
29 ║    # then predict for whole sample
30 ║    mtreat = sm.wls('''responded ~ medianhhincom +
31 ║    blackpercent + leg_democrat''',
32 ║    weights =  br.loc[br['leg_black'] == 1, 'ipw'],
33 ║    data = br.loc[br['leg_black'] == 1])
34 ║    mcontrol = sm.ols('''responded ~ medianhhincom +
35 ║    blackpercent + leg_democrat''',
36 ║    weights =  br.loc[br['leg_black'] == 0, 'ipw'],
37 ║    data = br.loc[br['leg_black'] == 0])
38 ║
39 ║    treat_predict = mtreat.fit().predict(exog = br)
40 ║    con_predict = mcontrol.fit().predict(exog = br)
41 ║
42 ║    # Compare means
43 ║    diff = np.mean(treat_predict) - np.mean(con_predict)
44 ║    return diff
45 ║
46 ║    # And a wrapper function to bootstrap
47 ║ def ipwra_b(br):
48 ║    n = br.shape[0]
49 ║    br = br.iloc[np.random.randint(n, size=n)]
50 ║    diff = ipwra(br)
51 ║    return diff
52 ║
53 ║ # Run once on the original data to get our estimate
54 ║ est = ipwra(br)
55 ║
56 ║ # And then a bunch of times to get the sampling distribution
57 ║ dist = [ipwra_b(br) for i in range(0,2000)]
58 ║
59 ║ # Our estimate
60 ║ est
61 ║ # and its standard error
62 ║ np.std(dist)
```

How about the other two methods I mentioned? I'll talk about these briefly. Just a tour of the space.

One is the Augmented Inverse Probability Weighted Estimator (AIPWE).[87] Like the Wooldridge approach, AIPWE has both a matching model to estimate, with treatment as a dependent variable, and a regression model, with the outcome as a dependent variable. The basic idea of AIPWE is that it includes an augmentation term (surprise) in the regression model that adjusts the results based on the degree of misspecification in the matching model.

[87] Zhiqiang Tan. Bounded, efficient and doubly robust estimation with inverse weighting. *Biometrika*, 97 (3):661–682, 2010.

AIPWE has some nice properties. It does okay in some settings even with mild misspecification in *both* models (although again, you would prefer at least one being correct). Plus, if it turns out that there *isn't* any misspecification in the model for treatment, AIPWE will produce lower standard errors in the outcome model. Neat!

The third and final method can be discussed briefly because I've already discussed it. Entropy Balancing, which showed up back in the Distance Matching section, is itself doubly robust.[88] Despite not directly including a regression, entropy balancing's focus on removing differences in means between treatment and control turns out to work a lot like regression's focus on removing differences in means. Double robustness!

[88] Qingyuan Zhao and Daniel Percival. Entropy balancing is doubly robust. *Journal of Causal Inference*, 5(1), 2017.

14.8   Matching and Treatment Effects

AS WITH ANY DESIGN AND ESTIMATION APPROACH, we have to ask what kind of treatment effect average (as in Chapter 10) we are getting.

When it comes to matching, that turns out to be a very interesting question. One of the coolest things about matching is that, unlike with regression, it's super easy to have a matching estimator give you just about any treatment effect average you want.[89]

How is that possible? Well, first let's start by thinking about the matching procedures I've been talking about this whole time throughout this chapter—what kind of treatment effect average do these give us?

We can think it through logically.[90] We're using matching to produce a group of control observations that are comparably similar to the treated group. They're supposed to be just like the treated group if the treated group hadn't been treated.

So... comparing our estimate of what *the treated group would have gotten without treatment* against *what the treated group actually got*? Using the rule-of-thumb logic from Chapter 10, that sounds like the average treatment on the treated to me. And indeed, that's what most of the methods I've described so far in this chapter will give you.[91]

OR AT LEAST PRETTY CLOSE TO WHAT THEY'LL GIVE YOU. Matching doesn't give you *exactly* the average treatment on the treated. Like with regression, it gives you an average that is affected by its approach to closing back doors. Just like regression gave variance-weighted treatment effects, matching

[89] Or at least it is in cases where you're identifying the effect by closing back doors. When combining matching with front-door isolating methods, it gets a bit more complex. Still more flexible than regression, though.

[90] This explanation doesn't apply to methods that also weight the treated group, like inverse probability weighting as I've described it—I'll get to that.

[91] At least most of the methods *as I've described them*. Some of the code examples give average treatment effects instead, because the defaults for those commands are to produce average treatment effects. Check the documentation! I talk about this in the next section as well.

gives *distribution-weighted* treatment effects, where each observation's treatment effect is weighted more heavily the more common its mix of matching variables is.

To give a basic example of this in action, let's imagine some types of people like we did back in Chapter 10. Table 14.2 shows two different types of people we might be sampling.

Name	Gender	Outcome Without Treatment	Outcome With Treatment	Treatment Effect
Alfred	Male	2	4	2
Brianna	Female	1	5	4

Table 14.2: Individual Hypothetical Outcomes

Now, imagine that we happen to end up with a sample with 500 untreated Alfreds and 500 untreated Briannas, as well as 1,000 treated Alfreds and 200 treated Briannas. And we're matching on Gender.

In the treated group we'd see an average outcome of $(1,000 \times 4 + 200 \times 5)/(1,200) \approx 4.17$. How about in the matched untreated group? Well, if we did one-to-one matching with replacement, for example, we'd end up with, just like in the treated group, 1,000 Alfreds and 200 Briannas, but they're untreated this time. So the average among the untreated group is $(1,000 \times 2 + 200 \times 1)/(1,200) \approx 1.83$. This gives us a treatment effect of $4.17 - 1.83 = 2.34$. This is much closer to the male treatment effect of 2 than the female treatment effect of 4 because there are far more men in the treated sample than women—it's weighted by how common a given observation's matching variable values are.[92]

14.8.1 Getting Other Treatment Effects from Matching

AH, BUT I HAVE BURIED THE LEDE.[93] The average treatment on the treated is not the only thing that matching can give you. We can get the average treatment on the treated, or even the average treatment on the *untreated*.[94] How can we do that?

It's actually fairly straightforward. If you want average treatment on the treated, as we've been getting, then construct a control group that is as similar to the treated group as possible.

Want average treatment on the *untreated*? Simple! Just reverse things. Instead of matching control observations to treated observations, instead match *treated* observations to *control* observations. Same procedure, but reverse which group is which. The result gives you a treated group that is as much

[92] This one is weighted by the distribution among the treated because it's an average-treatment-on-the-treated method. But distribution weighting applies to matching variable distributions among other relevant groups when doing average treatment effect, average treatment on the untreated, and so on.

[93] Yes, that's how it's spelled.

[94] Granted, subject to the distribution-weighting thing I just talked about.

like the untreated group as possible. Comparing that to the untreated group compares what you actually got from no treatment against what we think that group *would have gotten* if treated. Average treatment on the untreated![95]

[95] This does all assume that the variables that predict *variation in the treatment effect* are included as matching variables. And if you're using a propensity score to match, it further assumes that the variables that predict variation in the treatment effect are the same ones that predict treatment itself.

From here you can get to the average treatment effect fairly easily. If the average among your treated group is the average treatment effect on the treated (ATT), and the average among your untreated group is the average treatment effect on the untreated (ATUT), then what's the overall average, also known as the average treatment effect? Well, if a proportion p of people are getting treated, then it's just $ATE = (p \times ATT) + ((1 - p) \times ATUT)$. You're just averaging the average from each "side," weighted by how big each side is, to get the overall average.

The use of matching to get average treatment effects is fairly common, and sometimes is even the default option in software commands designed to do treatment effects by matching. Stata's `teffects` is an example of this—the `teffects` examples in this chapter produce average treatment effects by default (and all of them have an `atet` option for getting average treatment on the treated).

HOW DOES THIS APPLY TO INVERSE PROBABILITY WEIGHTING? After all, when you're doing inverse probability weighting you aren't really matching *to* anything, you're just weighting on the basis of the inverse propensity weight.

It turns out that the specific way you choose to turn the propensity score into an inverse propensity weight determines the kind of treatment effect average you get. Weighting to make the control group look like the treatment? Average treatment on the treated! Doing the reverse? Average treatment on the untreated! Doing both? Average treatment effect!

In fact, the approach to inverse probability weighting I've been talking about this whole time—with a weight of $1/p$ for the treated group and $1/(1 - p)$ for the untreated group, where p is the propensity score—serves to both upweight the members of the untreated group that are most like the treated group *and* upweight the members of the treated group that are most like the untreated group. This actually gives us the average treatment effect, not the average treatment on the treated.

What if we do want one of those other averages? We just use different weights.

What if we want average treatment on the treated? Intuitively here we want to make the untreated group like the treated group. So first we give *all* of the treated group the same weight of 1. We don't need to adjust them to be like anyone else—we want to adjust *everyone else to be like them* so the others can serve as a comparison. Sounds like average treatment on the treated to me.

The treated group gets a weight of 1. How about the untreated group? Instead of a weight of $1/(1-p)$, they get $p/(1-p)$. The $1/(1-p)$, which we did before, sort of brings the representation of the untreated group back to neutral, where everyone is counted equally. But we don't want equal. We want the untreated group to be *like the treated group*, which means weighting them *even more* based on how like the treated group they are. And thus $p/(1-p)$, which is more heavily responsive to p than $1/(1-p)$ is.

From there, average treatment on the untreated is pretty easy to figure out, and works for largely the same reason. The untreated group in this case gets a weight of 1, and the treated group, instead of the $1/p$ that gave us the ATE, gets $(1-p)/p$, which makes things match the untreated group properly and gives us the ATUT we want.

15

Simulation

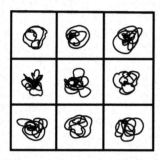

15.1   It is Proven

UNLIKE STATISTICAL RESEARCH, WHICH IS COMPLETELY MADE
UP OF THINGS THAT ARE AT LEAST SLIGHTLY FALSE, STATIS-
TICS ITSELF IS ALMOST ENTIRELY TRUE. Statistical methods
are generally developed by *mathematical proof.* If we assume A,
B, and C, then based on the laws of statistics we can prove that
D is true.

For example, if we assume that the true model is $Y = \beta_0 +
\beta_1 X + \varepsilon$, and that X is unrelated to ε,[1] then we can prove
mathematically that the ordinary least squares estimate of $\hat{\beta}_1$
will be the true β_1 on average. That's just *true,* as true as
the proof that there are an infinite number of prime numbers.
Unlike the infinite-primes case we can (and do) argue endlessly
about whether those underlying true-model and unrelated-to-ε
assumptions *apply in our research case.* But we all agree that
if they *did* apply, then the ordinary least squares estimate of $\hat{\beta}_1$
will be the true β_1 on average. It's been proven.

Pretty much all of the methods and conclusions I discuss in
this book are supported by mathematical proofs like this. I

[1] And some other less well-
discussed assumptions.

DOI: 10.1201/9781003226055-15

have chosen to omit the actual proofs themselves,[2] but they're lurking between the lines, and in the citations I can only imagine you're breezing past.

Why go through the trouble of doing mathematical proofs about our statistical methods? Because they let us know **how our estimates work.** The proofs underlying ordinary least squares tell us how its sampling variation works, which assumptions must be true for our estimate to identify the causal effect, what will happen if those assumptions *aren't* correct, and so on. That's super important!

SO WE HAVE AN OBVIOUS PROBLEM. I haven't taught you how to do econometric proofs. There's a good chance that you're not going to go out and learn them, or at least you're unlikely to go get good at them, even if you do econometric research. Proof-writing is really a task for the people developing methods, not the consumers of those methods. Many (most?) of the people doing active statistical research have forgotten all but the most basic of statistical proofs, if they ever knew them in the first place.[3]

But we still must know how our methods work! Enter *simulation.* In the context of this chapter, simulation refers to the process of *using a random process that we have control over* (usually, a data generating process) to produce data that we can evaluate with a given method. Then, we do that over and over and over again.[4] Then, we look at the results over all of our repetitions.

WHY DO WE DO THIS? By using a random process that we have control over, we can *choose the truth.* That is, we don't need to worry or wonder about which assumptions apply in our case. We can *make* assumptions apply, or not apply, by just varying our data generating process. We can choose whether there's a back door, or a collider, or whether the error terms are correlated, or whether there's common support.

Then, when we see the results that we get, we can *see what kind of results the method will produce given that particular truth.* We know the true causal effect since we *chose* the true causal effect—does our method give us the truth on average? How big is the sampling variation? Is the estimate more or less precise than another method? If we do get the true causal effect, what happens if we rerun the simulation but make the error terms correlated? Will it still give the true causal effect?

It's a great idea, when applying a new method, or applying a method you know in a context where you're not sure if it will

---

[2] Whether this is a blessing or a deprivation to the reader is up to you (or perhaps your professor). In my defense, there are already plenty of proof-based econometrics books out there for you to peruse. Those proof-based books are valuable, but I'm not sure how badly we need *another* one. Declining marginal value, comparative advantage, and all that. I'm still an economist, you know.

[3] Whether this is a problem is a matter of opinion. I think doing proofs is highly illuminating, and getting good at doing econometric proofs has made me understand how data works much better. But is it *necessary* to know the proof behind a method in order to use that method well as a tool? I'm skeptical.

[4] And then a few more times.

work, to just go ahead and try it out with a simulation. This chapter will give you the tools you need to do that.

15.1.1   The Anatomy of a Simulation

To give a basic example, imagine we have a coin that we're going to flip 100 times, and then estimate the probability of coming up heads. What estimator might we use? We'd probably just calculate the proportion of heads we get in our sample of $N = 100$.

What would a simulation *look like*?

1. Decide what our random process/data generating process is. We have control! Let's say we choose to have a coin that's truly heads 75% of the time.

2. Use a random number generator to produce our sample of 100 coin flips, where each flip has a 75% chance of being heads.

3. Apply our estimator to our sample of 100 coin flips. In other words, calculate the proportion of heads in that sample.

4. Store our estimate somewhere.

5. Iterate steps 2—4 a whole bunch of times, drawing a new sample of 100 flips each time, applying our estimator to that sample, and storing the estimate (and perhaps some other information from the estimation process, if we like).

6. Look at the distribution of estimates across each of our iterations.

That's it! There's a lot we can do with the distribution of estimates. First, we might look at the mean of our distribution. The *truth* that we decided on was 75% heads. So if our estimate isn't close to 75% heads, that suggests our estimate isn't very good.

We could use simulation to rule out bad estimators. For example, maybe instead of taking the proportion of heads in the sample to estimate the probability of heads, we take the *squared* proportion of heads in the sample for some reason. You can try it yourself—applying this estimator to each iteration won't get us an estimate of 75% heads.

We can also use this to *prove the value of good estimators to ourselves*. Don't take your stuffy old statistics textbook's word that the proportion of heads is a good estimate of the probability of heads. Try it yourself! You probably already *knew* your textbook was right. But I find that doing the simulation can also help you to *feel* that they're right, and understand *why* the estimator works, especially if you play around with it and try a few variations.[5]

On a personal note, simulation is far and away the first tool that I reach for when I want to see if an approach I'd like to take makes sense. Got an idea for a method, or a tweak to an existing method, or just a way you'd like to transform or incorporate a particular variable? Try simulating it and see whether the thing you want to do *works*. It's not too difficult, and it's certainly easier than *not* checking, finishing your project, and finding out at the end that it didn't work when people read it and tell you how wrong you are.

ASIDE FROM LOOKING AT WHETHER OUR ESTIMATOR CAN NAIL THE TRUE VALUE, we can also use simulation to understand *sampling variation*. Sure, on average we get 75% heads, but what proportion of samples will get 50% heads? Or 90%? What proportion will find the proportion to be statistically significantly different from 75% heads?

Simulation is an important tool for understanding an estimator's sampling distribution. Sampling distribution (and the standard errors it gives us that we want *so badly*) can be really tricky to figure out, especially if analysis has multiple steps in it, or uses multiple estimators. Simulation can help to figure out the degree of sampling variation without having to write a proof.

Sure, we don't need simulation to understand the standard error of our coin-flip estimator. The answer is written out pretty plainly on Wikipedia if you head for the right page. But what's the standard error if we first use matching to link our coin to other coins based on its observable characteristics, apply a smoothing function to account for surprising coin results observed in the past, and then drop missing observations from when we forgot to write down some of the coin flips? Uhhh... I'm lost. I really don't want to try to write a proof for that standard error. And if someone has already written it for me I'm not sure I'd be able to find it and understand it.[6] Simulation can help us get a sense of sampling variation even in complex cases.

Indeed, even the pros who know how to write proofs use simulation for sampling variation—see the bootstrap standard errors and power analysis sections in this chapter.

[5] It's harder to use simulation to prove the value of a *good* estimator to *others*. That's one facet of a broader downside of simulation—we can only run so many simulations, so we can only show it works for the particular data generating processes we've chosen. Proofs can tell you much more broadly and concretely when or where something works or doesn't.

[6] Or, similarly, what if the complexity is in the data rather than the estimator? How does having multiple observations per person affect the standard error? Or one group nested within another? Or what if treatments are given to different groups at different times, does that affect standard errors?

LASTLY, SIMULATION IS CRUCIAL IN UNDERSTANDING HOW ROBUST an estimation method can be to violations of its assumptions, and how sensitive its estimates are to changes in the truth. Sure, we know that if a true model has a second-order polynomial for X, but we estimate a linear regression without that polynomial, the results will be wrong. But how wrong will it be? A little wrong? A lot wrong? Will it be more wrong when X is a super strong predictor or a weak one? Will it be more wrong when the slope is positive or negative, or does it even matter?

It's very common, when doing simulations, to run multiple variations of the same simulation. Tweak little things about the assumptions—make X a stronger predictor in one and a weaker predictor in the other. Add correlation in the error terms. Try stuff! You've got a toy; now it's your job to break it so you know exactly when and how it breaks.

15.1.2   Why We Simulate

TO SUM UP, simulation is the process of using a random process that we control to repeatedly apply our estimator to different samples. This lets us do a few very important things:

- We can test whether our estimate will, when applied to the random process / data generating process we choose, give us the true effect

- We can get a sense of the sampling distribution of our estimate

- We can see how sensitive an estimator is to certain assumptions. If we make those assumptions no longer true, does the estimator stop working?

Now that we know *why* simulation can be helpful,[7] let's figure out how to do it.

15.2   Alas, We Must Code

THERE'S NO WAY AROUND IT. Modern simulation is about using code. And you won't just have to use prepackaged commands correctly—you'll actually have to write your own.[8]

Thankfully, not only are simulations relatively straightforward, they're also a good entryway into writing more complicated programs.

A typical statistical simulation has only a few parts:

[7] And I promise you it is. Once you get the hang of it you'll be doing simulations all the time! They're not just useful, they're addictive.

[8] On one hand, sorry. On the other hand, if this isn't something you've learned to do yet, it is one thousand million billion percent worth your time to figure out, even if you never run another simulation again in your life. As I write this, it's the 2020s. Computers run everything around you, and are probably the single most powerful type of object you interact with on a regular basis. Don't you want to understand how to tell those objects what to do? I'm not sure how to effectively get across to you that *even if you're not that good at it, learning how to program a little bit, and how to use a computer well, is **actual, tangible power** in this day and age.* That must be appealing. Right?

1. A process for generating random data

2. An estimation

3. A way of iterating steps 1—2 many times

4. A way of storing the estimation results from each iteration

5. A way of looking at the distribution of estimates across iterations

Let's walk through the processes for coding each of these.

HOW CAN WE CODE A DATA GENERATING PROCESS? This is a key question. Often, when you're doing a simulation, the whole point of it will be to see how an estimator performs under a particular set of data generating assumptions. So we need to know how to make data that satisfies those assumptions.

The first question is how to produce random data in general.[9] There are, generally, functions for generating random data according to all sorts of distributions (as described in Chapter 3). But we can often get by with a few:

- Random uniform data, for generating variables that must lie within certain bounds. `runif()` in R, `runiform()` in Stata, and `numpy.random. uniform()` in Python.[10]

- Normal random data, for generating continuous variables, and in many cases mimicking real-world distributions. `rnorm()` in R, `rnormal()` in Stata, and `numpy.random.normal()` in Python.

- Categorical random data, for categorical or binary variables. `sample()` in R, `runiformint()` in Stata, or `numpy.random. randint()` in Python. The latter two will only work if you want an equal chance of each category. If you want *binary* data with a probability p of getting a 1, there's `rbinomial(1,p)` in Stata and `numpy.random.binomial(1,p)` in Python.[11]

You can use these simple tools to generate the data generating processes you like. For example, take the code snippets a few paragraphs after this in which I generate a data set with a normally distributed ε error variable I call *eps*,[12] a binary X variable which is 1 with probability .2 and 0 otherwise, and a uniform Y variable.

[9] If you want to get picky, this chapter is all about *pseudo*-random data. Most random-number generators in programming languages are actually quite predictable since, y'know, *computers* are pretty predictable. They start with a "seed." They run that seed through a function to generate a new number and also a new seed to use for the next number. For our purposes, pseudo-random is fine. Perhaps even better, because it means we can *set our own seed* to make our analysis reproducible, i.e., everyone should get the same result running our code.

[10] Importantly, pretty much any random distribution can be generated using random uniform data, by running it through an inverse conditional distribution function. Or for categorical data, you can do things like "if $X > .6$ then $A = 2$, if $X <= .6$ and $X >= .25$ then $A = 1$, and otherwise $A = 0$," where X is uniformly distributed from 0 to 1. This is the equivalent of a creating a categorical A, which is 2 with probability .4 $(1 - .6)$, 1 with probability .35 $(.6 - .25)$, and 0 with probability .25.

[11] Want unequal probabilities with more than two categories in Stata or Python? Try carving up uniform data, as in the previous note.

[12] Yes, we can actually see the error term, usually unobservable! The beauty of creating the truth ourselves. This even lets you do stuff like, say, check how closely the residuals from your model match the error term.

```
1  # R CODE
2  library(tidyverse)
3
4  # If we want the random data to be the same every time, set a seed
5  set.seed(1000)
6
7  # tibble() creates a tibble; it's like a data.frame
8  # We must pick a number of observations, here 200
9  d <- tibble(eps = rnorm(200), # by default mean 0, sd 1
10     Y = runif(200), # by default from 0 to 1
11     X = sample(0:1, # sample from values 0 and 1
12               200, # get 200 observations
13               replace = TRUE, # sample with replacement
14               prob = c(.8, .2))) # Be 1 20% of the time
```

```
1  * STATA CODE
2  * Usually we want a blank slate
3  clear
4  * How many observations do we want? Here, 200
5  set obs 200
6
7  * If we want the random data to be the same every time, set a seed
8  set seed 1000
9
10 * Normal data is by default mean 0, sd 1
11 g eps = rnormal()
12 * Uniform data is by default from 0 to 1
13 g Y = runiform()
14 * For unequal probabilities with binary data,
15 * use rbinomial
16 g X = rbinomial(1, .2)
```

```
1  # PYTHON CODE
2  import pandas as pd
3  import numpy as np
4
5  # If we want the results to be the same
6  # every time, set a seed
7  np.random.seed(1000)
8
9  # Create a DataFrame with pd.DataFrame. The size argument of the random
10 # functions gives us the number of observations
11 d = pd.DataFrame({
12         # normal data is by default mean 0, sd 1
13         'eps': np.random.normal(size = 200),
14         # Uniform data is by default from 0 to 1
15         'Y': np.random.uniform(size = 200),
16         # We can use binomial to make binary data
17         # with unequal probabilities
18         'X': np.random.binomial(1, .2, size = 200)
19 })
```

[13] And others, if relevant. Read the documentation to learn about others. It's help(Distributions) in R, help random in Stata, and help(numpy.random) in Python. I keep saying to read the documentation because it's really good advice!

Using these three random-number generators,[13] we can then mimic causal diagrams. This is, after all, a causal inference textbook. We want to know how our estimators will fare in the face of different causal relationships!

15.2.1   Creating Data

HOW CAN WE CREATE DATA THAT FOLLOWS A CAUSAL DIA-
GRAM? EASY! If we use variable A to create variable B, then
$A \to B$. Simple as that. Take the below code snippets, for
example, which modify what we had above to add a back-door
confounder W (and make Y normal instead of uniform). These
code snippets, in other words, are representations of Figure 15.1.

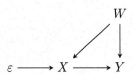

Figure 15.1: Causal Dia-
gram to be Replicated by
Simulation Code

As in Figure 15.1, we have that W causes X and Y, so we
put in that W makes the binary X variable more likely to be
1, but reduces the value of the continuous Y variable. We also
have that ε is a part of creating X, since it causes X, and
that X is a part of creating Y, with a true effect of X on Y
being that a one-unit increase in X causes a 3-unit increase
in Y.[14] Finally, it's not listed on the diagram,[15] but we have
an additional source of randomness, ν, in the creation of Y.
If the only arrows heading into Y were truly X and W, then
Y would be *completely determined* by those two variables—no
other variation once we control for them. You can imagine ν
being added to Figure 15.1 with an arrow $\nu \to Y$.

But presumably there's something else going into Y, and
that's ν, which is also the error term if we regress Y on X
and W. Where's ν in the code? It's that random normal data
being added in when Y is created. You'll also notice that *eps*
has left the code, too. It's actually still in there—it's the uniform random data! We create X here by checking if some uniform random data (ε) is lower than .2 plus W—higher W values
make it more likely we'll be under $.2 + W$, and so more likely
that $X = 1$. "Carving up" the different values of a uniform
random variable can sometimes be easier than working with binomial random data, especially if we want the probability to be
different for different observations.

[14] By the way, if you want to
include heterogeneous treatment effects in your code,
it's easy. Just create the
individual's treatment effect
like a regular variable. Then
use it in place of your single true effect (i.e., here, in
place of the 3) when making
your outcome.
[15] Additional sources of pure
randomness are often left
off of diagrams because they
add a lot of ink and we
tend to assume they're there
anyway.

```
1    # R CODE
2    library(tidyverse)
3
4    # If we want the random data to be the same every time, set a seed
5    set.seed(1000)
6
7    # tibble() creates a tibble; it's like a data.frame
```

```
 8 | # We must pick a number of observations, here 200
 9 | d <- tibble(W = runif(200, 0, .1)) %>% # only go from 0 to .1
10 |     mutate(X = runif(200) < .2 + W) %>%
11 |     mutate(Y = 3*X + W + rnorm(200)) # True effect of X on Y is 3
```

```
 1 | * STATA CODE
 2 | * Usually we want a blank slate
 3 | clear
 4 | * How many observations do we want? Here, 200
 5 | set obs 200
 6 |
 7 | * If we want the random data to be the same every time, set a seed
 8 | set seed 1000
 9 |
10 | * W only varies from 0 to .1
11 | g W = runiform(0, .1)
12 | g X = runiform() < .2 + W
13 | * The true effect of X on Y is 3
14 | g Y = 3*X + W + rnormal()
```

```
 1 | # PYTHON CODE
 2 | import pandas as pd
 3 | import numpy as np
 4 |
 5 | # If we want the results to be the same every time, set a seed
 6 | np.random.seed(1000)
 7 |
 8 | # Create a DataFrame with pd.DataFrame. The size argument of the random
 9 | # functions gives us the number of observations
10 | d = pd.DataFrame({
11 |     # Have W go from 0 to .1
12 |     'W': np.random.uniform(0, .1, size = 200)})
13 |
14 | # Higher W makes X = 1 more likely
15 | d['X'] = np.random.uniform(size = 200) < .2 + d['W']
16 |
17 | # The true effect of X on Y is 3
18 | d['Y'] = 3*d['X'] + d['W'] + np.random.normal(size = 200)
```

We can go one step further. Not only do we want to randomly create this data, we are going to want to do so *a bunch of times* as we repeatedly run our simulation. Any time we have code we want to run a bunch of times, it's often a good idea to put it in its own *function*, which we can repeatedly call. We'll also give this function an argument that lets us pick different sample sizes.

```
 1 | # R CODE
 2 | library(tidyverse)
 3 |
 4 | # Make sure the seed goes OUTSIDE the function. It makes the random
 5 | # data the same every time, but we want DIFFERENT results each time
 6 | # we run it (but the same set of different results, thus the seed)
 7 | set.seed(1000)
 8 |
 9 | # Make a function with the function() function. The "N = 200" argument
10 | # gives it an argument N that we'll use for sample size. The "=200" sets
11 | # the default sample size to 200
12 | create_data <- function(N = 200) {
```

```
13          d <- tibble(W = runif(N, 0, .1)) %>%
14          mutate(X = runif(N) < .2 + W) %>%
15          mutate(Y = 3*X + W + rnorm(N))
16
17          # Use return() to send our created data back
18          return(d)
19  }
20
21  # Run our function!
22  create_data(500)
```

```
1   * STATA CODE
2   * Make sure the seed goes OUTSIDE the function. It makes the random data
3   * the same every time, but we want DIFFERENT results each time we run it
4   * (but the same set of different results, thus the seed)
5   set seed 1000
6
7   * Create a function with program define. It will throw an error if there's
8   * already a function with that name, so we must drop it first if it's there
9   * But ugh... dropping will error if it's NOT there. So we capture the drop
10  * to ignore any error in dropping
11  capture program drop create_data
12  program define create_data
13      * `1': first thing typed after the command name, which can be our N
14      local N = `1'
15
16      clear
17      set obs `N'
18
19      g W = runiform(0, .1)
20      g X = runiform() < .2 + W
21      * The true effect of X on Y is 3
22      g Y = 3*X + W + rnormal()
23      * End ends the program
24      * Unlike the other languages we aren't returning anything,
25      * instead the data we've created here is just... our data now.
26  end
27
28  * Run our function!
29  create_data 500
```

```
1   # PYTHON CODE
2   import pandas as pd
3   import numpy as np
4
5   # Make sure the seed goes OUTSIDE the function. It makes the random data
6   # the same every time, but we want DIFFERENT results each time we run it
7   # (but the same set of different results, thus the seed)
8   np.random.seed(1000)
9
10  # Make a function with def. The "N = 200" argument gives it an argument N
11  # that we'll use for sample size. The "=200" sets the default to 200
12  def create_data(N = 200):
13      d = pd.DataFrame({
14      'W': np.random.uniform(0, .1, size = N)})
15      d['X'] = np.random.uniform(size = N) < .2 + d['W']
16      d['Y'] = 3*d['X'] + d['W'] + np.random.normal(size = N)
17      # Use return() to send our created data back
18      return(d)
19
```

```
20   # And run our function!
21   create_data(500)
```

WHAT IF WE WANT SOMETHING MORE COMPLEX? We'll just have to think more carefully about how we create our data!

A common need in creating more-complex random data is in mimicking a panel structure. How can we create data with multiple observations per individual? Well, we create an individual identifier, we give that identifier some individual characteristics, and then we apply those to the entire time range of data. This is easiest to do by merging an individual-data set with an individual-time data set in R and Python, or by use of **egen** in Stata.

```
1    # R CODE
2    library(tidyverse)
3    set.seed(1000)
4
5    # N for number of individuals, T for time periods
6    create_panel_data <- function(N = 200, T = 10) {
7        # Create individual IDs with the crossing()
8        # function, which makes every combination of two vectors
9        # (if you want an unbalanced panel, drop some observations)
10       panel_data <- crossing(ID = 1:N, t = 1:T) %>%
11       # And individual/time-varying data
12       # (n() means "the number of rows in this data"):
13       mutate(W1 = runif(n(), 0, .1))
14
15       # Now an individual-specific characteristic
16       indiv_data <- tibble(ID = 1:N,
17                            W2 = rnorm(N))
18
19
20       # Join them
21       panel_data <- panel_data %>%
22           full_join(indiv_data, by = 'ID') %>%
23           # Create X, caused by W1 and W2
24           mutate(X = 2*W1 + 1.5*W2 + rnorm(n())) %>%
25           # And create Y. The true effect of X on Y is 3
26           # But W1 and W2 have causal effects too
27           mutate(Y = 3*X + W1 - 2*W2 + rnorm(n()))
28
29       return(panel_data)
30   }
31
32   create_panel_data(100, 5)
```

```
1    * STATA CODE
2    * The first() egen function is in egenmore; ssc install egenmore if necessary
3    set seed 1000
4
5    capture program drop create_panel_data
6    program define create_panel_data
7            * `1': the first thing typed after the command name, which can be N
8            local N = `1'
9            * The second thing can be our T
10           local T = `2'
```

```
11
12              clear
13              * We want N*T observations
14              local obs = `N'*`T'
15              set obs `obs'
16
17              * Create individual IDs by repeating 1 T times, then 2 T times, etc.
18              * We can do this with floor((_n-1)/`T') since _n is the row number
19              * Think about why this works!
20              g ID = floor((_n-1)/`T')
21              * Now we want time IDs by repeating 1, 2, 3, ..., T a full N times.
22              * We can do this with mod(_n,`T'), which gives the remainder from
23              * _n/`T'. Think about why this works!
24              g t = mod((_n-1),`T')
25
26              * Individual and time-varying data
27              g W1 = rnormal()
28
29              * This variable will be constant at the individual level.
30              * We start by making a variable just like W1
31              g temp = rnormal()
32              * But then use egen to apply it to all observations with the same ID
33              by ID, sort: egen W2 = first(temp)
34              drop temp
35
36              * Create X based on W1 and W2
37              g X = 2*W1 + 1.5*W2 + rnormal()
38              * The true effect of X on Y is 3 but W1 and W2 cause Y too
39              g Y = 3*X + W1 - 2*W2 + rnormal()
40      end
41
42      create_panel_data 100 5
```

```
1       # PYTHON CODE
2       import pandas as pd
3       import numpy as np
4       from itertools import product
5       np.random.seed(1000)
6
7       # N for number of individuals, T for time periods
8       def create_panel_data(N = 200, T = 10):
9               # Use product() to get all combinations of individual and
10              # time (if you want some to be incomplete, drop later)
11              p = pd.DataFrame(
12              product(range(0,N), range(0, T)))
13              p.columns = ['ID','t']
14
15              # Individual- and time-varying variable
16              p['W1'] = np.random.normal(size = N*T)
17
18              # Individual data
19              indiv_data = pd.DataFrame({
20                      'ID': range(0,N),
21                      'W2': np.random.normal(size = N)})
22
23              # Bring them together
24              p = p.merge(indiv_data, on = 'ID')
25
26              # Create X, caused by W1 and W2
27              p['X'] = 2*p['W1'] + 1.5*p['W2'] + np.random.normal(size = N*T)
```

```
28
29           # And create Y. The true effect of X on Y is 3
30           # But W1 and W2 have causal effects too
31           p['Y'] = 3*p['X'] + p['W1']- 2*p['W2'] + np.random.normal(size = N*T)
32           return(p)
33
34    create_panel_data(100, 5)
```

Another pressing concern in the creation of random data is
in creating correlated error terms. If we're worried about the
effects of non-independent error terms, then simulation can be a
great way to try stuff out, if we can make errors that are exactly
as dependent as we like. There are infinite varieties to consider,
but I'll consider two: heteroskedasticity and clustering.

Heteroskedasticity is relatively easy to create. Whenever
you're creating variables that are meant to represent the error term, we just allow the variance of that data to vary based
on the observation's characteristics.

```
1     # R CODE
2     library(tidyverse)
3     set.seed(1000)
4
5     create_het_data <- function(N = 200) {
6             d <- tibble(X = runif(N)) %>%
7             # Let the standard deviation of the error
8             # Be related to X. Heteroskedasticity!
9             mutate(Y = 3*X + rnorm(N, sd = X*5))
10
11            return(d)
12    }
13
14    create_het_data(500)
```

```
1     * STATA CODE
2     set seed 1000
3
4     capture program drop create_het_data
5     program define create_het_data
6             * `1`: the first thing typed after the command name, which can be N
7             local N = `1'
8             clear
9             set obs `N'
10
11            g X = runiform()
12
13            * Let the standard deviation of the error
14            * Be related to X. Heteroskedasticity!
15            g Y = 3*X + rnormal(0, 5*X)
16    end
17
18    create_het_data 500
```

```
1     # PYTHON CODE
2     import pandas as pd
3     import numpy as np
```

```
4  np.random.seed(1000)
5
6  def create_het_data(N = 200):
7        d = pd.DataFrame({
8              'X': np.random.uniform(size = N)})
9        # Let the standard deviation of the error
10       # Be related to X. Heteroskedasticity!
11       d['Y'] = 3*d['X'] + np.random.normal(scale = 5*d['X'])
12
13       return(d)
14
15 create_het_data(500)
```

If we plot the result of `create_het_data` we get Figure 15.2. The heteroskedastic errors are pretty apparent there! We get much more variation around the regression line on the right side of the graph than we do on the left. This is a classic "fan" shape for heteroskedasticity.

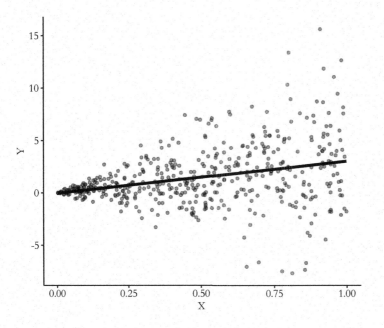

Figure 15.2: Simulated Data with Heteroskedastic Errors

Clustering is just a touch harder. First, we need something to cluster *at*. This is often some group-identifier variable like the ones we talked about earlier in this section when discussing how to mimic panel structure. Then, we can create clustered standard errors by creating or randomly generating a single "group effect" that can be shared by the group itself, adding on individual noise

In other words, generate an individual-level error term, then an individual/time-level error term, and add them together in

some fashion to get clustered standard errors. Thankfully, we
can reuse a lot of the tricks we learned for creating panel data.

```r
# R CODE
library(tidyverse)
set.seed(1000)

# N for number of individuals, T for time periods
create_clus_data <- function(N = 200, T = 10) {
        # We're going to create errors clustered at the
        # ID level. So we can follow our steps from making panel data
        panel_data <- crossing(ID = 1:N, t = 1:T) %>%
            # Individual/time-varying data
            mutate(W = runif(n(), 0, .1))

        # Now an individual-specific error cluster
        indiv_data <- tibble(ID = 1:N,
                             C = rnorm(N))

        # Join them
        panel_data <- panel_data %>%
            full_join(indiv_data, by = 'ID') %>%
            # Create X, caused by W1 and W2
            mutate(X = 2*W + rnorm(n())) %>%
            # The error term has two components: the individual
            # cluster C, and the individual-and-time-varying element
            mutate(Y = 3*X + (C + rnorm(n())))

        return(panel_data)
}

create_clus_data(100,5)
```

```stata
* STATA CODE
* The first() egen function is in egenmore; ssc install egenmore if necessary
set seed 1000

capture program drop create_clus_data
program define create_clus_data
        * `1': the first thing typed after the command name, which can be N
        local N = `1'
        * The second thing can be our T
        local T = `2'

        clear
        * We want N*T observations
        local obs = `N'*`T'
        set obs `obs'

        * We'll be making errors clustered by ID
        * so we can follow the same steps from when we made panel data
        g ID = floor((_n-1)/`T')
        g t = mod((_n-1),`T')

        * Individual and time-varying data
        g W = rnormal()

        * An error component at the individual level
        g temp = rnormal()
        * Use egen to apply the value to all observations with the same ID
```

```
28            by ID, sort: egen C = first(temp)
29            drop temp
30
31            * Create X
32            g X = 2*W + rnormal()
33            * The error term has two components: the individual cluster C,
34            * and the individual-and-time-varying element
35            g Y = 3*X + (C + rnormal())
36 end
37
38 create_clus_data 100 5
```

```
1  # PYTHON CODE
2  import pandas as pd
3  import numpy as np
4  from itertools import product
5  np.random.seed(1000)
6
7  # N for number of individuals, T for time periods
8  def create_clus_data(N = 200, T = 10):
9          # We're going to create errors clustered at the
10         # ID level. So we can follow our steps from making panel data
11         p = pd.DataFrame(
12         product(range(0,N), range(0, T)))
13         p.columns = ['ID','t']
14
15         # Individual- and time-varying variable
16         p['W'] = np.random.normal(size = N*T)
17
18         # Now an individual-specific error cluster
19         indiv_data = pd.DataFrame({
20                 'ID': range(0,N),
21                 'C': np.random.normal(size = N)})
22
23         # Bring them together
24         p = p.merge(indiv_data, on = 'ID')
25
26         # Create X
27         p['X'] = 2*p['W'] + np.random.normal(size = N*T)
28
29         # And create Y. The error term has two components: the individual
30         # cluster C, and the  individual-and-time-varying element
31         p['Y'] = 3*p['X'] + (p['C'] + np.random.normal(size = N*T))
32         return(p)
33
34 create_clus_data(100, 5)
```

HOW ABOUT OUR ESTIMATOR? No causal-inference simulation is complete without an estimator. Code-wise, we know the drill. Just, uh, run the estimator we're interested in on the data we've just randomly generated.

Let's say we've read in our textbook that clustered errors only affect the standard errors of a linear regression, but don't bias the estimator. We want to confirm that for ourselves. So we take some data with clustered errors baked into it, and we estimate a linear regression on that data.

The tricky part at this point is often figuring out how to get the estimate you want (a coefficient, usually), *out* of that regression so you can return it and store it. If what you're going after *is* a coefficient from a linear regression, you can get the coefficient on the variable `varname` in R with `coef(m)['varname']` for a model m, in Stata with `_b[varname]` just after running the model, and in Python with `m.params['varname']` for a model m.[16]

Since this is, again, something we'll want to call repeatedly, we're going to make it its own function, too. A function that, itself, calls the data generating function we just made! We'll be making reference to the `create_clus_data` function we made in the previous section.

[16] To extract something else you might have to do a bit more digging. But often an Internet search for "how do I get STATISTIC from ANALYSIS in LANGUAGE" will do it for you.

```
1    # R CODE
2    library(tidyverse)
3    set.seed(1000)
4
5    # A function for estimation
6    est_model <- function(N, T) {
7            # Get our data. This uses create_clus_data from earlier
8            d <- create_clus_data(N, T)
9
10           # Run a model that should be unbiased
11           # if clustered errors themselves don't bias us!
12           m <- lm(Y ~ X + W, data = d)
13
14           # Get the coefficient on X, which SHOULD be true value 3 on average
15           x_coef <- coef(m)['X']
16
17           return(x_coef)
18   }
19
20   # Run our model
21   est_model(200, 5)
```

```
1    * STATA CODE
2    set seed 1000
3
4    capture program drop est_model
5    program define est_model
6            * `1: the first thing typed after the command name, which can be N
7            local N = `1'
8            * The second thing can be our T
9            local T = `2'
10
11           * Create our data. This uses create_clus_data from earlier
12           create_clus_data `N' `T'
13
14           * Run a model that should be unbiased
15           * if clustered errors themselves don't bias us!
16           reg Y X W
17   end
18
19   * Run our model
```

```
20 || est_model 200 5
21 || * and get the estimate
22 || display _b[X]
```

```
 1 || # PYTHON CODE
 2 || import pandas as pd
 3 || import numpy as np
 4 || from itertools import product
 5 || import statsmodels.formula.api as smf
 6 || np.random.seed(1000)
 7 ||
 8 || def est_model(N = 200, T = 10):
 9 ||         # This uses create_clus_data from earlier
10 ||         d = create_clus_data(N, T)
11 ||
12 ||         # Run a model that should be unbiased
13 ||         # if clustered errors themselves don't bias us!
14 ||         m   = smf.ols('Y ~ X + W', data = d).fit()
15 ||
16 ||         # Get the coefficient on X, which SHOULD be true value 3 on average
17 ||         x_coef = m.params['X']
18 ||
19 ||         return(x_coef)
20 ||
21 || # Estimate our model!
22 || est_model(200, 5)
```

NOW WE MUST ITERATE. We can't just run our simulation once—any result we got might be a fluke. Plus, we usually want to know about the *distribution* of what our estimator gives us, either to check how spread-out it is or to see if its mean is close to the truth. Can't get a sampling distribution without multiple samples. How many times should you run your simulation? There's no hard-and-fast number, but it should be a bunch! You can run just a few iterations when testing out your code to make sure it works, but when you want results you should go for more. As a rule of thumb, the more detail you want in your results, or the more variables there are in your estimation, the more you should run—want to see the bootstrap distribution of a mean? A thousand or so should be more than you need. A correlation or linear regression coefficient? A few thousand. Want to precisely estimate a bunch of percentiles of a sampling distribution? Maybe go for a number in the *tens* of thousands.

Iteration is one thing that computers are super good at. In all three of our code examples, we'll be using a *for loop* in slightly different ways. A for loop is a computer-programming way of saying "do the same thing a bunch of times, once for each of these values."

For example, the (generic-language/pseudocode) for loop:

```
for i in [1, 4, 5, 10]: print i
```

would do the same thing (`print i`) a bunch of times, once for each of the values we've listed (1, 4, 5, and 10). First it would do it with `i` set to 1, and so print the number 1. Then it would do it with `i` set to 4, and print the number 4. Then it would print 5, and finally it would print 10.

In the case of simulation, our for loop would look like:

`for i in [1 to 100]: do my simulation`

where we're not using the iteration number `i` at all,[17] but instead just repeating our simulation, in this case 100 times.

The way this loop ends up working is slightly different in each language. Python has the purest, "most normal" loop, which will look a lot like the generic language. R *could* also look like that, but I'm going to recommend the use of the **purrr** package and its `map` functions, which will save us a step later. For Stata, the loop itself will be pretty normal, but the fact that we're creating new data all the time, plus Stata's limitation of only using one data set at a time, is going to require us to `clear` our data out over and over (plus, it will add a step later).

All of these code examples will refer back to the `create_clus` `_data` and `est_model` functions we defined in earlier code chunks.

```
1   # R CODE
2   library(tidyverse)
3   library(purrr)
4   set.seed(1000)
5
6   # Estimate our model 1000 times (from 1 to 1000)
7   estimates <- 1:1000 %>%
8       # Run the est_model function each time
9       map_dbl(function(x) est_model(N = 200, T = 5))
10  # There are many map functions in purrr. Since est_model outputs a
11  # number (a "double"), we can use map_dbl and get a vector of estimates
```

```
1   * STATA CODE
2   set seed 1000
3
4   * Use forvalues to loop from 1 to 1000 to estimate 1000 times
5   forvalues i = 1(1)1000 {
6       * We don't need to see the whole regression, so use quietly
7       quietly est_model 200 5
8
9       * But let's print out the result!
10      display _b[X]
11  }
```

```
1   # PYTHON CODE
2   import pandas as pd
3   import numpy as np
4   from itertools import product
5   import statsmodels.formula.api as smf
6   np.random.seed(1000)
7
```

```
 8 │ # This runs est_model once for each iteration as it iterates through
 9 │ # the range from 0 to 999 (1000 times total)
10 │ estimates = [est_model(200, 5) for i in range(0,1000)]
```

NOW A WAY OF STORING THE RESULTS. We already have a function that runs our estimation, which itself calls the function that creates our simulated data. We need a way to store those estimation results.

In both R and Python, our approach to iteration is enough. The `map()` functions in R, and the standard for-loop structure in Python, give us back a vector of results over all the iterations. So for those languages, we're already done. The last code chunk already has the instructions you need.

Stata is a bit trickier given its preference for only having one data set at a time. How can you store the randomly-generated data and the result-of-estimation data at once? We can use the `preserve`/`restore` trick—store the results in a data set, `preserve` that data set so Stata remembers it, then clear everything and run our estimation, save the estimate as a `local`, `restore` back to our results data set, and assign that `local` into data to remember it.[18]

Now, I will say, there *is* a way around this sort of juggling, and that's to use the `simulate` command in Stata. With `simulate`, you specify a single function for creating data and estimating a model, and then pulling all the results back out. `help simulate` should be all you need to know. I find it a bit less flexible than coding things explicitly and not much less work, so I don't use it in this book.[19] But you might want to check it out—you may find yourself preferring it.

[18] Stata 16 has ways of using more than one data set at once, but I don't want to assume you have Stata 16, and also I never learned this functionality myself, and suspect most Stata users out there are just `preserve`/`restoring` to their heart's content. We could also try storing the results in a matrix, but if you can get away with avoiding Stata matrices you'll live a happier life.

[19] And heck, I've been running simulations in Stata since like 2008 and I managed to get by without it until now.

```
 1 │ * STATA CODE
 2 │ set seed 1000
 3 │
 4 │ * Our "parent" data set will store our results
 5 │ clear
 6 │ * We're going to run 1000 times and so need 1000
 7 │ * observations to store data in
 8 │ set obs 1000
 9 │ * and a variable to store them in
10 │ g estimates = .
11 │
12 │ * Estimate the model 1000 times
13 │ forvalues i = 1(1)100 {
14 │     * preserve to store our parent data
15 │     preserve
16 │     quietly est_model 200 5
17 │     * restore to get parent data back
18 │     restore
19 │     * and save the result in the ith row of data
20 │     replace estimates = _b[X] in `i'
21 │ }
```

FINALLY, WE EVALUATE. We can now check whatever characteristics of the sampling estimator we want! With our set of results across all the randomly generated samples in hand, we can, say, look and see whether the mean of the sampling variation is the true effect we decided!

Let's check here. We know from our data generation function that the true effect of X on Y is 3. Do we get an average near 3 in our simulated analysis? A simple `mean()` in R, `summarize` or `mean` in Stata, or `numpy.mean()` in Python, gives us a mean of 3.00047. Pretty close! Seems like our textbook was right: clustered errors alone won't bias an otherwise unbiased estimate.

How about the standard errors? The distribution of estimates across all our random samples should mimic sampling distribution. So, **the standard deviation of our results is the standard error of the estimate.** `sd()` in R, `summarize` in Stata, or `numpy.std()` in Python show a standard deviation of .045.

So an honest linear regression estimator should give standard errors of .045. Does it? We can use simulation to answer that. We can take the simulation code we already have but swap out the part where we store a coefficient estimate to instead store the standard error. This can be done easily with `tidy()` from the **broom** package in R,[20] `_se[X]` in Stata, or `m.bse['X']` in Python, where X is the name of the variable you want the standard errors of. When we simulate *that* a few hundred times, the mean reported standard error is .044. Just a hair too small. If we instead estimate heteroskedasticity-robust standard errors, we get .044. Not any better! If we treat the standard errors as is most appropriate, given the clustered errors we know we created, and use standard errors clustered at the group level, we get .045. Seems like that worked pretty well!

[20] If you prefer, you could skip **broom** and get the standard errors with `coef(summary(m))['X', 'Std. Error']`. But **broom** makes extracting so many other parts of the regression object so much easier I recommend getting used to it.

15.3   The Old Man and the C

AS I MENTIONED,[21] one of the most powerful things that simulation can do in the context of causal inference is let you know if a given estimator *works* to provide, on average, the *true* causal effect. Bake a true causal effect into your data generating function, then put the estimator you're curious about in the estimation function, and let 'er rip! If the mean of the estimates is near the truth and the distribution isn't too wide, you're good to go.

[21] None of the code examples in this book use the C language, but I hope you'll excuse me.

This is a fantastic way to test out *any* new method or statistical claim you're exposed to. We already checked whether OLS still works on average with clustered errors. But we could do plenty more. Can one back door cancel another out if the effects on the treatment have opposite signs? Does matching really perform better than regression when one of the controls has a nonlinear relationship with the outcome? Does controlling for a collider *really* add bias? Make a data set with a back-door path blocked by a collider, and see what happens if you control or don't control for it in your estimation function.[22]

This makes simulation a great tool for three things: (1) trying out new estimators, (2) comparing different estimators, and (3) seeing what we need to break in order to make an estimator stop working.

We can do all of these things without having to write econometric proofs. Sure, our results won't generalize as well as a proof would, but it's a good first step. Plus, it puts the power in our hands!

We can imagine an old man stuck on a desert island somewhere, somehow equipped with a laptop and his favorite statistical software, but without access to textbooks or experts to ask. Maybe his name is Agus. Agus has his own statistical questions—what works? What doesn't? Maybe he has a neat idea for a new estimator. He can try things out himself. I suggest you do too. Trying stuff out in simulation is not just a good way to get answers on your own, it's also a good way to get to know how data works.

[22] By the way, making data with a collider in it is a great way to intuitively convince yourself that their uncontrolled presence doesn't bias you. To make data with, for example, a collider C with the path $X \rightarrow C \leftarrow Y$, you have to first make X, and then make Y with whatever effect of X, and then finally make C using both X and Y. You could have stopped before even *making* C and still run a regression of Y on X. Surely just *creating* the variable doesn't change the coefficient.

15.3.1   Trying out New Estimators

PROFESSORS TEND TO SHOW YOU THE ESTIMATORS THAT WORK WELL. But that can give the false illusion that these estimators fell from above, all of them working properly. Untrue! Someone had to think them up first, and try them out, and prove that they worked.

There's no reason you can't come up with your own and try it out. Sure, it probably won't work. But I bet you'll learn something.

Let's try our own estimator. Or Agus' estimator. Agus had a thought: you know how the standard error of a regression gets smaller as the variance of X gets bigger? Well... what if we only pick the extreme values of X? That will make the variance of X

get very big. Maybe we'll end up with smaller standard errors as a result.

So that's Agus' estimator: instead of regressing Y on X as normal, first limit the data to just observations with extreme values of X (maybe the top and bottom 20%), and then regress Y on X using *that* data.[23] We think it might reduce standard errors, although we'd also want to make sure it doesn't introduce bias.

Agus has a lot of ideas like this. He writes them on a palm leaf and then programs them on his old sandy computer to see if they work. For the ones that don't work, he takes the palm leaf and floats it gently on the ocean, giving it a push so that it drifts out to the horizon, bidden farewell. Will this estimator join the others at sea?

Let's find out by coding it up ourselves. Can you really code up your own estimator? Sure! Just write code that tells the computer to do what your estimator does. And if you don't know how to write that code, search the Internet until you do.

There's a tweak in this code, as well. We're interested in whether the effect is unbiased, but also whether standard errors will be reduced. So we may want to collect both the coefficient *and* its standard error. Two things to return. How can we do it? See the code.

[23] You might be anticipating at this point that this isn't going to work well. At the very least, if it did I'd probably have already mentioned it in this book or something, right? But do you know *why* it won't work well? Just sit and think about it for a while. Then, once you've got an idea of why you think it might be... you can try that idea out in simulation! How would you construct a simulation to test whether your explanation why it doesn't work is the right one?

```
1  # R CODE
2  library(tidyverse); library(purrr); library(broom)
3  set.seed(1000)
4
5  # Data creation function. Let's make the function more flexible
6  # so we can choose our own true effect!
7  create_data <- function(N, true) {
8          d <- tibble(X = rnorm(N)) %>%
9          mutate(Y = true*X + rnorm(n()))
10
11         return(d)
12 }
13
14 # Estimation function. keep is the portion of data in each tail
15 # to keep. So .2 would keep the bottom and top 20% of X
16 est_model <- function(N, keep, true) {
17         d <- create_data(N, true)
18
19         # Agus' estimator!
20         m <- lm(Y~X, data = d %>%
21             filter(X <= quantile(X, keep) | X >= quantile(X, (1-keep))))
22
23         # Return coefficient and standard error as two elements of a list
24         ret <- tidy(m)
25         return(list('coef' = ret$estimate[2],
26         'se' = ret$std.error[2]))
27 }
```

```
28
29    # Run 1000 simulations. use map_df to stack all the results
30    # together in a data frame
31    results <- 1:1000 %>%
32        map_df(function(x) est_model(N = 1000,
33                                     keep = .2,
34                                     true = 2))
35    mean(results$coef); sd(results$coef); mean(results$se)
```

```
1    * STATA CODE
2    set seed 1000
3
4    * Data creation function
5    capture program drop create_data
6    program define create_data
7            clear
8            local N = `1'
9            * The second argument will let us be a bit more flexible -
10           * we can pick our own true effect!
11           local true = `2'
12
13           set obs `N'
14
15           g X = rnormal()
16           g Y = `true'*X + rnormal()
17    end
18
19    * Model estimation function
20    capture program drop est_model
21    program define est_model
22           local N = `1'
23           * Our second argument will be  the proportion to keep in
24           * each tail so .2 would mean keep the bottom 20% and top 20%
25           local keep = `2'
26           local true = `3'
27
28           create_data `N' `true'
29
30           * Since X has no missings, if we sort by X, the top and bottom X%
31           * of X will be in the first and last X% of observations
32           * Agus' estimator!
33           sort X
34           drop if _n > `keep'*`N' & _n < (1-`keep')*`N'
35           reg Y X
36    end
37
38    * To return multiple things, we need only to have
39    * Multiple "parent" variables to store them in
40    clear
41    set obs 1000
42    g estimate = .
43    g se = .
44
45    forvalues i = 1(1)1000 {
46           preserve
47           quietly est_model 1000 .2 2
48           restore
49
50           replace estimate = _b[X] in `i'
51           replace se = _se[X] in `i'
```

```
52  }
53
54  summ estimate se
```

```
1   # PYTHON CODE
2   import pandas as pd
3   import numpy as np
4   import statsmodels.formula.api as smf
5   np.random.seed(1000)
6
7   # Data creation function. Let's also make the function more
8   # flexible - we can choose our own true effect!
9   def create_data(N, true):
10          d = pd.DataFrame({'X': np.random.normal(size = N)})
11          d['Y'] = true*d['X'] + np.random.normal(size = N)
12          return(d)
13
14  # Estimation function. keep is the portion of data in each tail
15  # to keep. So .2 would keep the bottom and top 20% of X
16  def est_model(N, keep, true):
17          d = create_data(N, true)
18
19          # Agus' estimator!
20          d = d.loc[(d['X'] <= np.quantile(d['X'], keep)) |
21          (d['X'] >= np.quantile(d['X'], 1-keep))]
22          m = smf.ols('Y~X', data = d).fit()
23
24          # Return the two things we want as an array
25          ret = [m.params['X'], m.bse['X']]
26          return(ret)
27
28  # Estimate the results 1000 times
29  results = [est_model(1000, .2, 2) for i in range(0,1000)]
30
31  # Turn into a DataFrame
32  results = pd.DataFrame(results, columns = ['coef', 'se'])
33
34  # Let's see what we got!
35  np.mean(results['coef'])
36  np.std(results['coef'])
37  np.mean(results['se'])
```

If we run this, we see an average coefficient of 2.00 (where the truth was 2), a standard deviation of the coefficient of .0332, and a mean standard error of .0339. See, told you the standard deviation of the simulated coefficient matches the standard error of the coefficient!

So the average is 2.00 relative to the truth of also-2. No bias! Pretty good, right? Agus should be celebrating. Of course, we're not done yet. We didn't just want no bias, we wanted lower standard errors than the regular estimation method. How can we get that?

15.3.2 Comparing Estimators

ONE THING THAT SIMULATION ALLOWS US TO DO is *compare the performance* of different methods. You can take the exact same randomly-generated data, apply two different methods to it, and see which performs better. This process is often known as a "horse race."

The horse race method can be really useful when considering multiple different ways of doing the same thing. For example, in Chapter 14 on matching, I mentioned a bunch of different choices that could be made in the matching process that, made one way, would reduce bias, but made another way would improve precision. Allowing a wider caliper brings in more, but worse matches, increasing bias but also making the sampling distribution narrower so you're less likely to be *way* off. That's a tough call to make—lower average bias, or narrower sampling distributions? But it might be easier if it just happens to turn out that the amount of bias reduction you get isn't very high at all, but the precision savings are massive! How could you know such a thing? Try a horse race simulation.[24]

IN THE CASE OF AGUS' ESTIMATOR, we want to know whether it produces lower standard errors than a regular regression. That was the whole idea, right? So let's check. These code examples all use the **create_data** function from the previous set of code chunks.

[24] Then, of course, the amount of bias and precision you're trading off would be sensitive to what your data looks like. So you might want to run that simulation with your actual data! You wouldn't know the "true" value in that case, but you could check balance to get an idea of what bias is. See the bootstrap section below for an idea of how simulation with actual data works.

```r
 1  # R CODE
 2  library(tidyverse); library(purrr);
 3  library(broom); library(vtable)
 4  set.seed(1000)
 5
 6  # Data creation function. Let's make the function more
 7  # flexible - we can choose our own true effect!
 8  create_data <- function(N, true) {
 9          d <- tibble(X = rnorm(N)) %>%
10                  mutate(Y = true*X + rnorm(n()))
11
12          return(d)
13  }
14
15  # Estimation function. keep is the portion of data in each tail
16  # to keep. So .2 would keep the bottom and top 20% of X
17  est_model <- function(N, keep, true) {
18          d <- create_data(N, true)
19
20          # Regular estimator!
21          m1 <- lm(Y~X, data = d)
22
23          # Agus' estimator!
24          m2 <- lm(Y~X, data = d %>%
25                  filter(X <= quantile(X, keep) | X >= quantile(X, (1-keep))))
```

```
26
27            # Return coefficients as a list
28            return(list('coef_reg' = coef(m1)[2],
29            'coef_agus' = coef(m2)[2]))
30  }
31
32  # Run 1000 simulations. Use map_df to stack all the
33  # results together in a data frame
34  results <- 1:1000 %>%
35            map_df(function(x) est_model(N = 1000,
36            keep = .2, true = 2))
37  sumtable(results)
```

```
1   * STATA CODE
2   set seed 1000
3
4   * Stata can make horse races a bit tricky since
5   * _b and _se only report the *most recent* regression results
6   * But we can store and restore our estimates to get around this!
7
8   * Model estimation function
9   capture program drop est_model
10  program define est_model
11            local N = `1'
12            local keep = `2'
13            local true = `3'
14
15            create_data `N' `true'
16
17            * Regular estimator!
18            reg Y X
19            estimates store regular
20
21            * Agus' estimator!
22            sort X
23            drop if _n > `keep'*`N' & _n < (1-`keep')*`N'
24            reg Y X
25            estimates store agus
26  end
27
28  * Now for the parent data and our loop
29  clear
30  set obs 1000
31  g estimate_agus = .
32  g estimate_reg = .
33
34  forvalues i = 1(1)1000 {
35            preserve
36            quietly est_model 1000 .2 2
37            restore
38
39            * Bring back the Agus model
40            estimates restore agus
41            replace estimate_agus = _b[X] in `i'
42
43            * And now the regular one
44            estimates restore regular
45            replace estimate_reg = _b[X] in `i'
46  }
47
```

```
48  * Summarize our results
49  summ estimate_agus estimate_reg
```

```
1   # PYTHON CODE
2   import pandas as pd
3   import numpy as np
4   import statsmodels.formula.api as smf
5   np.random.seed(1000)
6
7   def est_model(N, keep, true):
8           d = create_data(N, true)
9
10          # Regular estimator
11          m1 = smf.ols('Y~X', data = d).fit()
12
13          # Agus' estimator!
14          d = d.loc[(d['X'] <= np.quantile(d['X'], keep)) |
15          (d['X'] >= np.quantile(d['X'], 1-keep))]
16
17          m2 = smf.ols('Y~X', data = d).fit()
18
19          # Return the two things we want as an array
20          ret = [m1.params['X'], m2.params['X']]
21
22          return(ret)
23
24  # Estimate the results 1000 times
25  results = [est_model(1000, .2, 2) for i in range(0,1000)]
26
27  # Turn into a DataFrame
28  results = pd.DataFrame(results, columns = ['coef_reg', 'coef_agus'])
29
30  # Let's see what we got!
31  np.mean(results['coef_reg'])
32  np.std(results['coef_reg'])
33  np.mean(results['coef_agus'])
34  np.std(results['coef_agus'])
```

What do we get? While both estimators get 2.00 on average for the coefficient, the standard errors are actually lower for the regular estimator. The standard deviation of the coefficient's distribution across all the simulated samples is .034 for Agus and .031 for the regular estimator. Regular wins![25] Looks like Agus will set his creation adrift at sea once again.

15.3.3   Breaking Things

STATISTICAL ANALYSIS IS ABSOLUTELY FULL TO THE BRIM WITH ALL KINDS OF ASSUMPTIONS. Annoying, really. Here's the thing, though—these assumptions all need to be true for us to *know* what our estimators are doing. But maybe there are a few assumptions that could be false and we'd *still have a pretty good idea* what those estimators are doing.

[25] A few reasons why Agus' estimator isn't an improvement: first, it tosses out data, reducing the effective sample size. Second, while the variance of X *is* in the denominator of the standard error, the OLS residuals are in the numerator! Anything that harms prediction, like picking tail data, will harm precision and increase standard errors. Also, picking tail data is just... noisier.

So... how many assumptions can be wrong and we'll still get away with it? Simulation can help us figure that out.

For example, if we want a linear regression coefficient on X to be an unbiased estimate of the causal effect of X, we need to assume that X is unrelated to the error term, a.k.a. there are no open back doors.

In the social sciences, this is rarely going to be *true*. There's always *some* little thing we can't measure or control for that's unfortunately related to both X and the outcome. So should we never bother with regression?

No way! Instead, we should ask "how big a problem is this likely to be?" and if it's only a small problem, then hey, close to right is better than far, even if it's not all the way there!

SIMULATION CAN BE USED TO PROBE THESE SORTS OF IS- SUES by making a very flexible data creation function, flexible enough to try different kinds of assumption violations or degrees of those violations. Then, run your simulation with different set- tings and see how bad the results come out. If they're pretty close to the truth, even with data that violates an assumption, maybe it's not so bad![26]

Perhaps you design your data generation function to have an argument called `error_dist`. Program it so that if you set `error_dist` to `"normal"` you get normally-distributed er- ror terms. `"lognormal"` and you get log-normally distributed error terms, and so on for all sorts of distributions. Then, you can simulate a linear regression with all different kinds of error distributions and see whether the estimates are still good.

How could we use simulation in application to a remaining open back door? We can program our data generation function to have settings for the strength of that open back door. If we have $X \leftarrow W \rightarrow Y$ on our diagram, what's the effect of W on X, and what's the effect of W on Y? How strong do those two arrows need to be to give us an unacceptably high level of bias?

Once we figure out how bad the problem would need to be to give us unacceptably high levels of bias,[27] you can ask yourself how bad the problem is likely to be in your case—if it's under that bar, you might be okay (although you'd certainly still want to talk about the problem—and why you think it's okay—in your writeup).[28]

Let's code this up! We're going to simply expand our `create_ data` function to include a variable W that has an effect on both our treatment X and our outcome variable Y. We'll then include settings for the strength of the $W \rightarrow X$ and $W \rightarrow Y$

[26] Do keep in mind when do- ing this that the importance of certain assumptions, as with everything, can be sen- sitive to context. If you find that an assumption violation isn't that bad in your simu- lation, maybe try doing the simulation again, this time with the same assumption violation but otherwise very different-looking data. If it's still fine, that's a good sign!

[27] Which is a bit subjec- tive, and based on context— a .1% bias in the effect of a medicine on deadly side ef- fects might be huge, but a .1% bias in the effect of a job training program on earn- ings might not be so huge.

[28] This whole approach is a form of "sensitivity analysis"—checking how sensitive results are to violations of assump- tions. It's also a form of partial identification, as discussed in Chapter 11. This sensitivity-analysis approach to thinking about remaining open back doors is in common use, and is used to identify a range of reasonable regression coefficients when you're not sure about open back doors. For example, see Cinelli and Hazlett (2020).

Carlos Cinelli and Chad Hazlett. Making sense of sensitivity: Extending omit- ted variable bias. *Jour- nal of the Royal Statistical Society: Series B (Statisti- cal Methodology)*, 82(1):39– 67, 2020.

relationships. There are, plenty of other tweaks we *could* add as well—the strength of $X \rightarrow Y$, the standard deviations of X and Y, the addition of other controls, nonlinear relationships, the distribution of the error terms, and so on and so on. But let's keep it simple for now with our two settings.

Additionally, we're going to take our *iteration*/loop step and put that in a function too, since we're going to want to do this iteration a bunch of times with different settings.

```r
1   # R CODE
2   library(tidyverse); library(purrr)
3   set.seed(1000)
4
5   # Have settings for strength of W -> X and for W -> Y
6   # These are relative to the standard deviation
7   # of the random components of X and Y, which are 1 each
8   # (rnorm() defaults to a standard deviation of 1)
9   create_data <- function(N, effectWX, effectWY) {
10          d <- tibble(W = rnorm(N)) %>%
11              mutate(X = effectWX*W + rnorm(N)) %>%
12              # True effect is 5
13              mutate(Y = 5*X + effectWY*W + rnorm(N))
14
15          return(d)
16  }
17
18  # Our estimation function
19  est_model <- function(N, effectWX, effectWY) {
20          d <- create_data(N, effectWX, effectWY)
21
22          # Biased estimator - no W control!
23          # But how bad is it?
24          m <- lm(Y~X, data = d)
25
26          return(coef(m)['X'])
27  }
28
29  # Iteration function! We'll add an option iters for number of iterations
30  iterate <- function(N, effectWX, effectWY, iters) {
31          results <-  1:iters %>%
32          map_dbl(function(x) {
33                  # Let's add something that lets us keep track
34                  # of how much progress we've made. Print every 100th iteration
35                  if (x %% 100 == 0) {print(x)}
36
37                  # Run our model and return the result
38                  return(est_model(N, effectWX, effectWY))
39          })
40
41          # We want to know *how biased* it is, so compare to true-effect 5
42          return(mean(results) - 5)
43  }
44
45  # Now try different settings to see how bias changes!
46  # Here we'll use a small number of iterations (200) to
47  # speed things up, but in general bigger is better
48  iterate(2000, 0, 0, 200) # Should be unbiased
```

```
49 | iterate(2000, 0, 1, 200) # Should still be unbiased
50 | iterate(2000, 1, 1, 200) # How much bias?
51 | iterate(2000, .1, .1, 200) # Now?
52 | # Does it make a difference whether the effect
53 | # is stronger on X or Y?
54 | iterate(2000, .5, .1, 200)
55 | iterate(2000, .1, .5, 200)

 1 | * STATA CODE
 2 | set seed 1000
 3 |
 4 | * Have settings for strength of W -> X and for W -> Y
 5 | * These are relative to the standard deviation
 6 | * of the random components of X and Y, which are 1 each
 7 | * (rnormal() defaults to a standard deviation of 1)
 8 | capture program drop create_data
 9 | program define create_data
10 |         * Our first, second, and third arguments
11 |         * are sample size, W -> X, and W -> Y
12 |         local N = `1'
13 |         local effectWX = `2'
14 |         local effectWY = `3'
15 |
16 |         clear
17 |         set obs `N'
18 |         g W = rnormal()
19 |         g X = `effectWX'*W + rnormal()
20 |         * True effect is 5
21 |         g Y = 5*X + `effectWY'*W + rnormal()
22 | end
23 |
24 | * Our estimation function
25 | capture program drop est_model
26 | program define est_model
27 |         local N = `1'
28 |         local effectWX = `2'
29 |         local effectWY = `3'
30 |
31 |         create_data `1' `2' `3'
32 |
33 |         * Biased estimator - no W control! But how bad is it?
34 |         reg Y X
35 | end
36 |
37 | * Iteration function! Option iters determined number of iterations
38 | capture program drop iterate
39 | program define iterate
40 |         local N = `1'
41 |         local effectWX = `2'
42 |         local effectWY = `3'
43 |         local iter = `4'
44 |
45 |         * We don't need all this output! Use quietly{}
46 |         quietly {
47 |                 clear
48 |                 set obs `iter'
49 |                 g estimate = .
50 |
51 |                 forvalues i = 1(1)`iter' {
52 |                         * Let's keep track of how far along we are
```

```
53                       * Print every 100th iteration
54                       if (mod(`i',100) == 0) {
55                               * Noisily to counteract the quietly{}
56                               noisily display `i'
57                       }
58
59                       preserve
60                       est_model `N' `effectWX' `effectWY'
61                       restore
62                       replace estimate = _b[X] in `i'
63                   }
64
65                   * We want to know *how biased* it is,
66                   *so compare to true-effect 5
67                   replace estimate = estimate - 5
68              }
69          summ estimate
70  end
71
72  * Now try different settings to see how bias changes!
73  * Here we'll use a small number of iterations (200) to
74  * speed things up, but in general bigger is better
75
76  * Should be unbiased
77  iterate 2000 0 0 200
78  * Should still be unbiased
79  iterate 2000 0 1 200
80  * How much bias?
81  iterate 2000 1 1 200
82  * Now?
83  iterate 2000 .1 .1 200
84  * Does it make a difference whether the effect
85  * is stronger on X or Y?
86  iterate 2000 .5 .1 200
87  iterate 2000 .1 .5 200
```

```python
1   # PYTHON CODE
2   import pandas as pd
3   import numpy as np
4   import statsmodels.formula.api as smf
5   np.random.seed(1000)
6
7   # Have settings for strength of W -> X and for W -> Y
8   # These are relative to the standard deviation
9   # of the random components of X and Y, which are 1 each
10  # (np.random.normal() defaults to a standard deviation of 1)
11  def create_data(N, effectWX, effectWY):
12      d = pd.DataFrame({'W': np.random.normal(size = N)})
13      d['X'] = effectWX*d['W'] + np.random.normal(size = N)
14      # True effect is 5
15      d['Y'] = 5*d['X'] + effectWY*d['W'] + np.random.normal(size = N)
16
17    return(d)
18
19  # Our estimation function
20  def est_model(N, effectWX, effectWY):
21      d = create_data(N, effectWX, effectWY)
22
23          # Biased estimator - no W control!
24          # But how bad is it?
```

```
25          m = smf.ols('Y~X', data = d).fit()
26
27          return(m.params['X'])
28
29
30  # Iteration function! Option iters determines number of iterations
31  def iterate(N, effectWX, effectWY, iters):
32          results = [est_model(N, effectWX, effectWY) for i in range(0,iters)]
33
34          # We want to know *how biased* it is, so compare to true-effect 5
35          return(np.mean(results) - 5)
36
37  # Now try different settings to see how bias changes!
38  # Here we'll use a small number of iterations (200) to
39  # speed things up, but in general bigger is better
40
41  # Should be unbiased
42  iterate(2000, 0, 0, 200)
43  # Should still be unbiased
44  iterate(2000, 0, 1, 200)
45  # How much bias?
46  iterate(2000, 1, 1, 200)
47  # Now?
48  iterate(2000, .1, .1, 200)
49  # Does it make a difference whether the effect
50  # is stronger on X or Y?
51  iterate(2000, .5, .1, 200)
52  iterate(2000, .1, .5, 200)
```

What can we learn from this simulation? First, as we'd expect, with either *effectWX* or *effectWY* set to zero, there's no bias—the mean of the estimate minus the truth of 5 is just about zero. We also see that if *effectWX* and *effectWY* are both 1, the bias is about .5—half a standard deviation of the error term—and about 50 times bigger than if *effectWX* and *effectWY* are both .1. Does the size of the bias have something to do with the product of *effectWX* and *effectWY*? Sorta seems that way... it's a question you could do more simulations about!

15.4 Power Analysis with Simulation

15.4.1 What is Power Analysis?

STATISTICS IS AN AREA WHERE THE LESSONS OF CHILDREN'S TELEVISION ARE MORE OR LESS TRUE: IF YOU TRY HARD ENOUGH, ANYTHING IS POSSIBLE. It's also an area where the lessons of violent video games are more or less true: if you want to solve a really tough problem, you need to bring a whole lot of firepower (plus, a faster computer can really help matters).

Power analysis is the process of trying to figure out if the amount of firepower you have is *enough*. While there are

Power analysis. An analysis that links the sample size, effect size, and the statistical precision of an estimate, often with a goal of figuring out whether the sample size is big enough that, if your hypothesized effect size is accurate, you're likely to reject the null.

calculators and analytical methods out there for performing power analysis, power analysis is very commonly done by simulation, especially for non-experimental studies.[29]

WHY IS IT A GOOD IDEA TO DO POWER ANALYSIS? Once we have our study design down, there are a number of things that can turn statistical analysis into a fairly weak tool and make us less likely to find the truth:

1. Really tiny effects are really hard to find—good luck seeing an electron without a super-powered microscope! (in $Y = \beta_1 X + \varepsilon$, β_1 is tiny)

2. Most statistical analyses are about looking at variation. If there's little variation in the data, we won't have much to go on ($var(X)$ is tiny)

3. If there's a lot of stuff going on other than the effect we're looking for, it will be hard to pull the signal from the noise ($var(\varepsilon)$ is big)

4. If we have really high standards for what counts as evidence, then a lot of good evidence is going to be ignored so we need more evidence to make up for it (if you're insisting on a really tiny $se(\hat{\beta}_1)$)

Conveniently, all of these problems can be solved by increasing our firepower, by which I mean sample size. Power analysis is our way of figuring out exactly how much firepower we need to bring. If it's more than we're willing to provide, we might want to turn around and go back home.

If you have bad power and can't fix it, you might think about doing a different study. The downsides of bad statistical power aren't just "we might do the study and not get a significant effect—oh darn." Bad power also makes it more likely that, if you *do* find a statistical effect, it's overinflated or not real! That's because with bad power we basically *only* reject the null as a fluke. Think about it this way—if you want to reject the null at the 95% level that a coin flip is 50/50 heads/tails, you need to see an outcome that's less than 5% likely to occur with a fair coin. If you flip six coins (tiny sample, bad power), that can only happen with all-heads or all-tails. Even if the truth is 55% heads so the null really is false, the *only* way we can reject 50% heads is if we estimate 0% heads or 100% heads. Both of those are actually way further from the truth than the 50% we just rejected! This is why, when you read about a study that

[29] Power analysis calculators have to be specific to your estimate and research design. This is feasible in experiments where a simple "we randomized the sample into two groups and measured an outcome" design covers thousands of cases. But for observational data where you're working with dozens of variables, some unmeasured, it's infeasible for the calculator-maker to guess your intentions ahead of time. That said, calculators work well for some causal inference designs too, and I discuss one in Chapter 20 that I recommend for regression discontinuity.

seems to find an enormous effect of some innocuous thing, it often has a tiny sample size (and doesn't replicate).

Power analysis can be a great idea no matter what kind of study you're running. However, it's especially helpful in three cases:

- If you're looking for an effect that you think is probably not that central or important to what's going on—it's a small effect, or a part of a system where a lot of other stuff is going on (β_1 times $var(X)$ is tiny relative to $var(Y)$) a power analysis can be a great idea—the sample size required to learn something useful about a small effect is often much bigger than you expect, and it's good to learn that now rather than after you've already done all the work

- If you're looking for *how an effect varies across groups* (in regression, if you're mostly concerned with the interaction term), then a power analysis is a good idea. Finding *differences between groups in an effect* takes a lot more firepower to find than the effect itself, and you want to be sure you have enough

- If you're doing a randomized experiment, a power analysis is a must—you can actually control the sample size, so you may as well figure out what you need before you commit to way too little!

15.4.2 What Power Analysis Does

POWER ANALYSIS BALANCES FIVE THINGS, like some sort of spinning-plate circus act. Using X to represent treatment and Y for the outcome, those things are:

1. The **true effect size** (coefficient in a regression, a correlation, etc.)

2. The **amount of variation in the treatment** ($var(X)$)

3. The **amount of other variation in Y** (either the variation from the residual after explaining Y with X, or just $var(Y)$)

4. **Statistical precision** (the standard error of the estimate or statistical power, i.e., the true-positive rate)

5. **The sample size**

A power analysis holds four of these constant and tells you what the fifth can be. So, for example, it might say "if we think the effect is probably A, and there's B variation in X, and there's C variation in Y unrelated to X, and you want to have statistical precision of D or better, then you'll need a sample size of at least E." This tells us the *minimum sample size* necessary to get sufficient statistical power.

Or we can go in other directions. "If you're going to have a sample size of A, and there's B variation in X, and there's C variation in Y unrelated to X, and you want to have statistical precision of D or better, then the effect must be at least as large as E." This tells us the *minimum detectable effect*, i.e., the smallest effect you can hope to have a chance of reasonably measuring given your sample size.

You could go for the minimum acceptable *any* one of the five elements. However, minimum sample size and minimum detectable effect are the most common ones people go for, followed closely by the level of statistical precision.

HOW ABOUT THAT "STATISTICAL PRECISION" THING? What do I mean by that, specifically? Usually, statistical precision in power analysis is measured by a target level of statistical power (thus the name "power analysis"). Statistical power, also known as the true-positive rate, is a concept specific to hypothesis testing. Consider a case where we know the null hypothesis is false. If we had an infinitely-sized sample, we'd definitely reject that false null. However, we don't have an infinite sample. So, due to sampling variation, we'll correctly reject the null sometimes, and falsely fail to reject it other times. The proportion of times we correctly reject the false null is the *true-positive rate*, also known as *statistical power*. If we reject a false null 73% of the time, that's the same as saying we have 73% power to reject that null.[30,31]

Statistical power is dependent on the kind of test you're running, too—if you are doing a hypothesis test at the 95% confidence level, you're more likely to reject the null (and thus will have higher statistical power) than if you're doing a hypothesis test at the 99% confidence level. The more careful you are about false positives, the more likely you are to get a false negative. So there's a tradeoff there.

Power analyses don't have to be run with statistical power in mind. In fact, you don't necessarily need to think about things in terms of "the ability to reject the null," which is what statistical power is all about. You could run your power analysis

Minimum sample size. The smallest sample size that produces a certain statistical precision given a certain effect size, treatment variation, and unrelated outcome variation.

Minimum detectable effect. The smallest effect size that produces a certain statistical precision given a certain sample size, treatment variation, and unrelated outcome variation.

Statistical power, or the true-positive rate. The probability of rejecting the null hypothesis, when the null hypothesis is actually false.

[30] Power is specific to the null we're trying to reject. It's pretty easy to reject a null that's completely ridiculous, but rejecting a null that's near the truth, but is not the truth, takes a lot more firepower.

[31] Power analysis isn't necessarily tied to hypothesis testing. Statistical power is just a way of measuring statistical precision. We could instead, for example, ask how much firepower we need to get the standard error below a certain level.

with any sort of statistical precision as a goal, like standard errors. Given A, B, and C, what sample size D do you need to make your standard errors E or smaller?

15.4.3 Where Do Those Numbers Come From?

IN ORDER TO DO POWER ANALYSIS, you need to be able to fill in the values for four of those five pieces, so that power analysis can tell you the fifth one. How do we know those things?

We have to make the best guesses we can. We can use previous research to try to get a sense of how big the effect is likely to be, or how much variation there's likely to be in X. If there's no prior research, do what you can—think about what is likely to be true, make educated guesses. Power analysis at a certain level requires some guesswork.

Other pieces aren't about guesses but about standards. How much statistical power do you want? The higher your statistical power, the less chance there is of missing a true effect. But that means your sample size needs to go up a lot. Often people use a rule of thumb here. In the past, a goal of 80% statistical power has been standard. These days I see 90% a lot more often.

In practical terms, power analysis isn't a super-precise calculation, it's guidance. It doesn't need to be exact. A little guesswork (although ideally as little as possible) is necessary. After all, even getting the minimum sample size necessary doesn't guarantee your analysis is good, it just gives you a pretty good chance of finding a result if it's there. Often, people take the results of their power analysis as a baseline and then make sure to overshoot the mark, under the assumption they've been too conservative. So don't worry about being accurate, just try to make the numbers in the analysis close enough to be useful.

15.4.4 Coding a Power Analysis Simulation

MOST OF THE PIECES for doing a power analysis simulation we already have in place from the rest of the chapter. The "Breaking Things" section is the most informative, since many power analysis simulations will want to try things while changing quite a few settings.

Let's consider three components we've already gone over:

Data creation. We need a data creation function that *mimics what we think the actual data generating process looks like.* That's right, it's back to drawing out the causal diagram of our research setting! You already did this, right?

You can simplify a bit—maybe instead of sixteen variables you plan to measure and control for, and eight you don't, you can combine them all into "controls" and "unmeasured variables." But overall you should be thinking about how to create data as much like the real world as possible. You'll also want to try to match things like how much variation there is—how much variation in treatment do you expect there to be?[32] How much variation in the outcome? In the other variables? You won't get it exactly, but get as close as you can, even if it's only an educated guess.

Estimation and storing the results. Once we have our data, we need to perform the analysis we plan to perform. This works the exact same way it did before—just perform the analysis you plan to perform in your study, but using the simulated data!

The only twist comes if you are planning to use statistical power as your measure of statistical precision. How can we pull statistical power out of our analysis?

Remember, statistical power is the proportion of results that are statistically significant (assuming the null is false, which we know it is as long as our data creation process has a nonzero effect). So in each iteration of the simulation, we need to pull out *a binary variable indicating whether the effect was statistically significant* and return that. Then, the proportion that *are* significant across all the simulations is your statistical power. 45% of iterations had a significant effect (and your null was false)? 45% statistical power!

Getting out whether an effect was significant is a touch harder than pulling out the coefficient or standard error. In R, for a model m you can load the **broom** package and do d <- tidy(m). Then, d$p.value[2] < .05 will be TRUE if the p-value is below .05 (for a 95% significance test). The [2] there assumes your coefficient of interest is the second one (take a look at the tidy() output for one of the iterations to find the right spot for your coefficient).

In Stata, you'll need to dip a bit into the Mata matrix algebra underlying many of Stata's commands. Don't worry—it won't bite. Following a regression, pull out the matrix of results with matrix R = r(results). Then, get the p-value out with matrix p = R["pvalue","X"]. Then, you can put that p-value where you want by referring to it with p[1,1].

In Python, mercifully, things are finally easy. For a model m from **statsmodels**, we can get the p-value for the coefficient on

[32] If you have observational data, maybe you can estimate this, and other features of the data, using old data from before your study setting.

X with `m.pvalues['X']`. Compare that to, say, .05 for a significance test. `m.pvalues['X'] < .05` is `True` if the coefficient is significant at the 95% level, and `False` otherwise.

Iteration. Running the simulation over multiple iterations works the same as it normally would! No changes there.

However, we might want to add *another* iteration surrounding that one![33] Say for example we want to know the minimum detectable effect that gives 90% power. How can we find it? Well, we might try a bunch of different effect sizes, increasing them a little at a time until we find one that produces 90% power. "Trying a bunch of different effect sizes" is a job for a loop.

[33] Who iterates the iterators?

Similarly, if we want to find the minimum necessary sample size to get 90% power, we'd try a bunch of different sample sizes until power reached 90%.

So LET'S DO THE THING. I mentioned earlier that one case where there's poor power is where we want to know the *difference in the effect between two groups,* in other words the coefficient on an interaction term in a regression (where one of the terms being interacted is binary).

Let's see how bad the power actually is! Let's imagine that we want to know the effect of a reading-training program on children's reading test scores. Also, we're interested in whether the effect differs by gender.

In the study we plan to do, we know that the test score is explicitly designed to be normally distributed (many tests are designed with this in mind) with a standard deviation of 10.[34]

Other stuff matters for test scores too—let's pack a bunch of stuff together into a single variable we'll include as a control W—a one-standard-deviation change in this control increases test scores by 4.

[34] I'm about to use this 10 as a reason for creating the test score by adding up all our other determinants and then adding on a normally-distributed error with a standard deviation of 9. Why 9 and not 10? Because the other components we add up to make the test score will increase the standard deviation. This won't hit a standard deviation of 10 exactly but it will be close enough.

The reading-training program, in truth, increases test scores by 2 for boys and 2.8 for girls. That's a pretty big difference—the program is 40% more effective for girls! The training program is administered in a nonrandom way, with higher W values making it less likely you get the training.

We can estimate the difference in the effect of training between boys and girls with a regression

$$TestScore = \beta_0 + \beta_1 Training + \beta_2 Girl +$$
$$\beta_3 Training \times Girl + \beta_4 W + \varepsilon \quad (15.1)$$

where we will focus on $\hat{\beta}_3$.

We want to know the minimum sample size necessary to get 90% power to detect the difference in the effect between boys and girls. Then, let's imagine we don't know the true difference in effects, but *do* know that there are only 2,000 children available to sample, and ask what the smallest possible difference in effects is that we could detect with 90% power with that sample of 2,000. Let's code!

```
1    # R CODE
2    library(tidyverse); library(purrr); library(broom)
3    set.seed(1000)
4
5    # Follow the description in the text for data creation. Since we want
6    # to get minimum sample size and minimum detectable effect, allow both
7    # sample size and effect to vary.
8    # diff is the difference in effects between boys and girls
9    create_data <- function(N, effect, diff) {
10           d <- tibble(W = rnorm(N),
11                  girl = sample(0:1, N, replace = TRUE)) %>%
12                  # A one-SD change in W makes treatment 10% less likely
13                  mutate(Training = runif(N) + .1*W < .5) %>%
14                  mutate(Test = effect*Training + diff*girl*Training +
15                         0*W + rnorm(N, sd = 9))
16
17           return(d)
18   }
19
20   # Our estimation function
21   est_model <- function(N, effect, diff) {
22           d <- create_data(N, effect, diff)
23
24           # Our model
25           m <- lm(Test~girl*Training + W, data = d)
26           tidied <- tidy(m)
27
28           # By looking we can spot that the interaction
29           # term is in the 5th row
30           sig <- tidied$p.value[5] < .05
31
32           return(sig)
33   }
34
35   # Iteration function!
36   iterate <- function(N, effect, diff, iters) {
37           results <-  1:iters %>%
38           map_dbl(function(x) {
39                  # To keep track of progress
40                  if (x %% 100 == 0) {print(x)}
41
42                  # Run our model and return the result
43                  return(est_model(N, effect, diff))
44           })
45
46           # We want to know statistical power,
47           # i.e., the proportion of significant results
48           return(mean(results))
49   }
50
```

```
51   # Let's find the minimum sample size
52   mss <- tibble(N = c(10000, 15000, 20000, 25000))
53   mss$power <- c(10000, 15000, 20000, 25000) %>%
54           # Before we had to do function(x) here, but now the argument we're
55           # passing is the first argument of iterate() so we don't need it
56           map_dbl(iterate, effect = 2, diff = .8, iter = 500)
57   # Look for the first N with power above 90%
58   mss
59
60   # Now for the minimum detectable effect
61   mde <- tibble(effect = c(.8, 1.6, 2.4, 3.2))
62   mde$power <-  c(.8, 1.6, 2.4, 3.2) %>%
63           map_dbl(function(x) iterate(N = 2000, effect = 2,
64       diff = x, iter = 500))
65   # Look for the first effect wth poewr above 90%
66   mde
```

```
1    * STATA CODE
2    set seed 1000
3
4    * Follow the description in the text for data creation. Since we want to
5    * get minimum sample size and minimum detectable effect, allow both
6    * sample size and effect to vary.
7    * diff is the difference in effects between boys and girls
8    capture program drop create_data
9    program define create_data
10           local N = `1'
11           local effect = `2'
12           local diff = `3'
13
14           clear
15           set obs `N'
16
17           g W = rnormal()
18           g girl = runiformint(0,1)
19           * A one-SD change in W makes treatment 10% less likely
20           g Training = runiform() + .1*W < .5
21           g Test = `effect'*Training + `diff'*girl*Training +
22      4*W + rnormal(0,9)
23   end
24
25   * Our estimation function
26   capture program drop est_model
27   program define est_model
28           local N = `1'
29           local effect = `2'
30           local diff = `3'
31
32           create_data `N' `effect' `diff'
33
34           * Out model
35           reg Test i.girl##i.Training W
36   end
37
38   * Iteration function!
39   capture program drop iterate
40   program define iterate
41           local N = `1'
42           local effect = `2'
43           local diff = `3'
```

```
44          local iters = `4'
45
46          quietly {
47          * Parent data
48          clear
49          set obs `iters'
50          g significant = .
51
52          forvalues i = 1(1)`iters' {
53                  * Keep track of progress
54                  if (mod(`i',100) == 0) {
55                          display `i'
56                  }
57
58                  preserve
59                  est_model `N' `effect' `diff'
60                  restore
61
62                  * Get significance of the interaction (1.girl#1.Training)
63                  matrix R = r(table)
64                  matrix p = R["pvalue","1.girl#1.Training"]
65                  local sig = p[1,1] < .05
66                  local sig = `p' < .05
67                  replace significant = `sig' in `i'
68          }
69
70          * Get proportion significant
71          summ significant
72          }
73          display r(mean)
74 end
75
76 * Let's find the minimum sample size
77 * Look for the first N with power above 90%
78 foreach N in 10000 15000 20000 25000 {
79          display "Sample size `N'. Power:"
80          iterate `N' 2 .8 500
81 }
82
83 * Now for the minimum detectable effect
84 * Look for the first effect with power above 90%
85 foreach diff in .8 1.6 2.4 3.2 {
86          display "Effect difference `diff'. Power:"
87          iterate 2000 2 `diff' 500
88 }
```

```python
1  # PYTHON CODE
2  import pandas as pd
3  import numpy as np
4  import statsmodels.formula.api as smf
5  np.random.seed(1000)
6
7  # Follow the description in the text for data creation. Since we want to
8  # get minimum sample size and minimum detectable effect, allow both sample
9  # size and effect to vary.
10 # diff is the difference in effects between boys and girls
11 def create_data(N, effect, diff):
12     d = pd.DataFrame({'W': np.random.normal(size = N),
13                       'girl': np.random.randint(2, size = N)})
14     # A one-SD change in W makes treatment 10% less likely
```

```
15      d['Training'] = np.random.uniform(size = N) + .1*d['W'] < .5
16      d['Test'] = effect*d['Training'] + diff*d['girl']*d['Training']
17      d['Test'] = d['Test'] + 4*d['W'] + np.random.normal(scale = 9, size = N)
18      return(d)
19
20  # Our estimation function
21  def est_model(N, effect, diff):
22      d = create_data(N, effect, diff)
23
24      # Our model
25      m = smf.ols('Test~girl*Training + W', data = d).fit()
26
27      # By looking we can spot that the name of the
28      # interaction term is girl:Training[T.True]
29      sig = m.pvalues['girl:Training[T.True]'] < .05
30      return(sig)
31
32  # Iteration function!
33  def iterate(N, effect, diff, iters):
34      results = [est_model(N, effect, diff) for i in range(0,iters)]
35
36      # We want to know statistical power,
37      # i.e., the proportion of significant results
38      return(np.mean(results))
39
40  # Let's find the minimum sample size
41  mss = [[N, iterate(N, 2, .8, 500)] for
42  N in [10000, 15000, 20000, 25000]]
43  # Look for the first N with power above 90%
44  pd.DataFrame(mss, columns = ['N','Power'])
45
46  # Now for the minimum detectable effect
47  mde = [[diff, iterate(2000, 2, diff, 500)] for
48  diff in [.8, 1.6, 2.4, 3.2]]
49  pd.DataFrame(mde, columns = ['Effect', 'Power'])
```

What do we get from these analyses? From the minimum-sample-size power analysis we get Table 15.1. In that table, we see that we find 88.2% power with a sample size of 20,000, and 93.8% power with a sample size of 25,000. So our sweet spot of 90% power must be somewhere between them. We could try increasing the number of iterations, or trying different Ns between 20,000 and 25,000 to get a more precise answer, but "a little more than 20,000 observations is necessary to get 90% power on that interaction term" is pretty good information as is.[35]

[35] And if *all* of them had 90% power (or none of them did) that would be an indication we'd need to try smaller (larger) sample sizes. That's how I found the range of sample sizes to iterate over here.

N	Power
10000	0.602
15000	0.788
20000	0.882
25000	0.938

Table 15.1:   Results of Minimum-Sample-Size Power Analysis Simulation

Wait, uh, twenty *thousand?* For that *massive* interaction term (how often do you think a treatment intended to apply to everyone is really 40% more effective for one group than another)? Yep! "40% more effective" isn't as important as ".8 relative to the standard deviation of 10 for *Test*."

It's no surprise that if we have only 2000 observations, we're nowhere near 90% power for the true interaction effect of .8. But what's the minimum detectable effect we *can* detect with our sample size of 2000? The results are in Table 15.2.

Effect	Power
0.8	0.164
1.6	0.528
2.4	0.838
3.2	0.976

Table 15.2: Results of Minimum-Detectable-Effect Power Analysis Simulation

In Table 15.2 we reach 90% power with an effect somewhere between 2.4 and 3.2. Again, we could try with more iterations and more effect sizes between 2.4 and 3.2 to get more precise, but this gives us a ballpark.

If we only have 2000 observations to work with, we'd need the effect for girls to be something like 220% of the effect for boys (or 40% of the effect but with an opposite sign) to be able to detect gender differences in the effect with 90% power. Do we think that's a reasonable assumption to make? If not, maybe we shouldn't estimate gender differences in the effect for this study, even if we really want to. That's what power analysis is good at—knowing when a fight is lost and you should really go home.[36]

15.5   Simulation with Existing Data: The Bootstrap

SERIOUSLY, THE THIRD CHAPTER IN A ROW WHERE I'M TALKING ABOUT THE BOOTSTRAP? Well, deal with it. The bootstrap is basically magic. Don't you want magic in your life?[37]

Bootstrapping is the process of repeating an analysis many times after *resampling with replacement*. Resampling with replacement means that, if you start with a data set with N observations, you create a new simulated data set by picking N random observations from that data set... but each time you draw an observation, you put it back so you might draw it again. If your original data set was 1, 2, 3, 4, then a bootstrap data set might be 3, 1, 1, 4, and another one might be 2, 4, 1, 4.

[36] There are some statistical issues inherent to interactions that worsen their power, but the *real* problem is that interaction terms are in truth often small. Treatment effects vary across groups, but that variation is often much smaller than the effect itself. So however many observations we need to find the main effect, we need many more to find those tiny interactions. Small effects are very hard to find with statistics!

[37] The imaginary person I'm responding to here is such an ingrate.

The resampling process mimics the sampling distribution in the actual data. On average, the means of the variables will match your means, as will the variances, the correlations between the variables, and so on. In this way, we can *run a simulation using actually-existing data.* The resampling-with-replacement process replaces the `create_data` function in our simulation. One of the toughest parts of doing a simulation is figuring out how to create data that looks like the real world. If we can just use real-world data, that's miles better.

THE ONLY DOWNSIDE IS THAT, unlike with a simulation where you create the data yourself, *you don't know the true data generating process.* So you can't use bootstrapping to, say, check for bias in your estimate. You can see what your bootstrap-simulated estimate is, but you don't know the truth to compare it to (since you didn't create the data yourself), so you can't check whether you're right on average!

This means that the bootstrap, as a simulation tool, is limited to cases where the question doesn't rely on comparing your results to the truth. Usually this means using the bootstrap to estimate standard errors, since by mimicking the whole re-sampling process, you automatically simulate the strange interdependencies between the variables, allowing any oddities in the *true* sampling distribution to creep into your bootstrap-simulated sampling distribution.

Standard errors aren't all they're good for, though! Machine learning people use bootstrapping for all sorts of stuff, including using it to generate lots of training data sets for their models. You can also use a bootstrap to look at the variation of your model more generally. For example, if you make a bootstrap sample, estimate a model, generate predicted values from that model, and then repeat, you can see how your *predictions for each observation* vary based on the sampling variation of the whole model, or just aggregate all the simulated predictions together to get your actual prediction.[38],[39]

HOW CAN YOU DO A BOOTSTRAP? Borrowing from Chapter 13:

1. Start with a data set with N observations.

2. Randomly sample N observations from that data set, allowing the same observation to be drawn more than once. This is a "bootstrap sample."

3. Estimate our statistic of interest using the bootstrap sample.

[38] This process has the name "bagging" which on one hand I like as a term, but on the other hand really sounds like some super-dorky-in-retrospect dance craze from, like, 1991.

[39] One important note is that if you're planning to use your bootstrap distribution to do a hypothesis test, you want to first shift the distribution so it has the *un*-bootstrapped estimate as the mean, before you compare the distribution to the null. This produces the appropriate comparison you're going for with a hypothesis test.

4. Repeat steps 1—3 a whole bunch of times.

5. Look at the distribution of estimates from across all the times we repeated steps 1—3.[40] The standard deviation of that distribution is the standard error of the estimate. The 2.5th and 97.5th percentiles show the 95% confidence interval, and so on.

[40] If your data or estimate follows a super-weird distribution, perhaps one with an undefined mean (see Chapter 22), you'll want to drop some of the tails of your bootstrap distribution before using it.

How many times is "a bunch of times" in step 4? It varies. When you're just testing out code you can do it a few hundred times to make sure things work, but you wouldn't trust those actual results. Really, you should do it as many times as you feasibly can given the computing power available to you. Be prepared to wait! For anything serious you probably want to do it at least a few thousand times.

Bootstrapping comes with its own set of assumptions to provide good estimates of the standard error. We need a reasonably large sample size. We need to be sure that our data is at least reasonably well behaved—if the variables follow extremely highly skewed distributions, bootstrap will have a hard time with that. We also need to be careful in accounting for how the observations might be related to each other. For example, if there are individuals observed multiple times (panel data), we may want a "cluster bootstrap" that samples *individuals* with replacement, rather than observations (getting the 2005 and 2007 observations for you, and the 2005 observation for me twice, might not make sense). Or if the error terms are correlated with each other—clustered, or autocorrelated—the bootstrap's resampling approach needs to be very careful to deal with it.

15.5.1   Coding a Bootstrap

IN THE INTEREST OF NOT REPEATING MYSELF TOO MUCH, this section will be short. There are already coding examples for using built-in bootstrap standard error tools in Chapter 13 on Regression, and for inverse probability weighting with regression adjustment in Chapter 14 on Matching.

What's left to do? In this section, I'll show you how to do two things:

1. A generic demonstration of how to perform a bootstrap "by hand." The example will provide bootstrap standard errors. For R and Stata, bootstrap standard errors are easier done by the built-in methods available for most models, but the

code here gives you a better sense of what's happening under the hood, can be easily manipulated to do something
besides standard errors, which those built-in methods can't
help with, and can work with commands that don't have
built-in bootstrap standard error methods.

2. After this short refresher, I'll go into a little more about some
    variants of the bootstrap

LET'S DO A BOOTSTRAP BY HAND! The process here is super
simple. We just copy all the stuff we have already been doing
with our simulation code. Then, we replace the `create_data`
function, which created our simulated data, with a function that
resamples the data with replacement.

Or, we can save ourselves a little work and skip making the
bootstrap process its own function—usually it only takes one
line anyway. If we have N observations, just create a vector
that's N long, where each element is a random integer from 1
to N (or 0 to $N-1$ in Python). Then use that vector to pick
which observations to retain. In R and Python this is simply:

```r
# R CODE
# Load packages
library(tidyverse)

# Get our data set
data(mtcars)

# Resampling function
create_boot <- function(d) {
        d <- d %>%
                  slice_sample(n = nrow(d), replace = TRUE)
     return(d)

        # Or more "by-hand"
        # index <- sample(1:nrow(d), nrow(d), replace = TRUE)
        # d <- d[index,]
}

# Get a bootstrap sample
create_boot(mtcars)
```

```python
# PYTHON CODE
import pandas as pd
import numpy as np
import seaborn as sns

# Get our data to bootstrap
iris = sns.load_dataset('iris')

def create_boot(d):
        N = d.shape[0]
        index = np.random.randint(0, N, size = N)
        d = d.iloc[index]
```

```
13           return(d)
14
15   # Create a bootstrap sample
16   create_boot(iris)
```

Easy! You can slot that right into your simulations as normal.

However, when it comes to bootstrapping "by hand" we might want to take a half-measure—doing the estimation method, and returning the appropriate estimate, by hand, but letting a function do the bootstrap sampling and iteration work. All three languages have built-in bootstrap-running commands. We already saw Stata's in Chapter 14.

For all of these, we need to specify an estimation function *first*. Unlike in our previous code examples, that estimation function won't call a data-creation function to get data to estimate with. Instead, we'll call that estimation function *with our bootstrapping function*, which will provide a bootstrapped sample to be estimated, take the estimates provided, and aggregate them all together.

```
1    # R CODE
2    # Load packages
3    library(tidyverse); library(boot)
4    set.seed(1000)
5
6    # Get our data set
7    data(mtcars)
8
9    # Our estimation function. The first argument must be the data
10   # and the second the indices defining the bootstrap sample
11   est_model <- function(d, index) {
12           # Run our model
13           # Use index to pick the bootstrap sample
14           m <- lm(hp ~ mpg + wt, data = d[index,])
15
16           # Return the estimate(s) we want to bootstrap
17           return(c(coef(m)['mpg'],
18               coef(m)['wt']))
19   }
20
21   # Now the bootstrap call!
22   results <- boot(mtcars, # data
23                   est_model, # Estimation function,
24                   1000) # Number of bootstrap simulations
25
26   # See the coefficient estimates and their bootstrap standard errors
27   results
```

```
1    * STATA CODE
2    set seed 1000
3    * Get data
4    sysuse auto.dta, clear
5
6    * Estimation model
7    capture program drop est_model
```

```
 8   program def est_model
 9       quietly: reg price mpg weight
10   end
11
12   * The bootstrap command is only a slight difference from the simulate and
13   * bstat approach in the Matching chapter. Just expanding the toolkit!
14   bootstrap, reps(1000): est_model
```

```
 1   # PYTHON CODE
 2   import pandas as pd
 3   import numpy as np
 4   from resample.bootstrap import bootstrap
 5   import statsmodels.formula.api as smf
 6   import seaborn as sns
 7
 8   # Get our data to bootstrap
 9   iris = sns.load_dataset('iris')
10
11   # Estimation - the first argument should be for the bootstrapped data
12   def est_model(d):
13           # bootstrap() makes an array, not a DataFrame
14           d = pd.DataFrame(d)
15           # Oh also it tossed out the column names
16           d.columns = iris.columns
17           # And numeric types
18           d = d.convert_dtypes()
19           print(d.dtypes)
20           m = smf.ols(formula = 'sepal_length ~ sepal_width + petal_length',
21           data = d).fit()
22           coefs = [m.params['sepal_width'], m.params['petal_length']]
23           return(coefs)
24
25   # Bootstrap the iris data, estimate with est_model, and do it 1000 times
26   b = bootstrap(sample = iris, fn = est_model, size = 1000)
27
28   # Get our standard errors
29   bDF = pd.DataFrame(b, columns = ['SW','PL'])
30   bDF.std()
```

NOT TOO BAD! Okay, not actually easier so far than doing it by hand, but there's a real benefit here, in that these purpose-built bootstrap commands often come with options that let you bootstrap in ways that are a bit more complex than simply sampling each observation with replacement. Or they can provide ways of analyzing the distribution of outputs in slightly more sophisticated ways than just taking the mean and standard deviation.

One of the nicest benefits you get from these commands is the opportunity to do some *bias correction*. Here bias doesn't refer to failing to identify our effect of interest, but rather how the shape of the data can bias our statistical estimates. Remember how I mentioned that bootstrap methods can be sort of sensitive to skewed data? Skewed data will lead to skewed sampling distributions, which will affect how you get an estimate and standard error from that sampling distribution.

But there are ways to correct this bias, the most common of which is "BCa confidence intervals." This method looks at the proportion of times that a bootstrap estimate is *lower* than the non-bootstrapped estimate (if the distribution isn't skewed, this should be 50%), and the skew of the distribution.[41] Then, the BCa confidence intervals you get account for the skewing of the bootstrap sampling distribution.[42] You can then use these confidence intervals for hypothesis testing.

In R, you can pass your results from `boot` to `boot.ci(type = 'bca')` to get the BCa confidence intervals. In Stata, add the `bca` option to your `bootstrap` command, and then once it's done do `estat bootstrap, bca` (although in my experience this process is a bit finicky). In Python, import and run `confidence_interval` instead of `bootstrap` from **resample. bootstrap**.

15.5.2 Bootstrap Variants

LIKE JUST ABOUT EVERY STATISTICAL METHOD, THE BOOT-STRAP STUBBORNLY REFUSES TO JUST BE ONE THING, and instead there are a jillion variations on it. The fact that I've been referring to the thing I've been showing you as "*the* boot-strap" instead of more precisely as "pairs bootstrap" has almost certainly angered *someone*. I'm not entirely sure who, though.[43]

There are far more variants of the bootstrap than I'll go into here, but I will introduce a few, specifically the ones that focus on *non-independent data*. Remember when I said that the boot-strap only works if the observations are unrelated to each other? It's true! Or at least it's true for the basic version. But what if we have heteroskedasticity? Or autocorrelation? Or if our observations are clustered together? It's not the case that the bootstrap can't work, but it *is* the case that our bootstrapping procedure has to account for the relationship between the ob-servations.

THE CLUSTER/BLOCK BOOTSTRAP SEEMS LIKE A GOOD PLACE TO START, since I've already talked about it a bit. The cluster bootstrap is just like the regular bootstrap, except in-stead of resampling individual observations, you instead resam-ple *clusters* of observations, also known as *blocks*. That's the only difference!

[41] They also use a cousin of bootstrapping and (from Chapter 13) cross-validation called the "jackknife," in which for a sample with N observations the simulation is run N times, each time omitting exactly one obser-vation, so each observation is omitted from one sample. This is used here to see how influential each observation is on the estimates. It can also be used to generate out-of-sample predictions for the currently-excluded observa-tion.

[42] Interestingly, they're not symmetric around the estimate—if, for example, skew means there are "too many" bootstrap estimates that are lower than (to the left of) the non-bootstrapped estimate, then we know a bit more about the left-hand side of the distribution, and the 95% confidence interval doesn't need to be very far out. But on the right we need to go farther out.

[43] The literature on exactly which format of bootstrap to use, with which resam-pling method, with which random generation process, with which adjustment, is *deep* and littered with caveats like "this particular method seemed to do well in this particular instance." Unfortunately the learning curve is steep. If you're really interested in diving in, good luck!

You'd want to think about using a cluster bootstrap any time you have *hierarchical* or *panel* data.[44] Hierarchical data is when observations are nested within groups. For example, you might have data from student test scores. Those students are within classrooms, which are within schools, which are within regions, which are within countries.

Depending on the analysis you're doing, that might not necessitate a cluster bootstrap (after all, all data is hierarchical in *some* way). But if the hierarchical structure is built into your model—say, you are incorporating classroom-level characteristics into your model, then an individual-observation bootstrap might not make much sense.

Not all hierarchical data sets call for a cluster bootstrap, but when it comes to panel data it's often what you want. In panel data we observe the same individual (person, firm, etc.) multiple times. We'll talk about this a lot more in Chapter 16. When this happens, it doesn't make too much sense to resample individual observations, since that would sort of split each person in two and resample only part of them. So we'd want to sample *individuals*, and all of the observations we have for that individual, rather than each observation separately.

Another place where the cluster/block bootstrap pops up is in the context of time series data.[45] In time series data, the same value is measured across many time periods. This often means there is autocorrelation. When data is correlated across time, if you just resample randomly you'll break those correlation structures. Also, if your model includes lagged values, you're in trouble if you happen not to resample the value just before yours!

So, when applying a bootstrap to time series data, the time series is first divided into blocks of continuous time. The blocks themselves are often determined by one of several different available an optimal-block-length algorithms.[46] If you have daily stock index prices from 2000 through 2010, maybe each month ends up its own block. Or every six weeks is a new block, or every year, or something, or maybe some blocks are longer than others. The important thing is that *within* each block you have a continuous time series. Then, you resample the blocks instead of individual observations.[47]

THE CLUSTER/BLOCK BOOTSTRAP CAN BE IMPLEMENTED in Stata by just adding a `cluster()` option onto your regular `bootstrap` command. In R, you can use the `cluster` option in the `vcovBS()` function in the **sandwich** package if

[44] In these cases it's called "cluster bootstrap" instead of "block bootstrap" by convention.

[45] This time it's "block bootstrap" instead of "cluster bootstrap" by convention. Go figure.

[46] Dimitris N Politis and Halbert White. Automatic block-length selection for the dependent bootstrap. *Econometric Reviews*, 23(1): 53–70, 2004.

[47] A common alternative to this is the "moving block boostrap," which sort of has a sliding window that gets resampled instead of splitting the data up into fixed chunks.

what you want is bootstrap standard errors on a regression model (pass the `vcovBS()` result to the `vcov` argument in your `modelsummary` table). There doesn't appear to be a more general implementation in R or Python that works with any function like `boot` does. There doesn't appear to be any cluster bootstrap implementation at all in Python, but several varieties of block bootstrap are in `arch.bootstrap` in the **arch** package.

Thankfully, the cluster bootstrap is not too hard to do by hand in R and Python. Just (1) create a data set that contains only the list of ID values identifying individuals, (2) do a regular bootstrap resampling on *that* data set, and (3) use a merge to merge that bootstrapped data back to your regular data, using a form of merge that only keeps matches (`inner_join()` in R, or `d.join(how = 'inner')` in Python). This will copy each individual the appropriate number of times. This same method *could* also be used to do a block bootstrap, but only if you're willing to define the blocks yourself instead of having an algorithm do it (probably better than you can).

```r
1  # R CODE
2  # Load packages
3  library(tidyverse)
4  set.seed(1000)
5
6  # Example data
7  d <- tibble(ID = c(1,1,2,2,3,3),
8                    X = 1:6)
9  # Now, get our data set just of IDs
10 IDs <- tibble(ID = unique(d$ID))
11
12 # Our bootstrap resampling function
13 create_boot <- function(d, IDs) {
14         # Resample our ID data
15         bs_ID <- IDs %>%
16             sample_n(size = n(), replace = TRUE)
17
18         # And our full data
19         bs_d <- d %>%
20             inner_join(bs_ID, by = 'ID')
21         return(bs_d)
22 }
23
24 # Create a cluster bootstrap data set
25 create_boot(d, IDs)
```

```python
1  # PYTHON CODE
2  # Load packages
3  import pandas as pd
4  import numpy as np
5  np.random.seed(100)0)
6
7  # Example data
8  d = pd.DataFrame({'ID': [1,1,2,2,3,3],
```

```
9                           'X': [1,2,3,4,5,6]})
10  # Now, get our data frame just of IDs
11  IDs = pd.DataFrame({'ID': np.unique(d['ID'])})
12
13  # Our bootstrap resampling function
14  def create_boot(d, IDs):
15          # Resample our ID data
16          N = IDs.shape[0]
17          index = np.random.randint(0, N, size = N)
18          bs_ID = IDs.iloc[index]
19
20          # And our full data
21          bs_d = d.merge(bs_ID, how = 'inner', on = 'ID')
22          return(bs_d)
23
24  # Create a cluster bootstrap data set
25  create_boot(d, IDs)
```

ANOTHER POPULAR TYPE OF BOOTSTRAP ACTUALLY WORKS
IN A VERY DIFFERENT WAY. In *residual resampling*, instead of
resampling *whole observations*, you instead resample the *residuals*. Huh?

Under residual resampling, you follow these steps:

1. Perform your analysis using the original data.

2. Use your analysis to get a predicted outcome value $\hat{Y}$, and
   also the residual $\hat{r}$, which is the difference between the actual
   outcome and the predicted outcome, $\hat{r} = Y - \hat{Y}$.

3. Perform bootstrap resampling to get *resampled residuals* r_b.

4. Create a new outcome variable by adding the residual to the
   actual outcome, $Y_b = Y + r_b$.

5. Estimate the model as normal using Y_b.

6. Store whatever result you want, and iterate steps 3—5.

The idea here is that *the predictors in your model never
change* with this process. So whatever kind of interdependence
they have, you get to keep that! A downside of residual resampling is that it doesn't perform well if the errors are *in any way*
related to the predictors—this is a stronger assumption than
even the OLS assumption that errors are on *average* unrelated
to predictors.

Another popular form of the bootstrap in this vein, which
both has a very cool name and gets around the unrelated-error
assumption, is the *wild bootstrap*. The wild bootstrap is popular because it works under heteroskedasticity—even when we
don't know the exact form the heteroskedasticity takes. It also

performs well when the data is clustered. We already have the cluster/block bootstrap for that. However, the cluster bootstrap doesn't work very well if the clusters are of very different sizes. If your classroom has 100 kids in it, but mine has five, then re-sampling classrooms can get weird real fast. The cluster-based version of the wild bootstrap works better in these scenarios. The cluster-based version of the wild bootstrap also works much better when there are a *small number* of clusters, where "small" could even be as large as 50.[48]

Without getting too into the weeds, let's talk about what the wild bootstrap is. For a wild bootstrap, you follow the general outline of residual resampling. Except you *don't resample the residual.* You keep each residual lined up with its original observation each time you create the bootstrap data.

Instead of resampling the residuals, you first transform them according to a transformation function, and then multiply them by a random value with mean 0 (plus some other conditions).[49] That's it! You take your random value for each observation, multiply by the transformed residual, use that to get your new outcome value Y_b. Estimate the model with that, then iterate.

Do note that there are quite a few options for which transformation function, and which random distribution to draw from. Different choices seem to perform better in different scenarios, so I can't give any blanket recommendations. If you want to use a wild bootstrap you might want to do a little digging about how it applies to your particular type of data. If you *do* want to get into the weeds, I can recommend a nice introduction by Martin (2017).[50]

It seems incomprehensible as to *why* the wild bootstrap actually works. But the short of it is that the random value you multiply the residual by does manage to mimic the sampling distribution. Plus, it does so in a way that respects the original model, making things a bit more cohesive for the coefficients you're trying to form sampling distributions for.

Wild bootstrap standard errors can be implemented in R us-ing the `type` argument of `vcovBS()` in the **sandwich** package. There are different options for `type` depending on the kind of distribution you want to use to generate your random multipli-ers, although the default `'wild'` option uses the Rademacher distribution, which is simply −1 or 1 with equal probability. In Stata you can follow many regression commands with `boottest` from the **boottest** package. There does not appear to be a ma-ture implementation in Python at the moment.

[48] Regular clustered stan-dard errors get their results from *asymptotic* assump-tions, i.e., they're proven to work as intended as the number of clusters gets somewhere near infinity. In practice, for clustered er-rors, 50 is probably near enough to infinity to be fine. But the wild cluster boot-strap makes no such assump-tion, and so works much bet-ter with smaller numbers of clusters.

[49] The mean-zero part means that many of these random values are negative. Some observations get the signs of their residuals flipped.

[50] Michael A. Martin. *Wild Bootstrap*, pages 1–6. John Wiley & Sons, Ltd., 2017. ISBN 9781118445112.

16

Fixed Effects

16.1 How Does It Work?

16.1.1 Controlling for Unobserved Confounders

A LOT OF THIS BOOK has discussed ways to identify the answer
to your research question by *controlling* for variables. If we have
the diagram, then we can figure out what we need to control for,
and control for it, perhaps by following the methods in Chapters
13 or 14.

There are two big problems we keep running into, though.
First, what if we can't figure out the correct causal diagram that
will tell us what to control for? And second, what if we can't
measure all the variables we need to control for? In either of
those cases, we can't identify our answer by controlling for stuff.

Or can we?

Fixed effects is a method of controlling for *all* variables, whether
they're observed or not, *as long as they stay constant within
some larger category.* How can we do that? Simple! We just
control for the larger category, and in doing so we control for
everything that is constant within that category.[1,2]

[1] If you prefer, we're controlling for a variable higher up
on the hierarchy of our hierarchical data.

[2] Depending on your field,
you may be more famil-
iar with the term "fixed effects" to mean "not random
effects," or "no individual-
level variation in the intercept," as opposed to what
I'm talking about in this
chapter, which is even *more*
variation in the intercept
than random effects. That's
okay. It's not your fault
your field is wrong.

DOI: 10.1201/9781003226055-16

What does it mean for a variable to be "constant within some larger category?" For example, let's say we're looking at the effect of rural towns getting electricity on their productivity. An obvious back door is geography. Rural towns up in the mountainous hillside will be more difficult to electrify, and also might be different in their productivity for other reasons.

However, geography doesn't change a whole lot. If you observe the same town multiple times, it's likely to have the same geography every single time. So what if, observing the same town in many different years, we control for town?[3] If we do that, we'll be removing any variation explained by town. And since there's no variation in geography that isn't explained by town, after we control for town we'll have controlled for geography—there will be no variation in geography left!

NOW IN THIS CASE, we probably could have measured and controlled for geography, no big deal. But what about the stuff we can't control for?

To give one prominent example, there are lots of social science contexts where we'd want to control for "person's background," but that's a high-dimensional thing with lots of aspects we can't possibly measure. It's unobserved. But add fixed effects for individual person? Boom! It's controlled for.[4]

FIXED EFFECTS CAN BE THOUGHT OF as taking a whole long list of variables on back doors that are constant within some category and collapsing them to just be that category. Then, controlling for that category.

To give an example, let's say we're interested in the effect of a visit from the German chancellor on a country's level of trade with Germany, as depicted in 16.1.

[3] We can "control for town" the same way we'd control for any categorical variable. In regression this means adding a set of binary indicators, one for each town. In matching it would mean using exact matching on town.

[4] Variables that are *nearly* constant will also be *nearly* controlled for. For example, gender is constant within individual for most people. Some people change their gender, so it's not perfectly constant within individual. But as long as this is very rare, we might still say that fixed effects for individual controls for gender.

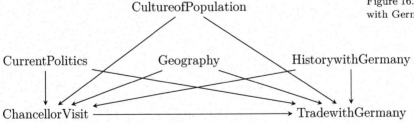

Figure 16.1: Causes of Trade with Germany

In the diagram, there are quite a few causes of trade with Germany, many of which would be difficult to keep track of or even measure. However, we can also note that several of these variables—the country's geography, the culture of the

population, and the history that country has with Germany—
are constant within country or at least (like the culture of the
population) are not likely to change much within the span of
any data we have. So we can redraw the diagram as in Figure
16.2.

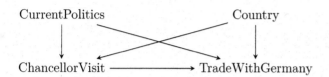

Figure   16.2:   Simplified
Causes   of   Trade   with
Germany

Now, with our simplified figure, we can identify the effect
without needing to control for each of the variables in Figure
16.1. We can just control for country instead.[5]

Keep in mind that we haven't collapsed everything into country. Anything likely to change regularly over time, like the current politics of a country, won't be addressed by fixed effects.
So if we wanted to identify this effect, we'd still need to control
for CurrentPolitics in addition to fixed effects for country.

16.1.2   The Variation Within

FIXED EFFECTS, IN ESSENCE, *controls for individual,* whether
"individual" in your context means "person," "company," "school,"
or "country," and so on.[6]

What this means is that it gets rid of any variation *between*
individuals. We know that because that's what it means to control for individual. We're taking all the variation in our data
explained by individual (i.e., all the variation between individuals) and getting rid of it.

What's left is variation *within* individual. What is variation
within individual? For example, let's say you're a fairly tall
person and we're tracking your height over time. The fact that
you're taller than other people is *between* variation. Comparing you to someone else, you're generally taller. However, we
also can observe that, as you grew up from a child to an adult,
your height changed over time. Comparing you to your shorter,
younger self is *within* variation. Something changed *within* individual and that's what we're picking up.

The idea of fixed effects is that by sweeping away all that
variation between individuals, we've controlled for any variables
that are fixed over time.[7] For example, no matter where I move
in life, I will always have been born in the same city.[8] The same

[5] This logic leads some people to conclude that you
can't identify *any* effect
without having access to
some sort of panel data /
repeated measurement over
time so you can use within
variation.   But that's not
true!   There are plenty of
methods in this book that
work fine without it. That's
why   we   worked   so   hard
on writing out our causal
diagrams—so we'd know if
we needed something like
fixed effects.
[6] More broadly, it controls
for group at some level of hierarchy. But for simplicity
let's say individual.

Within variation. Variation of a variable that occurs
within an individual across
(usually) different periods of
time.
Between variation. Variation of a variable that occurs *between* different individuals, usually at the same
period of time, or comparing
over-time averages.
[7] Technically the *effect* of
those variables must also be
fixed over time for this to
work, too.
[8] Holler to Arcata, California. Not Arcadia. Easy to
mix up.

goes for everybody else. So all the variation in "city of birth" is *between* variation. If we get rid of all between variation, we've gotten rid of all variation in city of birth, thus controlling for it.

One thing to note about this is that we really have gotten rid of that between variation. So if between variation is what we're interested in—say, we *want* to know the effect of city of birth—we can't do that with fixed effects.[9]

LET'S WALK THROUGH two technical examples of this, one with tables and one with graphs. Let's say we're interested in the effect of exercise on the number of times a year you get a cold. For subjects in our study we have you and me. We also suspect there are some variables like "genetics" that are fixed over time that we can adjust for with fixed effects. Let's start by looking at the data we have on exercise.

Individual	Year	Exercise
You	2019	5
You	2020	7
Me	2019	4
Me	2020	3

[9] We *can* do that with random effects—see later in the chapter.

Table 16.1:   Exercise and Colds

Table 16.1 shows our raw data, with exercise measured in hours per week. To get our within and between variation, we'll need to get the within-individual means, shown in Table 16.2.

Individual	Year	Exercise	MeanExercise	WithinExercise
You	2019	5	6	-1
You	2020	7	6	1
Me	2019	4	3.5	0.5
Me	2020	3	3.5	-0.5

Table 16.2:   Exercise with Within Variation

Table 16.2 shows how we can take the mean of exercise for each individual, and then subtract out that mean. The *difference between individuals in the means* is the *between variation*. So the between variation for exercise comes from comparing your average 6 hours per week against my 3.5 hours per week.

If we subtract out those individual means, we're left with the way in which we vary from time period to time period relative to our own averages. This is the within variation, looking at how things vary within individual. So my within variation in exercise compares the fact that in 2019 I exercised half an hour per week more than average, and in 2020 I exercised half and hour per week less than average.

WE CAN SEE THIS GRAPHICALLY as well. The top-left of
Figure 16.3 shows what the data might look like if we gathered
it for a few more years.

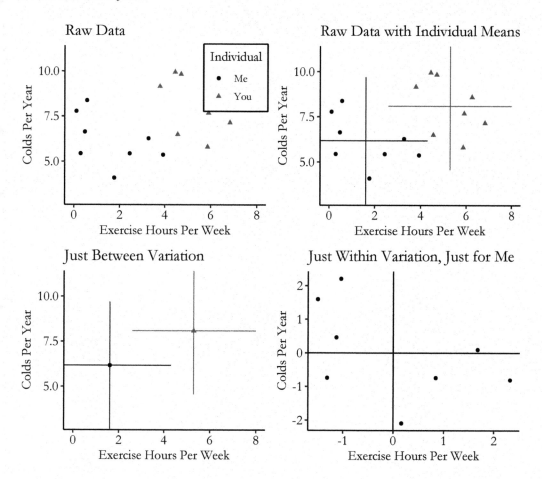

The top-right takes that raw data and adds some individual-
level means. Sort of like each of us is getting their own set of
axes. We find the mean exercise and mean number of colds for
you, and draw your set of axes at that point. That's $(0,0)$ on
your graph. Similarly, we draw my set of axes where my means
are. That's $(0,0)$ on *my* graph. Of course, on our shared graph,
it's not $(0,0)$ for anyone, it's our individual means on both the
x and y axes.

The bottom-left looks just at those axes. We put a point
at the center of those axes and ignore everything *but* those individual means. If we then compare them, say, by drawing a
line from one to the other, then we're looking at the *between*
variation.

Figure 16.3: Between and
Within Variation of Exercise
and Colds for You and Me

In the bottom right, we forget about you for a second, and just focus on me.[10] We zoom in to my set of axes and call those axes $(0,0)$. Now we can see that everything is relative to my own mean. Take that point in the bottom-left, for example. In the top-right graph that point was about .3 exercise hours per week and about 5.5 colds per year. But in this bottom-right graph, everything is relative to my mean. So instead of .3 and 5.5, it's about 1.3 *fewer hours of exercise per week than my average*, and about .8 *fewer colds per year than my average*.

To actually perform fixed effects, we'd then take your set of axes and slide it over on top of mine! Let's see how that works.

16.1.3 Fixed Effects in Action

WHAT DOES FIXED EFFECTS actually do to data? Well, it's basically the same as controlling for a categorical variable as described in Chapter 13. We're isolating within variation. So we find the mean differences between groups (individuals) and subtract them out. That's basically it.

We begin in Figure 16.4 with our raw data, noting that we have several individuals in there. In particular, we are interested in the effects of healthy-eating reminders on actual healthy eating. Our four subjects have each downloaded an app that, at random times, reminds them to eat healthy. They've each chosen how frequently they want reminders, but don't control exactly on which days they come. We've recorded how often they get reminders, and also scored them on how healthy they actually eat.

So we have variation over time in how intensively each individual gets reminders, and variation in their healthy eating. There might be some individual-level back doors that led them to choose their frequency and also their general healthy-eating level, so we will want to use fixed effects to close those back doors.

In Figure 16.4 there's a small positive correlation of 0.111 between reminders and healthy eating—more reminders means eating healthier! Maybe the messages are effective, if only a tiny little bit.[11] But of course, we haven't identified the answer to our question yet, since we know there are individual-based back doors. So we start our fixed effects estimation by calculating the mean reminders and mean healthy eating scores for each of our individuals.

Figure 16.5 shows the individual means. There's clearly a negative relationship between reminders and healthy eating for

[10] Try this line on a first date sometime.

[11] Although, looking at the graph itself, this isn't the most convincing finding.

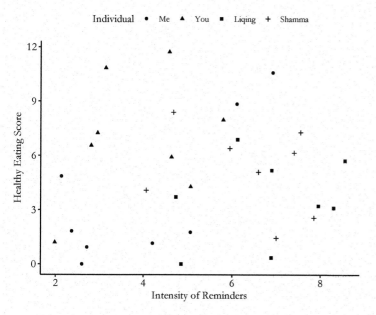

Figure 16.4: Healthy Eating Reminders and Healthy Eating for Four Individuals

the individual means. If you just look at the big crosses, *they* seem to follow a downward slope. The *between* part of the relationship is negative—the people who choose more frequent messages are the people who tend to eat more poorly on average. The *between* correlation is -0.440.

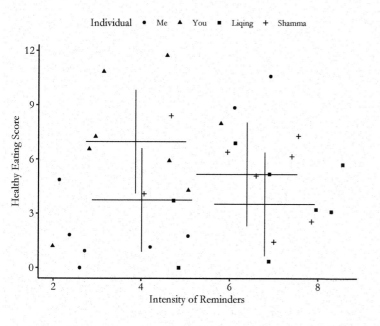

Figure 16.5: Healthy Eating Reminders and Healthy Eating for Four Individuals

Our next task is to remove any variation between individuals. So we're effectively going to take all four of those +'s and slide them on top of each other. First let's zoom in on each of the four individuals' within-variation alone, in Figure 16.6.

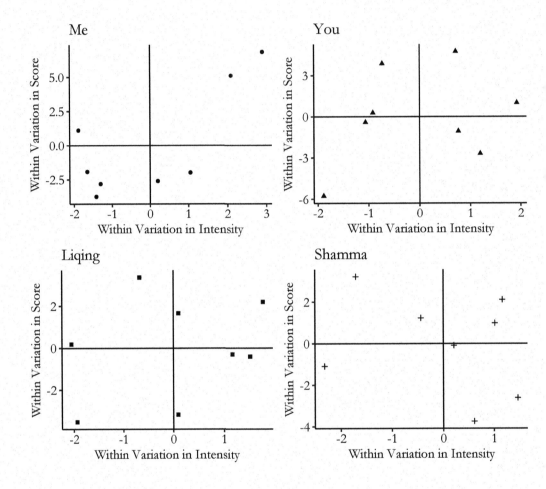

Figure 16.6: Healthy Eating Reminders and Healthy Eating for Four Individuals

Notice that for each of the four graphs in Figure 16.6, the data is centered at 0 on both the x- and y-axes. So now when we put all four on top of each other, they'll be hanging around the same part of the graph.

We put them all on top of each other, aligning all the +'s, in Figure 16.7.

Now, with the different individuals moved on top of each other, we see a more clear picture emerging, with a positive relationship between the intensity of reminders and the scores. While the correlation in the raw data was very small, the correlation in the within variation is 0.363. Much more positive.[12]

[12] Although still nothing to write home about. I doubt this reminder app is about to get a flurry of interested investors.

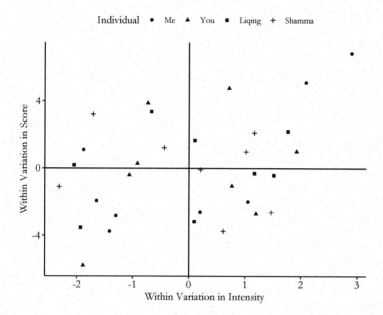

And since I created this data generating process myself I can tell you that a positive relationship is indeed correct.

16.2   How Is It Performed?

16.2.1   Regression Estimators

FIXED EFFECTS REGRESSIONS GENERALLY START with a regression equation that looks like this:

$$Y_{it} = \beta_i + \beta_1 X_{it} + \varepsilon_{it} \qquad (16.1)$$

This looks exactly like our standard regression equation from Chapter 13 with two main exceptions. The first is that X has the subscript it, indicating that the data varies both between individuals (i) and over time (t).

The second is that the intercept term β now has a subscript i instead of an 0, making it β_i. Thinking of regression as a way of fitting a straight line, this means that all the individuals in the data are constrained to have the same *slope* (there's no i subscript on β_1), but they have different *intercepts*.

Three obvious questions arise.

QUESTION 1. How does allowing the intercept to vary give you a fixed effects estimate where you use only within variation?

Varying intercepts.   A regression model in which, even if the slope (coefficient) on each independent variable is restricted to be the same for everyone, the intercept (constant term) is allowed to be different for each individual.

There are two ways to think about this intuitively. The first is procedural. By allowing the intercept to vary it's sort of like we're adding a control for each person. Let's take a study of countries for example. If India is in our sample, then one of the intercepts is β_{India}. This intercept *only* applies to India, not to any of the other countries.

That's a lot like adding a "This is India" binary control variable to our regression, and β_{India} is the coefficient on that control variable. We have one of these controls for each country in our data. So we're controlling for each different country. And we know what happens when we control for country. We get within-country variation.

The second way to think about why this works intuitively is to think in terms of a graph. Figure 16.8 uses data on GDP per capita and life expectancy from the Gapminder Institute.[13] Here we've isolated just two countries—India and Brazil, and are looking at their data since 1952.

[13] Gapminder Institute. Gapminder. https: //www.gapminder.org/, 2020. Accessed: 2020-03-09.

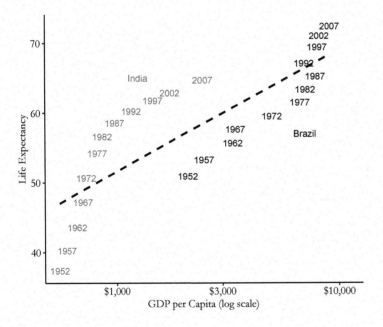

Figure 16.8: GDP per Capita and Life Expectancy in Two Countries

We can see a few things clearly in Figure 16.8. First, things in both countries are clearly getting better over time.[14] However, there are also some clear between-country differences we want to get rid of. And that regression line, with a single intercept for both, is clearly underestimating the slope of the lines that each of the two of them have.

[14] Neat!

By breaking up the intercept, we can instead use *two* lines—
parallel lines, since they have the same slope—and move them
up and down until we hit each country. This lets us not care
how far up or down each country's cluster of points is since we
can just move up and down to hit it (we get within variation
in the y-axis variable), and also lets us not care how far left or
right each country's cluster is, since by moving up and down,
the same slope can hit the cluster no matter how far out to the
right it is (we get within variation in the x-axis variable). And
thus we have within variation.

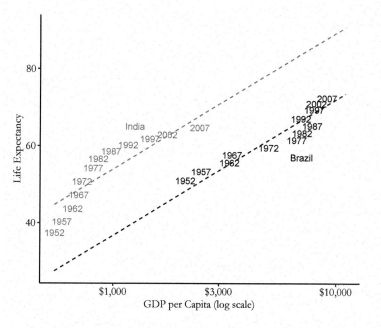

Figure 16.9:   GDP per
Capita and Life Expectancy
in Two Countries

We do exactly this—the same slope but with different
intercepts—in Figure 16.9. The lines are a lot closer to those
points now. Those lines capture the relationship that's clearly
there a lot better than the single regression line did.

QUESTION 2. How do we estimate a regression with individual intercepts?

There are actually a bunch of ways to do this. Some of them
we will cover in later parts of this chapter.[15] But within that
long list of ways, there are two standard methods, both of which
are dead-simple and only use tools we already know.

The first method is just to extract the within variation ourselves, and work with that.[16] Just calculate the mean of the
dependent variable for each individual ($\bar{Y}_i$) and subtract that
mean out ($Y_{it} - \bar{Y}_i$). Then do the same for each of the independent variables ($X_{it} - \bar{X}_i$). Then run the regression

[15] Some of them don't even
isolate only the within-
variation. Scandalous!

[16] This approach gets way
trickier if you have more
than one set of fixed effects.

$$Y_{it} - \bar{Y}_i = \beta_0 + \beta_1(X_{it} - \bar{X}_i) + \varepsilon_{it} \qquad (16.2)$$

This method is known as "absorbing the fixed effect."[17] Let's run this approach using the Gapminder data once again.

[17] β_0 is sometimes left out of the equation when doing this method, because it will be 0 anyway. Think about why that should be.

```r
# R CODE
library(tidyverse); library(modelsummary)
gm <- causaldata::gapminder

gm <- gm %>%
    # Put GDP per capita in log format since it's very skewed
    mutate(log_GDPperCap = log(gdpPercap)) %>%
    # Perform each calculation by group
    group_by(country) %>%
    # Get within variation by subtracting out the mean
    mutate(lifeExp_within = lifeExp - mean(lifeExp),
    log_GDPperCap_within = log_GDPperCap - mean(log_GDPperCap)) %>%
    # We no longer need the grouping
    ungroup()

# Analyze the within variation
m1 <- lm(lifeExp_within ~ log_GDPperCap_within, data = gm)
msummary(m1, stars = c('*' = .1, '**' = .05, '***' = .01))
```

```stata
* STATA CODE
causaldata gapminder.dta, use clear download

* Get log GDP per capita since GDP per capita is very skewed
g logGDPpercap = log(gdppercap)

* Get mean life expectancy and log GDP per capita by country
by country, sort: egen lifeexp_mean = mean(lifeexp)
by country, sort: egen logGDPpercap_mean = mean(logGDPpercap)

* Subtract out that mean to get within variation
g lifeexp_within = lifeexp - lifeexp_mean
g logGDPpercap_within = logGDPpercap - logGDPpercap_mean

* Analyze the within variation
regress lifeexp_within logGDPpercap_within
```

```python
# PYTHON CODE
import numpy as np
import statsmodels.formula.api as sm
from causaldata import gapminder
gm = gapminder.load_pandas().data

# Put GDP per capita in log format since it's very skewed
gm['logGDPpercap'] = gm['gdpPercap'].apply('log')

# Use groupby to perform calculations by group
# Then use transform to subtract each variable's
# within-group mean to get within variation
gm[['logGDPpercap_within','lifeExp_within']] = (gm.
groupby('country')[['logGDPpercap','lifeExp']].
transform(lambda x: x - np.mean(x)))

# Analyze the within variation
```

```
18 │ m1 = sm.ols(formula = 'lifeExp_within ~ logGDPpercap_within',
19 │ data = gm).fit()
20 │ m1.summary()
```

The second simple method is to add a binary control variable for every country (individual). Or rather, every country *except one*—if our model also has an intercept/constant in it, then including *every* country will make the model impossible to estimate, so we leave one out, or leave the intercept out.[18] That one left-out country is still in the analysis, it just doesn't get its own coefficient. It's stuck with the lousy constant. Thankfully, most software will do this automatically.

I should point out that this is rarely the way you want to go. Including a set of binary indicators in the model is fine when you have a few different categories, but if you're controlling for *each individual* you may have hundreds or *thousands* of terms. This can get very, very slow. And c'mon, were you really going to interpret all those coefficients anyway? But for demonstrative purposes, we'll run this version to see what we get.

[18] Why? Imagine we just have two people—me and you—and are estimating height. Say my average height over time is 66 inches and yours is 68, and we don't leave anyone out so we have $Height_{it} = \beta_0 + \beta_1 Me_i + \beta_2 You_i$. The values $\beta_0 = 0, \beta_1 = 66, \beta_2 = 68$ gives exactly the same fit as $\beta_0 = 1, \beta_1 = 65, \beta_2 = 67$. There are infinite ways to get the exact same fit. There's no way it can choose! So the model can't be estimated.

```
1  │ # R CODE
2  │ library(tidyverse); library(modelsummary)
3  │ gm <- causaldata::gapminder
4  │
5  │ # Simply include a factor variable in the model to get it turned
6  │ # to binary variables. You can use factor() to ensure it's done.
7  │ m2 <- lm(lifeExp ~ factor(country) + log(gdpPercap), data = gm)
8  │ msummary(m2, stars = c('*' = .1, '**' = .05, '***' = .01))
```

```
1  │ * STATA CODE
2  │ causaldata gapminder.dta, use clear download
3  │ g logGDPpercap = log(gdppercap)
4  │
5  │ * Stata requires variables to be numbers before automatically
6  │ * automatically making binary variables. So we encode.
7  │ encode country, g(country_number)
8  │ regress lifeexp logGDPpercap i.country_number
```

```
1  │ # PYTHON CODE
2  │ import numpy as np
3  │ import statsmodels.formula.api as sm
4  │ from causaldata import gapminder
5  │ gm = gapminder.load_pandas().data
6  │ gm['logGDPpercap'] = gm['gdpPercap'].apply('log')
7  │
8  │ # Use C() to include binary variables for a categorical variable
9  │ m2 = sm.ols(formula = '''lifeExp ~ logGDPpercap +
10 │ C(country)''', data = gm).fit()
11 │ m2.summary()
```

Let's take a look at our results from the two methods, using the R output, in Table 16.3. With our second method we also estimated a bunch of country coefficients, but those aren't on the table (there are a lot of them).

	Life Expectancy (within)	Life Expectancy
Log GDP per Capita		9.769***
		(0.297)
Log GDP per Capita (within)	9.769***	
	(0.284)	
Num.Obs.	1704	1704
R2	0.410	0.846

* p < 0.1, ** p < 0.05, *** p < 0.01

Table 16.3: Two Different Regression Approaches to Estimating a Model with Fixed Effects

There are some minor differences between the two in terms of their standard errors, and some big differences in R^2, but the coefficients are the same.[19] So now that we have the coefficients, how can we interpret them?

QUESTION 3. How do we interpret the results of this regression once we have estimated it?

The interpretation of the slope coefficient in a fixed effects model is the same as when you control for any other variable— within the same value of country, how is variation in log GDP per capita related to variation in life expectancy?

Put another way, since we have a coefficient of 9.769, we can say that for a given country, in a year where log GDP per capita is one unit *higher than it typically is for that country*, then we'd expect life expectancy to be 9.769 years *longer than it typically is for that country*.[20]

What else do we have on the table?

We can see that the R^2 values are quite different in the two columns. Why is this?

Remember that R^2 is a measure based on how much variation there is in our residuals relative to our dependent variable.[21] In the first column, our dependent variable is *lifeExp_within*, not *lifeExp*. So the R^2 is based on how much variation there is in the residuals relative to the the *within* variation, not the overall variation in *lifeExp*. This is called the "within R^2." On the contrary, the second column tells us how much variation there is in (those very same) residuals relative to all the variation in *lifeExp*. It counts the parts of *lifeExp* explained by between variation (our binary variables).

Lastly, let's think about something that's *not* on the table— the fixed effects themselves. That is, those coefficients on the country binary variables we got in the second regression. For example, India's is 13.971 and Brazil's is 6.318. These are the intercepts in the country-specific fitted lines, like in Figure 16.9.

[19] These code snippets are designed to be illustrative to show how both methods are equivalent. Most of the time, researchers will actually use one of the commands shown in the "Multiple Sets of Fixed Effects" section below, even if they only have one set of fixed effects.

[20] As you'll recall from Chapter 13, a one-unit increase in log GDP per capita is a percentage increase in GDP per capita. Specifically, a 171% increase. But for smaller increases, the value and percentage are very close, so we might more often say a log GDP per capita .1 above the typical level, or GDP per capita roughly 10% above, is associated with an life expectancy of .9769 years above the typical level.

[21] $R^2 = 1 - $ var(residuals) / var(dependent variable), where var is short for variance.

Within R^2 One minus the proportion of variance in the residuals relative to the variance in the dependent variable's within variation.

What can we make of these fixed effects? Not too much in absolute terms,[22] but they make sense *relative to each other.* So India's intercept is $13.971 - 6.318 = 7.653$ *higher* than Brazil's.

We can interpret that to mean that, if India and Brazil had the same GDP per capita, we'd predict that India's life expectancy would be 7.653 higher than Brazil's. We're still looking at the effect of one variable controlling for another—just now, we're looking at the effect of *country* controlling for *GDP per capita* rather than the other way around.

Sometimes researchers will use these fixed effects estimates to make claims about the individuals being evaluated. They might say, for example, that India has especially high life expectancy *given its level of GDP per capita.*

This can be an interesting way to look at differences between individuals. However, keep in mind that fixed effects is a good way of *controlling for a long list of unobserved variables that are fixed over time* to look at the effect of a few time-varying variables. Those individual effects don't have the same luxury of controlling for a bunch of unobserved time-varying variables. We can rarely think of these individual fixed effects estimates as causal.[23]

ONE LAST THING to keep in mind about fixed effects. It focuses on *within* variation. So the treatment effect (Chapter 10) estimated will focus a lot more heavily on people with a lot of within variation. In that life expectancy and GDP example, the coefficient on GDP we got is closer to estimating the effect of GDP on life expectancy in countries where GDP changed a *lot* over time. If your GDP was fairly constant, you count less for the estimate! You can address this by using weights to address the problem.[24,25]

16.2.2 *Multiple Sets of Fixed Effects*

WE'VE DISCUSSED FIXED EFFECTS as being a way of controlling for a categorical variable. This ends up giving us the variation that occurs *within* that variable. So if we have fixed effects for individual, we are comparing that individual to themselves over time. And if we have fixed effects for city, we are comparing individuals in that city only to other individuals in that city.

So why not include multiple sets of fixed effects? Say we observe the same individuals over multiple years. We could include fixed effects for individual—comparing individuals to themselves at different periods of time, or fixed effects for year—

[22] We can't consider them independently because they're all defined relative to whichever country didn't get its own coefficient—remember that? So Brazil's 6.318 doesn't mean its intercept is 6.318, it means its intercept is 6.318 *higher than the intercept of the country that was left out.*

[23] In addition, individual fixed effects estimates are estimated based on a relatively small number of observations—just the number we have per individual. So they tend to be a little wild and vary too much. Random effects—discussed later in the chapter—addresses this problem by "shrinkage" of the individual effects.

[24] Charles E. Gibbons, Serrato Juan Carlos Suárez, and Michael B. Urbancic. Broken or fixed effects? *Journal of Econometric Methods,* 8(1):1–12, 2019.

[25] The Gibbons, Serrato, and Urbancic (2019) paper suggests two solutions, the simplest of which is to calculate the variance of the treatment variable within individual, and weight by the inverse of that. This is an inverse variance weight, like in Chapter 13.

comparing individuals to each other at the same period of time. What if we include fixed effects for both individual and year?

This approach—including fixed effects for both individual and for time period—is a common application of multiple sets of fixed effects, and is commonly known as *two-way fixed effects*. The regression model for two-way fixed effects looks like:

$$Y_{it} = \beta_i + \beta_t + \beta_1 X_{it} + \varepsilon_{it} \qquad (16.3)$$

Two-way fixed effects. A model that includes fixed effects for both individual and for year.

What does this get us? Well, we are looking at variation *within* individual as well as *within* year at this point. So we can think of the variation that's left as being variation *relative to what we'd expect given that individual, and given that year*.[26]

FOR EXAMPLE, let's say we have data on how much each person earns per year (Y_{it}), and we also have data on how many hours of training they received that year (X_{it}). We want to know the effect of experience of in-job training on pay at that job.

We do happen to know that the year 2009 was a particularly bad year for earnings, what with the Great Recession and all. Let's say that earnings in 2009 were $10,000 below what they normally are. And let's say that we're looking at Anthony, who happens to earn $120,000 per year, a lot more money than the average person ($40,000).

In 2009 Anthony only earned $116,000. That's way above average, but that can be explained by the fact that it's Anthony and Anthony earns a lot. So *given that it's Anthony*, it's $4,000 below average. But *given that it's 2009*, most people are earning $10,000 less, but Anthony is only earning $4,000 less. So *given that it's Anthony, and given that it's 2009*, Anthony in 2009 is $6,000 more than you'd expect.

Or at least that's a *rough* explanation. Two-way fixed effects actually gets a fair amount more complex than that, since the individual fixed effects affect the year fixed effects and vice versa. Just like with one set of fixed effects, the variation that actually ends up getting used to estimate the treatment effect (Chapter 10) *focuses more heavily* on individuals that have a lot of variation over time. So if Anthony's level of in-job training is pretty steady over time, but some other person, Kamala, has a level of training that changes a lot from year to year, then the two-way fixed effects estimate will represent Kamala's treatment effect a lot more than Anthony's.[27]

TWO-WAY FIXED EFFECTS, WITH INDIVIDUAL AND TIME AS THE FIXED EFFECTS, AREN'T THE ONLY WAY to have multiple

[26] Crucially, this is not the same as *given that individual that year*, as we only observe that individual in that year once. There's no variation. Each of the "relative to"s is one at a time.

[27] Clément De Chaisemartin and Xavier d'Haultfoeuille. Two-way fixed effects estimators with heterogeneous treatment effects. *American Economic Review*, 110 (9):2964–96, 2020.

sets of fixed effects, of course. As described above, for example, you could have fixed effects for individual and also for city. Neither of those is time!

The interpretation here is a lot closer to thinking of just including controls for individual and controls for city.

One thing to remember, though. We're isolating variation *within* individual and also *within* city. So in order for you to have any variation at all to be included, you need to show up in multiple cities. This is a more common problem here (two sets of fixed effects, neither of them is time) than with two-way fixed effects (individual and time), since generally each person shows up in each time period only once anyway.

If we ran our income-and-job-training study with individual and city fixed effects, the treatment effect would *only* be based on people who are observed in different cities at different times. Never move? You don't count!

ESTIMATING A MODEL WITH MULTIPLE FIXED EFFECTS can be done using the same "binary controls" approach as for the regular ("one-way") fixed effects estimator with only one set of fixed effects. No problem!

However, this can be difficult if there are lots of different fixed effects to estimate. The computational problem gets thorny real fast. Unless your fixed effects only have a few categories each (say, fixed effects for left- or right-handedness and for eye color), it's generally advised that you use a command specifically designed to handle multiple sets of fixed effects.

These methods usually use something called *alternating projections*, which is sort of like our original method of calculating within variation and using that, except that it manages to take into account the way that the first set of fixed effects gives you within-variation in the other set, and vice versa.

Let's code this up, continuing with our Gapminder data.

```
1  # R CODE
2  library(tidyverse); library(modelsummary); library(fixest)
3  gm <- causaldata::gapminder
4
5  # Run our two-way fixed effects model (TWFE).
6  # First the non-fixed effects part of the model
7  # Then a /, then the fixed effects we want
8  twfe <- feols(lifeExp ~ log(gdpPercap) | country + year,
9            data = gm)
10 # Note that standard errors will be clustered by the first
11 # fixed effect by default. Set se = 'standard' to not do this
12 msummary(twfe, stars = c('*' = .1, '**' = .05, '***' = .01))
```

```
1 │ * STATA CODE
2 │ causaldata gapminder.dta, use clear download
3 │ g logGDPpercap = log(gdppercap)
4 │
5 │ * We will use reghdfe which must be installed with
6 │ * ssc install reghdfe
7 │ reghdfe lifeexp logGDPpercap, a(country year)
```

```
1  │ # PYTHON CODE
2  │ import linearmodels as lm
3  │ from causaldata import gapminder
4  │ gm = gapminder.load_pandas().data
5  │ gm['logGDPpercap'] = gm['gdpPercap'].apply('log')
6  │
7  │ # Set our individual and time (index) for our data
8  │ gm = gm.set_index(['country','year'])
9  │
10 │ # Specify the regression model
11 │ # And estimate with both sets of fixed effects
12 │ # EntityEffects and TimeEffects
13 │ # (this function can't handle more than two)
14 │ mod = lm.PanelOLS.from_formula(
15 │ '''lifeExp ~ logGDPpercap +
16 │               EntityEffects +
17 │               TimeEffects''',gm)
18 │
19 │ twfe = mod.fit()
20 │ print(twfe)
```

16.2.3   Random Effects

AS A RESEARCH DESIGN, fixed effects is all about isolating *within variation*. We think that there are some back doors floating around in that between variation, so we get rid of it and focus on the within variation. But as a *statistical method*, fixed effects can really be thought of as a model in which the intercept varies freely across individuals. As I've put it before, here's the model:

$$Y_{it} = \beta_i + \beta_1 X_{it} + \varepsilon_{it} \tag{16.4}$$

It turns out that fixed effects is, statistically, a relatively weak way of letting that intercept vary. After all, we're only allowing within variation. And how many observations per individual do we really have? We may be estimating those β_is with only a few observations, making them noisy, which in turn could make our estimate of β_1 noisy too.

Random effects takes a slightly different approach. Instead of letting the β_is be anything, it puts some structure on them. Specifically, it assumes that the β_is come from a known random distribution, for example assuming that the β_is follow a normal distribution.

This does a few things:

- It makes estimates more precise (lowers the standard errors).

- It improves the estimation of the individual β_i effects themselves.[28]

- Instead of just using within variation, it uses a weighted average of within and between variation.

- It solves the same back door problem that fixed effects does *if* the β_i terms are unrelated to the right-hand-side variables X_{it}.

THAT LAST ITEM is a bit of a doozy. Sure, better statistical precision is nice. But we're only doing fixed effects in the first place to solve our research design problem. Random effects only solves the same problem if the individual fixed effects are unrelated to our right-hand-side variable, including our treatment variable.

That seems unlikely. Consider our Gapminder example. For random effects to be appropriate, the "country effect" would need to be unrelated to GDP. That is, all the stuff about a country that's fixed over time and determines its life expectancy *other* than GDP per capita needs to be completely unrelated to GDP per capita. And that list would include things like industrial development and geography. Those seem likely to be related to GDP per capita.

For this reason, fixed effects is almost always preferred to random effects,[29] except when you can be pretty sure that the right-hand-side variables are unrelated to the individual effect. For example, when you've run a randomized experiment, X_{it} is a truly randomly-assigned experimental variable and so is unrelated to any individual effect.

One common way around this problem is the use of the *Durbin-Wu-Hausman test*. The Durbin-Wu-Hausman test is a broad set of tests that compare the estimates in one model against the estimates in another and sees if they are different. In the context of fixed effects and random effects, if the estimates are found to not be different (you fail to reject the null hypothesis that they're the same), then the relationship between β_i and X_{it} probably isn't too strong, so you can go ahead and use random effects with its nice statistical properties.

However, I do not recommend the use of the Durbin-Wu-Hausman test for this purpose.[30] For one thing, if we don't have

[28] Why? Because we assume that they come from the same distribution. We can use *all* the data to estimate that distribution, which means each estimate of β_i gets a lot more information to work off of.

[29] Or at least *this version* of random effects... keep reading.

[30] I do not like this testing plan, I do not like it Jerr Hausman.

a strong *theoretical* reason to believe that the unrelated-β_i-and-X_{it} assumption holds, it's hard to really believe that failing to reject a null hypothesis can justify it for you.

For another, the Durbin-Wu-Hausman test compares fixed effects to the simplest version of random effects, which doesn't really use random effects to its full advantage. Most studies that use random effects use them in more useful ways that get around that doozy of an assumption. For that, you'll want to look at "Advanced Random Effects" in the "How the Pros Do It" section.

16.3   How the Pros Do It

16.3.1   Clustered Standard Errors

ONE OF THE ASSUMPTIONS of the regression model is that the error terms are independent of each other. That is, the parts of the data generating process for the outcome variable that *aren't* in the model are effectively random. This assumption is necessary to correctly calculate the regression's standard errors.

However, we might imagine that this assumption is a tough sell when it comes to fixed effects. After all, we have multiple observations from the same individual or group. We might expect that some of the parts of the data generating process that are left out are shared across all of that individual or group's observations, making them correlated with each other, and making the standard errors wrong.

For this reason, it's not uncommon to hear economists say that you should almost *always* use clustered standard errors when you use fixed effects,[31] specifically standard errors clustered at the same level as the fixed effect. So for person fixed effects, for example, using standard errors clustered at the person level. Or for city fixed effects, using standard errors clustered at the city level. Clustered standard errors calculate the standard errors while allowing some level of correlation between the error terms.

HOWEVER, THIS COMMON WISDOM goes a bit too far. For clustering with fixed effects to be necessary (and a good idea), several conditions need to hold. First, there needs to be *treatment effect heterogeneity*. That is, the treatment effect must be quite different for different individuals.

If that is true, there's a second condition. Either the fixed effect groups/individuals in your data need to be a non-random

[31] See the introduction to clustered standard errors in Chapters 13 and 15.

sampling of the population,[32] or, within fixed effect groups/individuals, your treatment variable is assigned in a clustered way.[33]

So before clustering, think about whether both conditions are likely to be true.[34] If it is, go ahead and cluster! If not, don't bother, as the clustering will make your standard errors larger than they're supposed to be.

CLUSTERED STANDARD ERRORS usually come packaged with the same commands that are used to perform fixed effects, since people so commonly use clustered errors with fixed effects. Using Gapminder data once again:

```
1  # R CODE
2  library(tidyverse); library(modelsummary); library(fixest)
3  gm <- causaldata::gapminder
4
5  # feols clusters by the first fixed effect by default
6  clfe <- feols(lifeExp ~ log(gdpPercap) | country, data = gm)
7  msummary(clfe, stars = c('*' = .1, '**' = .05, '***' = .01))
```

```
1  * STATA CODE
2  causaldata gapminder.dta, use clear download
3  g logGDPpercap = log(gdppercap)
4
5  * We will use reghdfe which must be installed with ssc install reghdfe
6  reghdfe lifeexp logGDPpercap, a(country) vce(cluster country)
```

```
1   # PYTHON CODE
2   import pandas as pd
3   import linearmodels as lm
4   from causaldata import gapminder
5   gm = gapminder.load_pandas().data
6   gm['logGDPpercap'] = gm['gdpPercap'].apply('log')
7
8   # Set our individual and time (index) for our data
9   gm = gm.set_index(['country','year'])
10
11  mod = lm.PanelOLS.from_formula(
12      '''lifeExp ~ logGDPpercap +
13              EntityEffects''',gm)
14
15  # Specify clustering when we fit the model
16  clfe = mod.fit(cov_type = 'clustered',
17              cluster_entity = True)
18  print(clfe)
```

16.3.2   Fixed Effects in Nonlinear Models

THE INTUITION BEHIND the fixed effects research design applies consistently to all sorts of statistical models. We think there's a back door that can be closed by getting rid of all the between variation, so we do that.

However, the methods we have for getting rid of that variation—adding a set of binary control variables for the fixed effects,

[32] That is, some groups are more likely to be included in your sample than others.
[33] For example, with city fixed effects, are certain individuals in that city more likely to be treated than others?
[34] Alberto Abadie, Susan Athey, Guido W Imbens, and Jeffrey M Wooldridge. When should you adjust standard errors for clustering? Technical report, NBER, 2017. NBER Working Paper No. 24003.

or subtracting out the individual means—do *not* extend to all sorts of statistical models. Both of those approaches assume that we're using a linear model. That will be a problem if we're not using a linear model. Perhaps we're using probit, or logit, or poisson, or tobit, or ordered logit, and so on and so on.

Why might this be a problem?

First let's consider the approach where we include binary control variables. Including a set of binary control variables is fine as long as there aren't too many of them. For example, fixed effects for left-or-right handedness, or fixed effects for eye color. There aren't too many variations of those, so including controls for each option isn't too bad. The problem comes when there are lots of groups, as there might be in many "fixed effects for individual" settings. For example, in our Gapminder analysis there are 142 different countries in the data. This entails adding 142 control variables and estimating 142 coefficients. That's fine in a linear model, but in a nonlinear model you run into the "incidental parameters problem," where it just can't handle estimating that much stuff. The estimates start getting noisy and bad.

Second, let's consider the version where we subtract out the within-group means. This no longer works because the outcome variable isn't a continuous variable any more, so the mean is no longer a good representation of "what I need to subtract out to get rid of the between variation." The intuition still works, but subtracting out the mean is no longer the right trick.

So WHAT CAN WE DO? We still might want to use a fixed effects design even if we have a nonlinear dependent variable.

There are two common approaches people take. One is to drop the "nonlinear" part and just estimate a linear model anyway with your nonlinear dependent variable. This is especially common when the dependent variable is binary or ordinal. Many researchers, especially economists, consider the downsides of the misspecified linear model (with heteroskedasticity-robust standard errors) to not be as bad as the downsides associated with trying to estimate an actual nonlinear model with fixed effects.[35]

Another common approach is to use a nonlinear model variant that *is* designed to handle fixed effects properly. Things are likely out of your hands at this point and in the hands of the econometricians. Look for a method (or a software command) designed to do what you need. It might not exist! But if it does,

[35] This is, unfortunately, a matter of taste. There's not a single right answer here.

make sure to read the documentation so you understand what you're doing, and then use that.

There are too many different methods to go into them here, but once you come to this point you are probably down to searching the Internet and reading about your particular application. And keep in mind that there are usually multiple different ways to do this, each of which means something different.

However, the most common applications here are for binary dependent variables and wanting to do probit or logit with fixed effects. In R, look at the generalized linear model options in the **fixest** package. In Stata use the commands `xtprobit` or `xtlogit`. In Python there is the **pylogit** package, which allows you to do *conditional logit* regression, which is similar but not quite the same.

16.3.3   Advanced Random Effects

IN THE RANDOM EFFECTS SECTION above, I mentioned that fixed effects is usually preferred to random effects unless you have a strong reason (like randomization) to think that your right-hand-side variables are unrelated to the individual effects.

However, this argument uses a bit of a straw-man version of random effects. Lots of researchers use random effects even when the individual effect is almost certainly related to the right-hand-side variables. Surely they can't all be unaware of the problem!

And in fact, they are not. Modern approaches to random effects are more than capable of handling this issue.[36] This generally works by seeing your model as containing multiple *levels*. This makes sense for the context—in any fixed effects setting our data is *hierarchical*. At the very least it has multiple observations over time *within* individual. The data has levels. Let's model those levels.

So for example, perhaps we have the model:

$$Y_{it} = \beta_i + \beta_1 X_{it} + \varepsilon_{it} \qquad (16.5)$$

That's one of our levels. We would then say that the β_i term itself follows its *own* equation:

$$\beta_i = \beta_0 + \gamma_1 Z_i + \mu_i \qquad (16.6)$$

where Z_i is a set of individual-level variables that determine the individual effect, and μ_i is an error term.

[36] Andrew Bell and Kelvyn Jones. Explaining fixed effects: Random effects modeling of time-series cross-sectional and panel data. *Political Science Research and Methods*, 3(1):133–153, 2015.

Multilevel model.   A model in which the coefficients themselves are allowed to vary over the sample, potentially with other variables explaining how they vary.

This allows us to deal with the correlation between the β_i and the X_{it} because we're modeling it directly. That individual-level time-consistent part of X_{it} that might be related to β_i—hopefully we account for that part of it with the Z_i.

This approach does a few nice things for us. First of all, it gets us the nice statistical properties that random effects gives us over fixed effects, but without imposing as many additional assumptions.

Second, it lets us look at the effect of variables that are constant within individual. Let's say we wanted to look at the effect of hometown on whether you become an inventor. Well, each person's hometown doesn't change over their lifetime, so fixed effects would never let us study that question. But multi-level random effects would.

Third, it lets us explicitly look at between and within effects separately. Fixed effects says "within effects only, please." Basic random effects says "some mix of between and within." Multi-level random effects, if done properly, says "here are the between effects, and here are the within effects. Do what you will."

THERE ARE TWO DIRECTIONS we can go with this multi-level intuition in mind.

One approach, which helps explicitly separate out the between and within effects, is just to... include both effects in the regression. Simple as that![37]

All you do is:

[37] This descends from, although is not exactly the same as, Mundlak (1978).

1. Calculate within-individual means for each predictor ($\bar{X}_i$).

2. Calculate the within variation for each predictor ($X_{it} - \bar{X}$).

3. Regress the dependent variable on the within-individual means ($\bar{X}_i$), the within variation ($X_{it} - \bar{X}$), and any other individual-level controls you like (Z_i), and use random effects for the intercept.

This is referred to as "correlated random effects." It is handy in that it separates out for you the "between variation effect" (the coefficients on the $\bar{X}_i$s) and the "within variation effect" (the coefficients on the $X_{it} - \bar{X}$s), and also lets you see the effect of individual-level variables Z_i. You wanted that within effect from fixed effects anyway, and now you get some bonus stuff on top.

This approach is easy to implement and easy to interpret. If you can calculate within-individual means and within variation (code in the "How Is It Performed" section), then you can just toss everything into a model with random effects (in R, see `lmer` in the **lme4** package; in Stata see `xtreg` with the `re` option; and in Python see `RandomEffects` in the **linearmodels** package).

THE OTHER WAY WE CAN GO with our multi-level intuition is *whole hog* into mixed models, a.k.a. hierarchical linear models.

Mixed models say we don't need to stop at the *intercept* following a random distribution. Why stop there? How about the slope coefficients on the other variables?

$$Y_{it} = \beta_{0i} + \beta_{1i} X_{it} + \varepsilon_{it} \qquad (16.7)$$

$$\beta_{0i} = \beta_0 + \gamma_1 Z_i + \mu_i \qquad (16.8)$$

$$\beta_{1i} = \beta_1 + \delta_1 Z_i + \eta_i \qquad (16.9)$$

These models can get complex, but they can be used to represent very complex interactions between variables. After all, this is social science. Everything is complex! We might well expect that there are a bunch of variables that don't just explain Y_{it}, but explain *how* X_{it} *affects* Y_{it} and we want to model that.

Mixed models are extremely common in many fields, and if you're from one of those fields you might be surprised to find them tucked away in a little subsection in one chapter of this book. Alas, as the focus of this book is design rather than estimation, it makes sense to talk about fixed effects first and include mixed models as an extension of that.[38]

Still, they are worth exploring outside the confines of this book. I recommend the textbook by Gelman and Hill (2006) for further reading. You can also try mixed models out yourself with `lmer` in the **lme4** package in R; `xtmixed` in Stata; or `mixedlm` in the **statsmodels** package in Python. More broadly, there is Stan, which is sort of its own language dedicated entirely to fast and flexible estimation of hierarchical models.[39] Stan can be used from other languages, and the intuitively-named **RStan**, **StataStan**, and **PyStan** packages are all available.

[38] I can practically *feel the presence* of Andrew Gelman reaching through time and space to disagree with this characterization. Maybe the only reason putting mixed models here makes sense to me is I'm an ignorant economist. Sorry, Andrew.

[39] https://mc-stan.org/

17

Event Studies

17.1 How Does It Work?

17.1.1 An Event as Old as Time

THE EVENT STUDY IS PROBABLY THE OLDEST AND SIMPLEST CAUSAL INFERENCE RESEARCH DESIGN. It predates experiments. It predates statistics. It probably predates human language. It might predate humans. The idea is this: at a certain time, an event occurred, leading a treatment to be put into place at that time. Whatever changed from before the event to after is the effect of that treatment. Like in Chapter 16 on fixed effects, we are using only the within variation. Unlike Chapter 16, we are looking only at treatments that switch from "off" to "on," and (usually) instead of using panel data we use time series data where we track only one "individual" across multiple periods.

So when someone has a late-night beer, immediately falls asleep, and concludes the next day that beer makes them sleepy, that's an event study. When you put up a "no solicitors" sign on your door, notice fewer solicitors afterwards, and conclude the sign worked, that's an event study. When your dog is hungry,

DOI: 10.1201/9781003226055-17

finds you and whines, and becomes fed and full, then concludes that whining leads to getting fed, that's an event study. When a rooster concludes they're responsible for the sun rising because it rises every morning right after they crow, that's an event study.

EVENT STUDIES ARE DESIGNED TO WORK in cases where the relevant causal diagram looks like Figure 17.1. At some given time period, we move from "before-event" to "after-event" and a treatment goes into effect.[1] "After" here doesn't mean "the treatment came and went" but rather "in the time after the event occurred," so the treatment could still be in effect. By comparing the outcome when the treatment is in place to the outcome when it's not, we hope to estimate the effect of $Treatment \rightarrow Outcome$.

[1] Some event studies distinguish three time periods— before, after, and *during*, or "event day/period." This is done when you expect the effect of treatment to be very short-lived.

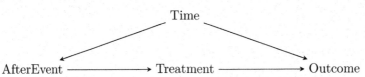

Figure 17.1: A Causal Diagram that Event Studies Work For

There is a back door to deal with, of course:

$$Treatment \leftarrow AfterEvent \leftarrow Time \rightarrow Outcome$$

Time here sort of collects "all the stuff that changes over time." We know the rooster isn't responsible for the crowing because it was going to rise anyway.[2] We have to close the back door somehow for an event study to work. We can't really control for *AfterEvent*—that would in effect control for *Treatment* too. The two variables are basically the same here— you're either before-event and untreated or after-event and treated. They're just separated in the diagram for clarity. However, because *Time* only affects *Treatment* through *AfterEvent*, there is the possibility that we can close the back door by controlling for some elements of *Time*, while still leaving variation in *AfterEvent*.

[2] This is a realization the rooster comes to in the 1991 film *Rock-a-Doodle*, which your professor now has an excuse to show in class instead of teaching for a day.

THINKING CAREFULLY ABOUT FIGURE 17.1 is the difference between a good event study and a bad one. And you can already imagine the bad ones in your head. In fact, many of the obviously-bad-causal-inference examples in this book are in effect poorly done event studies. Many *superstitions* are poorly done event studies. Maybe you have a friend who wore a pair of green sweatpants and then their sports team won, and they concluded that the sweatpants helped the team. Or maybe a new character is introduced to a TV show that has seen declining ratings for years. The show is canceled soon after, and the fans blame the new character. Those are both event studies.

But they're event studies done poorly! Thinking carefully about the causal diagram means two things: (1) thinking hard about whether the diagram actually applies in this instance—is the treatment the *only* thing that changed from before-event to after event, or is there something else going on at the same time that might be responsible for the outcomes we see? Event studies aren't bad, but they only apply in certain situations. Is your research setting one of them? (2) thinking carefully about how the back door through *Time* can be closed.

Fans of the canceled TV show have failed in step (2)—they needed to account for the fact that ratings were already on the decline in order to understand the impact of the new character. The sweatpants-wearing sports fan has probably fallen afoul more of estimation than the causal diagram—figuring out how to handle sampling variation properly can be tricky given that we only have a single repeated measurement.

17.1.2   Prediction and Deviation

THE REAL TRICKY PART ABOUT EVENT STUDIES IS THE BACK DOOR. The whole design of an event study is based around the idea that something changes over time—the event occurs, the treatment goes into effect. Then we can compare before-event to after-event and get the effect of treatment. However, we need that treatment to be the *only* thing that's changing. If the outcome would have changed anyway for some other reason and we don't account for it, the event study will go poorly.

This is the tricky part. After all, things are *always* changing over time. So how can we make sure we've removed all the ways that *Time* affects the outcome *except* for that one extremely important part we want to keep, the moment we step from before-event to after-event?

In short, we have to have some way of *predicting the counterfactual*. We have plenty of information from the before-event times. If we can make an assumption that whatever was going on before would have continued doing its thing if not for the treatment, then we can use that before-event data to predict what we'd expect to see without treatment, and then see how the actual outcome *deviates* from that prediction. The extent of the deviation is the effect of treatment.

For example, see Figure 17.2. There are two graphs here. In each case we have a dark black line representing the actual time series data we see. Things seem pretty well-behaved up to the date when the event occurs, at which point the time series veers off course—in Figure 17.2 (a) it juts up, and in Figure 17.2 (b) it drops down.

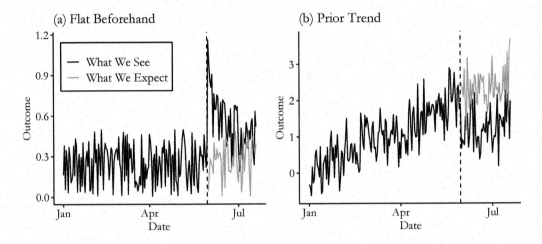

Figure 17.2: Two Examples of Graphs Representing Event Studies

For our event study to work, we need to have a decent idea of what would have happened without the event. Generally, we'll assume that whatever pattern we were seeing before the event will continue. In the case of Figure 17.2 (a), the time series is jumping up and down on a daily basis, but doesn't seem to be trending generally up or down or really changing at all. Seems like a safe bet that this would have continued if no event occurred. So we get the gray line representing what we might predict would have happened if that trend continued. To get the event study estimate, just compare the black line of real data to the gray line of predicted counterfactual data after the event goes into place. Looks like the treatment had an immediate positive effect on the outcome, with the black line far above the gray right after the treatment, but then that effect fades out over time and the lines come back together.

The same idea applies in graph Figure 17.2 (b). Here we have a clear upward trend before the event occurs. So it seems reasonable to assume that the trend would have continued. So this time, the gray line representing what we would have expected continues on that upward trend. The actual data on the black line drops down. We'd estimate a negative effect of treatment on the outcome in this case. Unlike in Figure 17.2 (a), the effect seems to stick around, with the black line consistently below the gray through the end of the data.

SO HOW CAN WE MAKE THOSE PREDICTIONS? While there are plenty of details and estimation methods to consider when it comes to figuring out the gray counterfactual line to put on the graph, on a conceptual level there are three main approaches.

Approach 1: Ignore It. Yep! You can just ignore the whole thing and simply compare after-event to before-event without worrying about changes over time at all. This is actually a surprisingly common approach. But, surely, a bad approach, right?

Whether this is bad or not comes down to context. Sure, some applications of the "ignore it" approach are just wishful thinking. But if you have reason to believe that there really isn't any sort of time effect to worry about, then why not take advantage of that situation?

There are two main ways that this can be justified. The first is if your data looks a lot like Figure 17.2 (a).[3] The time series is extremely flat before the event occurs, with no trend up or down, no change, no nothing. Just a nice consistent time series at roughly the same value, and ideally not too noisy. If this is the case, and also you have no reason to believe that anything besides the treatment changed when the event occurred, then you could reasonably assume that the data would have continued doing what it was already doing.

The second common way to justify ignoring the time dimension is if you're looking only at an *extremely tiny* span of time. This approach is common in finance studies that have access to super high-frequency data, where their time series goes from minute to minute or second to second. Sure, a company's stock may trend upwards or downwards over time, or follow some other weird pattern, but that trend isn't going to mean much if you're zooming way, way, way in and just looking at the stock's price change over the course of sixty seconds.

Approach 2: Predict After-Event Data Using Before-Event Data. With this approach, you take a look at the outcome data you have leading up to the event, and use the patterns in that data to predict what the outcome would be afterwards. You can see this in both Figure 17.2 (a) and (b)—there's an obvious trend in the data. By extrapolating that trend, we end up with some predictions that we can compare to the actual data.

[3] Or if you understand the data generating process *very* well and can say with confidence that there's absolutely no reason the outcome could have changed without the treatment—there is no back door.

Simple! Simple in concept anyway. In execution, there are many different ways to perform such a prediction, and some ways will work better than others. I'll discuss this later in the chapter.

Approach 3: Predict After-Event Data Using After-Event Data. This approach is very similar to Approach 2. Like in Approach 2, you use the before-event data to establish some pattern in the data. Then, you extrapolate to the after-event period to get your predictions.

The only difference here is that instead of only looking for things like trends in the outcome, you also look for *relationships* between the outcome and some other variables in the before-event period. Then, in the after-event period, you can bolster your prediction by also seeing how those other variables change.

A common application of this is in the stock market. Researchers doing an event study on stock X will look at how stock X's price relates to the rest of the market, using data from the before-event period. Maybe they find that a 1% rise in the market index is related to a .5% rise in stock X's price.

Then, in the after period, they extrapolate on any trend they found in stock X's price, like in Approach 2. But they also check the market index. If the market index happened to rise by 5% in the after-event period, then we'd expect stock X's price to rise by 2.5% for that reason, regardless of treatment. So we'd subtract that 2.5% rise out before looking for the effect of treatment.

SOMETHING THAT ALL THREE of these methods should make clear is that event studies are generally intended for a fairly short post-event window, or a "short horizon" in the event study lingo. The farther out you get from the event, the worse a job your pre-event-based predictions are going to do, and the more likely it is that the *Time* back door will creep in, no matter your controls. So keep it short, unless you're in a situation where you're sure that in your case there really *is* nothing else going on besides your treatment and business-as-usual in the long term, or you have a good model for handling those changes in the long term. Going for a long horizon is possible, but the time series design does a lot less for you, and you'll need to bolster the validity of your design, be more careful about the possibility of unrelated trends, and swat away the back doors and general noisiness that will start creeping back in.[4]

THAT'S ABOUT IT. The details are a bit more thorny than that, but event studies are ancient for a reason. They really are

[4] Sagar P Kothari and Jerold B Warner. Econometrics of event studies. In *Handbook of Empirical Corporate Finance*, pages 3–36. Elsevier, 2007.

conceptually simple. The complexities come when trying to get the estimation right, which I'll cover in the How Is It Performed section.

17.1.3   A Note on Terminology

IN THIS CHAPTER, I'll be covering what I call "event studies." However, as you might expect for such an intuitively obvious method, event studies were developed across many fields independently and so terminology differs, sometimes with some slight variation in meaning.

For example, in health fields, you might hear about "statistical process control," which is only intended to be used if there are no apparent pre-event trends (Approach 1 in the previous section). I'm just gonna call these, and the wide family of other similar methods, a form of event study.

Another confusing term is "interrupted time series." This *could* just refer to any event-study method where you account for a before-event trend. But in practice it usually *either* refers to an event study that uses time series econometrics, *or* to a regression-based approach that fits one regression line before-event and another after-event. And, naturally, users of either method will become scared and confused if the term is used to refer to the other one.[5]

There are also variants of event studies that control for *Time* by introducing a *control group* that is unaffected by the event. I won't cover this concept in the event study chapter,[6] but you'll see the idea of combining before-event vs. after-event comparisons with a control group later in the chapters on difference-in-differences (Chapter 18) and synthetic control (Chapter 21). Those cover the concept, anyway. In practice, studies calling themselves "event study with a control" or "interrupted time series with a control" will often include a bit more of the time series methodology I cover in the How the Pros Do It section than difference-in-differences or synthetic control approaches will.

Even more confusing, some economists use the term "event study" **only** to refer to studies that use multiple treated groups, all of which are treated at different times, whether or not there's also a control group. Oy. We'll similarly cover that in Chapter 18.

For now, just before-and-after event studies.

[5] If this happens, encourage deep breaths and tell them to imagine their hypothesis is correct.

[6] Aside from applications of Approach 3, which kinda does this.

17.2   How Is It Performed?

IT'S A BIT STRANGE to talk about "how event studies are performed." Because it's such a general idea of a method, and because so many different fields have their own versions of it, ways of performing event studies are highly varied.

So, instead of trying to take you on a tour of every single way that event study designs are applied, I'll focus on three popular ones. First, I'll take a look at event studies in finance, where a common approach is to explicitly calculate a prediction for each after-event period, and compare that to the data. Second, I'll look at a regression-based approach.[7] Third, in the "How the Pros Do It" section, I'll get into methods that take the time series nature of the data a bit more seriously.

[7] Economists would call this an "interrupted time series," although see my note about terminology above.

17.2.1   Event Studies in the Stock Market

EVENT STUDIES ARE HIGHLY POPULAR in studies of the stock market. Specifically, they're popular when the outcome of interest is a stock's return. Why? Because in any sort of efficient, highly-traded stock market, a stock's price should already reflect all the public knowledge about a company. Returns might vary a bit from day to day, or might rise if the market goes up overall, but the only reason the price should have any sort of sharp change is if there's surprising new information revealed about the company. Y'know, some sort of event!

Even better, the fact that stock prices reflect *information* about a company means that we can see the effect of that information immediately and so can look at a pretty short time window. This is good both for supporting the "nothing else changes at the same time" assumption, and for pinpointing exactly *when* an event is.[8] Say a company announces that they're planning a merger. That merger won't actually happen for quite a long time, but because investors buy and sell based on what they *anticipate* will happen, we'll be able to see the effect on the stock price right away, rather than having to wait months or years for the actual merger, allowing the time back door to creep back in.

The process for doing one of these event studies is as follows:

[8] Pinpointing the time of an event can be hard sometimes. Say you wanted to do an event study to look at the effect of someone being elected to office. When does that event occur? When polls indicate they'll win? When they actually win? When they take office? When they pass their first bill?

1. Pick an "estimation period" a fair bit before the event, and an "observation period" from just before the event to the period of interest after the event.

2. Use the data from the estimation period to estimate a model that can make a prediction $\hat{R}$ of the stock's return R in each period. Three popular ways of doing this are:

   (a) **Means-adjusted returns model.** Just take the average of the stock's return in the estimation period, $\hat{R} = \bar{R}$.

   (b) **Market-adjusted returns model.** Use the market return in each period, $\hat{R} = R_m$.

   (c) **Risk-adjusted returns model.** Use the data from the estimation period to fit a regression describing how related the return R is to other market portfolios, like the market return: $R = \alpha + \beta R_m$.[9] Then, in the observation period, use that model and the actual R_m to predict $\hat{R}$ in each period.

3. In the observation period, subtract the prediction from the actual return to get the "abnormal return." $AR = R - \hat{R}$.

4. Look at AR in the observation period. Nonzero AR values before the event imply the event was anticipated, and nonzero AR values after the event give the effect of the event on the stock's return. For the same efficient-market reason that stock returns are a good candidate for event studies in the first place, effects will generally spike and then fade out quickly.

Let's put this into action with an example. One of the nice things about event studies in finance is that you can apply them just about *anywhere*. So instead of borrowing an event study from a published paper, I'll make my own.

On August 10, 2015, Google announced that they would be rearranging their corporate structure. No longer would "Google" own a bunch of other companies like FitBit and Nest, but instead there would be a new parent company called "Alphabet" that would own Google along with all that other stuff.

How did the stock market feel about that? To find out, I downloaded data on the GOOG stock price (the stock symbol for Google before and Alphabet now) from May 2015 through the end of August 2015. I also downloaded the price of the S&P 500 index to use as a market-index measure.

Then, I calculated the daily return for GOOG and for the S&P 500 (which is the price divided by the previous day's price, all minus 1). Why use returns instead of the price? Because we're interested in seeing how the new information *changes* the

[9] Two notes: (1) I use α and β here instead of β_0 and β_1 because that's how they do it in finance. (2) You could easily add other portfolios here, perhaps setting up a Fama-French model or something.

price, so we may as well isolate those changes so we can look at
them. Looking at returns also makes it easier to compare stocks
that have wildly different prices.

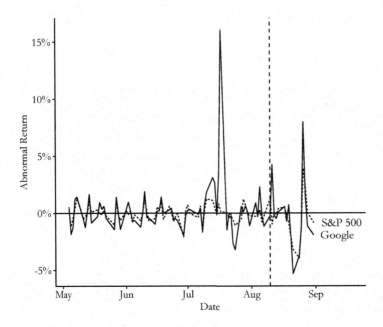

Figure 17.3: Stock Returns
for Google and the S&P 500

You can see those daily returns in Figure 17.3. Something
worth noticing in the graph is that Google's returns and the
S&P 500 returns tend to go up and down at the same time.
This is one reason why two of our three prediction methods
make reference to a market index—we don't want to confuse a
market move with a stock move and think it's just all about the
stock! Also, the returns are very flat a lot of the time. There
are some spikes—a big one in mid-July, for example. But in
general returns can be expected to stay at fairly constant levels
a lot of the time, until something happens.

Let's do the event study. We'll first pick an estimation period
and an observation period. For estimation I'll go May through
July,[10] and then for observation I'll start a few days before the
announcement, from August 6 through August 24.

Next, we'll use the data in the observation period to construct
our prediction models, and compare. The code for doing so is:

[10] You could reasonably argue that I should be leaving out July and just do-
ing May and June because
of that huge spike for Google
in July. This may well be
right. Think carefully about
whether you agree with this
critique.

```
1   # R CODE
2   library(tidyverse); library(lubridate)
3   goog <- causaldata::google_stock
4
5   event <- ymd("2015-08-10")
6
```

```
 7   # Create estimation data set
 8   est_data <- goog %>%
 9    filter(Date >= ymd('2015-05-01') &
10    Date <= ymd('2015-07-31'))
11
12   # And observation data
13   obs_data <- goog %>%
14    filter(Date >= event - days(4) &
15    Date <= event + days(14))
16
17   # Estimate a model predicting stock price with market return
18   m <- lm(Google_Return ~ SP500_Return, data = est_data)
19
20   # Get AR
21   obs_data <- obs_data %>%
22    # Using mean of estimation return
23    mutate(AR_mean = Google_Return - mean(est_data$Google_Return),
24    # Then comparing to market return
25    AR_market = Google_Return - SP500_Return,
26    # Then using model fit with estimation data
27    risk_predict = predict(m, newdata = obs_data),
28    AR_risk = Google_Return - risk_predict)
29
30   # Graph the results
31   ggplot(obs_data, aes(x = Date, y = AR_risk)) +
32    geom_line() +
33    geom_vline(aes(xintercept = event), linetype = 'dashed') +
34    geom_hline(aes(yintercept = 0))
```

```
 1   * STATA CODE
 2   causaldata google_stock.dta, use clear download
 3
 4   * Get average return in estimation window and calculate AR
 5   summ google_return if date <= date("2015-07-31","YMD")
 6   g AR_mean = google_return - r(mean)
 7
 8   * Get difference between Google and the market
 9   g AR_market = google_return - sp500_return
10
11   * Estimate a model predicting stock price with market return
12   reg google_return sp500_return if date <= date("2015-07-31","YMD")
13   * Get the market prediction for AR_risk
14   predict risk_predict
15   g AR_risk = google_return - risk_predict
16
17   * Graph the results in the observation window
18   g obs = date >= date("2015-08-06", "YMD") & date <= date("2015-08-24", "YMD")
19
20   * Format the date nicely for graphing
21   format date %td
22
23   * Use the tsline function for time series plots
24   tsset date
25   tsline AR_risk if obs, tline(10aug2015)
```

```
 1   # PYTHON CODE
 2   import pandas as pd
 3   import numpy as np
 4   import statsmodels.formula.api as smf
 5   import seaborn as sns
```

```
 6 │ import matplotlib.pyplot as plt
 7 │ from causaldata import google_stock
 8 │ goog = google_stock.load_pandas().data
 9 │
10 │ # Create estimation data set
11 │ goog['Date'] = pd.to_datetime(goog['Date'])
12 │ est_data = goog.loc[(goog['Date'] >= pd.Timestamp(2015,5,1)) &
13 │ (goog['Date'] < pd.Timestamp(2015,7,31))]
14 │
15 │ # And observation data
16 │ obs_data = goog.loc[(goog['Date'] >= pd.Timestamp(2015,8,6)) &
17 │ (goog['Date'] < pd.Timestamp(2015,8,24))]
18 │
19 │ # Estimate a model predicting stock price with market return
20 │ m = smf.ols('Google_Return ~ SP500_Return', data = est_data).fit()
21 │
22 │ # Get AR
23 │ # Using mean of estimation return
24 │ goog_return = np.mean(est_data['Google_Return'])
25 │ obs_data['AR_mean'] = obs_data['Google_Return'] - goog_return
26 │ # Then comparing to market return
27 │ obs_data['AR_market'] = obs_data['Google_Return'] - obs_data['SP500_Return']
28 │ # Then using model fit with estimation data
29 │ obs_data['risk_pred'] = m.predict(obs_data)
30 │ obs_data['AR_risk'] = obs_data['Google_Return'] - obs_data['risk_pred']
31 │
32 │ # Graph the results
33 │ sns.lineplot(x = obs_data['Date'],y = obs_data['AR_risk'])
34 │ plt.axvline(pd.Timestamp(2015,8,10))
35 │ plt.axhline(0)
```

If we plot all three lines together, we get Figure 17.4. What do we have? We certainly have a pretty big spike just after August 10, no matter which of the three methods we use. However, it's only a one-day spike. As we might expect in an efficient market, Google reaches its new Alphabet-inclusive price quickly, and then the daily returns dip back down to zero (even though the price itself is higher than before). We also don't see a whole lot of action *before* the event, telling us that this move was not highly anticipated.

As we look out to the right in Figure 17.4, we can spot a few quirks of this methodology that remind us to keep on our toes. First, notice how around August 20, the Mean line takes a dive, while the Market and Risk lines stay near zero? That's a spot where the Google price dipped, but so did the market. Probably not the result of the Alphabet switch, since it happened to the whole market. This is one good reason to use a method that incorporates market movements.

The graph, and the movement around August 20, also nudge us towards using a narrow window and not trying to get an event study effect far after the event. Given that the Alphabet change had a one-day effect that then immediately went away,

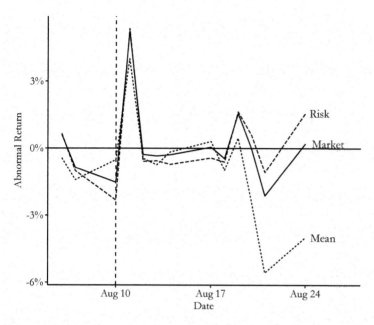

Figure 17.4: Event Study Estimation of the Impact of Google's Alphabet Announcement on its Stock Return

the changes a week and a half later probably aren't *because* of the Alphabet change. And yet they're showing up as effects! In the stock market, a week and a half is plenty of time for something *else* to change, and we're confusing whatever that is for being an effect of the Alphabet change because our horizon is too long.

NOW THAT WE HAVE THE DAILY EFFECTS, there are a few things we can do with them.

The first thing we might do is first ask whether a given day shows an effect by performing a significance test on the abnormal return that day. The most commonly-applied test is pleasingly simple. Just calculate AR for the *whole* time window, not just the observation window. Then get the standard deviation of AR. That's your standard error. Done! Divide the AR on your day of interest by the standard error to get a t-statistic, and then check if the AR on your day of interest is greater (or less than) than some critical value, usually 1.96 or -1.96 for a 95% significance test.

If I calculate the standard deviation of GOOG's AR_{risk} for the whole time window, I get .021. The AR_{risk} the day after the announcement is .054. The t-statistic is then $.054/.021 = 2.571$. This is well above the critical value 1.96, and so we'd conclude that the Alphabet announcement did increase the returns for GOOG, at least for that one day.

We might also be interested in the effect of the event on the *average* daily return, or the *cumulative* daily return. Calculating these is straightforward. For the average, just average all the *AR*s over the entire (post-event) observation period. For the cumulative, add them up.

Getting a standard error or performing a test for these aggregated measures, though, is another issue. There are lots and lots of different ways you can devise a significance test for them. These approaches generally rely on calculating the average or cumulative *AR* across lots of different stocks and using the standard deviation across the different stocks.

The simplest is the "cross-sectional test," which constructs a *t*-statistic by taking the average or cumulative *AR* for each firm, dividing by the standard deviation of those average or cumulative *AR*s, respectively, and multiplying by the square root of the number of firms. Easy!

Of course, this simple approach leaves out all sorts of stuff—time series correlation, increased volatility, correlation between the firms, and so on. There are, as I mentioned, a zillion alternatives. These include a number of tests descended from Patell (1976).[11] Patell's test accounts for variability in whether the returns come early or late in the observation window. The Adjusted Patell Test accounts also for the correlation of returns between firms. There's the Standardized Cross-Sectional Test, which also accounts for time series serial correlation and the possibility that the event increased volatility. And the Adjusted Standardized Cross-Sectional Test adds in the ability to account for correlation between firms. And that's before we even get into the non-parametric tests based on rankings of the abnormal returns. Oy, so many options.

These additional tests can be implemented in R in the **estudy2** package, and in Stata in the **eventstudy2** package. In Python you may be on your own.

[11] James M Patell. Corporate forecasts of earnings per share and stock price behavior: Empirical test. *Journal of Accounting Research*, 14 (2):246–276, 1976.

17.2.2   Event Studies with Regression

THE METHOD USED IN FINANCE is purpose-built for effects that spike and then, usually, quickly disappear. Not uncommon in finance! But what about other areas? What if you're interested in an event that changes the time series in a long-lasting way? In that case, one good tool is an event study design implemented with regression.

The basic idea is this: estimate one regression of the outcome on the time period before the event. Then estimate another time series of the outcome on the time period *after* the event. Then see how the two are different.

This approach can be implemented with the simple use of an interaction term (or "segmented regression" in the event study lingo):

$$Outcome = \beta_0 + \beta_1 t + \beta_2 After + \beta_3 t \times After + \varepsilon \quad (17.1)$$

where t is the time period and $After$ is a binary variable equal to 1 in any time after the event. This is just a basic linear example, of course, and forces the time trend to be linear. You could easily add some polynomial terms to allow the time series to take other shapes, as discussed in Chapter 13.

There are some obvious upsides and downsides to this method. As an upside, you can get a more-precise estimate of the time trend than going day by day (or second by second, or whatever level your data is at). As a downside, you're going to be limited in seeing the exact *shape* that the effect takes—seeing something like the one-day Google stock price bump in the previous section would be hard to do. And if you have the shape of the time trend wrong overall, your results will be bad. But if you can add the right polynomial terms to get a decent representation of the time series on either side of the event, you're good to go on that front.

Another *big* downside is that you need to be very careful with any sort of statistical significance testing. If your data is autocorrelated—each period's value depends on the previous period's value—then you're *extremely* likely to find a statistically significant effect of the event, even if it truly had no effect.[12] This occurs because the data tends to be "sticky" over time, with similar values clustering in neighboring time periods, which can make the regression a little too confident in where the trend is going. In data with a meaningful amount of autocorrelation, you can pick a fake "event" day at random, run a segmented regression, and get a statistically significant effect about half the time—much more than the 5% of the time you *should* get when doing a 95% significance test on a true-zero effect.

There are ways to handle this autocorrelation problem, though. Using heteroskedasticity- and autocorrelation-robust standard errors, as discussed in Chapter 13, definitely helps. Even better would be to use a model that directly accounts for the autocorrelation, as discussed in the How the Pros Do It section of this chapter.

[12] Autocorrelation is discussed further both in Chapter 13 and in the How the Pros Do It section in this chapter.

THERE'S NO NEW CODE for the segmented regression method; just run a regression of the outcome on some-order polynomial of the time period, with *After* as an interaction term. Then, since we're dealing with time series data, use heteroskedasticity- and autocorrelation-consistent (HAC) standard errors. Both interaction terms and HAC standard errors are discussed in Chapter 13.

Instead of new code, let's see how this might work in real life with an example. Taljaard, McKenzie, Ramsay, and Grimshaw (2014) use a regression-based approach to event studies to evaluate the effect of a policy intervention on health outcomes.[13] Specifically, they looked at an English policy put in place in mid-2010 to improve quality of health care received in the ambulance on the way to the hospital on the chances of heart attack and stroke afterwards. A bunch of quality-improvement teams were formed, they all collaborated, they shared ideas and they informed staff. So did it help with heart attack care?

That's what Taljaard et al. (2014) look at. They run a regression of heart attack performance (*AMI*, or Acute Myocardial Infarction performance) on $Week - 27$ (subtracting 27 "centers" $Week$ at the event period, which allows the coefficient on $Week - 27$ to represent the jump in the line), *After* (an indicator variable for being after the 27-week mark of the data where the policy was introduced), and an interaction term between the two:[14]

$$AMI = \beta_0 + \beta_1(Week - 27) + \beta_2 After +$$
$$\beta_3(Week - 27) \times After + \varepsilon \quad (17.2)$$

Their results for heart attack can be summarized by Figure 17.5. You can see the two lines that are fit to the points on the left and right sides of the event's starting period. That's the interaction term at work. The line to the left of 27 weeks is $\beta_0 + \beta_1(Week - 27)$, and the line to the right is $(\beta_0 + \beta_2) + (\beta_1 + \beta_3)(Week - 27)$.

So what do we see here? We certainly don't see any sort of jump at the cutoff. So no immediate effect of the policy on outcomes. That's not super surprising, given that you might expect that sort of collaborative behavior-changing policy to take time to work. The jump is represented by the change in prediction at the event time, $Week = 27$. Comparing the lines on the left and the right at that time, we get a prediction of β_0 on the left and $\beta_0 + \beta_1$ on the right (if $Week = 27$, then $Week - 27 = 0$ and the β_2 and β_3 terms in both lines fall out).

[13] Monica Taljaard, Joanne E McKenzie, Craig R Ramsay, and Jeremy M Grimshaw. The use of segmented regression in analysing interrupted time series studies: An example in pre-hospital ambulance care. *Implementation Science*, 9(1):1–4, 2014.

[14] They actually use logit regression instead of ordinary least squares. But they report the results in odds-ratio terms, so the lines are still nice and straight, if harder to interpret for my tiny little brain.

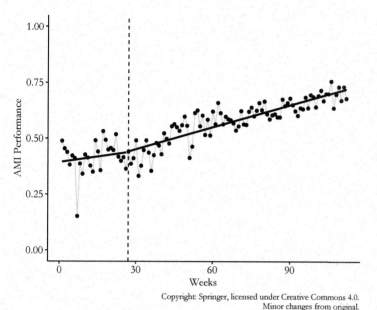

Figure 17.5: Effect of an Ambulance Quality Control Cooperative Policy on Heart Attack Care in Taljaard et al. (2014)

Copyright: Springer, licensed under Creative Commons 4.0.
Minor changes from original.

The difference is $(\beta_1 + \beta_0) - \beta_0 = \beta_1$. β_1 is the jump in the prediction at the moment the treatment goes into effect. In other words, that's the event study estimate.

As you might expect given the lack of a visible jump on the graph, the logit regression in Table 17.1 that goes along with this result shows that the coefficient on *After* (β_1) is not significantly different from 1. As with most medical studies, coefficients are in their odds-ratio form, meaning that the effect is multiplicative instead of additive. A null effect is 1, not 0, since multiplying by 1 makes no difference.

Variable	Coefficient (Odds Ratio)	95% C.I.	p-value
Week − 27	1.02	.97 to 1.07	.362
After	1.16	.93 to 1.44	.199
(*Week* − 27)× *After*	1.02	.97 to 1.08	.346

Odds ratios are multiplicative, so "no effect." is 1 not 0.

Copyright: Springer

Table 17.1: Event-Study Logit Regression of Heart Attack Performance on Ambulance Policy from Taljaard et al. (2014)

It does look in Figure 17.5 like we have a change in slope, though. This is the kind of thing that a regression model can pick up well on that the finance approach would have trouble with. Maybe the policy didn't change anything right away, but made things better slowly over time? It looks that way, except that when we look at the regression results, the change in slope is insignificant as well. The slope goes from β_2 to $\beta_2 + \beta_3$, for a change of $(\beta_2 + \beta_3) - \beta_2 = \beta_3$. So the slope changes by β_3, which is the coefficient on the $(Week - 27) \times After$ interaction, which is insignificant in Table 17.1. Ah well, so much for that policy!

17.3 How the Pros Do It

17.3.1 Event Studies with Multiple Affected Groups

ALL THE METHODS SO FAR in this chapter have assumed that we're talking about an event that affects only *one* group, or at least we're trying to get an effect only for that group. We're interested in the effect of the Alphabet announcement on Google's stock, not anyone else's, and we wouldn't really expect there to be much of an effect anyway.

However, there are plenty of events that apply to *lots* of groups. The Alphabet announcement really only affects Google's stock. But how about the announcement of something like the Markets in Financial Instruments Directive—a regulatory measure for stock markets in the European Union that should affect *all* the stocks traded there? How can we run an event study in a case like that?

There are a few ways we can go.

WE CAN PRETEND THAT WE ACTUALLY DON'T HAVE A BUNCH OF GROUPS. We have a method designed for a single time series, but we have lots of different time series that matter. Just squish 'em all together! If you have 10 different groups that might be affected by the event, average together their values in each time period to produce a single time series. Then do your event study estimation on that one time series.

Not really much to explain further. You're just averaging together a bunch of time series, and then doing what you already would do for a single group. There are some obvious downsides to this simple approach. You're losing a lot of information when you do this. Your estimate is going to be less precise than if you take advantage of all the variation you have, rather than throwing it out.

WE CAN TREAT EACH GROUP SEPARATELY. We have event study methods for single groups. And we have lots of groups. So... just run lots of event studies, one for each group. This will give us a different effect for each group.[15]

At this point you have an effect for each group—and this could be an immediate effect from a single post-event time

[15] Getting all the individual effects is a nice benefit of this approach. As Chapter 10 on treatment effects points out, the effect is likely to be truly different between groups, and having an effect for each group lets us see that.

period or the coefficient on *After* in a regression, or an average effect over the entire post-event period, or a cumulative effect. From there you can look at the distribution of effects across all the groups. You can also aggregate them together to get an overall immediate/average/cumulative effect.

WE CAN AGGREGATE WITH REGRESSION. The regression approach to event study estimation described in the How Is It Performed section will happily allow you to have more than one observation per time period. So stack those event studies up! In fact, this is actually what the Taljaard et al. ambulance study was doing—it had several different hospitals' worth of data to work with.

The regression model is nearly the same as when there's just one time series:

$$Outcome = \beta_i + \beta_1 t + \beta_2 After + \beta_3 t \times After + \varepsilon \qquad (17.3)$$

Just as before, there's a line being fit before the event ($\beta_i + \beta_1 t$) and a line after the event ($(\beta_i + \beta_2) + (\beta_1 + \beta_3)t$) The only difference is the switch from β_0 to a group fixed effect of β_i (as in Chapter 16). This isn't always necessary, but if the different groups you're working with have different baseline levels of *Outcome*, the fixed effect will help deal with that. Also, as before, this equation allows you to fit a *straight* line on either side of the event, but you could easily add polynomial terms to make it curvy.[16]

This regression will produce very similar results to what you'd get if you first averaged all the data together and then estimated a regression on that single time series. But because it has all the individual data points to work with, it will know how much variation there is in each time period and will do a better job estimating standard errors.

Of course, this regression method puts us in a box of having to fit a particular shape. What if we are looking at an event that should only matter for a day? Or we want to know the full shape? In that case we can use a different kind of regression:[17]

$$Outcome = \beta_0 + \beta_t + \varepsilon \qquad (17.4)$$

What's this stubby little equation? This is a regression of the outcome on *just* a set of time-period fixed effects. And we do the fixed effects in a particular way—we specifically select the *last time period before the event should have an effect* as the reference category (remember: any time we have a set of binary indicators we have to leave one out as the reference).

[16] Or even use some of the nonparametric methods discussed in Chapter 4—go nuts!

[17] Which should be estimated with standard errors that account for the time series nature of the data, like clustering at the group level (allowing for correlation across time within the same group).

The fixed effect for a given period is then just an estimate of the mean outcome in that period *relative* to the period just before the event. If we plot out the time-period fixed effects themselves, it will be a sort of single time series, just like if we'd mashed everything together ourselves by just taking the mean in each period. The only difference is that we now have standard errors for each of the periods, and we've made everything relative to the period just before the event.

Because we've made the period just before the event the reference group, we can more easily spot the event study effects. If the first post-event period is, say, period 4, and $\hat{\beta}_4 = .3$, then we'd say that the event made the outcome increase from period 3 to period 4 by .3, and the one-day effect is .3. And if $\hat{\beta}_5 = .2$, then we'd say that the event made the outcome increase from period 3 to period 5 by .2. You can also get average or cumulative effects by averaging, or adding up, the coefficients.

The process for doing this in code, using some fake data, is:

```
1   # R CODE
2   library(tidyverse); library(fixest)
3   set.seed(10)
4
5   # Create data with 10 groups and 10 time periods
6   df <- crossing(id = 1:10, t = 1:10) %>%
7     # Add an event in period 6 with a one-period positive effect
8     mutate(Y = rnorm(n()) + 1*(t == 6))
9
10  # Use i() in feols to include time dummies,
11  # specifying that we want to drop t = 5 as the reference
12  m <- feols(Y ~ i(t, ref = 5), data = df,
13    cluster = 'id')
14
15  # Plot the results, except for the intercep,# and add a line joining
16  # them and a space and line for the reference group
17  coefplot(m, drop = '(Intercept)',
18    pt.join = TRUE, ref = c('t:5' = 6), ref.line = TRUE)
```

```
1   * STATA CODE
2   * Ten groups with ten time periods each
3   clear
4   set seed 10
5   set obs 100
6   g group = floor((_n-1)/10)
7   g t = mod((_n-1),10)+1
8
9   * Add an event in period 6 with a one-period effect
10  g Y = rnormal() + 1*(t == 6)
11
12  * ib#. lets us specify which # should be the reference
13  * We want 5 to be the reference, right before the event
14  * And exclude the regular constant (nocons)
15  reg Y ib5.t, vce(cluster group) nocons
16
17  * Plot the coefficients
```

```
18 || coefplot, ///
19 ||     drop(_cons) /// except the intercept
20 ||     base /// include the 5 reference
21 ||     vertical /// rotate, then plot as connected line:
22 ||     recast(line)
```

```
 1 || # PYTHON CODE
 2 || import pandas as pd
 3 || import numpy as np
 4 || import statsmodels.formula.api as smf
 5 || import matplotlib.pyplot as plt
 6 || np.random.seed(10)
 7 ||
 8 || # Ten groups with ten periods each
 9 || id = pd.DataFrame({'id': range(0,10), 'key': 1})
10 || t = pd.DataFrame({'t': range(1,11), 'key': 1})
11 || d = id.merge(t, on = 'key')
12 || # Add an event in period 6 with a one-period effect
13 || d['Y'] = np.random.normal(0,1,100) + 1*(d['t'] == 6)
14 ||
15 || # Estimate our model using time 5 as reference
16 || m = smf.ols('Y~C(t, Treatment(reference = 5))', data = d)
17 ||
18 || # Fit with SEs clustered at the group level
19 || m = m.fit(cov_type = 'cluster',cov_kwds={'groups': d['id']})
20 ||
21 || # Get coefficients and CIs
22 || # The original table will have an intercept up top
23 || # But we'll overwrite it with our 5 reference
24 || p = pd.DataFrame({'t': [5,1,2,3,4,6,7,8,9,10],
25 ||        'b': m.params, 'se': m.bse})
26 || # And add our period-5 zero
27 || p.iloc[0] = [5, 0, 0]
28 ||
29 || # Sort for plotting
30 || p = p.sort_values('t')
31 || # and make CIs by scaling the standard error
32 || p['ci'] = 1.96*p['se']
33 ||
34 || # Plot the estimates as connected lines with error bars
35 || plt.errorbar(x = 't', y = 'b',
36 ||  yerr = 'ci', data = p)
37 || # Add a horizontal line at 0
38 || plt.axhline(0, linestyle = 'dashed')
39 || # And a vertical line at the treatment time
40 || plt.axvline(5)
```

Random number generation makes the results for each of these a bit different, but the R results look like Figure 17.6.[18] You can see that we've fixed the effect at the last pre-event period, 5, to be 0 so that everything is relative to it. And we do see the expected jump at period 6 where the effect is. The confidence interval doesn't come anywhere near 0—the effect is statistically significant on that day. We do also see significant effects in periods 2 and 4, though, where there isn't actually any effect. Oops! That's small samples for you.

[18] The results actually vary quite a bit if you run them multiple times without setting the seed—event studies are pretty noisy unless you have *lots* of groups or *lots* of time periods, or super well-behaved data.

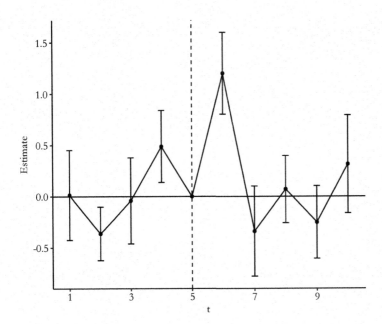

Figure 17.6:   Event Study
with Multiple Groups

When doing an event study this way it's important to keep in
mind exactly what it's doing. This approach isn't incorporating
any sort of trend—it's treating each time period separately. The
predicted counterfactual in each period is purely determined by
period 5—everything is relative to that. This approach really
only works if there's no time trend to worry about.[19] And even
then we should carefully check the pre-event periods.

This approach assumes that all the pre-event effects should
be zero. If they're not, that implies some sort of problem. This
is actually a neat bonus effect—we want to be able to spot if
there's a problem in our design, and this approach lets us do
that. It could just be that it's sampling variation (as it is here,
given what we know about how the data was generated), but it
could just as well be that there are some time-based back doors
we aren't accounting for properly. Hopefully, if you find nonzero
effects in the pre-event period, they follow some sort of obvious
trend, and then you can just build that trend into your model.
But if not, it might be an indication that your research design
just doesn't work!

WHICHEVER APPROACH TO MULTIPLE-GROUP EVENT STUD-
IES WE TAKE, we are a little limited in how we construct our
counterfactual predictions. Any approach that uses information from other groups to get a counterfactual—such as methods that use general market trends to predict what would have
happened to an affected group if there were no treatment—

[19] Or at least it doesn't handle the trend itself.   You
could estimate a trend, or a
relationship with some other
variable, and subtract out
the trend before doing this
method.

obviously don't work as counterfactuals any more, since *lots* of groups are treated. However, any method that doesn't rely on using other groups to predict counterfactuals is fine—the regression methods still work, as does the "mean" AR method for stock returns that takes the mean of pre-event returns as the post-event prediction.

There is always the possibility of finding still further groups that aren't treated and using them as a control group. But this is a task for other chapters, including Chapter 18 on difference-in-differences and Chapter 21 on synthetic control.

17.3.2   *Forecasting Properly with Time Series Methods*

THE CONCEPT BEHIND AN EVENT STUDY IS that you have some actual data in the after-event period and need, as a comparison, a prediction of what would have happened in the after-event period if the event hadn't occurred. Our prediction is invariably based on the idea that, without the event, whatever was going on before the event would have continued.

In other words, we're using the time series from before the event (and maybe some additional predictors) to predict beyond the event. This is the kind of thing that *time series forecasting* was designed for.

Technically, all the stuff we've done so far in this chapter is a form of time series forecasting, but only technically. Real time series forecasting uses methods that carefully consider the time dimension of the data. In a time series, observations are related across time. If a change happens in one period, the effects of that change can affect not just that period, but the whole pathway afterwards. This messes up both statistical estimation and prediction in an incalculable number of ways.

So if we want to get serious about predicting that counterfactual, and we aren't in a convenient setting like a well-behaved stock return, we're probably going to want to use a forecasting model that takes the time series element seriously.

That said, time series estimation and forecasting isn't just an entire field of statistics, it's a *huge* field in statistics and econometrics... and machine learning, and finance. And also it's an industry. And there's a whole parallel Bayesian version. I'm not even going to attempt to do the whole thing justice. Instead I will give you one peek into a single standard time series model used for forecasting, the ARMA, and then I'll point you towards a resource like Montgomery, Jennings, and Kulahci (2015) if you want to get started on your very own time series estimation journey.[20]

[20] Douglas C Montgomery, Cheryl L Jennings, and Murat Kulahci. *Introduction to Time Series Analysis and Forecasting*. John Wiley & Sons, 2015.

THE ARMA MODEL IS AN ACRONYM FOR two common features we see in time series data. AR stands for *autoregression*, meaning that the value of something today influences its value tomorrow. If I put a dollar in my savings account today, that increases the value of my account by a dollar today but also by a dollar tomorrow (plus interest), and the next day (plus a little more interest), and on into the future forever. That's autoregression. My savings account would be a model with an AR(1) process - the 1 meaning that the the value of Y in the single (1) most recent period is the only one that matters for predicting Y in *this* period. If I want to predict my bank balance *today*, then knowing what it was *yesterday* is very important. And if I already know what yesterday was, then learning about the day before that won't tell me anything new (if day-before-yesterday held additional information that yesterday didn't have, we'd have an AR(2), not an AR(1)). An AR(1) model looks like

$$Y_t = \beta_0 + \beta_1 Y_{t-1} + \varepsilon \qquad (17.5)$$

where the t and $t-1$ indicate that the predictor for Y in a given period is Y in the previous period, i.e., the "lag" of Y.

The MA stands for *moving-average*. Moving-average means that *transitory* effects on something last a little while and then fade out. If I lend my friend ten bucks from my savings account and she pays me back over the next few weeks, that's a ten-dollar hit to my account today, a little less next week as she pays some back, a little less the week after that as she pays more back, and then when she pays it all back, the amount in my account is back to normal as though nothing had ever happened. That's a moving-average process. A MA(1) process—where only the most recent transitory effect matters, would look like

$$Y_t = \beta_0 + (\varepsilon_t + \theta \varepsilon_{t-1}) \qquad (17.6)$$

Of course, since we can't actually observe ε_t or ε_{t-1}, we can't estimate θ by regular regression. There are alternate methods for that.

Given how I've just described it, since the amount in my savings account clearly has both AR and MA features, we'd say that the time series of my savings balance follows an ARMA process, and we'd want to use an ARMA model to forecast my savings in the future. An ARMA(2,1) model, for example, would include two autoregressive terms and one moving-average term, such that the current value of the time series depends on the previous two values, and the transitory effect in one previous period:

$$Y_t = \beta_0 + \beta_1 Y_{t-1} + \beta_2 Y_{t-2} + (\varepsilon_t + \theta \varepsilon_{t-1}) \qquad (17.7)$$

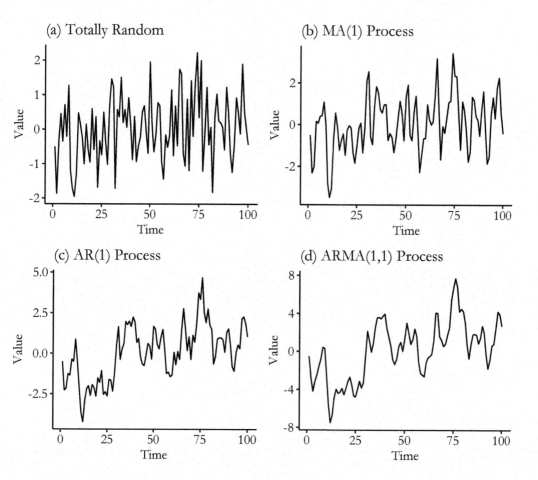

Figure 17.7: White Noise, and Different Time Series Processes

You can see how these things work out in Figure 17.7. In Figure 17.7 (a) we have some completely random white noise. Then in the other graphs I start taking that *exact same white noise* and make it those ε terms in the above processes. First you can see the MA(1) process in Figure 17.7 (b). You can see how each period in time is a bit closer to its neighbors than before. In (c) we have an AR(1) process. Things look a lot smoother— that's because each value directly influences the next. And finally in ARMA(1,1) the data looks more related still. Keep in mind, each of these four graphs has the *exact same amount of randomness* to start with. It's just made successively more strongly organized *in time*.

AR and MA are by no means the only features we can account for in a model, and time series analysts like to stack on all sorts of stuff. You can add an Integration term (I), which *differences* the data (i.e., instead of evaluating the amount in my account, you evaluate the daily *change* in the amount) and get an ARIMA.[21] You can add other time-varying eXogenous (X) predictors, like how much I get paid at my job, and get an ARIMAX. You can account for Seasonality (S), like how my savings dips every Christmas, and get a SARIMAX. And so on.[22]

THE CORE PART OF TIME SERIES FORECASTING IS finding the right kind of model to fit, fitting it, and then forecasting. Very similar in concept to finding the right kind of non-time-series model. Except this time, instead of thinking about the right functional form for the relationship between variables, and which controls to include, you're concerned with the features of that one time series. Is it seasonal? Do you need to integrate? Does it have autocorrelation, and if so, how many terms? Does it have a moving-average process, and if so, again, how many terms? How can you even tell how many terms you need?

Answering all these questions is, as I previewed before, an entire field in itself. Usually this is where I'd provide a code example of how to do this. But while I've introduced the concepts here, I can't pretend to have given you enough information to actually be able to do all this properly, so I'll refrain from leaving out the sharp knives in the form of a deficient demonstration. Go find a textbook all about time series.

I will leave you with just a few pointers, the first of which is to note that most packages will skip an ARMA function and just have ARIMA (which becomes an ARMA if you set the I term to 0). In R, estimation and forecasting for a few key models like ARIMA can be done easily in the **fable** package. In Stata, standard models like ARIMA are built-in with easily guessable names (`arima`), and you can do forecasting by adding a `dynamic()` option to `predict`. In Python, plenty of time series models are in **statsmodels.tsa**, and they come with `.predict()` methods for forecasting.

17.3.3   The Joint-Test Problem

EVENT STUDY DESIGNS ARE HIGHLY RELIANT on making an accurate prediction of the counterfactual.[23] That prediction is

[21] Differencing is done in an attempt to get a "stationary" process. To simplify, adding a dollar to my savings account, because of interest, affects tomorrow's balance *more* than it affects today's. Left to its own devices, that's an upward spiral that heads to infinity over a long enough period! You can imagine why a statistical model would have trouble dealing with infinite feedback loops spiraling out. So it differences in hopes that the differenced data has no such spiral.

[22] Oh, and don't forget the same sort of increasingly-complex model chain that allows for conditional heteroskedasticity (CH), where the variance of the error rises and falls over time. ARCH, GARCH, EGARCH...

[23] To some extent, all causal inference designs are highly reliant on making accurate counterfactual predictions. For that reason, you'll see similar placebo-test solutions like this one pop up in plenty of other chapters.

the only thing making the event study work. If you expect that, say, stock returns would have stayed the same if no treatment had occurred, but in fact they would have increased by .1%, then your event study estimates are all off by .1%, which can be a lot if you're talking about daily portfolio returns.

This means that the results we get from an event study are a combination of two things—the actual event-study effect, and the model we used to generate the counterfactual prediction. And if you, say, do a significance test of that effect, you're not just testing the true event-study effect. You're doing a *joint* test of both the true effect and whether the predictive model is right. Can't separate 'em!

This is worrying, especially since we can never really know for sure whether we have the right model. Sure, maybe we've picked a model that fits the pre-event data super duper well. But that's the rub with counterfactuals—you can't see them. Maybe that super great fit *would* have stopped working at the exact same time the treatment went into effect.

WHAT CAN YOU DO to deal with this joint-test problem? The joint-test problem never really goes away. But we can help make it less scary by doing the best we can to make sure our counterfactual-prediction model is as good as possible. There is, however, something we can do to try to jimmy off one part of that joint test.

The application of placebo testing in event studies has been around a long time and the approach seems to have been largely codified by the time of Brown and Warner (1985).[24],[25]

The idea is this: when there's supposed to be an effect, and we test for an effect, we can't tease apart what part of our findings is the effect, and what part is the counterfactual model being wrong. But what if we *knew* there were no effect? Then, anything we estimated could only be the counterfactual model being wrong. So if we got an effect under those conditions, we'd know our model was fishy and would have to try something else.

So, get a *bunch* of different time series to test, especially those unaffected by the event. In the context of stock returns, for example, don't just include the firm with the great announcement on a certain day, also include stocks for a bunch of other firms that had no announcement.

Then, start doing your event study over and over again. Except each time you do it, pick a random time series to do it on, and pick a random day when the "event" is supposed to have happened.[26]

[24] Stephen J Brown and Jerold B Warner. Using daily stock returns: The case of event studies. *Journal of Financial Economics*, 14(1): 3–31, 1985.

[25] In the context of causal inference, a placebo test is a test performed under conditions where there should be zero effect. So if you find an effect, something is wrong.

[26] All the code you need to do this yourself is in Chapter 15 on Simulation.

On average, there shouldn't be any true effect in these randomly picked event studies. Sure, sometimes by chance you'll pick a time series and day where something *did* happen to make it jump, but you're just as likely to pick one where something made it drop, or nothing happened at all. It will average out to 0. Do a bunch of these and your average should be 0. If it's not, that means that you're getting an effect when no effect is there. The culprit must be your counterfactual model. Better try a different one.

18
Difference-in-Differences

18.1 How Does It Work?

18.1.1 Across Within Variation

THERE ARE PLENTY OF EXAMPLES OF TREATMENTS that *occur* at a particular time. We can see the world before the treatment is applied, and after. We want to know how much of the change in the world is due to that treatment. That's the causal inference task we have set before us.

This sounds like I'm setting myself up to do Chapter 17 on event studies again. And in a sense, event studies will be our jumping-off point. Just like with an event study, we will be identifying a causal effect by comparing a group that received treatment before they received the treatment to after. We are focusing on the within variation here.

Also just like with an event study, the obvious back door we have to deal with can be summed up as "time." As in Figure 18.1, identifying the effect of *Treatment* on *Outcome* requires us to close the back door that goes through *Time*. But we can't do this entirely, because *all* of the variation in *Treatment* is

DOI: 10.1201/9781003226055-18

explained by *Time*. You're either in a before-treatment time and untreated, or in an after-treatment time and treated.

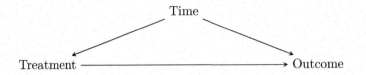

Figure 18.1: A Basic Time Back Door

EVENT STUDIES GET AROUND THIS PROBLEM by trying to use before-treatment information to construct a counterfactual after-treatment untreated prediction. Difference-in-differences (DID)[1] takes a different approach. Instead, it brings in *another* group that is *never* treated. So now in the data we have both the group that receives treatment at a certain point, and another group that never receives treatment. At first this seems counterintuitive—that untreated group may be different from the treated group! We have introduced a second back door, as in Figure 18.2.

[1] Some people say "difference-in-difference" instead of "difference-in-differences." Some people abbreviate it DD or Diff-in-Diff instead of DID. Let me be clear that I do not care.

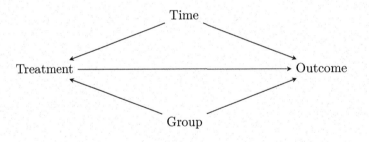

Figure 18.2: A Causal Diagram Suited for Difference-in-differences

Seems like we've made things worse by introducing the control group. The key is this, though: now that we have that untreated group, even though we've added a new back door, *we can now close both back doors.* How is this possible?

1. Isolate the within variation for both the treated group and untreated group. Because we have isolated within variation, we are controlling for group differences and closing the back door through *Group* (the "differences")

2. Compare the within variation in the treated group to the within variation in the untreated group. Because the within variation in the untreated group *is* affected by time, doing this comparison controls for time differences and closes the back door through *Time* (the "difference" in those differences)

In other words, we are looking for *how much more the treated group changed than the untreated group* when going from before to after. What we want is (treated group after − treated group before) − (untreated group after − untreated group before). The change in the untreated group represents *how much change we would have expected* in the treated group if no treatment had occurred. So any *additional* change beyond that amount must be the effect of the treatment.

18.1.2 Difference-in-Differences and Dirty Water

LET'S WALK THROUGH AN EXAMPLE OF DIFFERENCE-IN-DIFFERENCES with data from probably its first, and almost certainly its most famous, application: John Snow's 1855 findings that demonstrated to the world that cholera was spread by fecally-contaminated water and not via the air.[2,3,4]

Before the germ theory of disease became standard, medical thinkers of the world had a wide variety of ideas as to how disease spread. A popular one in Europe in the 19th century was "miasma theory" which held that disease spread through bad air coming from rotting material. This included cholera, which had routine outbreaks in many European cities. Other explanations for cholera besides miasma also abounded—bad breeding, low elevation, poverty, bad ground.[5]

John Snow, however, had reason to believe that cholera instead spread by dirty drinking water. He had a few ways of providing evidence, one of which is very similar to a modern-day difference-in-differences research design, and can be easily discussed in those terms.[6]

Snow's "before" and "after" periods were 1849 and 1854, respectively. London's water needs were served by a number of competing companies, who got their water intake from different parts of the Thames river. Water taken in from the parts of the Thames that were *downstream* of London contained everything that Londoners dumped in the river, including plenty of fecal matter from people infected with cholera. Between those two periods of 1849 and 1854, a policy was enacted—the Lambeth Company was required by an Act of Parliament to move their water intake upstream of London.

Lambeth moving their intake source gives us the Treated group: anyone in an area where the water came from the Lambeth company, and an Untreated group: anyone in an area without Lambeth.[7]

[2] John Snow. *On the Mode of Communication of Cholera.* John Churchill, 1855.

[3] His underlying theory was not widely accepted at the time, but they did buy the *result* and shut down the guilty water pump. Also his method certainly wouldn't have been called by the name "difference-in-differences" back then.

[4] My apologies if you're a fan of causal inference already, in which case you've almost certainly heard this story before. Sort of a cliché. But, hey, we tell it because it works.

[5] You can imagine how they might reasonably come to a conclusion like miasma theory. Rotting stuff does spread disease and also smells bad! And they were right that some diseases are airborne. Of course, they were wrong that masking those smells with nice smells would protect you. We can only imagine the Febreze commercials we'd get to see if we all still believed that. In any case, the *correlations* were definitely there. All of these things, including bad smell, were definitely correlated with cholera outbreak. The miasma theory is another reason why proper causal inference to understand the underlying data generating process is important.

[6] Thomas Coleman. Causality in the time of cholera: John Snow as a prototype for causal inference. *Available at SSRN 3262234*, 2019.

So then the question is: *did areas getting water from Lambeth see their Cholera numbers go down from 1849 to 1854 relative to areas getting no water from Lambeth?*

We can see the death rates in these areas in Table 18.1. First, we can see that in the pre-treatment period, the cholera death rates were fairly similar in the Lambeth and non-Lambeth areas. This isn't necessary for difference-in-differences, but does lend a bit of credibility to the assumption that these groups are comparable. Then, we can see that from 1849 to 1854, the cholera problem in non-Lambeth areas got *worse*, rising from 135 to 147, while the problem in the Lambeth areas got *better*, dropping from 130 to 84.9.

[7] It's a bit more complex because the areas with Lambeth water were really a mix of Lambeth with other companies. Since the "Treated" group is really a mix of treated and untreated people, our estimate will actually be an *understatement* of the effect of the clean water (why an understatement instead of an overstatement? An exercise for you...).

Region Supplier	Death Rates 1849	Death Rates 1854
Non-Lambeth Only (Dirty)	134.9	146.6
Lambeth + Others (Mix Dirty and Clean)	130.1	84.9

Death rates are deaths per 10,000 1851 population.

Table 18.1: London Cholera Deaths per 10,000 from Snow (1855)

Pretty convincing. Of course, looking at the Lambeth areas alone wouldn't be convincing—maybe cholera just happened to be going away at the time. We really need the non-Lambeth comparison to drive it home. The specific DID estimate we can get here is the Lambeth difference minus the non-Lambeth difference, or $(84.9 - 130) - (147 - 135) = -57.1$. The movement of the Lambeth pump reduced cholera mortality rates by 57.1 per 10,000 people. That's quite a lot!

18.1.3   How DID Does?

LET'S WORK THROUGH THE MECHANICS OF DIFFERENCE-IN-DIFFERENCES using a slightly more modern example, albeit one still on the topic of health. Specifically we'll be looking at a 2014 paper by Kessler and Roth,[8] which studies the rate at which people sign up to be organ donors.[9]

In the United States, people are not signed up to be organ donors by default. In most states, you are assumed to *not* be an organ donor. When you sign up for a driver's license, you can choose to *opt in* to the organ donation program. Check the organ donation box and—poof!—you're a donor. It's probably not surprising that organ donation rates in the US are considerably lower than in other countries where organ donation is opt-*out*— you're assumed to be a donor unless you actively choose not to be.

[8] Judd B Kessler and Alvin E Roth. Don't take "no" for an answer: An experiment with actual organ donor registrations. Technical report, National Bureau of Economic Research, 2014.
[9] Alvin Roth, the second author on that paper, is pretty well-known for talking about organ donation, as economists go. Won an econ Nobel for it, in fact! Check out his book *Who Gets What and Why*. It's not about causal inference but it is deeply interesting despite that one major flaw.

Outside of the opt-in and opt-out varieties of organ donation, there's also "active choice." Under active choice, when you sign up for a driver's license, you are asked to choose *whether or not* to be a donor. You can choose yes or no, but now the "no" option is *actively checking the "no" box* rather than skipping the question entirely as you can with opt-in approaches. Some policymakers have been advocating for active choice, with a goal of increasing donation rates, and active choice is the way things work in many states.

So does active choice work? In July 2011, the state of California switched from opt-in to active choice. Kessler and Roth decided to compare California against the twenty-five states that either have opt-in or a verbally given question with no fixed response (difference). Specifically, they compared the states on the basis of how their organ donation rates changed from before July 2011 to after (in differences).

We can see the kernel of the idea in Figure 18.3, which shows the raw data on organ donation rates in each state in each quarter.

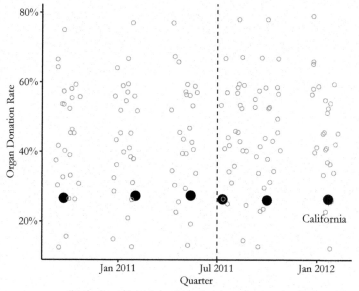

Figure 18.3: Organ Donation Rates in California and Other States

Jitter has been added to the x-axis to make points easier to see, since data is quarterly.

What can we see in the raw data? First off, we can see that California already doesn't have a great organ donation rate, sitting near the bottom of the pack. Second, you can see that California's rate didn't rise much after the policy went into effect— in fact, it seems to have dropped slightly. But maybe it just

dropped because *everyone's* rates were dropping at that time?
Nope—if anything, the other states seem to increase slightly.

Off the bat this already isn't looking too good for active
choice. But how would difference-in-differences actually handle
the data to tell us that?

We can see the steps in Figure 18.4. First, we calculate four
averages: before-treatment in the treated group (California),
after-treatment in the treated group, before-treatment in the
untreated group, and after-treatment in the untreated group.
We can see these averages in Figure 18.4 (a).

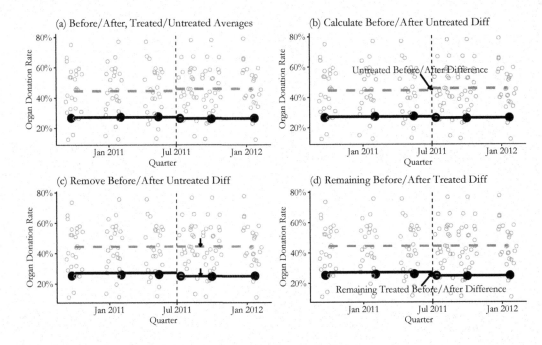

Second, we figure that any pre-/post-difference in the *untreated* group is the time effect. So we look at how that average changed from before to after for the untreated group (from
44.5% to 45.9%, an increase of 1.4 percentage points), in Figure
18.4 (b). We want to get rid of that time effect, so in Figure
18.4 (c) we subtract it out. Importantly, we subtract it out of
both the untreated *and* the treated group, lowering the treated
after-treatment values by 1.4 percentage points.

Finally, in Figure 18.4 (d), any remaining before/after difference in the treated (California) group is the difference-in-
difference effect. The raw difference is 26.3% − 27.1%, or a
reduction of .8 percentage points. Take out the 1.4-percentage-
point reduction from the untreated group and we see a DID

Figure 18.4: The Difference-
in-Differences Effect of
Active-Choice Organ
Donation Phrasing

effect of -2.2 percentage points of the active-choice phrasing on organ donor rates. Not great!

In this particular example, the before/after difference we see for the untreated group isn't that large, meaning there's not actually much of a time effect at all. This is actually nice—if there's a huge time effect we have to wonder if that time effect really *should* affect the treated and untreated groups differently. In any case, we can see how DID uses the data to come to its conclusion.

18.1.4   Untreated Groups and Parallel Trends

FOR ALL OF THIS TO WORK, WE HAVE TO HAVE THAT UNAF-FECTED GROUP, which we call the untreated group.[10] Can't do difference-in-differences without them. So what do we need in a untreated group?

I can talk about a lot of good features we can look for in an untreated group. But all of these are just observable pieces of the unobservable thing we really want to be true: We want our untreated group to be something that satisfies the *parallel trends assumption* with the treated group.

The parallel trends assumption says that, *if no treatment had occurred,* the difference between the treated group and the untreated group would have stayed the same in the post-treatment period as it was in the pre-treatment period.

Parallel trends is inherently *unobservable*. It's about the counterfactual of what would have happened if treatment had not occurred.

As an example of a clear failure of parallel trends, imagine you're looking at the effect of building additional roads on the popularity of its downtown restaurants. You find that Chicago built a bunch of additional roads in 2018, but Los Angeles did not. So you use Los Angeles as your untreated group.

You look at Chicago and Los Angeles in 2017 (pre-roads) and 2018 (post-roads), and use difference-in-differences to find that the roads somehow made downtown restaurants *less* popular in Chicago. What happened? Well, you might find that in 2018, a bunch of new highly-hyped restaurants opened in Los Angeles' downtown. So the 2017/2018 change in the Chicago/LA gap reflects both the new Chicago roads and the new Los Angeles restaurants. Obviously, we can't take this as a good estimate of the impact of the roads alone. We haven't properly identified

[10] You might also see them called the "comparison" group or the "control" group. I'll use "comparison" group occasionally in the chapter.

Parallel trends assumption. In a difference-in-differences design, the assumption that, had no treatment occurred, the gap between treated and untreated groups would have remained constant.

the effect of the roads. We should have picked a city that didn't build a bunch of new restaurants at the time.

Unfortunately, any other city we may pick as our untreated comparison group for Chicago may have *other* things changing. There may not even be anything obvious to pin it on. Maybe we pick New York as a comparison, and find that the roads *really* improved traffic to Chicago restaurants. Then we look at the New York data and notice that restaurant popularity has been trending down for years. Nothing special about 2018 in particular, but given the existing trend, the Chicago/New York gap likely would have grown in Chicago's favor even without the roads. The effect we get is clearly a combination of the long-term trend in New York with the roads in Chicago. Again, not identified.[11]

Remember, the entire plan behind a difference-in-differences design is to *use the change in the untreated group to represent all non-treatment changes in the treated group*. That way, once we subtract the untreated group's change out, all we're left with is the treated group's change. Parallel trends is necessary for us to assume that works. If, without a treatment, the gap between the two groups would have changed from the pre-period to the post-period *for any reason, or for no reason at all*, then that non-treatment-related change will get mixed up with the treatment-related change, and we won't be able to tell them apart.

We can put this in mathematical terms:

- The difference between pre-treatment and post-treatment in the treated group is $EffectofTreatment + OtherTreatedGroupChanges$

- The difference between pre-treatment and post-treatment in the untreated group is $OtherUntreatedGroupChanges$

- Difference-in-difference subtracts one from the other, giving us

$$EffectofTreatment$$
$$+ OtherTreatedGroupChanges$$
$$- OtherUntreatedGroupChanges \quad (18.1)$$

For DID to identify just $EffectofTreatment$, it has to be the case that $OtherTreatedGroupChanges$ exactly cancels out with $OtherUntreatedGroupChanges$. That's what parallel trends is really about. This is the assumption that we need to identify the effect. So think carefully about whether it's true in your case.

[11] Parallel trends is pretty poorly-named, making it easy to get confused. The problem with this New York example isn't that the Chicago and New York trends were diverging pre-treatment, but rather that *the trends would have continued to diverge post-treatment*. It's easy to look at how pre-treatment trends are changing, see that they're parallel or non-parallel, and think of that as the "parallel trends" you're interested in. This mistake is extra easy to make because it's the basis some suggestive tests for parallel trends use observed trends in the pre-treatment period. But you're really interested in the *counterfactual* trend from pre-treatment to post-treatment in the absence of treatment. If Chicago and New York had suddenly stopped trending apart in 2018 (other than because of the roads), that would have satisfied parallel trends despite the divergence before. Similarly, parallel pre-treatment trends could be broken at the moment of treatment, as in the Los Angeles example.

So if what we want is parallel trends, how should
we pick an untreated comparison group? What we want
in an untreated group is for it to change by the same amount as
the treated group (if treatment had occurred) from before the
treatment is applied to afterwards.

This means that there are a few good signs we can look for.
While none of these things are *requirements*, exactly, they are
all things someone would look for when thinking about whether
your DID design is believable:

1. There's no particular reason to believe the untreated group
   would suddenly change around the time of treatment.

2. The treated group and untreated groups are *generally similar
   in many ways*.

3. The treated group and untreated groups had *similar trajec-
   tories for the dependent variable before treatment*.

The first tip—that there's no reason to believe the untreated
group would suddenly change at the time of treatment—we've
already covered with the Chicago/Los Angeles roads example.
If there's something obviously changing in the untreated group
at the same time, DID will mix up the effects of the treatment
and whatever was changing in the untreated group.

Looking for an untreated group that is *generally similar in
many ways* makes sense. We are relying on an assumption that,
in the absence of treatment, the treated and untreated groups
would have changed over time in the same way. Groups that are
similar seem likely to have changed in similar ways over time.
Say we're looking at the impact of an event that considerably
increased immigration to Miami, Florida, in the United States,
as in the classic DID Mariel Boatlift study,[12] where a policy
change in Cuba led to a huge wave of immigrants coming to
Miami all at the same time, allowing us to look at the effects
of immigration on the labor market. In looking at how immigration affected the Miami labor market, our results would be
more plausible if using a demographically or geographically similar city like, say, Atlanta or Tampa, as opposed to something
far-off and very different like Reykjavik in Iceland.[13]

We may also want to look for an untreated group that has
a *similar trajectory for the dependent variable before treatment*.
That is, the outcome variable was growing or shrinking at about
the same rate in both the treated and untreated groups in the
pre-treatment period. If the two groups were trending similarly

[12] David Card. The im-
pact of the Mariel boatlift
on the Miami labor market.
ILR Review, 43(2):245–257,
1990.

[13] There would be noth-
ing technically wrong with
Reykjavik as an untreated
group. Any differences that
were *consistent over time*
would already be handled by
the within variation, as described in Chapter 16. But
we do need the changes over
time to be the same, which is
more plausible with a more-
similar city.

before treatment went into effect, that's a good clue that they would have continued to trend similarly if no treatment had occurred.

HOW CAN WE CHECK IF OUR UNTREATED GROUP IS AP-PROPRIATE? There are a few common ways to evaluate the parallel trends assumption, so core to our use of DID, and see whether it's plausible. I want to emphasize that *these are not tests of whether parallel trends is true.* "Passing" these tests does not mean that parallel trends is true. In fact, no test of the data could possibly confirm or disprove the parallel trends assumption, since it's based on a counterfactual we can't see. These tests are more along the lines of *suggestive evidence.* If these tests fail, that makes the parallel trends assumption *less plausible.* And that's about it.

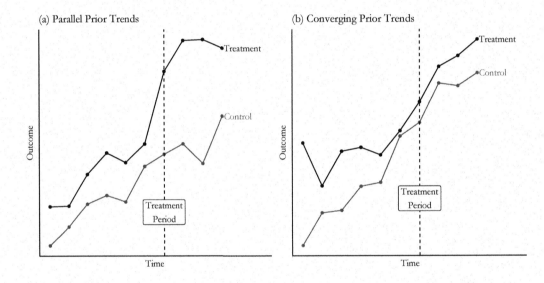

Figure 18.5: A Graph Where the Prior Trends Test Looks Good for DID, and a Graph Where It Doesn't

Hedging aside, there are some things we can do. One of them is the test of *prior trends.* This test simply looks to see whether the treated and untreated groups were trending similarly before treatment. For example, Figure 18.5 shows an example of a treated and untreated group that were heading in the same direction, and a pair that weren't. On the left, the distance between the treated and untreated group stays roughly constant in the leadup to treatment, even though both are trending upwards. This implies that, had the treatment not occurred, they likely would have continued having similar trends, lending more credibility to the parallel trends assumption. On the right, the fairly large gap between the two has already shrunk by the time treatment goes into effect, with the untreated group starting lower but gaining on the treated group. That trend likely would have continued without treatment, and so parallel trends is un-likely to hold.

Finding that the trends weren't identical doesn't necessar-ily disprove that DID works in your instance, but you'll defi-nitely have some explaining to do as to why you think that the gap from just-before to just-after treatment didn't change even though it did change from just-just-before to just-before![14]

The second test we can perform is the *placebo test*. In a placebo test for difference-in-differences, we'd take a situation where a treatment was applied in, for example, March 2019. Then, we'd use data only from *before* 2019, ignoring all the data from the periods where treatment was actually applied.

Then, using the pre-March 2019 data, we'd pick a few differ-ent periods and *pretend* that the treatment was applied at that time. We'd estimate DID using that pretend treatment date. If we consistently find a DID "effect" at those pretend treatment dates, that gives us a clue that something may be awry about the parallel trends assumption.

Certainly, a nonzero DID effect at a period where there is no actual treatment tells us that the non-treatment changes in the treated group don't exactly cancel out the non-treatment changes in the untreated group at the pretend-treatment time. So again, you'd have some explaining to do as to why we should believe that they exactly cancel out at the *actual* treatment time.

ONE FINAL NOTE ABOUT PARALLEL TRENDS that is too often overlooked: parallel trends means we have to think *very carefully* about how our dependent variable is measured and transformed. Because parallel trends isn't just an assumption about causality, it's an assumption about the size of a gap remaining constant, which means something different depending on how you measure that gap.

The most common way this pops up is when thinking about a dependent variable with a logarithm transformation. If parallel trends holds for dependent variable Y, *then it doesn't hold for* $\ln(Y)$, and vice versa—if it holds for $\ln(Y)$ it doesn't hold for Y.

For example, say that in the pre-treatment period Y is 10 for the control group and 20 for the treated group. In the post-treatment period, in the counterfactual world where treatment never happened, Y would be 15 for the control group and 25 for the treated group. Gap of $20 - 10 = 10$ before, and $25 - 15 = 10$ after. Parallel trends holds!

[14] One nice thing about looking at prior trends is it encourages you to look at how both treated *and* untreated are changing. If you get a big, significant difference-in-difference effect, but it's because your untreated group dropped down at the treatment time while the treated group just kept on-trend... well... *maybe* parallel trends holds, but it doesn't seem likely.

Placebo test. Applying a research design to a case where treatment was actu-ally not applied, to (ideally) show that the design already eliminates the influence of non-treatment influences.

What about for $\ln(Y)$? The gap before treatment is $\ln(20) -$ $\ln(10) = .693$, but the gap after treatment is $\ln(25) - \ln(15) =$ $.511$. Parallel trends doesn't hold![15]

This is the kind of thing that's obvious if you think about it for a second, but many of us never think about it for a second. So think carefully about *exactly what form of the dependent variable* you think parallel trends holds for, and use that form of the dependent variable.

18.2   How Is It Performed?

18.2.1   Two-Way Fixed Effects

THE CLASSIC APPROACH TO ESTIMATING DIFFERENCE-IN-DIFFERENCES IS VERY SIMPLE. The goal here is to control for group differences, and also control for time differences. So... easy. Just control for group differences and control for time differences.[16] The regression is

$$Y = \alpha_g + \alpha_t + \beta_1 Treated + \varepsilon \qquad (18.2)$$

where α_g is a set of fixed effects for the group that you're in—in the simplest form, just "Treated" or "Untreated"—and α_t is a set of fixed effects for the time period you're in—in the simplest form, just "before treatment" and "after treatment." *Treated*, then, is a binary variable indicating that you are being treated *right now*—in other words, you're in a treated group in the after-treatment period. The coefficient on *Treated* is your difference-in-differences effect.[17]

How about getting some control variables in that equation? Well, maybe. Any control variables that vary over group but don't change over time are unnecessary and would drop out— we already have group fixed effects (remember Chapter 16?). But what about control variables that do change over time? We may well think that parallel trends only holds conditional on some variables—perhaps the untreated group dropped relative to treatment because some predictor W unrelated to treatment just happened to drop at the same time treatment went into effect, but we can control for W. However, the inclusion of time-varying controls imposes some statistical problems related to whether those controls impact treated and untreated similarly, and, importantly, the assumption that treatment doesn't affect later values of covariates. If you need to include covariates, it's

[15] This isn't specific to the logarithm. If parallel trends holds, then any sort of non-linear transformation (or removing a nonlinear transformation) will break parallel trends.

[16] *Please only use this equation if treatment occurs at the same time across all your treated groups.* If that's not true... read on later in the chapter.

[17] This interpretation is specific to OLS. *Treated* is an interaction term (as will become clear in the next equation), and the meaning of interaction terms is a bit different in nonlinear models like logit or probit. So the difference-in-differences effect isn't just the coefficient on *Treated* in logit or probit. See Puhani (2012) for detail on the calculation. This complexity, by the way, is one reason why you often see people use OLS for DID even when they have a binary outcome (plus, OLS isn't so bad with a model that's just a bunch of binary variables).

often a good idea to show your results both with and without them.[18]

Another way to write the same difference-in-difference equation if you have only two groups and two time periods is

$$Y = \beta_0 + \beta_1 TreatedGroup + \beta_2 AfterTreatment +$$
$$\beta_3 TreatedGroup \times AfterTreatment + \varepsilon \quad (18.3)$$

where $TreatedGroup$ is an indicator that you're in the group being treated (whether it's before or after treatment is actually implemented), and $AfterTreatment$ is an indicator that you're in the "post"-treatment period (whether or not *your* group is being treated).[19] The third term is an interaction term, in effect an indicator for being in the treated group AND in the post-treatment period, i.e., you're actually being treated right now. This third term is equivalent to $Treated$ in the last equation, and $\hat{\beta}_3$ is our difference-in-differences estimate. This interaction-term version of the equation is attractive because it makes clear what's going on. By standard interaction-term interpretation, β_3 tells us *how much bigger* the $TreatedGroup$ effect is in the $AfterTreatment$ than in the before-period. That is, how much bigger the treated/untreated gap grows after you implement the treatment. Difference-in-differences!

Whichever way you write the equation, this approach is called the "Two-way fixed effects difference-in-difference estimator" since it has two sets of fixed effects, one for group and one for time period. This model is generally estimated using standard errors that are clustered at the group level.[20]

Two-way fixed effects, or TWFE, has some desirable properties in the sense that it is highly intuitive—we want to control for group and time differences, so we, uh... do exactly that. It also lets us apply what we already know about fixed effects. It gives us the exact same results as directly calculating (treated group after − treated group before) − (untreated group after − untreated group before). It also lets us account for multi-group designs where we have multiple groups, some of which are treated and some are not, rather than just one treated and untreated group.

There are some downsides of the TWFE approach, though. In particular, it doesn't work very well for "rollout designs," also known as "staggered treatment timing," where the treatment is applied at different times to different groups. Researchers used TWFE for these cases for a long time, but it turns out to not work very well—more on that later in the chapter. But if you

[18] Or, if you have both multiple treatment times *and* covariates, spring ahead to the How the Pros Do It section.

[19] Notice that these are, in effect, fixed effects for group and time, as in the last equation. But since there are only two groups and two time periods, we only need one coefficient for each.

[20] Clustering at the group level accounts for the fact that we expect errors to be related within group over time, which can make standard errors a little overconfident (i.e., too small) if not accounted for. Other ways of dealing with this, discussed in Betrand, Duflo, and Mullainathan (2004), include pre-aggregating the data to just one pre-treatment and one post-treatment period per group, or using a cluster bootstrap, as discussed in Chapter 15.

Marianne Bertrand, Esther Duflo, and Sendhil Mullainathan. How much should we trust differences-in-differences estimates? *The Quarterly Journal of Economics*, 119(1):249–275, 2004.

have a single treatment period, TWFE can be an easy way to estimate difference-in-differences.

HOW CAN WE ESTIMATE DIFFERENCE-IN-DIFFERENCES WITH TWO-WAY FIXED EFFECTS IN CODE? The following code chunks apply the same fixed effects code we learned in Chapter 16 to the Kessler and Roth organ donation study discussed earlier, with clustered fixed effects applied at the state level.[21] Following the code, we'll look at a regression table and interpret the results.

[21] In Stata there is also the **didregress** suite of functions that could replace several of the by-hand Stata code chunks in this chapter. But I will skip it as it is only available in Stata 17+. Also be aware that as of this writing, these functions do not handle rollout designs.

```r
1  # R CODE
2  library(tidyverse); library(modelsummary); library(fixest)
3  od <- causaldata::organ_donations
4
5  # Treatment variable
6  od <- od %>%
7      mutate(Treated = State == 'California' &
8             Quarter %in% c('Q32011','Q42011','Q12012'))
9
10 # feols clusters by the first
11 # fixed effect by default, no adjustment necessary
12 clfe <- feols(Rate ~ Treated | State + Quarter,
13          data = od)
14 msummary(clfe, stars = c('*' = .1, '**' = .05, '***' = .01))
```

```stata
1  * STATA CODE
2  causaldata organ_donations.dta, use clear download
3
4  * Create treatment variable
5  g Treated = state == "California" & ///
6      inlist(quarter, "Q32011","Q42011","Q12012")
7
8  * We will use reghdfe which must be installed with
9  * ssc install reghdfe
10 reghdfe rate Treated, a(state quarter) vce(cluster state)
```

```python
1  # PYTHON CODE
2  import linearmodels as lm
3  from causaldata import organ_donations
4  od = organ_donations.load_pandas().data
5
6  # Create Treatment Variable
7  od['California'] = od['State'] == 'California'
8  od['After'] = od['Quarter_Num'] > 3
9  od['Treated'] = 1*(od['California'] & od['After'])
10 # Set our individual and time (index) for our data
11 od = od.set_index(['State','Quarter_Num'])
12
13 mod = lm.PanelOLS.from_formula('''Rate ~
14 Treated + EntityEffects + TimeEffects''',od)
15
16 # Specify clustering when we fit the model
17 clfe = mod.fit(cov_type = 'clustered',
18 cluster_entity = True)
19 print(clfe)
```

Table 18.2 shows the result of this two-way fixed effects regression, with the fixed effects themselves excluded from the

	Organ Donation Rate
Treatment	-0.022***
	(0.006)
Num.Obs.	162
FE: Quarter	X
FE: State	X

* p < 0.1, ** p < 0.05,*** p < 0.01

Standard errors clustered at the state level.

Table 18.2: Difference-in-differences Estimate of the Effect of Active-Choice Phrasing on Organ Donor Rates

table and only the coefficient on the *Treated* variable ("treated-group" and "after-treatment" interacted) shown. Notice at the bottom of the table a row each for the state and quarter fixed effects. The "X" here just indicates that the fixed effects are included. It's fairly common to skip reporting the actual fixed effects—there are so many of them!

The coefficient is $-.022$ with a standard error of 0.006. From this we can say that the introduction of active-choice phrasing in California saw a *reduction* in organ donation rates that was .022 (or 2.2 percentage points) larger in California than it was in the untreated states. The standard error is .006, so we have a t-statistic of $-.022/.006 = 3.67$, which is high enough to be considered statistically significant at the 99% level. We can reject the null that the DID estimate is 0.

18.2.2 Treatment Effects in Difference-in-Differences

AS WITH ANY RESEARCH DESIGN, WE WANT TO THINK CARE-FULLY ABOUT WHAT OUR RESULT ACTUALLY MEANS. When we estimate difference-in-differences, *who* are we getting the effect for?

The chapter on treatment effects, Chaper 10, gives us some clues. What are we comparing here? Difference-in-differences compares what we *see* for the treated group after treatment against *our best guess at what the treatment group would have been without treatment.*

We are specifically isolating the difference between being treated and not being treated *for the group that actually gets treated.* So, we are getting an average treatment effect among that group. In other words it's the "average treatment on the treated."

So, the estimate that standard difference-in-differences gives us is all about how effective the treatment was for the groups that actually got it. If the untreated group would have been affected differently, we have no way of knowing that.[22]

18.2.3   Supporting the Parallel Trends Assumption

THE PARALLEL TRENDS ASSUMPTION SAYS that, if no treatment had in fact occurred, then the difference in outcomes between the treated and untreated groups would not have changed from before the treatment date to afterwards. It's okay that there *is* a difference,[23] but that difference can't change from before treatment to after treatment for any reason *but* treatment.

This assumption, though it's completely crucial for what we're doing with difference-in-differences, must remain an assumption. It relies directly on a counterfactual observation—what would have happened without the treatment.

We can't *prove* or even really *test* parallel trends, but in the last section I discussed two tests that can provide some evidence that at least makes parallel trends look more plausible as an assumption. Those are the *test of prior trends* and the *placebo test*.

THE TEST OF PRIOR TRENDS LOOKS AT whether the treated and untreated groups already had differing trends in the leadup to the period where treatment occurred. There are two good ways to actually perform this test. The first is to graph the average outcomes over time in the pre-treatment period and see if they look different, as we already did in Figure 18.5.

The second is to perform a statistical test to see if the trends are different, and if so, how much different. The simplest form of this uses the regression model

$$Y = \alpha_g + \beta_1 Time + \beta_2 Time \times Group + \varepsilon \qquad (18.4)$$

estimated using only data from before the treatment period, where $\beta_2 Time \times Group$ allows the time trend to be different for each group. A test of $\beta_2 = 0$ provides information on whether the trends are different.[24] This is the simplest specification, and you could look for more complex time trends by adding polynomial terms or other nonlinearities to the model.

If you do find that the trends are different, you'll want to look at how different, exactly. Are the trends barely different, but the difference is statistically significant because of a large sample? Are the trends different because they were mostly consistent

[22] Another way of coming to this same conclusion is to notice that we never see any variation in treatment for the untreated group, since they're never treated. How could we possibly see the effect of treatment on them if there's no variation in treatment in that group?

[23] Although one does wonder whether we have a good comparison group if the outcomes are very different.

[24] Finding that $\beta_2 = 0$ is unlikely shows that the trends are different

with each other but there was a brief period of deviation a few years back? Think about it! Don't just use the significance test to answer the question.

When failing a prior trends test, some researchers will see this as a reason to add "controls for trends" to salvage their research design by including the $Time$ variable in their difference-in-differences model directly, rather than the time fixed effects α_t. However, this can have the unfortunate effect of controlling away some of the actual treatment effect, especially for treatments with effects that get stronger or weaker over time.[25] There are also ways to control only for $prior$ trends, sort of like running an event study for the treated and untreated groups, but this can make things worse in its own way unless it's done precisely (and you're definitely at the "How the Pros Do It" stage there).[26]

NEXT WE CAN CONSIDER THE PLACEBO TEST. Placebo tests are good ways of evaluating untestable assumptions in a number of different research designs, not just difference-in-differences, and they pop up a few times in this book.

For the difference-in-differences placebo test, we can follow these steps:

1. Use only the data that came before the treatment went into effect.

2. Pick a fake treatment period.[27]

3. Estimate the same difference-in-differences model you were planning to use (for example $Y = \alpha_t + \alpha_g + \beta_1 Treated + \varepsilon$), but create the $Treated$ variable as equal to 1 if you're in the treated group and after the $fake$ treatment date you picked.

4. If you find an "effect" for that treatment date where there really shouldn't be one, that's evidence that there's something wrong with your design, which may imply a violation of parallel trends.

Another way to do this if you have multiple untreated groups is to use all of the data, but drop the data from the treated groups. Then, assign different untreated groups to be fake treated groups, and estimate the DID effect for them. This approach is less common since it doesn't address parallel trends quite as directly (and it's not really a problem if parallel trends fails $among\ your\ untreated\ groups$), but this is a very common

[25] Justin Wolfers. Did unilateral divorce laws raise divorce rates? A reconciliation and new results. *American Economic Review*, 96 (5):1802–1820, 2006.

[26] Jonathan Roth. Should we adjust for the test for pre-trends in difference-in-difference designs? *arXiv preprint arXiv:1804.01208*, 2018.

[27] There are many ways to do this, depending on context. It could be a specific alternative that makes sense, or it could be at random, or you could try a BUNCH of random treatment periods and see where the true treatment fits in the sampling distribution of the fake effect. This latter approach is called "randomization inference."

placebo test for the synthetic control method (which will be discussed later in this chapter, and in Chapter 21).

We can put the fake-treatment-period placebo method to work in code form in the following examples. We will continue to use our organ donation data, although this process tends to work better when you have a lot of pre-treatment periods, rather than just the three we have here:

```r
# R CODE
library(tidyverse); library(modelsummary); library(fixest)
od <- causaldata::organ_donations %>%
    # Use only pre-treatment data
    filter(Quarter_Num <= 3)

# Create our fake treatment variables
od <- od %>%
    mutate(FakeTreat1 = State == 'California' &
            Quarter %in% c('Q12011','Q22011'),
            FakeTreat2 = State == 'California' &
            Quarter == 'Q22011')

# Run the same model we did before but with our fake treatment
clfe1 <- feols(Rate ~ FakeTreat1 | State + Quarter,
        data = od)
clfe2 <- feols(Rate ~ FakeTreat2 | State + Quarter,
    data = od)

msummary(list(clfe1,clfe2), stars = c('*' = .1, '**' = .05, '***' = .01))
```

```stata
* STATA CODE
causaldata organ_donations.dta, use clear download

* Use only pre-treatment data
keep if quarter_num <= 3

* Create fake treatment variables
g FakeTreat1 = state == "California" & inlist(quarter, "Q12011","Q22011")
g FakeTreat2 = state == "California" & quarter == "Q22011"

* Run the same model as before
* But with our fake treatment
reghdfe rate FakeTreat1, a(state quarter) vce(cluster state)
reghdfe rate FakeTreat2, a(state quarter) vce(cluster state)
```

```python
# PYTHON CODE
import linearmodels as lm
from causaldata import organ_donations
od = organ_donations.load_pandas().data

# Keep only pre-treatment data
od = od.loc[od['Quarter_Num'] <= 3]

# Create fake treatment variables
od['California'] = od['State'] == 'California'
od['FakeAfter1'] = od['Quarter_Num'] > 1
od['FakeAfter2'] = od['Quarter_Num'] > 2
od['FakeTreat1'] = 1*(od['California'] & od['FakeAfter1'])
od['FakeTreat2'] = 1*(od['California'] & od['FakeAfter2'])
```

```
15
16    # Set our individual and time (index) for our data
17    od = od.set_index(['State','Quarter_Num'])
18
19    # Run the same model as before
20    # but with our fake treatment variables
21    mod1 = lm.PanelOLS.from_formula('''Rate ~
22    FakeTreat1 + EntityEffects + TimeEffects''',od)
23    mod2 = lm.PanelOLS.from_formula('''Rate ~
24    FakeTreat2 + EntityEffects + TimeEffects''',od)
25
26    clfe1 = mod1.fit(cov_type = 'clustered',
27    cluster_entity = True)
28    clfe2 = mod1.fit(cov_type = 'clustered',
29    cluster_entity = True)
30
31    print(clfe1)
32    print(clfe2)
```

Table 18.3: Placebo DID Estimates Using Fake Treatment Periods

	Second-Period Treatment	Third-Period Treatment
Treatment	0.006	-0.002
	(0.005)	(0.003)
Num.Obs.	81	81
FE: Quarter	X	X
FE: State	X	X

* $p < 0.1$, ** $p < 0.05$, *** $p < 0.01$
Standard errors clustered at the state level.

In Table 18.3, we see that if we drop all data after the actual treatment (which occurs between the third and fourth period in the data), and then pretend that the treatment occurred either between the first and second, or second and third periods, we find no DID effect. That's as it should be! There wasn't actually a policy change there, so there shouldn't be a DID effect.

18.2.4   Long-Term Effects

THE WAY WE'VE BEEN TALKING ABOUT TIME SO FAR with difference-in-differences has basically assumed we're dealing with two time periods—"before treatment" and "after treatment." Sure, the two-way fixed effect model allows for as many time periods as you like, but in talking about it I've lumped all those time periods into those two big buckets: before and after, and we've only estimated a single effect that's implied to apply to the entire "after" period.

This can leave out a lot of useful detail. We're interested in the effect of a given treatment. Certain treatments become more or less effective over time, or take a while for the effect

Dynamic treatment effect.   A treatment effect that is allowed to vary over time, either in absolute terms or relative to the time the treatment was implemented.

to show up. And if you think about it, "after" is sort of an arbitrary time period. If we *were* looking at only one "after" period, when is that? The day after treatment? The month? The year? Four years?[28] Well, dang, why not just check *all* those "after" periods?

We can do that! Difference-in-differences can be modified just a bit to allow the effect to differ in each time period. In other words, we can have *dynamic treatment effects*. This lets you see things like the effect taking a while to work, or fading out.[29]

A common way of doing this is to first generate a *centered* time variable, which is just your original time variable minus the treatment period. So time in the last period before treatment is $t = 0$, the first period with treatment implemented is $t = 1$, the second-to-last period before treatment is $t = -1$, and so on.

Then, interact your *Treatment* variable with a set of binary indicator variables for each of the time periods. Done!

$$Y = \alpha_g + \alpha_t +$$
$$\beta_{-T_1} Treated + \beta_{-(T_1-1)} Treated + ... + \beta_{-1} Treated +$$
$$\beta_1 Treated + ... + \beta_{T_2} Treated + \varepsilon \quad (18.5)$$

Where there are T_1 periods before the treatment period, and T_2 periods afterwards. Do note that there's no β_0 coefficient for the last period before treatment here—that needs to be dropped or else you get perfect multicollinearity.[30]

THIS SETUP DOES A FEW THINGS FOR YOU. First, you *shouldn't* find effects among the before-treatment coefficients $\beta_{-T_1}, \beta_{-(T_1-1)}, ..., \beta_{-1}$. These should be close to zero (and insignificant, if doing statistical significance testing). This is a form of placebo test—it gives difference-in-differences an opportunity to find an effect before it should be there, and hopefully you find nothing.[31]

Second, the after-treatment coefficients $\beta_1, ... \beta_{T_2}$ show the difference-in-difference estimated effect in the relevant period: the effect one period after treatment is β_1, and so on.

This approach is additionally easy to implement with a good ol' interaction term. Adjust your time variable `time` such that the treatment period (or last period before treatment, if treatment occurs between periods) is 0. Then just add the interaction to your model. In R this is `factor(time)*treatedgroup` (or, if you're using **fixest**, `i(time, treatedgroup)`). In Stata it's `i.time##i.treatedgroup`. In Python, `C(time)*treatedgroup`. The code may have to be a bit fancier if you want to pick which

[28] If you include all the time periods, you get the average effect over all the post-treatment periods. That's not precisely accurate if you also have control variables in the model, but that's the general idea.

[29] You'll notice many similarities to Chapter 17, and indeed this is often referred to as "event study difference-in-differences."

[30] While more broadly used, application of this method often refers to Autor (2003).

David H Autor. Outsourcing at will: The contribution of unjust dismissal doctrine to the growth of employment outsourcing. *Journal of Labor Economics*, 21(1):1–42, 2003.

[31] Do keep in mind that if you have a lot of pre-treatment periods, then by sheer chance alone *some* might show effects anyway, even if things are actually fine. So if you have a lot of pre-treatment periods and you find the rare occasional large effect, don't freak out.

time period to drop, or graph the results. See the below code
for that.

THERE ARE A FEW IMPORTANT THINGS TO KEEP IN MIND
when using a dynamic difference-in-differences approach.

First, regular difference-in-differences takes advantage of all
the data in the entire "after" period to estimate the effect. As
you might guess, each period's effect estimate in the dynamic
treatment effects approach relies mostly on data from that one
period. That's a lot less data. So you can expect much less
precision in your estimates. And don't be surprised if your confidence intervals exclude 0 in the overall difference-in-differences
estimate but don't do so here (i.e., the overall effect is statistically significant but the individual-period effects aren't).

Second, when interpreting the results, everything is relative
to that omitted time-0 effect. As always when we have a categorical variable, everything is relative to the omitted group. So
the β_2 coefficient, for example, means that the effect two periods
after treatment is β_2 *higher* than the effect in the last period
before treatment. Of course, there *should* be no actual effect
in period 0. But if there was (oops), it will make your results
wrong but you'll have a hard time spotting the problem.

Third, a good way to present the results from a dynamic estimate like this is usually graphically, with time across the x-axis,
and with the difference-in-difference estimates and (usually) a
confidence interval on the y-axis. This lets you see at a glance
much easier than a table how the effect evolves over time, and
how close to 0 those pre-treatment effects are.

The following code examples show how to run the dynamic
treatment effect model and then produce these graphs, again
using our organ donation data.

```
1   # R CODE
2   library(tidyverse); library(fixest)
3   od <- causaldata::organ_donations
4
5   # Treatment variable
6   od <- od %>% mutate(California = State == 'California')
7
8   # Interact quarter with being in the treated group using
9   # the fixest i() function, which also lets us specify
0   # a reference period (using the numeric version of Quarter)
1   clfe <- feols(Rate ~ i(Quarter_Num, California, ref = 3) |
2                          State + Quarter_Num, data = od)
3
4   # And use coefplot() for a graph of effects
5   coefplot(clfe)
```

```
1   * STATA CODE
2   causaldata organ_donations.dta, use clear download
```

```stata
 3 | g California = state == "California"
 4 |
 5 | * Interact being in the treated group with Qtr,
 6 | * using ib3 to drop the third quarter (the last one before treatment)
 7 | reghdfe rate California##ib3.quarter_num, ///
 8 |     a(state quarter_num) vce(cluster state)
 9 | * There's a way to graph this in one line using coefplot
10 | * But it gets stubborn and tricky, so we'll just do it by hand
11 | * Pull out the coefficients and SEs
12 | g coef = .
13 | g se = .
14 | forvalues i = 1(1)6 {
15 |         replace coef = _b[1.California#`i'.quarter_num] if quarter_num == `i'
16 |         replace se = _se[1.California#`i'.quarter_num] if quarter_num == `i'
17 | }
18 |
19 | * Make confidence intervals
20 | g ci_top = coef+1.96*se
21 | g ci_bottom = coef - 1.96*se
22 |
23 | * Limit ourselves to one observation per quarter
24 | keep quarter_num coef se ci_*
25 | duplicates drop
26 |
27 | * Create connected scatterplot of coefficients
28 | * with CIs included with rcap
29 | * and a line at 0 from function
30 | twoway (sc coef quarter_num, connect(line)) ///
31 |        (rcap ci_top ci_bottom quarter_num) ///
32 |        (function y = 0, range(1 6)), xtitle("Quarter") ///
33 |        caption("95% Confidence Intervals Shown")
```

```python
 1 | # PYTHON CODE
 2 | import pandas as pd
 3 | import matplotlib as plt
 4 | import linearmodels as lm
 5 | from causaldata import organ_donations
 6 | od = organ_donations.load_pandas().data
 7 |
 8 | # Create Treatment Variable
 9 | od['California'] = od['State'] == 'California'
10 |
11 | # Create our interactions by hand,
12 | # skipping quarter 3, the last one before treatment
13 | for i in [1, 2, 4, 5, 6]:
14 |         name = 'INX'+str(i)
15 |         od[name] = 1*od['California']
16 |         od.loc[od['Quarter_Num'] != i, name] = 0
17 |
18 | # Set our individual and time (index) for our data
19 | od = od.set_index(['State','Quarter_Num'])
20 |
21 | mod = lm.PanelOLS.from_formula('''Rate ~
22 | INX1 + INX2 + INX4 + INX5 + INX6 +
23 | EntityEffects + TimeEffects''',od)
24 |
25 | # Specify clustering when we fit the model
26 | clfe = mod.fit(cov_type = 'clustered',
27 | cluster_entity = True)
28 |
```

```
29   # Get coefficients and CIs
30   res = pd.concat([clfe.params, clfe.std_errors], axis = 1)
31   # Scale standard error to CI
32   res['ci'] = res['std_error']*1.96
33
34   # Add our quarter values
35   res['Quarter_Num'] = [1, 2, 4, 5, 6]
36   # And add our reference period back in
37   reference = pd.DataFrame([[0,0,0,3]],
38   columns = ['parameter',
39                           'lower',
40                           'upper',
41                           'Quarter_Num'])
42   res = pd.concat([res, reference])
43
44   # For plotting, sort and add labels
45   res = res.sort_values('Quarter_Num')
46   res['Quarter'] = ['Q42010','Q12011',
47                     'Q22011','Q32011',
48                     'Q42011','Q12012']
49
50   # Plot the estimates as connected lines with error bars
51
52   plt.pyplot.errorbar(x = 'Quarter', y = 'parameter',
53                                   yerr = 'ci', data = res)
54   # Add a horizontal line at 0
55   plt.pyplot.axhline(0, linestyle = 'dashed')
```

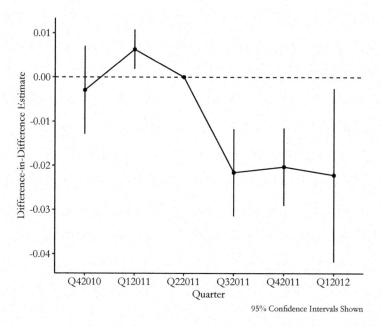

Figure 18.6: The Dynamic Effect of Active-Choice Phrasing on Organ Donation Rates

95% Confidence Intervals Shown

From Figure 18.6 we can see effects near zero in the three pre-treatment periods—always good, although the confidence interval for the first quarter of 2011 is above zero. That's not ideal, but as I mentioned, a single dynamic effect behaving badly isn't a reason to throw out the whole model or anything, especially when the deviation is fairly small in its actual value. We also see three similarly negative effects for the three periods after treatment goes into effect. The impact appears to be immediate and consistent, at least within the time window we're looking at.

18.2.5 Rollout Designs and Multiple Treatment Periods

AND NOW WE COME TO THE SECRET SHAME OF ECONOMETRICS,[32] which concerns the issue of *rollout designs.* It's one thing to have a difference-in-difference setup with multiple groups in the "treated" category. That's fine. A rollout design is when you have all that, but also *the groups get treated at different times.*

[32] The most recent one, at least.

For example, say we wanted to know the impact of having access to high-speed Internet on the formation of new businesses. We know that King County got broadband in 2001, Pierce County got it in 2002, and Snohomish County got it in 2003. They each have a before and after period, but those treatment times aren't all the same.

What's the problem? Well, from a *research design* perspective, there's no problem. You're just tossing a bunch of valid difference-in-difference designs together. But from a *statistical* perspective, this makes our two-way fixed effects regression not work any more.[33] The problem can actually be so bad that, in some rare cases, you can get a negative difference-in-differences estimate *even if the true effect is positive for everyone in the sample.* And thus the secret shame: for *decades* researchers were basically unaware of this problem and used two-way fixed effects anyway. Only very recently is the tide beginning to turn on this.

[33] Andrew Goodman-Bacon. Difference-in-differences with variation in treatment timing. Technical report, National Bureau of Economic Research, 2018.

Why doesn't two-way fixed effects work when we have multiple treatment periods? It's a little complex, but the real problem occurs because this setup *leads already-treated groups to get used as an untreated group.* Think about what fixed effects *does*—it makes us look at variation within group.[34] And in a sense, "no treatment last period, no treatment this period" is the same amount of within-group variation in treatment as "treatment last period, treatment this period." No change either way. So groups that stick with "still treated" get used as comparisons just as groups that stick with "not treated" do.

[34] As per Chapter 16.

And why is *that* a problem? Sure, it's a little weird to use a continuously-treated group as your comparison, but why wouldn't parallel trends hold there anyway? It might, but also, if *the effect itself is dynamic*, as I discussed in the previous section, or if *the treatment effect varies across groups*, as in Chapter 10, then you've set your estimation up in a way so that parallel trends won't hold! If you have a treatment effect that gets stronger over time, for example, then the "treated comparison group" should be trending upwards over time in a way that the "just-now treated group" shouldn't. Parallel trends breaks and the identification fails.[35]

If you are using a balanced panel (each group is observed in every time period) and you're not using control variables, you can check how much of a problem the use of two-way fixed effects is in a difference-in-differences given study using the Goodman-Bacon decomposition, as described in Goodman-Bacon (2018). The decomposition shows how much weight the just-treated vs. already-treated comparisons are getting relative to the cleaner treated vs. untreated comparisons. The decomposition can be performed in R or Stata using their respective **bacondecomp** packages.

What to do, then, when we have a nice rollout design? Don't use two-way fixed effects, but also don't despair. You're not out of luck, you're just moving into the realm of what the pros do.

[35] The problems can get even stranger when it comes to things like the dynamic treatment effects estimator discussed in the last section. The effects in the different periods start "contaminating" each other. This is described in Sun and Abraham (2020).

Liyang Sun and Sarah Abraham. Estimating dynamic treatment effects in event studies with heterogeneous treatment effects. *Journal of Econometrics*, 2020. Forthcoming.

18.3   How the Pros Do It

18.3.1   Doing Multiple Treatment Periods Right!

AN AREA OF ACTIVE RESEARCH PRESENTS TO THE TEXTBOOK AUTHOR both the purest terror and the sweetest relief. Whatever I say will almost certainly be outdated by the time you read it. But the inevitability of failure, dang, that's some real freedom.

When it comes to ways of handling multiple treatment periods in difference-in-differences, where some groups are treated at different times than others (rollout designs), "active area of research" is right! Because concern over the failure of the two-way fixed effects model for rollout designs is relatively recent, at least on an academic time scale, the approaches to solving the problem are fairly new, and it's not yet clear which will become popular, or which will be proven to have unforeseen errors in them.

I will show two ways of addressing this problem. First, I will show how our approach to dynamic treatment effects can help us fix the staggered rollout problem. Then I'll discuss the method described in Callaway and Sant'Anna (2020).[36] More technical details on all of these, as well as discussion of some additional fancy-new estimators, are in Baker, Larcker, and Wang (2021), which also discusses a third approach called "stacked regression."[37] But there is more coming out regularly about all of this. Hey, maybe even a version of the random-effects Mundlak estimator from Chapter 16 could fix the problem![38] So you'll probably want to check in on new developments in this area before getting too far with your staggered rollout study.

MODELS FOR DYNAMIC TREATMENT EFFECTS, modified for use with staggered rollout, can help in the case of staggered difference-in-differences in a few ways.

First, they separate out the time periods when the effects take place. Since our whole problem is overlapping effects in different time periods, this gives us a chance to separate things out and fix our problem.

Second, they're just plain a good idea when it comes to difference-in-differences with multiple time periods. As described in the Long-Term Effects section, we can check the plausibility of prior trends, and also see how the effect changes over time (and most effects do).

Third, because they *do* let us see how the treatment effect evolves, and because treatment effects evolving is one of the problems with two-way fixed effects, that gives us another opportunity to separate things out and fix them.

What do I mean by "modified for use with staggered rollout," then? A few things, all described in Sun and Abraham (2020).[39]

First, we need to center each group relative to its own treatment period, instead of "calendar time." This helps make sure that already-treated groups don't get counted as comparisons. But as pointed out in the Long-Term Effects section, dynamic treatment effects can still get in the way here. So Sun and Abraham (2020) go further. They don't just use a set of time-centered-on-treatment-time dummies, they *interact those dummies* with group membership. This really allows you to avoid making any comparisons you don't want to make, since now your regression model is barely comparing anything. It's giving each group and time period its own coefficient. The comparisons are then up to you, after the fact. You can average those coefficients together in a way that gives you a time-varying treatment effect.[40]

[36] Brantly Callaway and Pedro HC Sant'Anna. Difference-in-differences with multiple time periods. *Journal of Econometrics*, 2020. Forthcoming.

[37] Andrew C. Baker, David F. Larcker, and Charles C. Y. Wang. How much should we trust staggered difference-in-differences estimates? Technical report, Social Science Research Network, 2021.

[38] Jeffrey M Wooldridge. Two-way fixed effects, the two-way Mundlak regression, and difference-in-differences estimators. Technical report, SSRN, 2021.

[39] Liyang Sun and Sarah Abraham. Estimating dynamic treatment effects in event studies with heterogeneous treatment effects. *Journal of Econometrics*, 2020. Forthcoming.

[40] Weighting by group size is standard here. See the paper, or its simpler description in Baker, Larcker, and Wang (2021).

The Sun and Abraham estimator can be estimated in R using the sunab function in the **fixest** package, or in Stata using the **eventstudyinteract** package.

THE FIRST THING CALLAWAY AND SANT'ANNA DO is focus closely on *when each group was treated*. What's one way to deal with all those different treatment periods giving your estimation a hard time? Consider them separately! They consider each treatment period just a little different, and instead of estimating an average treatment-on-the-treated for the whole sample, they estimate "group-time treatment effects," which are average treatment effects on the group treated *in a particular time period*, so you end up with a bunch of different effect estimates, one for each time period where the treatment was new to *someone*.

Now dealing with the treated groups separately by when they were treated, they compare Y between each treatment group and the untreated group, and use propensity score matching (as in Chapter 14) to improve their estimate. So each group-time treatment effect is based on comparing *the post-treatment outcomes of the groups treated in that period* against *the never-treated groups that are most similar to those treated groups*.

Once you have all those group-time treatment effects, you can summarize them to answer a few different types of questions. You could carefully average them together to get a single average-treatment-on-the-treated.[41] You could compare the effects from earlier-treated groups against later-treated groups to estimate dynamic treatment effects. Plenty of options.

The Callaway and Sant'Anna method can be implemented using the R package **did**. In Stata you can use the **csdid** package.

[41] I say "carefully" because you'll want to do a weighted average with some specially-chosen weights, as described in the paper.

18.3.2   Picking an Untreated Group with Matching

DIFFERENCE-IN-DIFFERENCES ONLY WORKS IF THE COMPARISON GROUP IS GOOD. You really need parallel trends to hold. And since you can't check parallel trends directly, you need to pick an untreated group (or a set of untreated groups) good enough that the assumption is as plausible as it can be. So however you're doing your difference-in-differences, you want to be sure that you can really *justify why your untreated group makes sense and parallel trends should hold*.

That said, what do you do when you have a bunch of potential untreated groups? You can choose between them (or aggregate them together) by *matching* untreated and treated groups, as Callaway and Sant'Anna did in the previous section.

The idea here is pretty straightforward. Pick a set of predictor variables X from the pre-treatment period,[42] and then use one of the matching methods from Chapter 14 to match each treated group with an untreated group, or produce a set of weights for the untreated groups based on how similar they are to the treated groups.

Then, run your difference-in-difference model as normal, with the matching groups/weights applied. Done! Make sure to adjust your standard errors for the uncertainty introduced by the matching process, perhaps using bootstrapped standard errors, as discussed in Chapter 14.

[42] Definitely don't use any data from the post-treatment period. Remember, matching is one way of closing a path, and picking data from after the treatment goes into place is a great way to close a front door we don't want to close.

You can go even further and look into the *synthetic control* method. In synthetic control, you match your treated group to a bunch of untreated groups based not just on prior covariates but also *prior outcomes*. If the synthetic control matching goes well, then prior trends are just about forced to be the same because you've specifically chosen weights for your untreated groups that have the same average outcomes as your treated group in each prior period. Why don't we just call this a form of difference-in-differences with matching? There are some differences, like the way the result is calculated and how the matches are made, and the fact that it's designed to work with only a single treated group. In any case, synthetic control will be discussed further in Chapter 21.

THIS COMBINED MATCHING/DIFFERENCE-IN-DIFFERENCES APPROACH satisfies a few nice goals. First, we have difference-in-differences. Since DID has group fixed effects, it already controls for any differences between treated and untreated groups that is *constant over time*. However, what DID *doesn't* do is say anything about *why certain groups come to be treated and others don't*.

If there's some back door between "becomes a treated group" and "evolution of the outcome in the post-treatment period," as it seems likely there would be, then we still aren't identified. That's where matching comes in. If we can pick a set of matching variables X that close the back doors between *which groups become treated and when* and the outcome, we get parallel trends back.

If you like, you can go even further and use doubly robust difference-in-differences, which applies both matching and regression in ways that identifies the effect you want even if one of those two methods has faulty assumptions (although you're still in trouble if both are faulty).[43] You can apply this method in R with the **DRDID** package.

WHEN THINKING ABOUT MATCHING WITH DIFFERENCE-IN-DIFFERENCES GENERALLY, THERE IS ONE THING to be concerned about, and that is *regression to the mean*. Regression to the mean is a common problem whenever you are looking at data that varies over time. The basic idea is this: if a variable is far above its typical average *this* period, then it's likely to go down *next* period, i.e., regress back towards the mean.[44] This is because a far-above-average observation is, well, far above average. In a random period, you're likely to be closer to the mean than far-above-average. So typically an extreme observation (either above or below) will be followed by something closer to the mean.

Another way to think about it is this: the day you win the ten-million-dollar lottery is likely to be followed by a day where you make way less than ten million dollars.

What does this have to do with difference-in-differences and matching? The problem arises if the *pre-period outcome levels* are related to the probability of treatment. For example, say there are two cities with similar covariates. Policymakers are planning to put a job training program in place and want to know the effects of the program on unemployment. They choose City A for the program since unemployment is currently really bad in City A. Then we use A as the treated group and B as the untreated group, since B has similar covariates.

What happens then? After the policy goes into effect, unemployment might get better in City A for two reasons: the effect of the policy, *or* regression to the mean, if A was just having an unusually bad period when policymakers were choosing where to put the training program. Difference-in-differences can't tell the two apart, so the estimate is wrong.[45]

Strangely, this is only a problem *because* we matched A and B. If we'd just used a bunch of untreated cities, or a random city from a set of potential comparisons, the bias wouldn't be there. That's because B was selected as a good match for an *unusually bad* time in A's history. Heck, maybe B's unemployment is usually *way* worse, and this is an unusually good time for them. The matching emphasizes comparisons that are especially subject to regression to the mean.[46]

That said, this applies specifically when *outcome levels* are strongly related to whether your group gets treated, and is worse

[43] Pedro HC Sant'Anna and Jun Zhao. Doubly robust difference-in-differences estimators. *Journal of Econometrics*, 219(1): 101–122, 2020.

[44] This has nothing to do with regression as in "running a regression" or OLS.

[45] Jamie R Daw and Laura A Hatfield. Matching and regression to the mean in difference-in-differences analysis. *Health Services Research*, 53(6):4138–4156, 2018.

[46] I've described this as though they're being matched on their pre-treatment *outcome* levels, which is something that some researchers do. But this also applies whenever you're matching on pre-treatment values of time-varying *covariates* for the same reason—those covariates are related to the prior outcomes.

the more different your treated and untreated group's typical outcome levels are. Treatment being related to outcome trends is less of an issue. This also isn't an issue if you're matching on covariates that don't change much over time. Plus, if this outcome-level-based-assignment thing *isn't* an issue, then matching can really help. So this isn't a reason to throw out matching in difference-in-differences, just a reason to think carefully about it.

18.3.3   The Unfurling Logic of Difference-in-Differences

AT ITS CORE, BEYOND ALL THE CALCULATION DETAILS, difference-in-differences is extremely simple. We see a difference— a gap in outcome levels—and see how that difference changes from before a policy change to afterwards. But then who says we have to be limited to looking at how a *gap in outcome levels* changes? We could look at how a gap in just about *anything* changes.

What do I mean? For example, maybe we want to see how a difference in a *relationship* changes from before to after. Let's consider a teacher training program that is introduced in some districts but not others. The goal of this training program is to help ease educational income disparities. That is, *the relationship between parental income and student test scores should be weaker with the introduction of the training program.*

So, what can we do? We already know how to get the effect of *Income* on *TestScores*:

$$TestScores = \beta_0 + \beta_1 Income + \varepsilon \qquad (18.6)$$

Now, we want to do difference-in-differences but on β_1 instead of on Y. In other words $(\beta_1^{Treated,After} - \beta_1^{Treated,Before}) - (\beta_1^{Untreated,After} - \beta_1^{Untreated,Before})$. That looks like difference-in-differences to me![47]

We can get at these sorts of effects with interaction terms. If we want to know the "within" variation in the effect, we interact *After* with what we have:

$$TestScores = \beta_0 + \beta_1 Income + \beta_2 After + \beta_3 Income \times After + \varepsilon \qquad (18.7)$$

Estimate that model for the treated group, and β_3 gives us

$$(\beta_1^{Treated,After} - \beta_1^{Treated,Before}) \qquad (18.8)$$

Estimate it for the untreated group and we get

[47] This logic is very similar to what we covered in the Oster paper way back in the Describing Relationships chapter, Chapter 4.

$$(\beta_1^{Untreated,After} - \beta_1^{Untreated,Before}) \qquad (18.9)$$

We can combine everything into one regression with a *triple-interaction* term:

$$TestScores = \beta_0 + \beta_1 Income + \beta_2 After + \beta_3 Income \times After +$$
$$\beta_4 Treated + \beta_5 Treated \times Income + \beta_6 Treated \times 2After +$$
$$\beta_7 Treated \times Income \times After + \varepsilon \quad (18.10)$$

Let's walk through the logic carefully here. β_7 looks at the effect of *Income*, and sees how that effect changes from before to after ($Income \times After$), and then sees how *that before-after change* is different between the treated and untreated groups ($Treated \times Income \times After$). This is difference-in-differences but on a relationship rather than the average of an outcome.

This approach works for basically any kind of research design. We've done it here for a basic correlation, but with the right setup you could see how an instrumental variables (Chapter 19) or regression discontinuity (Chapter 20) estimate changes from before to after. As long as you want to know the effect of some policy *on the strength of some other effect*, you can do what you like.

Aside from applying DID to effects themselves (relationships), you can also apply difference-in-differences logic to other kinds of summary descriptions of a single variable rather than the mean (like DID would do). One application of this is in using DID with quantile regression—a form of regression that looks at how predictors affect *the entire distribution* of a variable. This lets you use a DID research design to see how a policy affects, say, values at the tenth percentile.

WE CAN EVEN APPLY THIS TO DIFFERENCE-IN-DIFFERENCES ITSELF and get the difference-in-difference-in-differences model, also known as triple-differences or DIDID![48] DIDID could be used to see how a newly implemented policy changes a DID-estimated effect. However, DIDID is also used to help strengthen the parallel trends assumption by finding a treated group that *shouldn't* be affected at all, and subtracting out their effect. This works using the same triple-interaction regression we just discussed.

For example, imagine a new environmental policy that increases funding to remove trash from marshes. The policy is implemented in some prefectures but not others. You want to run DID to see if the policy actually reduced trash levels. But

[48] Really! You could even do DIDIDID or DIDIDIDID or DIDLDIDI or...

what if you ran DID on *parks* instead of marshes and found an effect? That tells you there's something going on that breaks parallel trends—trash levels in the treated and untreated groups are trending apart even in places that shouldn't be affected by the policy.[49] What can we do? Well, just run DID on marshes, and also run DID on parks, and take the difference of them (difference-in-DID or DIDID). This hopefully subtracts out the violation of parallel trends we saw, leaving us with a better estimate of the effect on marshes.

To see how this might work in action, take Collins and Urban (2014).[50] This paper looks at a policy enacted in the state of Maryland in 2008 intended to help reduce home foreclosure rates. When someone misses payments on their mortgage, the loan servicer *could* foreclose on them, *or* they could do nothing and just wait to see if they start paying again, *or* they could help the mortgage holder modify their loan so they might be able to get back on track with payment. All the policy did was require mortgage servicers to report what sorts of loan-modification activities they'd done to help people. Collins and Urban wanted to know whether the policy got loan servicers to modify more loans and reduce foreclosures.

They *could* have just run DID. Simply compare loan modifications and foreclosures in Maryland and some untreated state before and after the policy. However, it might be tricky to believe that parallel trends would hold. After all, 2008 was heading right into the Great Recession, and all sorts of things were changing all over the place, *especially* in regards to loan delinquencies and foreclosures. Maryland and whatever untreated state got picked might trend apart for no reason other than pure unbridled financial chaos.

However, Collins and Urban were in luck: the policy only applied to *some* loan servicers but not others. So they added that third difference in: see how different the DID effect is between servicers subject to the new policy and those that weren't. After all, the servicers *not* subject to the policy shouldn't be affected at all, so any DID effect we *do* see is more on the "financial chaos" side of things rather than the "effect of the policy" side, and we can subtract that out.

What did they find? They found that the policy *did* get loan servicers to modify more loans to make them easier to pay, even though the underlying financial reality of the loan stayed exactly the same. However, they *also* found that the policy led to more foreclosures, contrary to the intent of the policy.[51] These results

[49] Okay, sure, maybe they *should* be affected if there are budget spillovers. But let's ignore that for now.

[50] J Michael Collins and Carly Urban. The dark side of sunshine: Regulatory oversight and status quo bias. *Journal of Economic Behavior & Organization*, 107:470–486, 2014. DOI: 10.1016/j.jebo.2014.04.003.

[51] How could this policy possibly lead to more foreclosures? Collins and Urban suggest that this may be because it encourages loan servicers away from a "don't do anything, wait and see" approach to the loan. Modifying a loan is one way to "do something," but, uh... foreclosing on the house is another.

can be seen in Figure 18.7—on the left, the number of loans that get modified is the dependent variable, and on the right, it's the foreclosure rate. Each point is the difference between Maryland and other states (difference), and the different lines represent the ESRR loan servicers affected and the not-ESRR servicers not affected by the policy (in difference). You can see how the gap grows by quite a bit after the policy goes into effect (in difference).

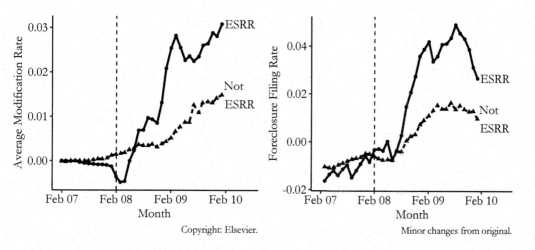

Copyright: Elsevier.

Minor changes from original.

Figure 18.7: Differences Between Maryland and Other States from Collins and Urban (2014)

19
Instrumental Variables

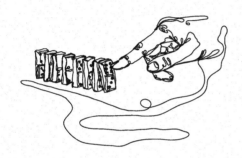

19.1 How Does It Work?

19.1.1 Isolating Variation

IF YOU WANT TO BUILD A STATUE, ONE WAY TO GO ABOUT IT
IS TO GET A BIG BLOCK OF MARBLE and chip away at it until
you reveal the desired form underneath. Another way is to take
something like concrete or molten steel, and pour it into a mold
that only allows the desired form to show through. Research
designs based on closing back doors and removing undesirable
variation are sort of like the former. Instrumental variables designs are like the latter.

Instrumental variables are more like using a mold because instead of trying to strip away all the undesirable variation using
controls, it finds a source of variation that allows you to isolate
just the front-door path you are interested in. This is similar
to how a randomized controlled experiment works. In a randomized experiment, you have control over who gets treatment
and who doesn't. You can choose to assign that treatment in
a random way. When you do so, the variation in treatment

DOI: 10.1201/9781003226055-19

among participants in your experiment has no back doors. By using *randomized assignment to treatment* instead of *real-world treatment,* you isolate *only* the covariation between treatment and outcome that's due to the treatment, and completely purge the variation that's due to those nasty back doors.

Instrumental variables designs seize directly on the concept of randomized controlled experiments, and are in effect an attempt to mimic a randomized experiment, but using statistics and opportune settings instead of actually being able to influence or randomize anything.[1]

IN A TYPICAL SETTING, if we have a *Treatment* and an *Outcome*, we want to know the effect of treatment on outcome, *Treatment → Outcome.* However, there is, to say the least, an *annoying* back door path. There could be a lot going on here but let's just say we also have *Treatment ← Annoyance → Outcome.* Further, because in the social sciences *Annoyance* represents *a bunch* of different things, it's unlikely we can control for it all. A randomized experiment shakes up this system by adding a new source of variation, *Randomization,* that's completely unrelated to *Annoyance.*[2] This is all shown in Figure 19.1.

[1] You will recognize a lot of the logic here from Chapter 9 on Finding Front Doors and isolating front-door paths.

[2] In all of this, we can refer back to the lingo from Chapter 13: *Treatment* is *endogenous* due to how it is determined by *Annoyance.* But *Randomization* is *exogenous,* or we might say it's "an exogenous source of variation," because it's not caused by anything else in the system.

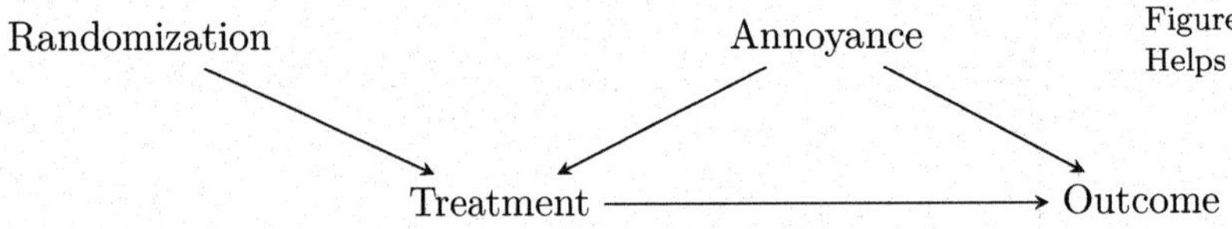

Randomization   Annoyance

Treatment ─────────────────→ Outcome

Figure 19.1: Randomization Helps Avoid Back Doors

The instrumental variables design works in the *exact same way.* The only difference is that, instead of randomizing the variable ourselves, we hope that something has already randomized it for us. *We look in the real world for a source of randomization that has no back doors,*[3] *and use that to mimic a randomized controlled experiment.*

An instrumental variables design does not remove the requirement to identify an effect by closing back doors. But it does *move the requirement,* hopefully to something easier! Instead of needing to close the back doors between *Treatment* and *Outcome,* which would require us to control for *Annoying,* we instead just need to close the back doors between *Randomization* and *Outcome* (if there are any), as well as any front door between *Randomization* and *Outcome* that doesn't pass through *Treatment.*

[3] Or at least has back doors that are easily closed by controlling.

THE MECHANICS FOR ACTUALLY IMPLEMENTING INSTRU-MENTAL VARIABLES are, in effect, a means of trying to do what an experiment does when you don't actually have perfect control of the situation—makes sense, as that's what is happening.

How can we mimic a randomized experiment? Well, in a randomized experiment, we generate some random variation and then just isolate the part of the treatment that is driven by that random variation—we only use the data from the experiment. So in instrumental variables we're going to isolate just the part of the treatment that is driven by the instrument, but *statistically*.

In other words, we're going to:

1. Use the instrument to explain the treatment

2. Remove any part of the treatment that is *not* explained by the instrument

3. Use the instrument to explain the outcome

4. Remove any part of the outcome that is *not* explained by the instrument[4]

5. Look at the relationship between the remaining, instrument-explained part of the outcome and the remaining, instrument-explained part of the treatment

This is, in effect, the *opposite* of controlling for a variable. When we get the effect of X on Y while controlling for W, we use W to explain X and Y and remove the explained parts, since we want to close any path that goes through W. But when we get the effect of X on Y using Z as an instrument, we use Z to explain X and Y and remove the *unexplained* parts, to give us *only* the paths that come from Z. The part that's explained by Z is the part with no back doors, which is what we want.

The whole story can be seen in Figure 19.2, which demonstrates what an instrumental variables design actually does to data to get its estimate. The figure uses a binary instrument—this isn't necessary for instrumental variables generally, but the figure is much easier to follow this way. We start with Figure 19.2 (a), graphing the raw data, which shows a slight positive relationship between X and Y. We can also see the different values of Z on different parts of the graph—the 0s to the bottom-right and the 1s to the top-left.

[4] Steps 3 and 4 here turn out to not actually make a difference for your estimate in the typical case—you end up getting the explained part of the outcome anyway if you use instrument-explained treatment to explain regular-ol' outcome. But I keep it in because I think it makes the concept clearer.

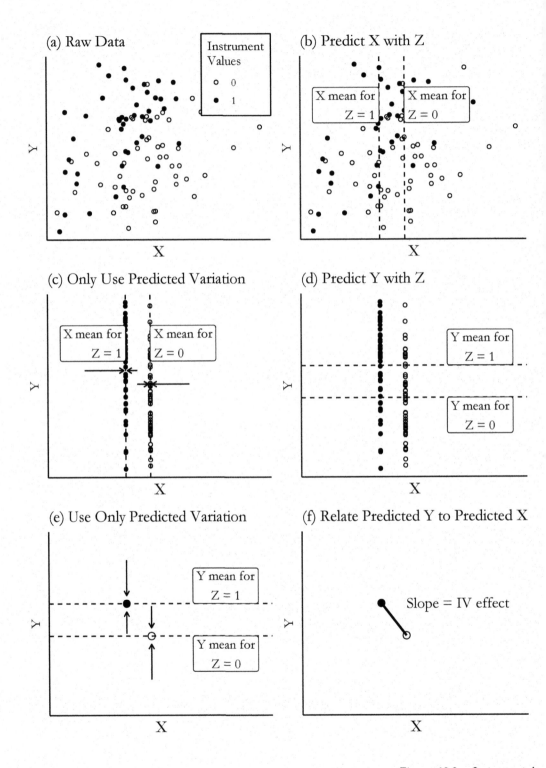

Figure 19.2:   Instrumental Variables with a Binary Instrument, Step by Step

We can move on to Figure 19.2 (b), which shows us taking the mean of X within the different Z values. Instrumental variables uses the part of X that is explained by Z. When Z is binary that's just the mean of X for each group. We only want to use the part of X that's explained by Z—that's the part without any back doors, so in Figure 19.2 (c) we get rid of all the variation in X that *isn't* explained by Z, leaving us with only those predicted (mean-within-Z) X values.

Then in Figure 19.2 Panels (d) and (e) we repeat the process with Y. We do this, again, to close the back doors between X and Y. That nice Z-driven variation is free of back doors, so we want to isolate just that part.

Finally, in Figure 19.2 (f), we look at the relationship between the predicted part of X and the predicted part of Y. Since Z is binary, each observation in the data gets only one of two predictions—the $Z = 0$ prediction for X and Y, or the $Z = 1$ prediction for X and Y. Only two predictions means only two dots left on the graph, and we can get the relationship between X and Y by just drawing a line between those two points. The slope of that line is the effect of X on Y, as estimated using Z as an instrumental variable, i.e., isolating only the part of the X/Y relationship that is predicted by Z. The slope we have here is negative, unlike the positive relationship from (a). And since I generated this data myself I can tell you that the true relationship between X and Y in this data is in fact negative, just like the slope in (f).

19.1.2   Assumptions for Instrumental Variables

FOR INSTRUMENTAL VARIABLES TO WORK, we must satisfy two assumptions: *relevance* of the instrument, and *validity* of the instrument.[5]

Relevance is fairly straightforward. The idea of instrumental variables is that we use the part of X, the treatment, that is explained by Z, the instrument. But what if no part of X is explained by Z? What if they're completely unrelated? In that case, instrumental variables doesn't work.

This follows pretty intuitively—we can't isolate the part of X explained by Z if there isn't a Z-explained part of X to isolate. Also, if we follow the steps described earlier in the chapter,[6] then instrumental variables in effect asks "for each Z-explained movement in X, how much Z-explained movement in Y was there?" In other words, in its basic form instrumental variables

[5] Rather, we must satisfy *at least* two assumptions. But these are the two with real star power that everyone talks about. Regular OLS assumptions also apply, as well as monotonicity— the assumption that the relationship between the IV and the endogenous variable is always of the same sign (or zero). We'll talk more about monotonicity in the How Is It Performed? section.

[6] And are using a standard linear one-treatment one-instrument version of IV.

gives us $Cov(Z,Y)/Cov(Z,X)$.[7] If Z doesn't explain X, then $Cov(Z,X) = 0$ and we get $Cov(Z,Y)/0$. And we can't divide by 0. The estimation simply wouldn't work.

Now in real-world settings, hardly any correlation is truly zero. But even if $Cov(Z,X)$ is just *small* rather than zero, we still have problems. If the covariance $Cov(Z,X)$ is small, we'd call Z a *weak instrument* for X. The estimate of $Cov(Z,X)$ is likely to jump around a bit from sample to sample—that's the nature of sampling variation. If $Cov(Z,X)$ is, for example, on average 1, and it varies a bit from sample to sample, we might get values from .95 to 1.05, changing our $Cov(Z,Y)/Cov(Z,X)$ instrumental variables estimate by maybe 10% across samples. But if $Cov(Z,X)$ follows the same range for a much lower baseline, maybe from .01 to .11, that changes our $Cov(Z,Y)/Cov(Z,X)$ estimate by as much as 1100% across samples. Big, swingy estimates! Big and swingy may be positive qualities in some areas of life but not in statistics.

So we need to be sure that Z actually relates to X. Thankfully, this is one of the few assumptions in this half of the book we can really confidently check. Simply look at the relationship between X and Z and see how strong it is. The stronger it is, the more confident you can be in the relevance assumption, and the less the estimate will jump around from sample to sample.[8]

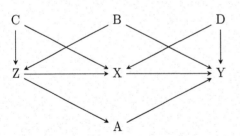

[7] Referring back to Figure 19.2 above, this is the slope you get after isolating the Z-explained parts of both variables. After all, slope is rise (how much Y increases for an increase in Z) over run (how much X increases for a given increase in Z), which is another way of thinking about a slope.

[8] How strong does it need to be? Good question. We'll talk about it in the next section.

Figure 19.3: A Diagram Amenable to Instrumental Variables Estimation... If We Pick Controls Carefully!

PERHAPS THE MORE FRAUGHT OF THE TWO ASSUMPTIONS is the assumption of *validity*. Validity is, in effect, the assumption that the instrument Z *is* a variable that has no open back doors of its own.[9]

Perhaps a bit more precisely, *any paths between the instrument Z and the outcome Y must either pass through the treatment X or be closed*. Remember, instrumental variables doesn't relieve us of the duty of closing all the back doors that are there, it just moves that responsibility from the treatment to the instrument, and hopefully the instrument is easier to close back doors for.

[9] Validity is sometimes also called an "exclusion restriction," because it is an assumption that Z can reasonably be *excluded* from the model of Y after the $Z \rightarrow X$ path is included.

Take Figure 19.3 for example. If we want to use Z as an instrumental variable to identify the effect of X on Y, what needs to be true for the validity assumption to hold?

There are a few paths from Z to Y we need to think about:

- $Z \rightarrow X \rightarrow Y$

- $Z \rightarrow X \leftarrow D \rightarrow Y$

- $Z \leftarrow C \rightarrow X \rightarrow Y$

- $Z \leftarrow C \rightarrow X \leftarrow D \rightarrow Y$

- $Z \leftarrow B \rightarrow Y$

- $Z \rightarrow A \rightarrow Y$

We want all open paths from Z to Y to contain X. The first four paths in the list contain X, so those are good to go. It's not even a problem that we have $Z \leftarrow C \rightarrow X$. If that path is left open, we can't identify $Z \rightarrow X$. But who cares? We don't have to identify $Z \rightarrow X$, we just need there to be no way to get from Z to Y except through X.[10,11]

How about D? That's the annoying source of endogeneity that we presumably couldn't control for that made $X \rightarrow Y$ unidentified and required us to use IV in the first place. But even though we have $Z \rightarrow X \leftarrow D \rightarrow Y$, that's fine. Once we isolate the variation in X driven by Z, any other arrow leading *to* X basically doesn't matter anymore. Those arrows might go to X, but they don't go to Z, and we're only using variation that starts with Z.

Putting the paths with X in them to the side, that leaves us with two: $Z \leftarrow B \rightarrow Y$ and $Z \rightarrow A \rightarrow Y$. These two paths *are* problems and must be shut down for Z to be valid. $Z \leftarrow B \rightarrow Y$ is an obvious one. That looks just like any normal back door we'd be concerned about.

But why is $Z \rightarrow A \rightarrow Y$ a problem, too? Because it gives Z another way to be related to Y other than through X. When we isolate the variation in X driven by Z, that variation will also be closely related to A.[12] So when we then look at the relationship between the variation in X driven by Z and Y, we'll mix together the effect of X and the effect of A.

So for the validity assumption to hold for Z, we need to close both the $Z \leftarrow B \rightarrow Y$ path and the $Z \rightarrow A \rightarrow Y$ path. If we can control for both A and B, then we're fine.[13]

[10] We probably *don't want* to control for C (although doing so wouldn't make Z invalid). The presence of C probably strengthens the *predictive* effect of Z on X, depending on whether those arrows are positive or negative. So controlling for C might make Z less *relevant*. That's a shocker—our identification of $X \rightarrow Y$ is actually *helped* in this instance by failing to identify $Z \rightarrow X$.

[11] C would also be a great instrument in its own right on this diagram.

[12] In fact, as long as we're talking about linear models, the variation in X driven by Z will be *perfectly correlated* with the variation in A driven by Z.

[13] Because we need to control for things to get validity to hold in this instance, we'd say that validity holds *conditional* on A and B.

WE NEED RELEVANCE AND VALIDITY. HOW REALISTIC IS VALIDITY, ANYWAY? We ideally want our instrument to behave just like randomization in an experiment. But in the real world, how likely is that to actually happen? Or, if it's an IV that requires control variables to be valid, how confident can we be that the controls really do everything we need them to?

In the long-ago times, researchers were happy to use instruments without thinking too hard about validity. If you go back to the 1970s or 1980s you can find people using things like parental education as an instrument for your own (surely your parents' education can't possibly affect your outcomes except through your own education!). It was the Wild West out there...

But these days, go to any seminar where an instrumental variables paper is presented and you'll hear no end of worries and arguments about whether the instrument is valid. And as time goes on, it seems like people have gotten more and more difficult to convince when it comes to validity.[14]

There's good reason to be concerned. Not only is it hard to justify that there exists a variable strongly related to treatment that somehow *isn't at all* related to all the sources of hard-to-control-for back doors that the treatment had in the first place, we also have plenty of history of instruments that we thought sounded pretty good that turned out not to work so well.

To pick two examples of many, we can first look at Acemoglu and Johnson (2007).[15] In this study, they want to understand the effect that a population's health has on economic growth in that country. We'd expect that a healthier population leads to better economic growth, right? They use the timing of when new medical technology (like vaccines for certain diseases) is introduced as an instrument for a country's health (specifically, mortality from certain diseases). They find that changes in health in a given year driven by changes in medical technology that year actually has a *negative* effect on a country's economic growth!

Bloom, Canning, and Fink (2014) have a different view on the instrument that Acemoglu and Johnson used.[16] Bloom, Canning, and Fink point out that the original study didn't account for the possibility that changes in health may have *long-term* effects on growth. Why might this matter? Because the healthier you *already are*, the less a new medical technology can improve your mortality rates even further. So the effectiveness of that medical technology is related to pre-existing health, which is related to pre-existing economic factors, which is related to

[14] This focus on validity is good, but sometimes comes at the expense of thinking about other IV considerations, like monotonicity (we'll get there) or even basic stuff like how good the data is.

[15] Daron Acemoglu and Simon Johnson. Disease and development: The effect of life expectancy on economic growth. *Journal of Political Economy*, 115(6):925–985, 2007.

[16] David E Bloom, David Canning, and Günther Fink. Disease and development revisited. *Journal of Political Economy*, 122(6):1355–1366, 2014.

current economic growth. A back door emerges. Bloom, Canning, and Fink find that when they add a control for pre-existing factors, presumably fixing the validity issue, the negative effect that Acemoglu and Johnson found actually becomes positive.[17]

A particularly well-known example of a popular IV falling into disrepute is rainfall, which was a commonly-used instrument in studies of modern agricultural societies.[18] Rainfall levels have been used as an instrument for all sorts of things, like personal income or economic activity.[19] Less rain (or too much rain) can harm agricultural output. If everyone's a farmer, well... that's no good! Since farmers don't have any control over rainfall, it seems reasonable to treat it as exogenous and without any back doors.

To follow one example of a study that uses rainfall as an instrument, Miguel, Satyanath, and Sergenti (2004) use rainfall as an instrument for economic growth in different countries in Africa,[20] and then look at how economic growth affects civil conflict. They find a pretty big effect: a negative change in income growth of five percentage points raises the chances of civil war by 50% in the following year.

But is rainfall a valid instrument for income? Sarsons (2015) looks at data from India and finds that the relationship between rainfall and conflict is similar in areas irrigated by dam water (where rainfall doesn't affect income much) and in areas without dams (where rainfall matters for income a lot).[21] This implies that rainfall must be affecting conflict via some mechanism other than rainfall.[22] Some of her potential explanations include rainfall spurring migration, rainfall affecting dam-fed areas by spillovers from nearby non-dam-fed areas, or rainfall making it difficult to riot or protest.

IF VALIDITY IS SO DIFFICULT, WHY BOTHER? You can't use data to prove that your instrument is valid, and convincing others it's valid on the basis that you think your causal diagram is right is difficult. That said, that doesn't mean we need to give up on instrumental variables entirely. We just may need to be choosy in when we apply it.

First, there really are some situations where the instrument is as-good-as-random. You just have to pay attention and get lucky to find one. Some force happens to act to assign treatment that *truly comes from outside the system* or is applied *almost completely randomly*. A really good instrument usually takes one of two forms. Either it represents real randomization, like an actual randomized experiment, or something like

[17] On a personal level I basically never believe the validity assumption when the data is on the level of a *whole country*—closing all the necessary back doors seems impossible for something so big and interconnected! "Another macro IV paper?" I chuckle and roll my eyes, tutting at the world. The fools. Then they get thousands of cites and probably a Nobel or something while I sit alone at 3AM in the dark wondering if anyone would *really* notice if I submitted to a predatory pay-to-publish journal.

[18] These days, saying you're going to use rainfall as an instrument is a sort of almost-funny in-joke among a certain set of causal inference nerds. The joke is completely incomprehensible to all normal people, just like the things causal inference nerds say that aren't jokes.

[19] In general, the same instrument being used as an instrument for multiple different treatments should be a cause for suspicion. After all, this implies that your instrument causes multiple things that also cause your outcome. So to get rid of all paths that don't go through *your* treatment, you're going to have to control for all the *other* treatments your instrument affects. This is often not possible.

[20] Edward Miguel, Shanker Satyanath, and Ernest Sergenti. Economic shocks and civil conflict: An instrumental variables approach. *Journal of Political Economy*, 112(4):725–753, 2004.

[21] Heather Sarsons. Rainfall and conflict: A cautionary tale. *Journal of Development Economics*, 115:62–72, 2015.

"Mendelian randomization," which uses the random process of combining parental genes to produce a child as an instrument.[23] Alternately (and more common in the social sciences) a good instrument is probably one that you would never think to include in a model of the outcome variable, and in fact you may be surprised to find that it ever had anything to do with assigning treatment.

To pick one example of this kind of thing, I'll toot my own horn and mention the only study I've ever worked on using a validity assumption.[24,25] The paper was about the teacher hiring process in a school district that used a scoring system to evaluate applicants before hiring them. When we looked at the data, we realized that the evaluation scores on the different criteria were added up to produce an overall score by hand. Sometimes this adding-up was done incorrectly. Addition errors were a cause of making it to the next stage of the hiring process, and sound pretty random to me.[26]

Second, there are some forms of IV analysis that allow for a little bit of failure of the validity assumption. So as long as you can make the validity assumption *close* to being true, you can get something out of it. We'll talk about that in the How the Pros Do It section.

Third, there are certain settings where randomness is really more believable. For one, you can apply instrumental variables to *actual randomized experiments,* and in fact this is commonly done when not everybody does what you've randomized them to do ("imperfect compliance").[27]

Did you randomize some people to take a new medication and others not to, but some people skipped their medication doses anyway? If you just analyze the experiment as normal, you'll underestimate the effect of the medication, since you'll have people who never actually took it mixed in there. This would be the "intent-to-treat" estimate, the effect of *assigning* people to take medication, rather than the effect of the medication itself. But if you use the random assignment as an instrument for *taking* the medication (treatment), you get the effect of treatment. And that random assignment sounds pretty darn valid. After all, it really was random. No alternate paths!

Another place where IV pops up with a really believable validity assumption is in application to regression discontinuity designs. We'll get to those in Chapter 20. But for now, let's get a little deeper into how we can actually use IV ourselves, if we do happen to believe that validity assumption.

[22] Other authors have other reasons for doubting the validity of rainfall, this is only one of them. Sarsons lists a few other studies in her paper if you're interested.

[23] Even with something really random like Mendelian randomization, validity is not assured. If a given gene affects your treatment of interest but *also* some other variable that affects the outcome (perhaps the same gene affects height and weight, both of which affect athleticism), that's a back door!

[24] We ran a "selection model" in this paper rather than using instrumental variables, but this method similarly required a validity assumption.

[25] Dan Goldhaber, Cyrus Grout, and Nick Huntington-Klein. Screen twice, cut once: Assessing the predictive validity of applicant selection tools. *Education Finance and Policy*, 12(2):197–223, 2017.

[26] Of course, maybe we missed something. Maybe evaluators were more likely to make errors if they didn't like someone for some reason. It's super important to think about what drives your instrument, just as you would think about what drives your treatment in a regular non-IV setting.

[27] If you recall, we already talked about this back in Chapter 10 on treatment effects.

19.1.3   Canonical Designs

THE USE OF INSTRUMENTAL VARIABLES MEANS you can find causal inference using all sorts of clever designs, and in all sorts of places that seem helplessly chock-full of back doors. "Clever" is probably the most common adjective you'll hear for a good instrument.[28]

One downside of super-clever instruments, though, is that they're often highly context-dependent. But some instrumental variables designs seem to be both pretty plausible, apply in lots of different situations, and get used over and over again.[29] Some of these I've already discussed—charter school lotteries from Chapter 9 are an example. And of course the use of instruments in the case of random experiments with imperfect compliance. Another is—uh-oh—rainfall as an instrument for agricultural productivity. But others have weathered years of criticism a bit better than rainfall did.[30]

One good example of a canonical instrumental variables design is *judge assignment*.[31] In many court systems, when you are about to go on trial, the process of assigning a judge to your trial is more or less random. This is important because some judges are harsher, and others are less harsh. This means that *JudgeHarshness* can act as an instrument for *Punishment*. Simply estimate the harshness of each judge using their prior rulings. Then, the harshness of *your* judge is an instrument for your punishment. This can be used anywhere you have randomly-assigned judges and want to know the impact of harsher punishments (or even being judged guilty) on some later outcome.

Another is the use of the military draft as an instrument for military service.[32] Military drafts are usually done semi-randomly, with the order in which people are drafted assigned in a random way. Whether you are drafted or not is a combination of where you are in the order and how many people they want to draft. So where you fall in the random draft order is a source of randomization for military service. It can also be a source of randomization for other things, like going to college, that may let you avoid the draft.[33]

Perhaps the most commonly used instrumental variables design ever is compulsory schooling as an instrument for years of education.[34,35] There are a zillion things that go into your decision of how long to continue in education and when to stop. But in many places, there is a compulsory schooling age—you can't

[28] Although whether that word implies excitement about the design or an indictment of the whole idea of instruments depends on who is talking.

[29] These are generally *pairs* of instruments and treatments that get reused. If you had one instrument used over and over again for different treatments, that would imply the instrument wasn't very good since for any one treatment, you're likely to have some open front doors from instrument to outcome that don't go through treatment—they go through *other* treatments too.

[30] You may notice that most of these surviving designs have some element of explicit randomization to them; take that as you will.

[31] Anna Aizer and Joseph J Doyle Jr. Juvenile incarceration, human capital, and future crime: Evidence from randomly assigned judges. *The Quarterly Journal of Economics*, 130(2):759–803, 2015.

[32] Joshua D. Angrist. Lifetime earnings and the Vietnam era draft lottery: Evidence from Social Security administrative records. *The American Economic Review*, pages 313–336, 1990.

[33] Uh-oh. An instrument for two treatments! This one is a bit less of a concern than rainfall though, since these activities are sort of alternatives to each other. An instrument for *any* treatment would also be an instrument for not-getting-treatment. You could instead think of draft order as an instrument for "what you do after high school."

legally drop out until a certain age. So if you'd like to drop out at fifteen, but the law keeps you in until sixteen, that's a little nudge of extra education for you. Variation between regions in these laws over time can act as an instrument for how much education you get.

Lastly, there's the "Bartik Shift-Share IV."[36] In this design, shared economy-wide trends are combined with the distribution of industries across different regions as an instrument for economic activity. In other words, if you live in Ice Cream Town, which makes a lot of ice cream, and the next town over is Popcorn, Town which makes a lot of popcorn, and over the past ten years ice cream has gotten way more popular on a national level, that's going to be a fairly random boon for your town. Easy! Recent work looks more closely at this design and finds a few changes in how we interpret it,[37] but the design lives on.

There are plenty of other canonical designs, of course. Settling locations of an original wave of immigrants as an instrument for further immigration, the birth of twins as an instrument for the number of children, housing supply elasticity as an instrument for housing prices, natural disasters as instruments for all sorts of things... I could go on! In the course of the chapter, I'll talk about three other canonical designs: the direction of the wind as an instrument for pollution, whether your first two children are the same sex as an instrument for having a third, and "Mendelian randomization," which uses someone's genetic code as an instrument for all sorts of things.

19.2   How Is It Performed?

19.2.1   Instrumental Variables Estimators

WE CAN START WITH OUR APPROACH TO ESTIMATING INSTRU-MENTAL VARIABLES BY TAKING IT PRETTY LITERALLY. Instrumental variables as a research design is all about isolating the variation in the treatment that is explained by the instrument. So let's just, uh, do that.

Two-stage least squares, or 2SLS,[38] is a method that uses two regressions to estimate an instrumental variables model. The "first stage" uses the instrument (and other controls) to predict the treatment variable. Then, you take the predicted (explained) values of the treatment variable from that first stage, and use that to predict the outcome in the second stage (along with the controls again).

[34] Joshua D. Angrist and Alan B. Keueger. Does compulsory school attendance affect schooling and earnings? *The Quarterly Journal of Economics*, 106(4): 979–1014, 1991.

[35] Boy that Angrist guy gets around, huh?

[36] Timothy J Bartik. *Who Benefits from State and Local Economic Development Policies?* WE Upjohn Institute for Employment Research, 1991.

[37] Paul Goldsmith-Pinkham, Isaac Sorkin, and Henry Swift. Bartik instruments: What, when, why, and how. *American Economic Review*, 110(8):2586–2624, 2020.

[38] Or TSLS, but that one just looks wrong if you ask me.

Given our instrument Z, treatment X, outcome Y, and controls W, we would estimate the models:

$$X = \gamma_0 + \gamma_1 Z + \gamma_2 W + \nu \qquad (19.1)$$

$$Y = \beta_0 + \beta_1 \hat{X} + \beta_2 W + \varepsilon \qquad (19.2)$$

where ν and ε are both error terms, $\hat{X}$ are the predicted values of X, predicted using an OLS estimation of the first equation, and γ are regression coefficients just like β, only given a different Greek letter to avoid confusing them with the βs.[39]

The procedure is quite easy to do by hand (although I wouldn't recommend it, for reasons I'll get to in a moment). Simply run OLS of X on Z (lm in R, regress in Stata, sm.ols().fit() in Python with **statsmodels**). Then, predict X using the results of that regression (predict() in R, predict in Stata, sm.ols().fit().predict() in Python). Finally, do a regression of Y on the predicted values. Don't forget to include any controls in both the first and second stages.

We're not entirely done yet—if we simply do this procedure as I've described it, our standard errors will be wrong. So there's a standard error adjustment to be done, changing them to account for the fact that we've *estimated* $\hat{X}$ rather than *measuring* it, and therefore there's more uncertainty in those values than OLS will pick up on.[40]

But with that under our belt, we have created our own instrumental variables estimate!

What is it doing, precisely, anyway? 2SLS produces a *ratio* of effects, dividing the effect of Z on Y by the effect of Z on X. It asks "for each movement in X we get by changing Z, how much movement in Y does that lead to?" The answer to this question, since Z has no back doors, should give us the causal effect of X on Y.

2SLS has some nice features—it's easy to estimate, it's flexible (adding more instruments to the first stage is super easy, although adding more treatment variables is less easy), and since it really just uses OLS it's easy to understand. These are the reasons why 2SLS is by far the most common way of implementing instrumental variables.

2SLS has some downsides too. It doesn't perform that well in small samples, for one. While the instrument *in theory* has no back doors, in an actual data set the relationship between Z and the non-X parts of Y is going to be at least a *little* nonzero,

[39] If the goal is to close the back doors associated with the parts of X not explained by Z, why don't we take the *residuals* instead of the predicted values, and then control for them in the second stage, alongside regular ol' X? Well... you could! This is called the *control function* approach. In standard linear IV this produces basically the same results as 2SLS. But it has some important applications for nonlinear IV, which I'll get to later in the chapter.

[40] For this reason, you generally will want to use a software command specifically designed for IV to run IV.

just by random chance. The smaller the sample is, the more often this "nonzero by random chance" is going to be not just nonzero but fairly large, driving Z to not be quite valid in a given sample and giving you bias. Additionally, 2SLS doesn't perform particularly well when the errors are heteroskedastic.[41]

WHILE TWO-STAGE LEAST SQUARES IS THE MOST LITERAL WAY OF THINKING ABOUT INSTRUMENTAL VARIABLES, it is only one estimator of many. And there's a good case to be made that it isn't even that great a pick among the different estimators, despite its popularity.

Here, I'll talk a bit about the generalized method of moments (GMM) approach to estimating instrumental variables. Some other methods, each with their own strengths, will pop up throughout the chapter.

GMM is an approach to estimation that's much broader than instrumental variables, but in this chapter at least we're just using it for IV.[42] The basic idea is this: based on your assumptions and theory, construct some statistical *moments* (means, variances, covariances, etc.) that *should* have certain values.

For example, if we want to use GMM to estimate the expected value (mean, basically) of a variable Y, we'd say that the difference between our sample estimation of the expected value μ and the population expected value $E(Y)$ should be zero. So we'd make $\mu - E(Y) = 0$ a condition of our estimation. Replace $E(Y)$ with its sample value $\frac{1}{N} \sum Y$ and solve the equation to get $\hat{\mu} = \frac{1}{N} \sum Y$. GMM will pick the μ that makes the moment condition true on average. Or for OLS, we assume that X is unrelated to the error term ε. So we use the condition that $Cov(X, \varepsilon) = 0$. In the actual data that works out to $\sum Xr = 0$, where r is the residual. Do a little math and you end up back at the same estimate for β_1 that we already had.[43,44]

You might be able to guess how GMM works with IV from the OLS example. We assume that Z is unrelated to the second-stage error term ε. So we use the condition $Cov(Z, \varepsilon) = 0$. Substitute in the sample-data version of the covariance and do a little math to end up with $\beta_1 = \sum(ZY)/\sum(ZX)$, which is what we had before.

So, same answer. Big whoop, right? Except that things get a bit more interesting when we either *have heteroskedasticity* (which we probably do) or *add more instruments*.

GMM and 2SLS are both capable of handling heteroskedasticity. GMM does so naturally, as it doesn't really make assumptions about the error terms in the same way that the OLS

[41] Heteroskedasticity-robust standard errors only do an okay job fixing this.

[42] Many good textbooks out there go into more detail. A lot of them are titled, surprisingly enough, "Generalized Method of Moments" with a few extra words tacked on.

[43] How?   Let's simplify by assuming everything is mean-zero so we don't need an intercept.   Start with $\sum Xr = 0$. Then plug in $r = Y - \beta_1 X$ to get $\sum X(Y - \beta_1 X) = \sum(XY - \beta_1 X^2) = 0$. Solve for β_1 to get $\beta_1 = \sum(XY)/\sum(X^2)$, which is what we had before.

[44] In the case of the mean and OLS, the *solution* it comes to is the same as we've already gotten—take the mean of the sample data. But the way we got there is different.

Overidentified.   An IV model with more instrumental variables than treatment/endogenous variables.

Just identified or **exactly identified.**   An IV model with the same number of instrumental variables as treatment/endogenous variables.

equations making up 2SLS do. But we can get there with 2SLS by simply using heteroskedasticity-robust standard errors.

But things get more interesting when the number of instruments is bigger than the number of treatment/endogenous variables you have, which is called being "overidentified."

When your model is overidentified, 2SLS and GMM diverge.[45] GMM is going to be more precise, at least if there's heteroskedasticity involved. GMM will have less sampling variation (and thus smaller standard errors) under heteroskedasticity than 2SLS, even if you add heteroskedasticity-robust standard errors. This continues to be true if you start applying clustered standard errors or corrections for autocorrelation.

Keep in mind—GMM isn't just a different way of adjusting the standard errors. The estimates themselves will actually be different when there's overidentification. The GMM standard errors are smaller not because we choose to claim we have more information and thus more precision, but because the method itself produces more precise estimates, and the smaller standard errors reflect that.

LET'S CODE UP SOME INSTRUMENTAL VARIABLES. For this exercise we'll be following along with the paper "Social Networks and the Decision to Insure" by Cai, De Janvry, and Sadoulet (2015).[46] The authors are looking into the decision that farmers make about whether to buy insurance against weather events. In particular, they're interested in whether information about insurance travels through social networks.

They look at a randomized experiment in rural China, where households were randomized into two rounds of different informational sessions about insurance. The question then is: how much does *what your friends learn about insurance* affect your own takeup of insurance? They look at people in the second round of sessions, and they look at what their friends did and saw in the first round of sessions, to see how the former is affected by the latter.[47]

Cai, De Janvry, and Sadoulet do in general find that farmers' decisions were affected by what their friends saw and the information they received. I'll be looking at a particular one of their analyses where they ask whether farmers' decisions were affected by what their friends *did*: does your friends *actually buying* insurance make you more likely to buy?

We want to identify the effect of *FriendsPurchaseBehavior* on *YourPurchaseBehavior* among people in the second-round informational sessions, looking at the average purchasing

[45] In overidentified cases, GMM can't actually satisfy all of the moment conditions exactly, so it needs to pick weights for the conditions to decide which are more important. Roughly, it weights the conditions by how difficult they are to satisfy.

[46] Jing Cai, Alain De Janvry, and Elisabeth Sadoulet. Social networks and the decision to insure. *American Economic Journal: Applied Economics*, 7(2):81–108, 2015.

[47] By structuring things in this way where they look only at the effect of the first round on the second, they avoid "reflection," a common problem with studies about how your friends affect what you do; these studies are also known as studies of "peer effects." The reflection problem, attributed to perennial Nobel overlookee Charles Manski, points out the identification problem you face if you want to learn about how *your friends affect you...* the problem being that *you affect your friends too!* There's a feedback loop, and we know that a causal diagram hates a feedback loop. By having a first round and a second round, we know which direction the causal arrow points.

behavior of their friends who were in the first-round informational sessions. This effect has some obvious back doors. Preferences for insurance may be higher or lower by region, or you may simply be more likely to have friends with preferences similar to yours, including on topics like insurance.

As an instrument for $FriendsPurchaseBehavior$ they use the variable $FirstRoundDefault$, which is a binary indicator for whether your friends were randomly assigned to a "default buy" informational session, where attendees were assigned to buy insurance by default, and had to specify their preference not to buy it, or a "default no buy" session, where attendees were assigned to not buy insurance by default, and had to specify their preference to buy it. Everyone had the same options and got the same information, but the defaults were different. People follow defaults! Those in the "default buy" sessions were twelve percentage points more likely to buy insurance than those in the "default no buy" sessions. The fact that $FirstRoundDefault$ is randomized makes the argument that it's a valid instrument pretty believable. Plus, a twelve percentage point jump seems like plenty to satisfy the relevance assumption.[48]

Okay, now, *finally* let's code the thing. We'll start with 2SLS, and then, using the same data, will do GMM, LIML, and IV with fixed effects ("panel IV"). Wait, I slipped "LIML" in there—what's that? That's limited-information maximum likelihood. I'll show how to code it up here, since it's easy enough to switch out methods, but I'll actually talk about the method later in the chapter.

[48] We're going to use the analysis they did on the subsample of second-round participants *who were told what purchasing decisions their first-round friends had made*. They did another analysis with second-round participants who were not told, and found no effect.

```
1   # R CODE
2   # There are many ways to run 2SLS;
3   # the most common is ivreg from the AER package.
4   # But we'll use feols from fixest for speed and ease
5   # of fixed-effects additions later
6   library(tidyverse); library(modelsummary); library(fixest)
7   d <- causaldata::social_insure
8
9   # Include just the outcome and controls first, then endogenous ~ instrument
10  # in the second part, and for this study we cluster on address
11  m <- feols(takeup_survey ~ male + age + agpop + ricearea_2010 +
12                 literacy + intensive + risk_averse + disaster_prob +
13                 factor(village) | pre_takeup_rate ~ default,
14                 cluster = ~address, data = d)
15
16  # Show the first and second stage, omitting all
17  # the controls for ease of visibility
18  msummary(list('First Stage' = m$iv_first_stage,
19          'Second Stage' = m),
20          coef_map = c(default = 'First Round Default',
21          fit_pre_takeup_rate = 'Friends Purchase Behavior'),
22          stars = c('*' = .1, '**' = .05, '***' = .01))
```

```stata
1  * STATA CODE
2  causaldata social_insure.dta, use clear download
3
4  * We want village fixed effects, but that's currently a string
5  encode village, g(villid)
6
7  * The order doesn't matter, but we need controls here
8  * as well as (endogenous = instrument)
9  * don't forget to specify the estimator 2sls!
10 * and we cluster on address
11 * and also show the first stage with the first option
12 ivregress 2sls takeup_survey (pre_takeup_rate = default) male age agpop ///
13     ricearea_2010 literacy intensive risk_averse disaster_prob ///
14     i.villid, cluster(address) first
```

```python
1  # PYTHON CODE
2  import pandas as pd
3  from linearmodels.iv import IV2SLS
4  from causaldata import social_insure
5  d = social_insure.load_pandas().data
6
7  # Create an control-variable DataFrame
8  # including dummies for village
9  controls = pd.concat([d[['male','age','agpop','ricearea_2010',
10 'literacy','intensive','risk_averse','disaster_prob']],
11 pd.get_dummies(d[['village']])],
12 axis = 1)
13
14 # Create model and fit separately
15 # since we want to cluster, and will use
16 # m.notnull to see which observations to drop
17 m = IV2SLS(d['takeup_survey'],
18 controls,
19 d['pre_takeup_rate'],
20 d['default'])
21 second_stage = m.fit(cov_type = 'clustered',
22 clusters = d['address'][m.notnull])
23
24 # If we want the first stage we must do it ourselves!
25 first_stage = IV2SLS(d['pre_takeup_rate'],
26 pd.concat([controls,d['default']],axis = 1),None,
27 None).fit(cov_type = 'clustered',
28 clusters = d['address'][m.notnull])
29
30 first_stage
31 second_stage
```

This gives us the result in Table 19.1 (keeping in mind this table doesn't show a bunch of coefficients for all the control variables).[49] The first stage regression has the endogenous variable (whether your friends purchased insurance) as the dependent variable, and a coefficient for our instrument. The coefficient is .118 and statistically significant. It's showing that your friends being assigned to the "default-purchase" experimental condition leads to a 11.8 percentage point increase in the probability that they'll buy insurance. We predict whether your friends bought

[49] Copyright American Economic Association; reproduced with permission of the *American Economic Journal: Applied Economics*.

insurance using that .118 bump (as well as the other predictors not shown on the table) and use those predicted values in the second stage.

	First Stage	Second Stage
First Round Default	0.118***	
	(0.034)	
Friends Purchase Behavior		0.791***
		(0.273)
Num.Obs.	1378	1378
R2	0.469	0.127
R2 Adj.	0.448	0.092
Std. Errors	Clustered (address)	Clustered (address)

Table 19.1: Instrumental Variables Regression from Cai, de Janvry, and Sadoulet (2015)

* p < 0.1, ** p < 0.05, *** p < 0.01

Controls for gender, age, agricultural proportion, farming area, literacy, intensiveness of assigned treatment, risk aversion, perceived disaster probability, and village excluded from table.

The second stage has the actual outcome of *you* buying insurance as the outcome. A one-unit increase in the rate at which your friends buy insurance, using only the random variation driven by the random experimental assignment and the controls, increases your chances of buying insurance by .791. That's a pretty strong spillover effect!

That's 2SLS. How about GMM and LIML? In Stata and Python the transition is easy. In Stata, run `ivregress gmm` or `ivregress liml` instead of `ivregress 2sls`. Unfortunately, there are some important LIML parameters α and κ we'll discuss later that `ivregress liml` won't let you set on your own. In Python, you can use `IVGMM` or `IVLIML` instead of `IV2SLS`, and here you do get control over α and κ.

And in R? In R we unfortunately have to switch packages, at least for now. The `ivmodel()` function in the **ivmodel** package is capable of doing LIML, with options for setting α and κ. How about GMM? For the moment, you'll have to set up the whole two-equation GMM model yourself using the `gmm()` function in the **gmm** package, although your options for standard error adjustments are a bit more limited. The syntax for this example is:

```
# R CODE
library(modelsummary); library(gmm)
d <- causaldata::social_insure
# Remove all missing observations ourselves
d <- d %>%
    select(takeup_survey, male, age, agpop, ricearea_2010,
           literacy, intensive, risk_averse, disaster_prob,
```

```
8              village, address, pre_takeup_rate, default) %>%
9        na.omit()
10
11   m <- gmm(takeup_survey ~ male + age + agpop + ricearea_2010 +
12           literacy + intensive + risk_averse + disaster_prob +
13           factor(village) + pre_takeup_rate,
14           ~ male + age + agpop + ricearea_2010 +
15           literacy + intensive + risk_averse + disaster_prob +
16           factor(village) + default, data = d)
17
18   # We can apply the address clustering most easily in msummary
19   msummary(m, vcov = ~address, stars = c('*' = .1, '**' = .05, '***' = .01))
```

Finally, what if we have a lot of fixed effects in our IV model? There are some technical adjustments that must be made in these cases. This time it's R that has the easiest transition. The `feols()` function we already used can easily incorporate fixed effects—the +`factor(village)` just becomes | `village`. Stata and Python aren't too hard though. In Stata you switch from `ivregress` to `xtivreg`. This comes with a few other changes— you must `xtset` your data to tell Stata what the panel structure is, it only does 2SLS rather than GMM or LIML, and you must tell it whether you want fixed effects (`fe`), random effects (`re`), or something else (see the help file). At the moment there is no Python equivalent for this, although our typical Python syntax will let you estimate IV models with fixed effects in them by adding sets of binary indicator variables.

19.2.2   *Instrumental Variables and Treatment Effects*

WHAT DOES INSTRUMENTAL VARIABLES ESTIMATE?   We can refer back to Chapter 10 when thinking about what kind of treatment effect average instrumental variables produces.

In general, we know that IV is a method all about isolating the variation in treatment that is explained by the instruments. This means we are looking at a local average treatment effect, where the individual treatment effects are weighted by how responsive that individual observation is to the instrument.[50]

In the case of a standard estimator like 2SLS or GMM with one treatment/endogenous variable and one instrument, the weights are what the individual effect of the instrument would be for you in the first stage.

For example, say there has been a recent set of television advertisements that encourage people to exercise more. You want to use exposure to the advertisements as an instrument for how much you exercise, and then will look at the effect of exercise on blood pressure.

[50] We aren't necessarily destined for a LATE, and there are estimation methods that will produce other averages. This will get some brief coverage in the How the Pros Do It section.

Consider three people in the sample: Jakeila, Kyle, and Li. The advertisements would make Jakeila exercise an additional half hour each week, and an additional hour of exercise each week would lower her blood pressure by 6 points. You can see these values, as well as the values for Kyle and Li, on Table 19.2.

Name	Effect of Ads on Exercise Hours	Effect of Exercise Hours on Blood Pressure
Jakeila	.5	−2
Kyle	.25	−8
Li	0	−10

Table 19.2: Effect Sizes for Three People in our Exercise Study

Keep in mind—those effects of ads on exercise hours are what the effect *theoretically would be* for those people. Obviously we can't see Jakeila both advertised to and not advertised to at the same time. But this is saying that Jakeila-with-ads exercises half an hour more each week than Jakeila-without-ads.

What will 2SLS tell us the effect of exercise hours on blood pressure is? Well, Jakeila responds the strongest to the ads, so the −2 effect of exercise that *she* gets will be more heavily weighted. Specifically, it gets the .5 weight she has on the effect of ads. Similarly, Kyle gets a weight of .25. Li, on the other hand, doesn't respond to the ads at all—they make no difference to him. So it turns out he makes no difference to the 2SLS estimate. He gets a weight of 0.

So the estimated LATE is $(.5 \times (-2) + .25 \times (-8) + 0 \times (-10))/(.5 + .25 + 0) = (-1 + -2)/(.75) = -4$. That's what 2SLS will give us. This is in contrast to the average treatment effect which is $((-2) + (-8) + (-10))/3 = -6.67$.

This immediately points us to a very important result: *2SLS will give different results depending on which IV is used.* If we had picked an instrument that was really effective at getting Li to exercise but not so effective for Jakeila, then 2SLS would estimate a much stronger effect of exercise on blood pressure.

What if there's more than one instrument? The same logic applies—the stronger the effect of the instrument for you, the more strongly your treatment effect will be weighted. But at this point the math gets a bit more complex, because it's going to be a mix of the different weights you'd have had from the different instruments. So if one instrument would give Jakeila a .5 weight, but a different one would give her .8, the weight she'd get from using both instruments would be some mix of the .5 and .8 (and not necessarily just an average of the two).

One thing to emphasize here is that the specific weights we get *are* dependent on the estimator. Different ways of estimating instrumental variables can produce different weighted average treatment effects. If you're not using 2SLS, do look into what precisely you are getting. Although outside of some corner cases you are generally still getting something that weights you more strongly the more you are affected by the instrument. The idea that we're estimating a LATE does hold up in most cases, although the specifics on *which LATE we're getting* change from estimator to estimator.

THERE'S A COMMON TERMINOLOGY USED IN INSTRUMENTAL VARIABLES WHEN THINKING ABOUT THESE WEIGHTS, AND IN PARTICULAR THE 0S. We can divide the sample into three groups:

- **Compliers:** For compliers, the effect of the instrument on the treatment is in the expected direction. Jakeila and Kyle were compliers in Table 19.2 because the ad telling them to exercise more got them to exercise more.

- **Always-takers/never-takers:** Always-takers/never-takers are completely unaffected by the instrument. Li was an always-taker/never-taker in Table 19.2 because the effect of the ads on his exercise level was zero.[51]

- **Defiers:** Defiers are affected by the instrument in the *opposite* of the expected direction.[52]

From this terminology we can get one result and one assumption we need to make:

First, the result: if all of the compliers are affected by the instrument to the same degree, then 2SLS gives the average treatment effect... among compliers.[53] Neat! Although it is unlikely that everyone is affected in the same way by the instrument.

Second, the assumption: for all of this to work, we need to assume that there are *no defiers*. Imagine we add a fourth person to Table 19.2: Macy, who is so annoyed by the ads that she decides to exercise .25 hours *less* if she sees them. An hour of weekly exercise would reduce her blood pressure by 8 points.

How do the LATE weights work out now? Macy gets a weight of $-.25$. The weighted-average calculation now gives us $(.5 \times (-2) + .25 \times (-8) + 0 \times (-10) + -.25 \times (-4))/(.5 + .25 + 0 + -.25) = (-1 + -2 + 2)/(.5) = -2$. Exercise was actually *more* effective for Macy than the effect we already estimated, but adding her made the effect *shrink*! This is because she had a negative weight,

[51] The terminology here comes from cases where the treatment variable is binary. "Always-takers" always get the treatment, regardless of the instrument, and "never-takers" never get the treatment, regardless of the instrument. The terms are a little odd applied to continuous treatments like we have here in exercise, but it's the same idea.

[52] If there are people affected in both directions in the sample, it's a bit arbitrary to call one of them "compliers" and the other "defiers," but the important thing really is that there *are* people in both directions.

[53] You will find no shortage of people telling you that the LATE *is the same thing* as the average treatment effect among compliers. But that's not really true, and relies on this assumption that the instrument is equally effective on all compliers.

which makes the math of weighted averages get all wonky so they're not really weighted averages any more.[54]

The assumption that there are no defiers is also known as the *monotonicity assumption*. Along with validity and relevance, this is another key assumption we need to make for instrumental variables, although this one tends to receive a lot less attention.

So if you have an instrument that has an effect *on average*, think carefully about whether that effect is likely to be in the same direction for everyone. There are plenty of cases where it wouldn't be. For example, if there are people out there who would be so annoyed by an intervention that it would have the opposite effect of what's intended.

Or, more broadly, people are just different and react in different ways. Angrist and Evans (1998) are a good example for thinking about this.[55] They observed that families appeared to have a preference for having both a boy *and* a girl. A family that happens to have two boys as their first two children, or two girls, would be more likely to have a third child so as to try for a mix. So, "your first two kids being the same gender" has been used as an instrument for "having a third child" in a whole bunch of studies, following on from Angrist and Evans.

But for this to work, there have to be no defiers. Even if *most* people would be more likely to have the third kid if the first two are the same gender (or not base their third-kid decision on that at all), if there are *some* people who would be *less* likely to have the third kid because the first two are the same gender, then monotonicity is violated. Maybe some parents are terrified by the possibility of three kids of the same gender? Whatever story we come up with, people have a lot of complex reasons for choosing to have more kids, and some of them might contradict what we need them to be. So studies using this instrument need to think carefully about whether monotonicity is likely to be satisfied, and what they can do about it.

[54] Importantly, the problem here isn't that Macy is affected *negatively*, it's that she's affected in a *different direction to the others*. Conformity is pretty helpful here. It wouldn't be a problem if *everyone* had a negative effect. Then, all the negatives on the weights would just factor out. It's like multiplying the original LATE calculation by $-1/-1 = 1$. Makes no difference.

Monotonicity. In the context of instrumental variables, the assumption that for everyone in the data, the instrument either affects them in the same direction (positively or negatively), or not at all (zero effect).

[55] Joshua D. Angrist and William N. Evans. Children and their parents' labor supply: Evidence from exogenous variation in family size. *American Economic Review*, 88(3):450–477, 1998.

19.2.3   Checking the Instrumental Variables Assumptions

FOR IV TO DO WHAT WE WANT IT TO, WE ARE RELYING ON THE RELEVANCE ASSUMPTION. Remember, the relevance assumption is the assumption that the instrumental variable Z and the treatment/endogenous variable X *are related to each other*.

We can go a bit further and say that we need to assume that X and Z are *strongly enough related to each other* that we don't run into the "weak instruments problem."

A *weak instrument* is one that is valid and does predict the treatment variable, but it only predicts the treatment variable a little bit. It predicts weakly. Keeping in mind our general intuition that IV gives us $Cov(Z,Y)/Cov(Z,X)$, then if $Cov(Z,X)$ is *small*, we're nearing a divide-by-zero problem! The estimate as a whole balloons up really big (since you're dividing by something tiny) and the sampling variation gets huge.

Weak instrument. An instrument that only weakly predicts the treatment/endogenous variable(s).

Thankfully, unlike a lot of the assumptions we cover, weak instruments (and by extension relevance) is fairly easy to test for.[56] After all, we just need to make sure that Z and X are related, and not in a weak way. We know how to measure and test relationships!

[56] Although maybe it's not the best way to deal with the weak-instrument problem... read to the end of the section.

By far the most common way to check for relevance is the *first-stage F-statistic test*.[57] Conveniently, it's also the easiest. All you have to do is:

[57] This is sometimes referred to as an underidentification test.

1. Estimate the first stage of the model (regress the treatment/endogenous variable on the controls and instruments)

2. Do a joint F test on the instruments

3. Get the F statistic from that joint F-test and use it to decide if the instrument is relevant

That's it! The calculation gets a bit trickier if there's more than one treatment/endogenous variable (since there's not really a single first stage in the same way) but the idea remains the same.

We covered the code for doing a joint F test back in Chapter 13 on regression, but because this is such a common test, your IV command will often do it for you. In R, if your `feols` regression is stored as `m`, `summary(m)` will report a "first-stage F statistic."[58] In Stata you can follow your `ivregress` command with `estat firststage`.[59] In Python with **statsmodels** you can do `IV2SLS().fit().first_stage` and look at the "partial F-stat."

[58] If you're using a different IV command this may not work. Many of them, such as ivreg from the **AER** package, work with summary(m, diagnostics = TRUE).

[59] If you've done 2SLS, this will also give you the different relevant critical values, which we'll discuss in a second.

WE'VE GOT OUR FIRST-STAGE F-STATISTIC NOW. HOW BIG DOES IT NEED TO BE? Checking if the instrument is statistically significant is not nearly enough—we aren't just concerned that the relationship is zero, we're concerned that it's *small*.

Since there's no single precise definition of "small," there's no single correct cutoff F-statistic to look for. Instead, we have a tradeoff. The bigger your F-statistic, the less bias you get. So the F-statistic you want will be based on how much bias you're willing to accept.

Wait—what bias? Who said anything about bias? Weak instruments lead to bias because, even if the instrument is truly valid, in an actual sample of data the instrument will have a nonzero relationship with the error term just by random chance, worsening validity and giving you bias. The weaker the instrument is, the worse this gets.

We can frame our choice of cutoff F-statistic in responding to this tradeoff. Stock and Yogo (2005) calculate the bias that you get with instrumental variables *relative to the bias you'd get by just running OLS on its own*.[60] The stronger the instrument is, the less the IV bias will be relative to the OLS bias. Their tables at the end of the paper will show you that, for example, if you have one treatment/endogenous variable and three instruments, you need an F-statistic above 13.91 to reduce IV bias to less than 5% of OLS bias in 2SLS, but only an F-statistic above 9.08 to reduce IV bias to less than 10% of OLS bias. What if you have four instruments? Then you need 16.85 or 10.27. Some first-stage test commands (like Stata's) will tell you the relevant cutoffs automatically.[61]

What to do if your F-statistic doesn't measure up? You could just give up at that point—more than a few researchers would advise ditching a project with a small first-stage F-statistic. But this practice is bad if everyone does it, especially if you're talking about projects that might end up getting published. If low-F-statistic projects get dropped based on an F-statistic cutoff, then the projects that *do* get published will be a mix of actually-strong IVs and actually-weak IVs where the researcher just happened to get a sample where the F-statistic was large. When an actually-weak IV happens to produce a large F-statistic, that's what gives us *the exact kind of weak-instrument bias we were worried about*. We're specifically selecting projects to complete based on getting samples where the instrument is invalid by random chance. So the F-statistic cutoff leads to a larger proportion of biased results in the published literature.

What to do? In the case of weak instruments, as long as you are indeed pretty certain that the instrument is valid, instead of ditching the project you might want to try an approach to IV that is more robust to weak instruments. A few of these methods, like Anderson-Rubin confidence intervals, are shown in the How the Pros Do It section.

[60] James H Stock and Motohiro Yogo. Testing for weak instruments in linear IV regression. In *Identification and Inference for Econometric Models: Essays in Honor of Thomas Rothenberg*, pages 80–108. Cambridge University Press, Cambridge, 2005.

[61] There is another rule of thumb that just says your F-statistic must be 10 or above in general. That's certainly much easier to remember than looking up a value in a table, but it's also very rough. Like many one-size-fits-all values in statistics, this is a tradition probably better left behind.

In fact, you may just want to go ahead and use those weak-instrument-robust methods from How the Pros Do It anyway. The problem with weak instruments is that they introduce a lot of sampling variation and mess with your standard errors. The idea with pre-testing is that you only pursue analyses where you know the problem isn't *too* big. But that doesn't mean the problem goes away. Lee, McCrary, Moreira, and Porter (2020) show that your confidence intervals will at least be a *teeny* bit wrong all the way up to a first-stage F statistic of 104.7.[62] Yikes! That doesn't mean toss out anything with F below 104.7. Instead it just means that maybe pre-testing isn't going to solve the real problem.

So WE CAN TEST FOR RELEVANCE AND WEAK INSTRUMENTS. HOW ABOUT VALIDITY? There are a few tests I should probably mention, despite the fact that they make me grumpy. In particular, there are several very well-known tests that are designed to test the validity assumption. While these tests have their uses, they are commonly applied in a way that I find thoroughly unconvincing, for reasons I'll mention. However, they are very common and so worth at least knowing enough about to know what's going on.

First off, what does it mean exactly to test for validity? We want to test whether there are any open back doors between the instrument Z and the outcome Y. We can reframe this as checking whether Z is related to the second-stage error term ε. We want Z and ε to be unrelated. If they're related, validity is violated.

Testing for validity has some obvious hurdles to overcome. First off, we can't actually observe the error term ε, and so can't just look at the relationship between Z and ε.

Why not just look at the relationship between Z and the residual r instead? We use r in place of ε all the time for calculating things like standard errors. However, if Z *is* invalid, then the second-stage estimates will be biased, which will make the residuals not a great representation of the error. Since the error is the difference between the *true-model* prediction and the outcome, and we know that the model is biased, the estimated model isn't even the true model on average, so the residuals don't represent the errors very well.

What, then, can we do?

One approach is to run the second-stage model but include the instrument as a control.

[62] David L Lee, Justin McCrary, Marcelo J Moreira, and Jack Porter. Valid t-ratio inference for IV. *arXiv preprint arXiv:2010.05058*, 2020.

$$Y = \beta_0 + \beta_1 X + \beta_2 Z + \varepsilon \qquad (19.3)$$

If the coefficient on the instrument Z is nonzero, that suggests a violation of validity. Why? Because all open paths from Z to Y should run through X, right? So if we look at the effect of Z on Y while controlling for X, that should close all pathways. If there's still a relationship evident in $\hat{\beta}_2$, then it seems that there are other, validity-breaking pathways still open![63]

Another approach uses the Durbin-Wu-Hausman test. The Durbin-Wu-Hausman test compares the results from two models, one of which may have inconsistent results if some assumptions are wrong, while the other doesn't rely on those assumptions but is less precise. If the results are similar, it suggests that the assumptions at least aren't *so* wrong that they mess up the results, and it means a green light to use the more precise model.

In the context of instrumental variables, Durbin-Wu-Hausman is used in two ways. First, it can be used to compare OLS (inconsistent if X is related to ε) to IV (less precise). If the results are different, that means that X really *does* have open back doors, and we should probably be using IV.

Second, Durbin-Wu-Hausman can be used to compare two different IV models in an overidentification test. If we have *more instruments than we need* (we are overidentified), then as long as we're *really certain* that we have as many valid instruments as we need, we can compare the IV model that uses all of our instruments (inconsistent if some of them are invalid) against an IV model that only uses the instruments we're really sure about (less precise). If they're different, that tells us that the additional instruments are likely invalid.

In practice, overidentification tests are more commonly performed for 2SLS using a Sargan test, which gets the residuals from the second stage of a 2SLS model and looks at the relationship between those residuals and the instruments. For GMM models, often a Hansen test is used, which I won't go into deeply here. For all of these tests there are many ways to go at them.

In R you get both Durbin-Wu-Hausman exogeneity tests and Sargan overidentification tests automatically when you `summary()` the IV model you get from `feols`. If you're using a different command you may need to use the `diagnostics = TRUE` option. In Stata you can follow your regression with `estat endog`

[63] This particular test is a bit ad-hoc and is used more because it makes sense than because it has any nice statistical properties. I bring it up here more so you know what the logic is when you see it, rather than really suggesting you use it.

to test OLS vs IV or `estat overid` to do the Sargan test. In Python you can get the Durbin-Wu-Hausman exogeneity test from your model with the `.wu_hausman()` method, or the Sargan test with `.sargan()`.

So those are the main tools in our belt for testing for validity of instruments. Why do they make me grumpy? Because people try to use them to justify iffy instruments. Imagine someone was showing you some research with an instrument that you didn't think was valid, or an OLS model with a treatment you thought had some open back doors remaining. But ah—no worries! They've done an endogeneity test and have failed to reject the null hypothesis that the iffy treatment/instrument is valid. No problem, right?

Still a problem! There's a lot that goes into whether a null hypothesis is rejected other than whether the underlying thing you're testing is true or not. Statistical power, sampling variation, and so on. Failing to find evidence of a validity violation *could* mean the instrument is valid, or it could mean one of those other things.

A more reasonable use of these tests is when you have an instrument that you *are* really certain about, but are worried that maybe you've just had a bad luck of the draw. Using these tests to disappoint yourself about an instrument you were pretty certain on is a much more justifiable use of them than using them to reassure yourself about an instrument you're uncertain on.[64]

There's another reason these tests, in particular overidentification tests, make me grumpy. It has to do with the local average treatment effect (LATE) estimate being produced by the instrumental variables model. When you have more than one instrument, those instruments will generate different LATE estimates, because they explain different parts of variation in the treatment. When you combine the two instruments you'll get yet again another LATE. So if your overidentification test finds that the model produces different results with different instruments, well... that doesn't necessarily mean that the instruments are invalid. It just means that the instruments don't produce the same results.[65]

[64] Good ol' statistics, always looking for a way to disappoint.

[65] Paulo MDC Parente and JMC Santos Silva. A cautionary note on tests of overidentifying restrictions. *Economics Letters*, 115(2): 314–317, 2012.

19.3   How the Pros Do It

19.3.1   Don't Just TEST for Weakness, Fix It!

WEAK INSTRUMENTS ARE A REAL PROBLEM. I've already discussed ways to try to detect weak instruments using an F-test. But this test has its own problems. We could, instead, just go ahead and use a method for estimating instrumental variables that is not as strongly affected by weak instruments. They do exist!

The first fix is also the easiest. In the case where we have one treatment/endogenous variable and one instrument,[66] we can adjust the standard errors to account for the possibility that the instruments are weak.

This solution is very old, as far back as 1949. Anderson-Rubin confidence intervals provide valid measures of uncertainty in our estimate of the effect even if the instruments are weak.[67],[68] This doesn't really *solve* the problem of having a weak instrument. That is, you still have a weak instrument and have the sampling variation issues that go along with it. But it does make sure that your results *reflect* that variation. In other words, they'll be more honest about your weak-instrument problems.

That's... that's really it. If you have one instrument and one treatment/endogenous variable, you may as well skip the whole process of testing for weak instruments and just report Anderson-Rubin confidence intervals whether your instrument is weak or not. It's kind of a shocker that this isn't a completely universal practice. As you'll see in the next paragraph, it isn't even included as an option in common IV commands. We've known about it for ages. Ah, well.

In R you can get Anderson-Rubin confidence intervals by first estimating your IV model using the `ivreg` function in the **AER** package, and then passing that to the `anderson.rubin.ci` function in the **ivpack** package. In Stata you can get Anderson-Rubin confidence intervals by following up your `ivregress` model with the `weakiv` function from the **weakiv** package. To my knowledge, there's no prepackaged way of getting Anderson-Rubin confidence intervals in Python.

WE CAN MORE HONESTLY REPORT OUR WEAK-INSTRUMENT PROBLEM, BUT CAN WE REALLY *FIX* IT? Maybe!

A common estimation method that attempts to perform better when the instrument is weak is limited-information maximum likelihood (LIML).[69]

[66] Let's be honest, this is by far the most common scenario. Why did we bother with those overidentification tests again?

[67] Theodore W Anderson and Herman Rubin. Estimation of the parameters of a single equation in a complete system of stochastic equations. *The Annals of Mathematical Statistics*, 20(1):46–63, 1949.

[68] This is not to be confused with the Anderson-Rubin *test* of weak instruments. Confusing, I know.

[69] Limited-information maximum likelihood is often supported by the same IV estimation software that supports 2SLS and GMM, and the code for estimation using LIML was discussed earlier in this chapter.

In short, instrumental variables uses the parts of the treatment/endogenous variable X and the outcome Y that are predicted by the instrument Z. LIML in the context of instrumental variables does the same thing, except that it scales down the prediction, making it weaker, using a parameter κ, which is generally estimated in the model based on the data. If $\kappa = 1$, we scale the prediction by 1, making no change, and ending up back with 2SLS. But if $\kappa < 1$, we bring back a little of the original endogenous variable we would have used in OLS, relying less on that weak instrument.

We can also make adjustments to κ using the parameter known as Fuller's α. Instead of $\kappa = \hat{\kappa}$ estimated on the basis of the data, α makes it even smaller, using instead $\kappa = \hat{\kappa} - \alpha/(N - N_I)$, where N is the number of observations and N_I is the number of instruments. $\alpha = 4$, for example, minimizes the mean squared error of the estimate across samples. It can also be estimated using the data. The LIML procedure reduces the bias that comes along with weak instruments, but this may come, surprise surprise, at the expense of worse precision.[70]

Anything else? One of the most promising avenues of research looks at the use of *lots* of weak instruments. Sure, no one of them means much for the treatment. But an army of them? This is the finale to a movie. I cannot possibly do the whole literature justice here, but one starting place is with Chao and Swason (2005).[71]

19.3.2   Way Past LATE

ARE WE STUCK WITH LOCAL AVERAGE TREATMENT EFFECTS? Not at all! A local average treatment effect is what 2SLS gives us by default, but there are other approaches. There's a wide literature out there on various approaches to instrumental variables estimation that produce average treatment effects (or other averages) instead of a local average treatment effect. Naturally, they all rely on different assumptions or apply in different contexts.

I'll mention only two here, briefly. These are only two papers representing two different general approaches to getting average treatment effects, and those two approaches are far from the only two.[72] These are starting points.

First, Heckman and Vytlacil (1999) look at instrumental variables that take lots of different values.[73] They realize that you can identify a bunch of effects by comparing different values—

[70] Sören Blomquist and Matz Dahlberg. Small sample properties of LIML and jackknife IV estimators: Experiments with weak instruments. *Journal of Applied Econometrics*, 14 (1):69–88, 1999.

[71] John C Chao and Norman R Swanson. Consistent estimation with a large number of weak instruments. *Econometrica*, 73(5): 1673–1692, 2005.

[72] And a third, in fact, will pop up in the next section.

[73] James J Heckman and Edward J Vytlacil. Local instrumental variables and latent variable models for identifying and bounding treatment effects. *Proceedings of the National Academy of Sciences*, 96(8): 4730–4734, 1999.

what's the treatment effect among those pushed from the very lowest value of the instrument to just above that? What's the effect from those pushed from the just-above-that value to just-above-*that*? And so on. So now you've got a whole bunch of effects for the range of the instrument. Under the condition that the probability of treatment is zero for some values of the instrument, and one for others, you can average out this whole distribution of effects to get an average treatment effect.

Second is the literature using correlated random effects (remember those from Chapter 16?) in combination with 2SLS. Another way to think about heterogeneous treatment effects is that the treatment interacts with some individual variation we can't observe. Correlated random effects try to model unobserved individual variation. Seems like a good starting place.

By interacting treatment with a set of control variables, and then estimating the model with correlated random effects, you can produce an average treatment effect. Or, heck, get the conditional average treatment effect for any group you like, defined by the control variables you've given the model. Wooldridge (2008) provides a readable overview of the method and when it does or doesn't work well.[74] One important note is that this method tends to work better the more continuous the treatment variable is, and things get pretty dicey for a binary treatment.

19.3.3   Nonlinear Instrumental Variables

WHAT CAN WE DO WITH INSTRUMENTAL VARIABLES AND A NONLINEAR MODEL? What we've talked about so far with instrumental variables, and *especially* with two-stage least squares, has all assumed that we've been in the realm of linear models. But what if we have a binary outcome or treatment/endogenous variable? Won't we want to run a probit or logit?

The obvious temptation in the common case where you have a binary treatment/endogenous variable is to run the "first stage" of 2SLS using probit or logit, take the predicted values, and use those to estimate the second stage. Maybe run a probit or logit in that second stage too if the outcome is binary. This approach has been termed by econometrician Jerry Hausman as the "forbidden regression" as it simply does not work the way you want it to.[75] The reason it doesn't work has to do with the fact that in a nonlinear regression, each variable's effect depends on the values of the other variables. So while the instrument Z and the second-stage error term ε are unrelated, the *fitted values*

[74] Jeffrey M Wooldridge. Instrumental variables estimation of the average treatment effect in the correlated random coefficient model. In *Modelling and Evaluating Treatment Effects in Econometrics*. Emerald Group Publishing Limited, Bingley, 2008.

[75] Jeffrey M Wooldridge. *Econometric Analysis of Cross Section and Panel Data*. MIT press, 2010.

$\hat{X}$ no longer get to borrow that nice unrelatedness from Z. The back door remains.

Another temptation is, upon learning that you can't just mix probit/logit and 2SLS all willy-nilly, is to simply run a linear model, that is, a linear probability model, ignoring the nonlinearity.[76] This isn't the worst idea—2SLS does *work* this way, unlike with the "forbidden regression." But on the other hand, you do get all the downsides of linear probability models relative to probit/logit. Don't forget the poor performance of OLS vs. probit/logit when the mean of the binary variable is near 0 or 1. Quite a few binary treatment variables we might be interested in that are very rare, or very common. Plus, this approach produces estimates that are less precise than models that do properly take into account the nonlinearity.[77]

So then what to do? Our options are many at this point, and depend on where it is you want to model the nonlinearity—the first or second stage. In this section we'll keep it simple and only focus on binary treatments/outcomes, rather than the infinite alternative forms of nonlinear modeling we might be worried about. I'll also keep from going into *every* way of estimating these models, as there is no shortage of approaches.

LET'S START WITH BINARY TREATMENTS. There is a handy and easy-to-implement method for incorporating a binary treatment into 2SLS popularized by Jeffrey Wooldridge.[78]

First, estimate the first stage using nonlinear regression of the treatment/endogenous variable on the instruments and controls. Run that probit! Then, get the predicted values.

Then, instead of sticking those predicted values into the second stage as in the forbidden regression, use those predicted values *in place of the instrument* in 2SLS.

That's it! Under this process, the nonlinearity no longer biases the estimate, and we get more precise estimates than if we'd run a linear probability model.

You may prefer a "treatment effect regression" approach, which avoids 2SLS entirely and directly models the actual binary structure of the data. These approaches basically estimate the probit first stage and the linear second stage *at the same time*,[79] allowing the instrument to influence the treatment/endogenous variable as usual. To estimate a treatment effect regression you can look into `treatReg` in the **sampleSelection** package in R, or the `etregress` function in Stata. These functions also have tools for helping to estimate treatment effect averages other

[76] Economists tend to do this. In general, economists are pretty likely to look at a research design that gives nonlinear models a hard time, like fixed effects or instrumental variables, figure that the additional "hard time" makes fixing the nonlinearity introduce more problems than it solves, and go with a linear model. I admit to being sympathetic to this view.

[77] And IV is already less precise than regular regression. Do we really need *less* precision?

[78] Jeffrey M Wooldridge. *Econometric Analysis of Cross Section and Panel Data*. MIT press, 2010.

[79] Or at least the maximum likelihood approach does. Two-stage approaches to these estimators are more similar to the control function approach described in the next section, but do some adjustment with the first-stage predicted values to scale them properly.

than the LATE. You'll have to do it yourself in Python, at least for now.

WHAT IF THE OUTCOME IS BINARY INSTEAD? In these cases a common fix is to use the *control function* approach to instrumental variables.

The control function approach is a lot like 2SLS, except that instead of *isolating the explained part of X* and using that in the second stage, you instead use regular ol' X, but also *control for the unexplained part of X*. In regular linear instrumental variables, 2SLS and the control function give the same point estimates.[80]

However, when you're looking at *nonlinear* instrumental variables, the two methods are no longer the same. And while the 2SLS approach produces a biased estimate, the control function approach where the residuals from a linear first stage are added as a control to a probit second stage does not.

That said, like with 2SLS there are a few adjustments that must be made to use the control function approach properly with a binary outcome variable, so you will generally want to use an estimation function designed for the job. In R this can be done in the **ivprobit** package. In Stata this is `ivprobit` with the `twostep` option.[81] To my knowledge, if you want to do this in Python you'll have to do it yourself. Walking through the code for the R **ivprobit** package is a good place to start on that project.

UH OH. WHAT IF THEY'RE BOTH BINARY? In these cases it's common to apply a *bivariate probit* model. Remember the treatment effect regression from a few paragraphs ago? In that model, we estimate the probit first stage and the linear second stage at the same time using maximum likelihood.

In the bivariate probit, we use maximum likelihood to estimate two probit models at the same time, where the dependent variables of the two models are correlated. So if we make one of those dependent variables a predictor of the other and also give it a predictor excluded from the other model, then that means we are estimating the probit first stage and the *probit second stage* at the same time. Instrumental variables! The technical details of the treatment effect regression and bivariate probit models are different, but the concept is the same.[82]

Bivariate probit is important as a 2SLS alternative because 2SLS can be especially imprecise when both stages are binary. As a bonus, bivariate probit can give you an average treatment

[80] It makes a lot of intuitive sense that you'd get the same answer whether you're isolating explained variation or controlling away unexplained variation. Sort of comforting that it actually works.

[81] For both the R and Stata versions listed here, they default to a non-control-function approach with maximum likelihood. Nothing wrong with that.

[82] Technically, the nonlinearity of the system would allow you to estimate the model without an actual instrument, which is different from linear instrumental variables, but skipping the instrument is generally considered a pretty bad idea.

effect, as opposed to a local average treatment effect. That can come in handy.[83]

Bivariate probit can be estimated in R using `gjrm` in the **GJRM** package with the options `Model = "B"` (bivariate model) and `margins = c('probit','probit')`. In Stata you can use the `biprobit` command. In Python you are unfortunately once again on your own.

[83] Richard C Chiburis, Jishnu Das, and Michael Lokshin. A practical comparison of the bivariate probit and linear IV estimators. *Economics Letters*, 117(3):762–766, 2012.

19.3.4   Okay, You Can Have a Little Validity Violation

ONCE AGAIN WE ASK OURSELVES, IS IT WORTH IT? We futz around with all these details—estimation approaches, tests for the plausibility of assumptions. IV with panel data. IV with nonlinear variables. Weak instruments. Monotonicity. But none of it matters if validity doesn't hold. And we are rightfully skeptical about whether it does outside of actually-randomized treatments.

Thankfully there is an ongoing literature on approaches to using IV that work even if validity doesn't perfectly hold.

To be clear, by "work" I mean that you can get useful inferences out of the analysis. You may have to provide more kinds of other information, or perhaps the inferences you get might not be quite as precise as when you do have a fully valid IV.

There are two popular approaches here. The first *tries to do what it can with a mostly-valid instrument*, and the second *makes up for invalidity by using a large number of instruments*.

AS LONG AS THE INSTRUMENT IS PUSHING AROUND THE TREATMENT, surely we can do *something* with it, right? Yes!

One thing we can do is, instead of treating an invalid instrument as a reason to throw out the analysis, instead think about *how bad* that validity violation is—how strongly the instrument is related to the second stage error term—and construct a *range* of plausible estimates based on the plausible range of violations.

This approach gives us something called "partial identification," since instead of giving you a *single* estimate (identifying an estimate exactly), it gives you a range, based on a range of different assumptions you might make. A key paper in this literature is Conley, Hansen, & Rossi (2012).[84] They show how, instead of assuming the instrument is completely valid, you can construct a range of plausible invalidity values, and produce a range of reasonable estimates that go with those values. Major validity violation? Then your range is big. But the stronger your instrument, the narrower your range as well.

[84] Timothy G Conley, Christian B Hansen, and Peter E Rossi. Plausibly exogenous. *Review of Economics and Statistics*, 94(1):260–272, 2012.

ANOTHER COMMON APPROACH is to replace your validity assumption with some *other* assumption about the instrument that may be more plausible than validity, and seeing how far that assumption can take you. There are many of these papers, each relying on different assumptions about the instrument. Maybe there's a paper out there using assumptions that are very plausible for your instrument!

One such paper is Nevo & Rosen (2012).[85] Here, the key additional assumption we have to make about the instrument is that the correlation between the instrument and the second-stage error term is *the same sign* as the correlation between the endogenous/treatment variable and the error term, but is smaller. This produces a partial-identification range of estimates based on where exactly the correlation with the instrument is—somewhere between 0 and the correlation with the endogenous/treatment variable.

Another is Flores & Flores-Lagunes (2013).[86] The assumption they rely on here is monotonicity—specifically, that the effect of the instrument has the same sign within groups. From this, they are once again able to produce partial-identification bounds on the estimate by taking advantage of how the instrument is always pushing in the same direction.

These approaches, of course, require us to check whether our instrument satisfies these alternative assumptions theoretically. And hopefully they do! But there might be some more firepower we can bring to the party.

ANOTHER LARGE AND GROWING LITERATURE ON INVALID INSTRUMENTS REALLY STRESSES THE "S" IN "INSTRUMENTS." In other words, it focuses on situations where many instruments are available. And yes, there *are* situations where many instruments are available. Often, this shows up in places where an instrument has many different subcomponents that can each be measured separately and treated as their own instrument. For example, perhaps the instrument is assignment to a particular group, which results in a whole bunch of binary-variable fixed effect instruments, like some of the designs from Canonical Designs earlier in the chapter: the randomly-assigned judges design, or Mendelian randomization, which uses genes as instruments—each part of the gene (of which there are many) can be its own instrument!

If you think back to the discussion of regularized regression, LASSO in particular, in Chapter 13, it may not come as much of a surprise that it gets applied here too. LASSO is a method that

[85] Aviv Nevo and Adam M Rosen. Identification with imperfect instruments. *Review of Economics and Statistics*, 94(3):659–671, 2012.

[86] Carlos A Flores and Alfonso Flores-Lagunes. Partial identification of local average treatment effects with an invalid instrument. *Journal of Business & Economic Statistics*, 31(4):534–545, 2013.

modifies OLS to tell it to minimize a sum of the absolute values of the coefficients, in addition to the sum of squared residuals, in effect encouraging it to toss variables out of the model entirely.

LASSO can be used as it normally is, here, selecting the most important predictors among both control variables and instruments.[87] However, LASSO can also be used to help *spot* invalid instruments. By including all of the instruments in the second stage and estimating the model with LASSO, you would expect that the *valid* instruments would have no predictive power once the treatment is controlled for, and so would be chucked out of the second stage (and back into the first stage where they belong). Windmeijer, Farbmacher, Davies, and Davey Smith refine this approach so that it can be expected to work as long as fewer than half of the instruments are invalid.[88]

The fact that we have lots of instruments (many of them valid, we think) might also just make our standard methods work all on their own. Kolesár, Chetty, Friedman, Glaeser, and Imbens show that in a many-instruments setting, 2SLS just plain works even with some invalid instruments, under some conditions.[89,90] Since the 2SLS first stage sorta squashes together the effects of all the instruments, the valid instruments can wash out the invalid instruments. This doesn't happen all the time, but they lay out the conditions under which it does, which rely on the relationship between the effect of the valid instruments and the invalid ones. Their result also doesn't come from regular-ol'-2SLS; they use a version that comes with a bias correction.

[87] Victor Chernozhukov, Christian Hansen, and Martin Spindler. Post-selection and post-regularization inference in linear models with many controls and instruments. *American Economic Review*, 105(5): 486–90, 2015.

[88] Frank Windmeijer, Helmut Farbmacher, Neil Davies, and George Davey Smith. On the use of the LASSO for instrumental variables estimation with some invalid instruments. *Journal of the American Statistical Association*, 114 (527):1339–1350, 2019.

[89] Although maximum likelihood doesn't.

[90] Michal Kolesár, Raj Chetty, John Friedman, Edward Glaeser, and Guido W Imbens. Identification and inference with many invalid instruments. *Journal of Business & Economic Statistics*, 33(4):474–484, 2015.

20

Regression Discontinuity

20.1 How Does It Work?

SAN DIEGO IS A LARGE CITY IN THE UNITED STATES THAT COVERS A FAIRLY WIDE GEOGRAPHIC AREA OF MORE THAN 300 SQUARE MILES. It's pretty well-off too, with an average annual household income above $85,000 as of 2019, about 50% above the national average. As you move south through the area, the scene changes a bit. There's a little less money in some southern parts of the city. By the time you get to San Ysidro, the district of the city just on the Mexican border, household income has dropped to more like $50,000 or $55,000. The further south you go, the lower you can expect incomes to be.

But our trip through the city has nothing on what happens when we *cross* that border into Tijuana, Mexico. It took us 16 miles to drive from downtown to San Ysidro and watch incomes drop by 25%. But you could get out of the car in an area with $50,000 household income and just walk a few feet over the border into Tijuana. Suddenly and sharply, household income drops to $20,000.

DOI: 10.1201/9781003226055-20

Sure, there may be something different about the people, the opportunities, or the geography that is different in south San Diego as opposed to the downtown area that could explain some of the difference in incomes. But there's no way that *just going south* explains what happens when we take one step over the border and see incomes crash. From downtown to San Ysidro there's continuous change. But at the border it *jumps*. Something is *fundamentally different*. We have a *cutoff*.

Whenever some treatment is assigned *discontinuously*—people just on one side of a line get it, and people just on the other side of the line don't, we can be pretty sure that the differences we see aren't just because there's something different about how far south they are. Without the line they might be a little different, but not *that* different. The big change we can attribute to that line separating Mexico and the US.

This is the idea behind *regression discontinuity*. Compare people *just* on either side of a cutoff. Without the cutoff they'd likely be pretty similar. So if they're different, we can probably attribute the difference to whatever it is that happens at that cutoff.[1]

REGRESSION DISCONTINUITY FOCUSES ON TREATMENT THAT IS ASSIGNED AT A CUTOFF. Just to one side, no treatment. Just to the other, treatment![2]

It's worth establishing some terminology here:

- **Running variable.** The running variable, also known as a **forcing variable.**, is the variable that determines whether you're treated or not. For example, if a doctor takes your blood pressure and will assign you a blood-pressure reducing medicine if your systolic blood pressure is above 135, then your blood pressure is the running variable.

- **Cutoff.** The cutoff is the value of the running variable that determines whether you get treatment. Using the same blood pressure example, the cutoff in this case is a systolic blood pressure of 135. If you're above 135, you get the medicine. If you're below 135, you don't.[3]

- **Bandwidth.** It's reasonable to think that people *just barely to either side* of the cutoff are basically the same other than for the cutoff. But people farther away (say, downtown San Diego vs. further inside of Mexico) might be different for reasons other than the cutoff. The bandwidth is how much area *around* the cutoff you're willing to consider comparable.

[1] In the San Diego/Tijuana case it's really multiple treatments since so many things change at the border. But the border is clearly doing *something!*

[2] Regression discontinuity is almost always used in reference to a binary treatment variable. This chapter will not bother hedging its wording to imply any other use, although regression discontinuity with non-binary treatment *is* out there, and you'll see a tiny hint of it in the Regression Kink section in How the Pros Do It below.

[3] In this example, and many examples, you get treatment if you're *above* the cutoff, and not if you're below. But in many scenarios you would get treatment if you were *below* the cutoff. The same logic works.

Ten feet to either side of the US/Mexico border? 1000 feet? 80 miles?

The core of the research design of regression discontinuity is to (a) account for how the running variable normally affects the outcome, (b) focus on observations right around the cutoff, inside the bandwidth, and (c) compare the just-barely-treated against the just-barely-didn't to get the effect of treatment.

THERE ARE PLENTY OF CUTOFFS LIKE THIS IN THE WORLD. Quite a few policies are designed with explicit cutoffs in mind. Make just slightly too much money? You might not qualify for some means-tested program. Or have a test score just slightly too low? You might not qualify for the gifted-and-talented program.

These cutoffs could be geographic—we already had the San Diego/Tijuana example. But also, live just to one side of a time zone border or another? You might be getting up an hour earlier. Or if you're just to one side of a police jurisdiction's border, you may be experiencing different policing policies.

Politics is another place where we have hard cutoffs. In an election and win 49.999% of the vote among the top two in your election? You don't get the office. But win 50.001%? Enjoy your office! Now we can see what the effect of *you* is.

In each of these cases, we can pretty reasonably imagine that cases that are *just barely* to either side of the cutoff are comparable, and any differences between them are really the fault of treatment. In other words, we've identified the effect of treatment.

The goal of this approach is to carefully select the variation that lets us ignore all sorts of back doors without actually having to control for them.

WHEN CAN WE APPLY REGRESSION DISCONTINUITY? What we're looking for is some sort of treatment that is assigned based on a cutoff. There's the running variable that determines treatment. And there's some cutoff value. If you're just to one side of the cutoff value, you don't get treated. If you're to the other side, you do get treated.[4] Plus, since our strategy is going to be assuming that people close to the cutoff are effectively randomly assigned,[5] there shouldn't be any obvious impediments to that randomness—people *shouldn't be able to* manipulate the running variable so as to choose their own treatment, and also the people who choose what the cutoff is *shouldn't be able to* make that choice *in response* to finding out who has which running variable values.

[4] Or at least your chances of treatment change dramatically—we'll get to this variant when we talk about fuzzy regression discontinuity.

[5] More precisely, we need *continuity* of the relationship between the outcome and the running variable. If, for example, there's a positive relationship between the outcome and the running variable, we'd expect a higher outcome to the right of the cutoff than to the left. That's totally fine, even though the two groups aren't exactly comparable. As long as *there wouldn't be a discontinuity in that positive relationship without the treatment.* Roughly, you can think of this as "after we adjust for the overall relationship between the outcome and the running variable, the two sides are basically randomly assigned."

This gives us a causal diagram that looks like Figure 20.1. Here we have a *RunningVariable* with a back door through *Z* (which perhaps we can't control for). *RunningVariable* also has a direct effect on the *Outcome*.

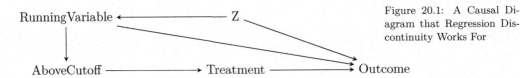

Figure 20.1: A Causal Diagram that Regression Discontinuity Works For

Now, we aren't actually interested in the effect of *Running Variable*. We're interested in the effect of *Treatment*. *Treatment* has a back door through *RunningVariable*. However, if we control for *all the variation in RunningVariable except for being above the cutoff*, then we can close that back door, identifying the effect of *Treatment* on *Outcome*.[6]

Based on the diagram it sort of seems like regression discontinuity is just a case of controlling for a variable (*RunningVariable*) and saying we've identified the effect. But it's a bit more interesting than that. We're actually *isolating a front-door path*, as in Chapter 9 (Finding Front Doors) or Chapter 19 (Instrumental Variables), not closing back doors. By only looking right around the cutoff we are getting rid of *any* variation that doesn't lie on the *AboveCutoff* → *Treatment* → *Outcome* path. Nothing else should really be varying once we limit ourselves to that cutoff, so all the other variables go away!

So, sure, in Figure 20.1 we could identify the effect by doing regular ol' statistical adjustment for *RunningVariable*. But what if there were other back doors? Perhaps even back doors for *Treatment* itself, for example if *Z* → *Treatment* were on the graph? Controlling for *RunningVariable* wouldn't solve the problem, but regression discontinuity would, by isolating *AboveCutoff* → *Treatment* → *Outcome* and letting us ignore any other arrows heading into *Treatment*.[7]

SO HOW DOES REGRESSION DISCONTINUITY DO ALL THIS COOL STUFF, EXACTLY? We can perform regression discontinuity by making only a few basic choices. Once we've done that, the process is pretty straightforward.

1. Choose a method for predicting the outcome on each side of the cutoff

2. Choose a bandwidth (optional)

[6] As has popped up a few times in this book, this diagram isn't exactly a kosher causal diagram. An arrow allows any sort of *shape* for the relationship, and so "*RunningVariable* affects *Treatment* only at the cutoff" could be handled by a *RunningVariable* → *Treatment* arrow, with the cutoff incorporated using a "limiting diagram" where *RunningVariable* is replaced by *RunningVariable* → *Cutoff* as a node in itself. I think this is more complex than the way I've done it, though, so here we are. If you keep working with causal diagrams, the more proper version will pop up.

[7] You may have noticed the similarity between Figure 20.1 and the causal diagram in Chapter 17 for event studies. They are, indeed, nearly the same, except event studies have *Time* as a running variable. You will notice other similarities to event studies, especially interrupted time series, through the chapter. The main difference is that a lot of the necessary assumptions become more plausible with a non-*Time* running variable, for reasons that hopefully will become evident.

That's it![8] Let's put these choices to work on some simulated data. You can see the process all worked out in Figure 20.2.

In part (a), we see a setup that's pretty typical for regression discontinuity: a running variable along the x-axis. The running variable does seem to be related to the outcome even when we're not around the cutoff, which is fine. We have a clear upward slope here. We also have a cutoff value at which treatment was applied. And we see a jump in the outcome at that cutoff! That's exactly what we want to see.

[8] Although the estimation will usually be more complex than I have it here—we'll get to it in the How Is It Performed section.

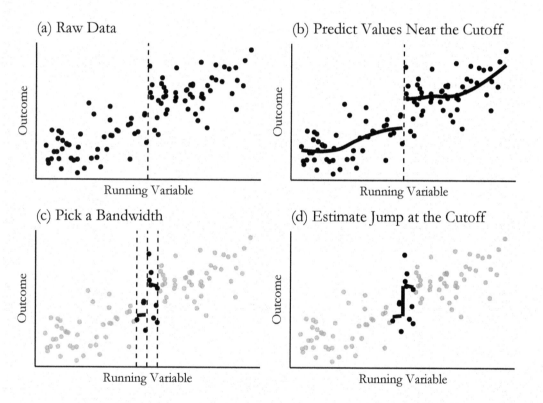

Figure 20.2: Regression Discontinuity, Step by Step

Moving on to (b), we need a way of predicting what the outcome will be just to either side of the cutoff. There are a bunch of different prediction methods we could use—regression, local means, and so on.[9] In this case, we're using local means with LOESS, which is just a smoothed line based on the prediction from a regression that uses only points within a certain range of each x-axis value.

In (c) we can see our bandwidth being applied (which will modify the predictions that our method in (b) makes). We are really only interested in the values right around the cutoff. Values far away from the cutoff still have value—we can use them

[9] Shoot, we could just guess what we think the values will be based on gut instinct. Journals tend to frown on this practice though. Ivory tower snobs, the lot of them.

to figure out trends and lines that let us better predict the values at the cutoff. But get too far away and you're allowing in a bunch of back-door variation that we're doing regression discontinuity to avoid in the first place. But maybe we plan to account for that problem by adjusting for the relationship between the outcome and the running variable and the regression, and we can skip the bandwidth step. We'll get to the details of this later,[10] but for now let's opt to use a bandwidth. Here we're using only data that's really close to the cutoff to calculate our prediction around the cutoff. Everything else gets ignored.

Finally, based on our bandwidth and prediction method, we have our predictions for what the outcome is just on either side of the cutoff. Our regression discontinuity estimate is then just how much the prediction jumps at the cutoff. We can see that in (d), where the estimate is the vertical line going from the prediction just-on-the-left to the prediction just-on-the-right.

So far, I've shown examples where the cutoff is the only determinant of treatment. Often this isn't the case. In a lot of regression discontinuity applications, being on one side or another of the cutoff only *changes the probability* of treatment. In these cases we have what's called a *fuzzy regression discontinuity*, as opposed to a "sharp" regression discontinuity where treatment rates jump from 0% to 100%.

[10] The basic struggle of the choice of bandwidth is this: the farther out your bandwidth, the less certain you can be about your identifying assumption that people are comparable on either side of the cutoff. But the closer-in your bandwidth, the fewer observations you have, making your estimate less precise.

Fuzzy regression discontinuity. A regression discontinuity design in which treatment is not completely determined by being on one side or another of the cutoff.

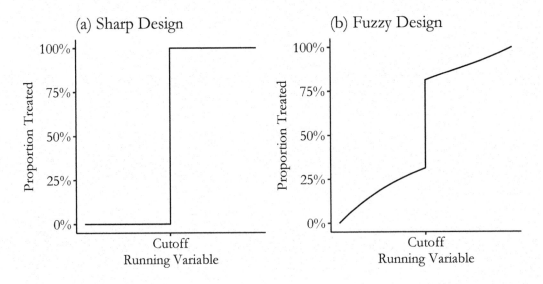

(a) Sharp Design (b) Fuzzy Design

Figure 20.3: Proportion Treated in Sharp and Fuzzy Regression Discontinuity Designs

Figure 20.3 shows how treatment rates might change in sharp and fuzzy designs.[11] While each shows a jump in the rate of treatment at the cutoff, the sharp design shows no action at all anywhere but the cutoff.

One example of a topic where fuzzy regression discontinuity is applied a lot is in looking at the impact of retirement. In many occupations and countries, there are certain ages, or number of years you've been at your job, where a new policy kicks in—pension income, access to retirement funds—and the retirement rate jumps significantly at these points. Not everyone retires as soon as they're able to, though, and some retire earlier, making this a fuzzy discontinuity.

Battistin, Brugiavini, Rettore, and Weber (2009) are one paper of many that apply fuzzy regression discontinuity to retirement.[12] In their paper, they look at *consumption* and how it changes at the point of retirement. Specifically, they want to know if retirement causes consumption to immediately drop.[13] And if it does drop, why?

They look at data from Italy in 1993—2004, where they have information on when people become eligible for their pension. At the exact age of eligibility there's a more than 30 percentage point jump in the proportion of men who are retired. That's not a 0% to 100% jump (and thus is fuzzy), but a 30 percentage point jump is something we can work with! We can see the jump in one of the years they studied in Figure 20.4. There's clearly *something* going on at that cutoff.

They want to look at how consumption changes at that cutoff, just as you normally would with a non-fuzzy RDD. But if they just treated it as normal, their results would be way off. They'd be acting as though the cutoff increased the retirement rate by 100 percentage points, when it only did it by 30 percentage points. So they'd be expecting *everyone* to change their consumption habits when only 30 percent of people would, and would way understate the effects.

So they scale by the effect of the cutoff. Roughly,[14] they scale the effects they see by dividing them by the 30 percentage point jump (dividing by .3) to see how big the change would have been if everyone got treated.

What do they find? They do indeed find a drop in consumption at the moment someone retires. But conveniently their regression discontinuity design doesn't just identify the effect of retirement on overall consumption, it identifies the effect of retirement on just about anything, in this case *specific types* of

[11] Pay attention to the fact that the y-axis on these graphs is *the proportion of observations that are treated*, not *the average outcome*.

[12] Erich Battistin, Agar Brugiavini, Enrico Rettore, and Guglielmo Weber. The retirement consumption puzzle: Evidence from a regression discontinuity approach. *American Economic Review*, 99(5):2209–2226, 2009.

[13] Finding such a drop is a "puzzle" to economists, as standard models predict that people would rather not experience big sudden drops in their consumption, and retirement is a case where they should be able to avoid that drop if they want to, by adjusting consumption *before* retirement. This is the kind of thing that feels like a very silly and unrealistic prediction on the part of economists but makes more sense the more you think about it.

[14] And we'll get into more specifics in the How Is It Performed? section.

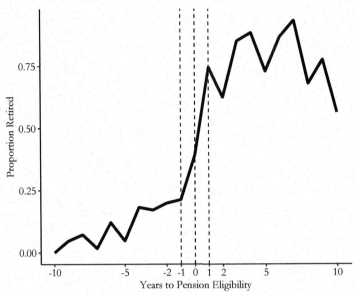

Figure 20.4: The Proportion of Household Heads Retired by Years to Pension Eligibility in 2000, from Battistin, Brugiavini, Rettore, and Weber (2009)

Copyright American Economic Association. Reproduced with permission of the *American Economic Review*.

consumption and the number of kids at home. They find that consumption drops resulting from retirement are largely due to things like using your extra leisure time to cook rather than order from a restaurant, not needing nice work clothes any more, and adult children tending to move out of the house. Not so bad as a drop in consumption might seem.

SURELY THERE'S A CATCH. Well, sort of. As with anything we need to ask ourselves what assumptions need to be made for all of this to work, what the pros and cons are, and what exactly we end up estimating.

First, as with any other method where we isolate just the part of the variation where we can identify the effect, we need to assume that nothing fishy is going on in that variation. We're estimating the effect of treatment, sure, but really we're estimating the effect *of the cutoff* and attributing that (or part of that, if it's fuzzy) to the treatment. So if anything else is changing at the cutoff, we're in trouble. This assumption is the assumption that the outcome is *smooth at the cutoff*. That is, if treatment status actually hadn't changed at the cutoff (if nobody near the cutoff had gotten treated, or everyone had, or everyone had the same chance of being treated), then there would be no jump or discontinuity to speak of.[15]

Say we want to know the effect of a gifted-and-talented school program that requires a score of 95 out of 100 on a math test to get in. Does getting a 95 or more on your math test both

[15] This is an assumption about a counterfactual—we can't really check it. We just need to understand the context we're studying very well and think carefully about what else might have changed at the cutoff.

get you into the gifted-and-talented program *and* the science-museum-field-trip program? Well, then you can't use regression discontinuity to get just the effect of the gifted-and-talented program. No way to separate it from those field trips!

This is a problem even if there isn't another explicit treatment happening at the cutoff. Maybe gifted-and-talented is the only program assigned with a 95-score cutoff, but if it just so happens that the proportion of tall people is way different for people with 94s as opposed to people with 96s. Even if getting a 94 vs. a 96 is completely random and the height difference was happenstance, it's hard to distinguish the effect of gifted-and-talented from the effect of being tall.

Second, the whole crux of this thing is that people are basically randomly assigned to either side of the cutoff if we zoom in close enough.[16,17] So we need to assume that the running variable is randomly assigned around the cutoff and isn't manipulated. If teachers want more students in gifted-and-talented, so they take kids with scores of 94 and secretly regrade their tests so they get 96s instead, then suddenly it's not really random whether you get a 94 or a 96. We can't say that a comparison between the 94s and 96s identifies the effect any more!

Plus, since this is based on being able to zoom in far enough that assignment to either side of the cutoff is random, we need to have a running variable that's precisely measured enough to let us zoom in to the necessary level. Sure, maybe the 94s and 96s are comparable. But what if the test data is only reported in 5-point bins, so we have to compare the 90—94s against the 95—100s? We might trust our random-assignment assumption a lot less then.

Moving on from the key assumptions, our third concern is about statistical power. Regression discontinuity by necessity focuses in on a tiny slice of the data—the people just around the cutoff. There simply *aren't that many people* with a 94, 95, or 96. Alternately, in Figure 20.2 (c) you can see how much data we're graying and throwing out. So, in comparing just the people on either side, we're severely limiting our sample size and so making our estimate noisier. We can fix this by incorporating more data from people farther away from the cutoff, but this reintroduces the bias we were trying to get rid of.[18]

FINALLY, IF WE DO THIS ALL, what kind of treatment effect are we estimating (Chapter 10)? Regression discontinuity is fairly easy to figure out on this front. We're using variation only from just around the cutoff. So we get the effect of treatment

[16] After correcting for "trend."

[17] There are ways to think about regression discontinuity without it actually needing to be "random around the cutoff." But this is such a simple way to think about it that it will remain the focus in this chapter.

[18] It's a good idea to do a *power analysis* (as in Chapter 15) before doing a regression discontinuity to see if you'll have sufficient observations to estimate anything precisely enough to care. In R and Stata, the **rdpower** package does this specifically for regression discontinuity. In R you can install the package as normal, but in Stata there's a different **rdpower** package that does something else, so you'll want to visit `https://rdpackages.github.io/rdpower/` to get installation instructions.

for people who are just around the cutoff. This is a local average treatment effect, getting the weighted average treatment effect for those just on the margin of being given treatment.

There are pros and cons to getting the local average treatment effect in regression discontinuity. As a pro, if we were to use the information from our study to expand treatment to more people, we'd probably do it by shifting the cutoff a bit to allow for more people. The newly-treated people are very similar to the at-the-cutoff people we just estimated the effect for. After all, knowing the effect of a gifted-and-talented program on someone who gets a 5 out of 100 on their math test isn't too useful—they're unlikely to ever be put in the program anyway. Knowing the local effect could be more useful than knowing the overall average treatment effect.

Of course, in other situations we might rather have the average treatment effect. In plenty of cases we're looking at the effect of a treatment that *could* be applied to everyone, but we just got lucky enough to be able to identify the effect in a place where it was assigned based on a cutoff. For example, imagine a relief program that sends $1,600 relief checks to major portions of the populace, with an income cutoff of $75,000. If we were to analyze the effect of these checks using regression discontinuity, we'd get the effect for people who earn very close to $75,000. But we'll never know what the effect would have been for people who earn $100,000. More concerning, we'd have no idea what the effect was for people who earn closer to $20,000—they're the ones who would probably need the most help. But even though the $20,000-earners *got* checks, we can't use regression discontinuity to see the effect of checks for them since they're not near the cutoff.

20.2   How Is It Performed?

20.2.1   Regression Discontinuity with Ordinary Least Squares

THUS FAR, I'VE WRITTEN ABOUT REGRESSION DISCONTINUITY IN TERMS OF DISCONTINUITIES BUT HAVE LEFT OUT THE RE-GRESSION. Seems inappropriate, so let's fix it. There are many ways to predict the outcome just on either side of the regression, but we'll start with a very simple regression-based approach:

$$Y = \beta_0 + \beta_1(Running - Cutoff) + \beta_2 Treated +$$
$$\beta_3(Running - Cutoff) \times Treated + \varepsilon \quad (20.1)$$

This is a simple linear approach to regression discontinuity, where *Running* is the running variable, which we've centered around the cutoff by using $(Running - Cutoff)$. $(Running - Cutoff)$ takes a negative value to the left of the cutoff, zero at the cutoff, and a positive value to the right. We're talking about a sharp regression discontinuity here, so *Treated* is both an indicator for being treated and an indicator for being above the cutoff—these are the same thing,[19] and you could easily write the equation with *AboveCutoff* instead of *Treated*. The model is generally estimated using heteroskedasticity-robust standard errors, as one might expect the discontinuity and general shape of the line we're fitting to exhibit heteroskedasticity in most cases.[20] We'll ignore the issue of a bandwidth for now and come back to it later—this regression approach can be applied whether you use all the data or limit yourself to a bandwidth around the cutoff.

Notice the lack of control variables in Equation 20.1. In most other chapters, when I do this it's to help focus your attention on the design. Here, it's very intentional. The whole idea of regression discontinuity is that you have nearly random assignment on either side of the cutoff. You shouldn't need control variables because the design itself should close any back doors. No open back doors? No need for controls. Adding controls implies you don't believe the assumptions necessary for the regression discontinuity method to work, and makes the whole thing becomes a bit suspicious to any readers.[21]

It's not that adding controls is *wrong*, but you should think carefully about whether you need them if you think that treatment is completely assigned by the cutoff—where's the back door you're closing in that case? You'd have to think there are some back doors just around the cutoff (which isn't impossible, just not a *given*). Controls may be more necessary in fuzzy regression discontinuity, where there definitely *are* determinants of treatment other than the cutoff, so maybe there are back doors to be worried about. And as always, you could still add controls to try to reduce the size of the residuals and so shrink standard errors.

We end up with two straight lines—one with the intercept β_0 and the slope β_1, to the untreated side of the cutoff, and another with the intercept $\beta_0 + \beta_2$ and the slope $\beta_1 + \beta_3$ to the treated side of the cutoff. Applying this regression to some simulated data gives us Figure 20.5.

[19] They no longer will be once we get to estimating a fuzzy regression discontinuity.

[20] *How exactly to do* heteroskedasticity-robust standard errors does get a bit trickier here. You might see papers out in the wild using standard errors that are *clustered on the running variable*—we used to think this was a good idea but no longer do. I'll get to that one in the How the Pros Do It section.

[21] That said, sometimes the addition of controls can help improve the precision of the estimate by reducing the amount of unexplained variation. The process of adding controls can get a bit tricky if you're doing local regression, though, as I'll describe soon. There are methods for it, though, such as Calonico, Cattaneo, Farrell, and Titiunik (2019), which are implemented in pre-packaged regression discontinuity commands.

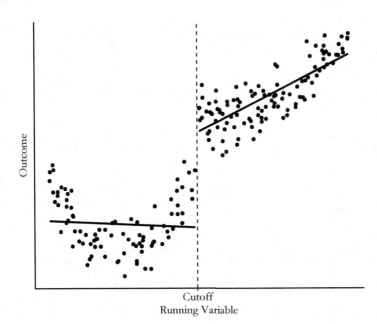

Figure 20.5: Regression Discontinuity Estimated with Linear Regression with an Interaction

Ah, but don't forget—since we centered the running variable to be 0 at the cutoff, the intercept of each of these lines is our prediction *at* the cutoff. So our estimate of how big the jump is from one line to the other at the cutoff is simply the change in intercepts, or $(\beta_0 + \beta_2) - (\beta_0) = \beta_2$. β_2 gives us our estimate of the regression discontinuity effect.

WHAT ARE SOME OF THE PROS AND CONS OF THIS APPROACH? An obvious pro is that this is easy to do, but that's probably not a great reason to pick an estimation method.

The real pros and cons are sort of wrapped up together, and it has to do with the fact that we're using a fitted shape to predict the outcome at the cutoff. There are some clear benefits to this, especially in the form of statistical precision. Using a shape lets us incorporate information from points further from the cutoff to inform the trajectory that the outcome is taking as you get closer and closer to the cutoff. See that clear positive trend to the right of the cutoff in Figure 20.5? There's a bit of a flat, or even downward-sloping area in the data before the trend becomes obvious—it would be difficult to pick that trend up cleanly without incorporating the data further from the cutoff, which could lead us to overestimate the value at the cutoff coming from the right, as shown in Figure 20.6.

However, applying a fitted shape only works if we are applying the *right* fitted shape. If we pick the wrong one, as we clearly have in Figure 20.5 to the left of the cutoff, the prediction we get from our shape will be wrong.

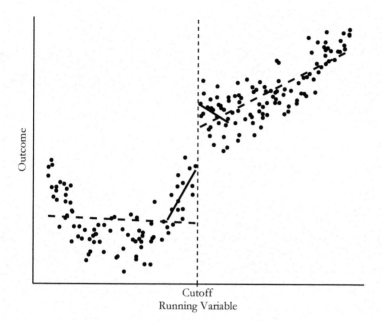

Figure 20.6: Regression Dis-
continuity Estimated with
Linear Regression with an
Interaction, both Without
and With a Bandwidth
Restriction

That's always the case when we use ordinary least squares and fitted shapes, of course. But the problem is especially bad here because *fitted shapes tend to be at their most wrong at the edges of the available data.* As you get to the very edge of your data, the prediction may just veer off in a strange direction because the shape you've chosen forces it to keep going in that direction. Again, we see this in Figure 20.6 on the left.

An obvious instinct is to just try a more flexible shape. We can do this using the exact same idea as Equation 20.1, but use some more flexible function for $(Running - Cutoff)$:

$$Y = \beta_0 + f(Running - Cutoff, Treated) + \beta_2 Treated + \varepsilon \quad (20.2)$$

where $f()$ is some nonlinear function of $Running - Cutoff$, with one version for $Treated = 0$ and another for $Treated = 1$. For example, this might be a 2nd-order polynomial:

$$Y = \beta_0 + \beta_1(Running - Cutoff) + \beta_2(Running - Cutoff)^2 + \\ \beta_3 Treated + \beta_4(Running - Cutoff) \times Treated + \\ \beta_5(Running - Cutoff)^2 \times Treated + \varepsilon \quad (20.3)$$

Equation 20.3 again produces two lines—a parabola to the left of the cutoff, and a parabola to the right. Like before, the coefficient on $Treated$ (here β_3) shows us how the prediction jumps at the cutoff, giving us the regression discontinuity estimate.

However, we need to be careful with more flexible OLS lines. It's hard to fix the "wrong shape" problem in RDD by making the shape more flexible, because more-flexible shapes make even *weirder* predictions and hard-left-turns-to-infinity than simpler shapes near the edges of the data. Figure 20.7 shows regression discontinuity estimated for the same data using (a) the 2nd-order polynomial from Equation 20.3, as well as (b) a 4th-order polynomial and (c) an 25th-order polynomial. Adding all those polynomials, we get stranger and stranger results, with huge swings near the cutoff that don't even seem to track the data all that well.

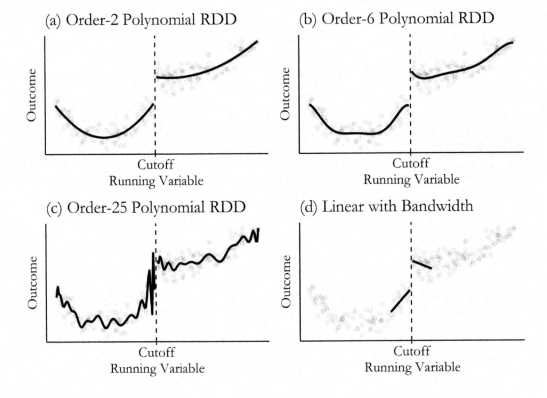

Figure 20.7: Regression Discontinuity with Different Polynomials

Because the data is simulated I can tell you that the true effect is .15 and the true model is an order-2 polynomial. The linear model gives an effect of .29. The 2nd, 6th, and 25th-order polynomial models give .17, .26, and... even if that 25th-order polynomial *happened* to get close to the true effect it would just be luck. Look how wild that curve is near the cutoff! Flexibility isn't always so good.

Because of this problem, it's not a great idea to go above the 2nd-order term when performing regression discontinuity with ordinary least squares.[22] If there's a complex shape that needs fitting, instead try limiting the range of the data with a bandwidth and use a simpler shape. All those twists and curves tend to go away when you zoom in close with a bandwidth, anyway, and a straight line may be just fine. This can be seen in Figure 20.7 (d), which gives an effect of .19, as compared to the true effect of .15. It did just about as well as the order-2 polynomial, but without having to use the knowledge that the true underlying model is order-2.

WE CAN THINK OF THIS "ZOOM-IN-AND-GO-SIMPLE" AP-PROACH as a *local regression*. A local regression is one that estimates the relationship between some predictor X and some outcome Y as normal, but allows that relationship to vary freely across the range of X.[23] Local regression of some kind is how most researchers choose to implement their regression discontinuity design, at least if they have a large sample.[24]

What local regression does in particular is: for each value of X, say, $X = 5$, it gets its estimate of Y by running its own special $X = 5$ regression. This $X = 5$ regression fits some sort of regression shape *specific to $X = 5$*. How is it specific? When estimating the regression, it weights observations more heavily the closer they are to $X = 5$. The specific form of weighting is called the *kernel*. Sometimes the kernel is a cutoff—the weight is 1 if you're close and 0 if you're too far away. Other times it's more gradual—you're weighted less and less the farther you get until finally, slowly, reaching zero. A very commonly-used kernel in regression discontinuity is the *triangular kernel*, which looks in Figure 20.8 just like it sounds—a full weight right at the value you're looking at, and a linearly declining weight as you move away, until you get to 0 at a certain point. That point is the "bandwidth"—i.e., how far away you can get and still count. This draws a neat lil' triangle.

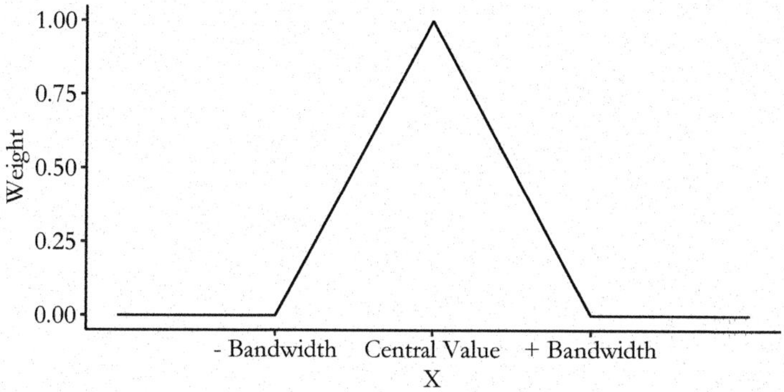

[22] Andrew Gelman and Guido W Imbens. Why high-order polynomials should not be used in regression discontinuity designs. *Journal of Business & Economic Statistics*, 37 (3):447–456, 2019.

Local regression. A regression that is fit differently for different values of the predictor variable, weighting observations based on how close they are to that value of the predictor.

[23] Local regression is also known as a form of "nonparametric regression" since it does not force a given shape on the entire range of data. However, I will point out that it *does* force a shape on subsets of the data. So it's more "less-parametric regression" if you ask me.

[24] The flexibility and focus on the cutoff is the upside, but (as with nearly any time you toss out data or drop assumptions), by imposing less structure on the data you get less precision in your estimates. If the assumptions you'd need to impose would be very wrong, that's a great tradeoff! But you'd better have lots of observations near the cutoff if you want to use local regression, or else the estimate isn't going to be precise enough to use.

Figure 20.8: Example of a Triangular Kernel Weight Function

Computationally, local regression takes a lot of computing firepower. But conceptually this is very simple. Just roll along that x-axis, making a bunch of new regressions along the way based on the observations closest at hand at the time. Then predict values from it and that's your curve.

What kind of regression should be run at each point, though?

We might just take the average, in which case we end up with *kernel regression*. Kernel regression can give us problems in the case of regression discontinuity, especially if there's any sort of trend in the outcome in the lead-up to the cutoff. Since it's just taking the average, it tends to flatten out the trend, leading to an at-the-cutoff prediction that attributes some of the trend that's really there to the jump at the cutoff. Figure 20.9 shows an exaggerated example of how this can occur. Notice that the prediction to the left of the cutoff is way lower than all the actual data, because it's missing the trend upwards.

Kernel. A function that describes how values are weighted. In the context of local regression, it describes how they are weighted based on how close they are to the local value of interest.

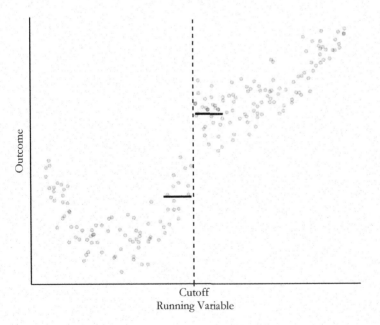

Figure 20.9: Regression Discontinuity with Means at the Cutoff

More commonly, in application to regression discontinuity, we'll do *local regression*, the most common form of which is LOESS, which we already discussed in Chapter 4 on Describing Relationships, and earlier in this chapter in Figure 20.2. Instead of taking a local mean, LOESS estimates a linear ("local linear regression") or 2nd-order polynomial ("local polynomial regression" or "local quadratic regression") regression at each point.

The local regression approach allows for the relationship between X and Y to not be totally flat. Depending on how curvy it is, the linear or 2nd-order polynomial function we've chosen might not be quite enough to capture the relationship. However, since we've zoomed in so far, it's unlikely that anything will be that highly nonlinear over such a narrow range of the x-axis.

Of course, given our plans to use local regression, we don't actually have to do anything special. While applying a local regression estimator across the whole range of the data will give us a great idea of what the overall relationship between running variable and outcome is, if we're just interested in estimating regression discontinuity, well... we're only really interested in two local points—just to the left of the cutoff, and just to the right. So if we estimate a regular-ol' interacted linear regression like in Equation 20.1 (for local linear regression) or Equation 20.3 (for local polynomial regression), but with a bandwidth (or perhaps some weighting for a slow-fade-out kernel), we've got it.

LET'S CODE UP SOME REGRESSION DISCONTINUITY! I'm going to do this in two ways. First I'm going to run a plain-ol' ordinary least squares model with a bandwidth and kernel weight applied, with heteroskedasticity-robust standard errors. However, there are a number of other adjustments we'll be talking about in this chapter, and it can be a good idea to pass this task off to a package that knows what it's doing. I see no point in showing you code you're unlikely to use just because we haven't gotten to the relevant point in the chapter yet. So we'll be using those specialized commands as well, and I'll talk later about some of the extra stuff they're doing for us.

For this example we're going to use data from Government Transfers and Political Support by Manacorda, Miguel, and Vigorito (2011).[25] This paper looks at a large poverty alleviation program in Uruguay which cut a sizeable check to a large portion of the population.[26] They are interested in whether receiving those funds made people more likely to support the newly-installed center-left government that sent them.

Who got the payments? You had to have an income low enough. But they didn't exactly use income as a running variable; that might have been too easy to manipulate. Instead, the government used a bunch of factors—housing, work, reported income, schooling—and *predicted* what your income would be from that. Then, the predicted income was the running variable, and treatment was assigned based on being below a cutoff. About 14% of the population ended up getting payments.

[25] Marco Manacorda, Edward Miguel, and Andrea Vigorito. Government transfers and political support. *American Economic Journal: Applied Economics*, 3 (3):1–28, 2011.

[26] The Plan de Atención Nacional a la Emergencia Sociál, or PANES. These were big payments, too—about $70 per month, which is about half what the typical beneficiary had been earning beforehand. On top of that, anyone with kids got a food card. There were some other goodies too.

The researchers polled a bunch of people near the income cutoff to check their support for the government afterwards. Did people just below the income cutoff support the government more than those just above?

The data set we have on government transfers comes with the predicted-income variable pre-centered so the cutoff is at zero. Then, support for the government takes three values: you think they're better than the previous government (1), the same (1/2) or worse (0). The data only includes individuals near the cutoff—the centered-income variable in the data only goes from −.02 to .02.[27]

Before we estimate our model, let's do a nearly-compulsory graphical regression discontinuity check so we can confirm that there does appear to be some sort of discontinuity where we expect it.[28] This is a "plot of binned means." There are some preprogrammed ways to do this like **rdplot** in R or Stata in the **rdrobust** package or **binscatter** in Stata from the **binscatter** package, but this is easy enough that we may as well do it by hand and flex those graphing muscles.

For one of these graphs, you generally want to (1) slice the running variable up into bins (making sure the cutoff is the edge between two bins), (2) take the mean of the outcome within each of the bins, and (3) plot the result, with a vertical line at the cutoff so you can see where the cutoff is. Then you'd generally repeat the process with treatment instead of the outcome to produce a graph like Figure 20.3.

[27] Some of the results here will differ slightly from the original paper so as to simplify the data-preparation process or highlight some potential things you *could* do that they didn't.

[28] Have you noticed how many graphs there are in this chapter? RDD is a lovingly visual design.

```
1   # R CODE
2   library(tidyverse)
3   gt <- causaldata::gov_transfers
4
5   # Use cut() to create bins, using breaks to make sure it breaks at 0
6   # (-15:15)*.02/15 gives 15 breaks from -.02 to .02
7   binned <- gt %>%
8       mutate(Inc_Bins = cut(Income_Centered,
9           breaks = (-15:15)*(.02/15))) %>%
10      group_by(Inc_Bins) %>%
11      summarize(Support = mean(Support),
12      Income = mean(Income_Centered))
13  # Taking the mean of Income lets us plot data roughly at the bin midpoints
14
15  ggplot(binned, aes(x = Income, y = Support)) +
16      geom_line() +
17      # Add a cutoff line
18      geom_vline(aes(xintercept = 0), linetype = 'dashed')
```

```
1   * STATA CODE
2   causaldata gov_transfers.dta, use clear download
3
4   * Create bins. 15 bins on either side of the cutoff from
```

```
5   * -.02 to .02, plus 0, means we want "steps" of\dots
6   local step = (.02)/15
7   egen bins = cut(income_centered), at(-.02(`step').02)
8
9   * Means within bins
10  collapse (mean) support, by(bins)
11
12  * And graph with a cutoff line
13  twoway (line support bins) || (function y = 0, horiz range(support)), ///
14      xti("Centered Income") yti("Support")
```

```
1   # PYTHON CODE
2   import pandas as pd
3   import numpy as np
4   import matplotlib.pyplot as plt
5   import seaborn as sns
6   from causaldata import gov_transfers
7   d = gov_transfers.load_pandas().data
8
9   # cut at 0, and 15 places on either side
10  edges = np.linspace(-.02,.02,31)
11  d['Bins'] = pd.cut(d['Income_Centered'], bins = edges)
12
13  # Mean within bins
14  binned = d.groupby(['Bins']).agg('mean')
15
16  # And plot
17  sns.lineplot(x = binned['Income_Centered'],
18              y = binned['Support'])
19  # Add vertical line at cutoff
20  plt.axvline(0, 0, 1)
```

A little extra theming gets us Figure 20.10.[29] There's definitely a break at the cutoff there, and some higher support to the left of the cutoff (treatment is to the left here since you need low income to get paid) that isn't what we'd expect by following the trend on the right of the cutoff.

Now that we have our graph in mind, we can actually estimate our model. We'll start doing it with OLS and a 2nd-order polynomial, and then we'll do a linear model with a triangular kernel weight, limiting the bandwidth around the cutoff to .01 on either side. The bandwidth in this case isn't super necessary— the data is already limited to .02 on either side around the cutoff—but this is just a demonstration.

It's important to note that the first step of doing regression discontinuity with OLS is to center the running variable around the cutoff. This is as simple as making the variable $RunningVariable - Cutoff$, translated into whatever language you use. The second step would then be to create a "treated" variable $RunningVariable < Cutoff$ (since treatment is applied below the cutoff in this instance). But in this case, the running variable comes pre-centered ($IncomeCentered$) and the

[29] Specifically from the R version, as with all the graphs in this book I added the `theme_pubr()` theme from **ggpubr**, and some additional text options in the theme function. It's very much worth your time and effort to learn how to use the graphing package of your choice to make graphs look *nice*.

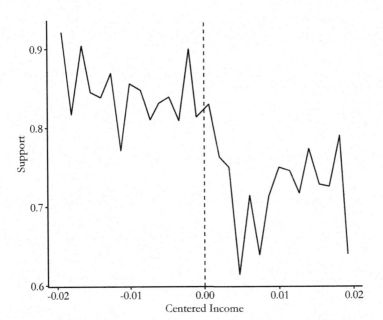

Figure 20.10: Government Support by Income with Policy Cutoff

below-cutoff treatment variable is already in the data (*Participation*) so I'll leave those parts out.

```
1  # R CODE
2  library(tidyverse); library(modelsummary)
3  gt <- causaldata::gov_transfers
4
5  # Linear term and a squared term with "treated" interactions
6  m <- lm(Support ~ Income_Centered*Participation +
7          I(Income_Centered^2)*Participation, data = gt)
8
9  # Add a triangular kernel weight
10 kweight <- function(x) {
11         # To start at a weight of 0 at x = 0, and impose a bandwidth of .01,
12         # we need a "slope" of -1/.01 = 100,
13         # and to go in either direction use the absolute value
14         w <- 1 - 100*abs(x)
15         # if further away than .01, the weight is 0, not negative
16         w <- ifelse(w < 0, 0, w)
17         return(w)
18 }
19
20 # Run the same model but with the weight
21 mw <- lm(Support ~ Income_Centered*Participation, data = gt,
22          weights = kweight(Income_Centered))
23
24 # See the results with heteroskedasticity-robust SEs
25 msummary(list('Quadratic' = m, 'Linear with Kernel Weight' = mw),
26     stars = c('*' = .1, '**' = .05, '***' = .01), vcov = 'robust')
```

```
1  * STATA CODE
2  causaldata gov_transfers.dta, use clear download
3
```

```
4  * Include the running variable, its square (by interaction with itself)
5  * and interactions of both with treatment, and robust standard errors
6  reg support i.participation##c.income_centered##c.income_centered, robust
7
8  * Create the triangular kernel weight. To start at a weight of 0 at x = 0,
9  * and impose a bandwidth of .01, we need a "slope" of -1/.01 = 100
10 * and to go in either direction use the absolute value
11 g w = 1 - 100*abs(income_centered)
12 * if further away than .01, the weight is 0, not negative
13 replace w = 0 if w < 0
14
15 reg support i.participation##c.income_centered [aw = w], robust
```

```
1  # PYTHON CODE
2  import numpy as np
3  import statsmodels.formula.api as smf
4  from causaldata import gov_transfers
5  d = gov_transfers.load_pandas().data
6
7  # Run the polynomial model
8  m1 = smf.ols('''Support~Income_Centered*Participation +
9  I(Income_Centered**2)*Participation''', d).fit()
10
11 # Create the kernel function
12 def kernel(x):
13     # To start at a weight of 0 at x = 0,
14     # and impose a bandwidth of .01, we need a "slope" of -1/.01 = 100
15     # and to go in either direction use the absolute value
16     w = 1 - 100*np.abs(x)
17     # if further away than .01, the weight is 0, not negative
18     w = np.maximum(0,w)
19     return w
20
21 # Run the linear model with weights using wls
22 m2 = smf.wls('Support~Income_Centered*Participation', d,
23 weights = kernel(d['Income_Centered'])).fit()
24
25 m1.summary()
26 m2.summary()
```

From this we get the results in Table 20.1. From the 2nd-order polynomial model our estimate is that receiving payments increased support for the government by 9.3 percentage points—not bad![30] The version with the kernel weight shows a statistically insignificant effect of 3.3 percentage points. The first model is more plausible to me—the bandwidth was already fairly narrow since the data only included observations near the cutoff anyway, so limiting the data further isn't necessary. You can also see the reduction in sample size from 1948 to 937—this occurs because it's not counting anyone who gets a weight of 0 as being a part of the sample. The potential for reduced bias (by taking fewer people farther from the cutoff) comes with the price of a reduced sample and less precision.

[30] Using the specific methods from the original paper, their estimates were more like 11 to 13 percentage points.

Table 20.1: The Effect of Government Payments on Support Regression Discontinuity Estimates

	Quadratic	Linear with Kernel Weight
(Intercept)	0.769***	0.819***
	(0.034)	(0.015)
Income (Centered)	-11.567	-23.697***
	(8.101)	(3.219)
Participation	0.093**	0.033
	(0.044)	(0.021)
(Income (Centered))^2	562.247	
	(401.982)	
Income (Centered) × Participation	19.300*	26.594***
	(10.322)	(4.433)
(Income (Centered))^2 × Participation	-101.103	
	(502.789)	
Num.Obs.	1948	937
R2	0.036	0.041
Std. Errors	Robust	Robust

* $p < 0.1$, ** $p < 0.05$, *** $p < 0.01$

Now that we have our by-hand version down, we can try things out using some prepackaged commands, both for estimating the model effects and for making regression discontinuity-style plots. For both R and Stata, there is the **rdrobust** package, which is a part of the "RD packages" universe that contains a bunch of useful tools I'll be including throughout this chapter.[31] Unfortunately, there's not a fantastic Python option at the moment, although a Python version of **rdrobust** is in development as of this writing. Check the **rdrobust** website for updates and code examples. In the meantime you can call R from Python using the **rpy2** package.

The rdrobust function in the **rdrobust** package in both R and Stata does a few things for us. First, we don't have to choose the bandwidth. It will do an optimal bandwidth selection procedure for us. More on how that works in the How the Pros Do It section.

Second, it will properly implement local regression with your polynomial order of choice (linear by default) and a triangular kernel.

Third, as implied by the name, it will apply heteroskedasticity-robust standard errors. Specifically, it uses a robust standard error estimator that takes into account the structure of the regression discontinuity problem. It looks for each observation's *nearest neighbors* along the running variable, and uses that

[31] See https://rdpackages.github.io/, and the many econometrics papers cited therein.

information to figure out the amount of heteroskedasticity and how to correct for it.

Fourth, it applies a *bias correction*. Methods for optimal bandwidth choice tend to select bandwidths that are wide, so as to help increase the effective sample size and improve precision. However, estimates of standard errors rely on our idea of what the distribution of the error term is. And those wide bandwidths futz up what those distributions look like. So there is a method here to correct for that bias, improving the estimate of the standard error. More on this in the How the Pros Do It section.

```
1  # R CODE
2  library(tidyverse); library(rdrobust)
3  gt <- causaldata::gov_transfers
4
5  # Estimate regression discontinuity and plot it
6  m <- rdrobust(gt$Support, gt$Income_Centered, c = 0)
7  summary(m)
8  # Note, by default, rdrobust and rdplot use different numbers
9  # of polynomial terms. You can set the p option to standardize them.
10 rdplot(gt$Support, gt$Income_Centered)
```

```
1  * STATA CODE
2  * If necessary: ssc install rdrobust
3  causaldata gov_transfers.dta, use clear download
4
5  * Run the RDD model and plot it. Note, by default,
6  * rdrobust and rdplot use different
7  * numbersof polynomial terms.
8  * You can set the p() option to standardize them.
9  rdrobust support income_centered, c(0)
10 rdplot support income_centered, c(0)
```

If you run this code you'll find a fairly different **rdrobust** output as opposed to when we did it by hand. Results are insignificant and in the opposite direction of the original paper.[32] This comes down to **rdrobust** selecting a narrow bandwidth— if you force the bandwidth to be the entire range, as might be reasonable given how narrow it already is—you get results that are nearly identical to the quadratic-model results from Table 20.1. Making these decisions is hard! Just because we have a premade package doesn't mean we can let it make all the decisions. Use that head of yours, and explore the options in the functions you use by reading the documentation.

[32] Coefficients are positive, but it's assuming that treatment is to the right of the cutoff, so this translates into a negative effect of being treated, which occurs here on the left of the cutoff.

20.2.2 Fuzzy Regression Discontinuity

HOW MANY TREATMENTS ARE THERE REALLY that are *completely* determined by a cutoff in a running variable? Some, certainly. But we live in the real world. Even in cases where a policy is *supposed* to be assigned based on a cutoff, surely there might be a few people above or below who manage to get the "wrong" treatment. Forgetting that, there are plenty of scenarios where the policy isn't necessarily *supposed to* be assigned based on a cutoff, but still for some reason the chances of treatment jump significantly at a cutoff. Perhaps you wanted to evaluate the effect of holding a high school degree on some outcome. The chances that you hold a high school degree jump significantly in the summer following your 18th birthday. But there are quite a few people who graduate at 17, or 19, or not in the summer, or never graduate at all. There's a *jump* in the treatment rate at that summer cutoff, but it's not from 0% to 100%. We can see how the treatment rate might look across the running variable in Figure 20.11.

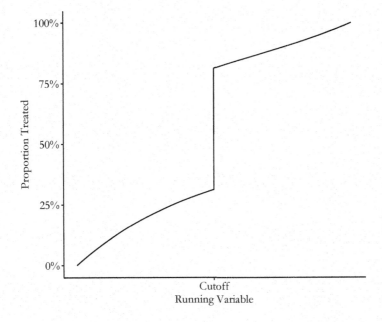

Figure 20.11: An Example of Treatment Rates in a Fuzzy Regression Discontinuity

I already covered the basic concept here in the How Does It Work section. But how can we actually implement a fuzzy regression discontinuity design?

We can figure it out by modifying the causal diagram we had in Figure 20.1. The only changes we need to account for is that there's some other way for *RunningVariable* and *Treatment* to be related *other* than the cutoff, which is why we see changes in the treatment rate at non-cutoff points in Figure 20.11. We also might have some non-*RunningVariable* determinants of *Treatment*.

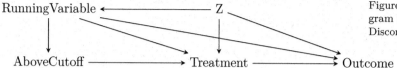

Figure 20.12: A Causal Diagram that Fuzzy Regression Discontinuity Works For

We get our modified causal diagram in Figure 20.12.[33] What can we get out of this diagram?

Well, one thing to note is that the idea for identifying the effect is still basically the same. If we control for *RunningVariable*, then there is no back door from *AboveCutoff* to *Outcome*. By cutting out any variation driven by *RunningVariable* anywhere *but* the cutoff, we're left only with variation that has no back doors and identifies the effect of *Treatment* on *Outcome*.

The only real change is this: we can no longer simply limit the data to the area around the cutoff and call that a day on controlling for *RunningVariable*. Doing that would lead us to understate the effect. If we're only getting a jump in treatment rates of, say, 15 percentage points, then we should only expect to see 15% of the effect. We have to scale it back to the full size.

Thankfully this can be done by applying instrumental variables, as in Chapter 19, using basically the same regression discontinuity equations as for the sharp RDD, like Equation 20.1 or Equation 20.3. Except these equations now become our second-stage equations. Our first stage uses *AboveCutoff* as an instrument for *Treated* (and interactions with *AboveCutoff* as instruments for the interactions with *Treated*).

This scales the estimate exactly as we want it to. In its basic form, instrumental variables divides the effect of the instrument on the outcome by the effect of the instrument on the endogenous/treatment variable.[34] In other words, we're scaling the effect of being above the cutoff on the outcome (i.e., what we'd get from a typical sharp-regression-discontinuity model) but dividing to account for the fact that being above the cutoff only led to a partial increase in treatment rates.

JUST LIKE WITH SHARP REGRESSION DISCONTINUITY, we can implement fuzzy regression discontinuity in code either using our standard tools (the same instrumental variables estimators we used in Chapter 19), or specialized regression-discontinuity specific tools. Let's do both here.

For this, we'll be using data from Fetter (2013).[35] Fetter also gives us an opportunity to see regression discontinuity

[33] In this diagram, Z might represent a bunch of different variables here—it might even be different variables with arrows to *RunningVariable* and to *Treatment*. The logic still holds up.

[34] Because of those interactions we're doing something not quite the same as the basic form, but the logic carries over very well.

[35] Daniel K Fetter. How do mortgage subsidies affect home ownership? Evidence from the mid-century GI bills. *American Economic Journal: Economic Policy*, 5(2):111–47, 2013.

performed in a slightly less-clean setting. Fetter's main question of interest is how much of the increase in the home ownership rate in the midcentury US was due to mortgage subsidies given out by the government.

How does he get at this question? He looks at people who were about the right age to be veterans of major wars like World War II or the Korean War. Anyone who was a veteran of these wars received special mortgage subsidies.

What does this have to do with regression discontinuity? There's an age requirement to join the military. If you were born one year too late to join the military to fight in the Korean War, then you won't get these mortgage subsidies (or at least far fewer veterans were eligible). Born just in time? You can join up and get subsidies! So there's a regression discontinuity here based on birth year. Born just in time, or a bit too late?

Of course, not everybody born in a given year joins the military.[36] So the "treatment" of being eligible for mortgage subsidies would only apply to some people born at the right time. Treatment rates jump from 0% to... above 0% but not 100%. That's fuzzy!

And why do I say this is "less-clean" than some of other examples? Because this isn't an explicitly designed cutoff. There's a lot more choice in terms of who gets treated, and, heck, some people were probably getting into the military underaged. We have to think very carefully here about whether we believe the assumptions necessary to use regression discontinuity. I think the design here is pretty reasonable, but my opinion on that is not as rock-solid as it might be for something for which the assignment of treatment had fewer moving parts.[37]

The author also has to go through some checks and tests that someone with a simpler story might not have to. For example, while I won't show it in the code below, he performs a placebo test by repeating the analysis using data from other eras, where veteran status wouldn't have an effect on receiving mortgage subsidies, finding no effect at those times. You're about to see the analysis include control variables for season of birth, race, and the US state in which they were born—controls generally aren't necessary for regression discontinuity, but it makes sense to include them here. If the design itself won't do all the work for you, then geez I guess you're gonna have to work a little harder.

Let's code up the analysis. First we'll do it ourselves using the instrumental variables code we used in Chapter 19. We're

[36] Some can't, or wouldn't be allowed. They weren't taking a whole lot of women at the time, for example. But even among eligible men, not everyone would choose to or end up drafted.

[37] But limiting ourselves only to the cleanest analyses would leave us with a lot of stuff we couldn't study at all.

using the running variable quarter of birth (qob), which has been centered on the quarter of birth you'd need to be to be eligible for a mortgage subsidy for fighting in the Korean War (qob_minus_kw). This determines whether you were a veteran of either the Korean War or World War II (vet_wwko). All of this is to see if veteran status (and its accompanying mortgage subsidies) affects your home ownership status (home_ownership)—remember, the goal here is looking for the effect of those mortgage subsidies. There are also controls for being white or non-white, and your birth state.

To keep things simple, we'll apply a bandwidth before doing estimation, we'll skip the kernel weighting, and we'll just do a linear model. The code we did before for incorporating these things still works here, though, if you want to add them.

```r
# R CODE
library(tidyverse); library(fixest); library(modelsummary)
vet <- causaldata::mortgages

# Create an "above-cutoff" variable as the instrument
vet <- vet %>% mutate(above = qob_minus_kw > 0)

# Impose a bandwidth of 12 quarters on either side
vet <- vet %>%  filter(abs(qob_minus_kw) < 12)

m <- feols(home_ownership ~
    nonwhite  | # Control for race
    bpl + qob | # fixed effect controls
    qob_minus_kw*vet_wwko ~ # Instrument our standard RDD
    qob_minus_kw*above, # with being above the cutoff
    se = 'hetero', # heteroskedasticity-robust SEs
    data = vet)

# And look at the results
msummary(m, stars = c('*' = .1, '**' = .05, '***' = .01))
```

```stata
* STATA CODE
* We will be using ivreghdfe,
* which must be installed with ssc install ivreghdfe,
* along with ftools, ranktest, reghdfe, and ivreg2
causaldata mortgates.dta, use clear download

* Create an above variable as an instrument
g above = qob_minus_kw > 0

* Impose a bandwidth of 12 quarters on either side
keep if abs(qob_minus_kw) < 12

* Regress, using above as an instrument for veteran status.
* Note that qob_minus_kw by itself doesn't need instrumentation
* so we separate it. Usually I advise against it,
* but in this case this is easier if we just make our
* own interaction with g (generate)
g interaction_vet = vet_wwko*qob_minus_kw
g interaction_above = above*qob_minus_kw
```

```
20 | ivreghdfe home_ownership nonwhite qob_minus_kw ///
21 |     (vet_wwko interaction_vet = above interaction_above), ///
22 |     absorb(bpl qob) robust
23 |
24 | * We can also use regular ol' ivregress but it will be slower
25 | encode bpl, g(bpl_num)
26 | ivregress 2sls home_ownership nonwhite qob_minus_kw i.bpl_num i.qob ///
27 |     (vet_wwko interaction_vet = above interaction_above), robust
```

```
 1 | # PYTHON CODE
 2 | import pandas as pd
 3 | from linearmodels.iv import IV2SLS
 4 | from causaldata import mortgages
 5 | d = mortgages.load_pandas().data
 6 |
 7 | # Create an above variable as an instrument
 8 | d['above'] = d['qob_minus_kw'] > 0
 9 | # Apply a bandwidth of 12 quarters on either side
10 | d = d.query('abs(qob_minus_kw) < 12')
11 |
12 | # Create an control-variable DataFrame
13 | # including dummies for bpl and qob
14 | controls = pd.concat([d[['nonwhite']],
15 |                 pd.get_dummies(d[['bpl']])], axis = 1)
16 |
17 | d['qob'] = pd.Categorical(d['qob'])
18 | # Drop one since we already have full rank with bpl
19 | # (we'd also drop_first with bpl if we did add_constant)
20 | controls = pd.concat([controls,
21 |                 pd.get_dummies(d[['qob']],
22 |                 drop_first = True)], axis = 1)
23 | # the RDD terms:
24 | # qob_minus_kw by itself is a control
25 | controls = pd.concat([controls, d[['qob_minus_kw']]], axis = 1)
26 |
27 | # we need interactions for the second stage
28 | d['interaction_vet'] = d['vet_wwko']*d['qob_minus_kw']
29 | # and the first
30 | d['interaction_above'] = d['above']*d['qob_minus_kw']
31 |
32 | # Now we estimate!
33 | m = IV2SLS(d['home_ownership'], controls,
34 |         d[['vet_wwko','interaction_vet']],
35 |         d[['above','interaction_above']])
36 |
37 | # With robust standard errors
38 | m.fit(cov_type = 'robust')
```

These show that veteran status at this margin increases home ownership rates by 17 percentage points. Not bad. We could also apply the exact same regression discontinuity plot code from last time (there's no special "fuzzy" plot, although now we'd call the typical regression discontinuity plot the "second stage" plot), and also apply that plot code using the treatment variable as an outcome (the "first stage"), to make sure that the assignment variable is working as expected. This gives us Figure 20.13.[38] It looks good! We definitely see a sharp change in the rate of

[38] Another way to think about it is grabbing hold of both graphs at the top and bottom and pulling them taller until the jump in Figure 20.13 (a) goes from 0% to 100%. However big the resulting jump is in Figure 20.13 (b) is the fuzzy RDD estimate—it's how much of a jump you'd expect if *everybody* went from untreated to treated (0% to 100%).

eligible veterans around the cutoff, and we see a discontinuity in the rate of home ownership at the same point. The fuzzy RDD estimate is the jump at the cutoff in Figure 20.13 (b) divided by the jump in the cutoff in Figure 20.13 (a).

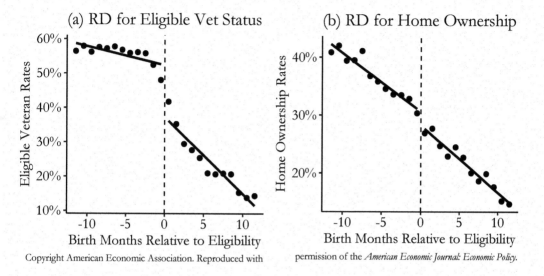

Copyright American Economic Association. Reproduced with permission of the *American Economic Journal: Economic Policy*.

Figure 20.13: Eligibility for Mortgage Subsidies for being a Korean War Veteran and Home Ownership from Fetter (2013)

Next we can, as before, use the **rdrobust** function in R and Stata, to do local polynomial regression, kernel weighting, easy graphs, all that good stuff. We just need to use **vet_wwko** as our "treatment status indicator" and give that to the **fuzzy** option.

```r
library(tidyverse); library(rdrobust)
vet <- causaldata::mortgages

# It will apply a bandwidth anyway, but having it
# check the whole bandwidth space will be slow. So let's
# pre-limit it to a reasonable range of one year
vet <- vet %>%
    filter(abs(qob_minus_kw) < 12)

# Create our matrix of controls
controls <- vet %>%
    select(nonwhite, bpl, qob) %>%
    mutate(qob = factor(qob))
# and make it a matrix with dummies
conmatrix <- model.matrix(~., data = controls)

# This is fairly slow due to the controls, beware!
m <- rdrobust(vet$home_ownership,
              vet$qob_minus_kw,
              fuzzy = vet$vet_wwko,
              c = 0,
              covs = conmatrix)

summary(m)
```

```
1   * STATA CODE
2   * We will be using ivreghdfe, which must be installed with ssc install ivreghdfe
3   * along with (all with ssc install) ftools, ranktest, reghdfe, and ivreg2
4   causaldata mortgages.dta, use clear download
5
6   * Create an above variable as an instrument
7   g above = qob_minus_kw > 0
8   * Impose a bandwidth of one year on either side
9   keep if abs(qob_minus_kw) < 12
10
11  * Regress, using above as an instrument for veteran status
12  * Note that qob_minus_kw by itself doesn't need instrumentation,
13  * so we separate it. Usually I advise against it, but in this case
14  * this is easier if we just make our own interaction
15  g interaction_vet = vet_wwko*qob_minus_kw
16  g interaction_above = above*qob_minus_kw
17  ivreghdfe home_ownership nonwhite qob_minus_kw ///
18      (vet_wwko interaction_vet = above interaction_above), ///
19      absorb(bpl qob) robust
20
21  * We can also use regular ol' ivregress
22  * but it will be slower
23  encode bpl, g(bpl_num)
24  ivregress 2sls home_ownership nonwhite qob_minus_kw i.bpl_num i.qob ///
25      (vet_wwko interaction_vet = above interaction_above), robust
```

It's not included in the code, but we could easily apply `rdplot` just like we did last time (and maybe apply it with the treatment as the dependent variable as well, just to make sure it looks like we expect).

The `rdrobust` estimate is quite different from the linear instrumental variables estimate we did earlier, with veteran status increasing home ownership rates by 51.8 percentage points, as opposed to 17.0! That's quite a difference. It's useful to know when estimation methods differ like this. Which should we trust? Well, let's think through why they differ: `rdrobust` uses a 2nd-order polynomial rather than a linear regression, it uses local regression, and it uses a narrower bandwidth (it picks 3.39 birth quarters on either side of the cutoff, rather than 12).

Our first step in thinking which to trust would have to ask if one of the models is making assumptions that are much more likely to be accurate. These are the sorts of things you'd have to defend when describing your research design. If we think linearity is reasonable, for example, then the linear model might be less noisy and a more precise estimate than the 2nd-order polynomial. Does Figure 20.13 look linear? If we think that the wider bandwidth is likely to contain too many non-comparable people, then we might think the narrower bandwidth is better. Do we think 12 months on either side of the cutoff is too far away to be comparable?

Often these days, researchers performing regression discontinuity will see how sensitive the estimate is to different bandwidths, running the model many times and reporting how the effect changes, letting the reader of the study think about how close to the cutoff things should be limited.[39] The Bana, Bedard, and Rossin-Slater (2020) paper I discuss in the Regression Kink section does this.

20.2.3   Placebo Tests

THE ASTONISHING THING ABOUT REGRESSION DISCONTINUITY is that it closes all back doors, even the ones that go through variables we can't measure. That's the whole idea—since we're isolating variation in such a narrow window of the running variable, it's pretty plausible to claim that the *only* thing changing at the cutoff is treatment, and by extension anything that treatment affects (like the outcome).

Even more convenient is that this extremely-nice result from regression discontinuity gives us a fairly straightforward way to test if it's true. Simply run your regression discontinuity model on something that the treatment *shouldn't* affect. Anything we might normally use as a control variable should serve for this purpose.[40] If we *do* find an effect, our original regression discontinuity design might not have been quite right. Perhaps things weren't as random at the cutoff as we'd like, for some reason.

There's not too much to say here. The equations, code, everything is the exact same for this test as for regular regression discontinuity. We just swap out the actual outcome variable for some other variable that the treatment shouldn't affect. In many cases we can try it with a long list of variables that might serve as control variables.

Figure 20.14 shows this test with two variables from the Government Transfers and Political Support paper. No effects here! Exactly what we were hoping for.

One important thing to keep in mind with these placebo tests is that, since you can run them on a long list of potential placebo outcomes, you're likely to find a few nonzero effects just by random chance. It happens. If you test a long list of variables and find a few differences, that's not a fatal problem with your design. In these cases, you may want to add the variables with the failed placebo tests to the model as control variables. Adding controls can be tricky if you're using local regression, but there are methods for doing it,[41] and pre-packaged regression discontinuity commands often incorporate them.

[39] In general, for any parameter you have to pick for your analysis, like a bandwidth, it's not a terrible idea to try a few different ones and report how the results change. Maybe it doesn't change much, which is reassuring. Or maybe it does, and that's still good to know, and honest to report.

[40] As long as it's not something that treatment should affect. But then, we know from Chapter 8 that we wouldn't usually want to control for something treatment would affect anyway.

[41] Sebastian Calonico, Matias D Cattaneo, Max H Farrell, and Rocio Titiunik. Regression discontinuity designs using covariates. *Review of Economics and Statistics*, 101(3):442–451, 2019.

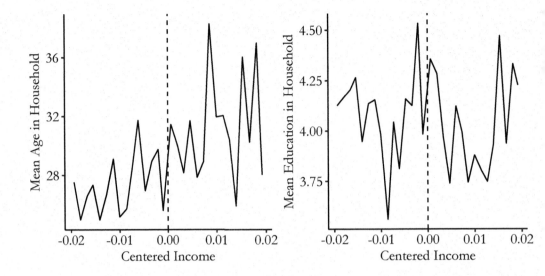

Figure 20.14: Performing
Regression Discontinuity
with Controls as Outcomes
for a Placebo Test

20.2.4   The Density Discontinuity Test

FOR REGRESSION DISCONTINUITY TO WORK, WE NEED A SIT-
UATION WHERE people just on either side of the cutoff are effectively randomly assigned to treatment. That won't happen
if someone's thumb is on the scale!

There are two ways manipulation could happen. First, who-
ever (or whatever) is in charge of setting the cutoff value might
do so with the knowledge of exactly who it will lead to getting
treated. Imagine the track and field coach running tryouts who
conveniently decides that anyone on the team must run a mile
in 5 minutes and 37 seconds or better *just after* his son clocks
in at 5 minutes and 35 seconds.

Second, individuals themselves likely have some control over
their running variable. Sometimes they have direct control—if
my local supermarket offers me a discounted price on tuna if I
buy 12 or more cans, then I can choose exactly what my running
variable (number of cans of tuna I buy) is, and it's not really
plausible to say that someone buying 12 cans is comparable to
someone buying 11 so I can identify the effect of the discount.

More often, we have indirect control—if I'm trying out for
that track and field team, I can decide to take it easy or push
myself really hard to run faster. Or maybe the person keeping
time really likes me and decides to shave a few seconds off of
the time they write down for me. This is a problem if we have
indirect control and also know what the cutoff is. If I know the

cutoff is 5:37 and I usually run a 5:40, I know I have to push extra hard to get a chance at the team and so will try very hard, whereas someone who normally runs 4:30 would just take it easy. Or maybe the person keeping time might see that I actually ran a 5:38, and figure I came *so close* it wouldn't hurt to cheat a little and write down 5:36 instead so I can get on the team. The person who runs 6:30 isn't likely to get a whole minute chopped off, though, that would be too much to ask.

In all of these cases, we want to be sure we understand how the running variable is determined. Are we in a cans-of-tuna situation? A slyly-choose-the-cutoff situation? A situation where we know the cutoff and may manipulate our behavior for it? For any of these, we may be worried that manipulation of the cutoff, or of the running variable near the cutoff, makes people on either side incomparable. If my friendly run-timer shaves off a few seconds for me so I get on the team, but not for Denny who also runs a 5:38 but the run-timer doesn't like,[42] then treatment assignment wasn't random, it was based on who the run-timer liked, which has all sorts of back doors.

IN THE CASE OF INDIRECT CONTROL we do have a test we can perform to check whether manipulation seems to be occurring at the cutoff. Conceptually it's very simple. We just look at the distribution of the running variable around the cutoff. It should be smooth, as we might expect if the running variable were being randomly distributed without regard for the cutoff.

But if we see a distribution that seems to have a dip just to one side of the cutoff, with those observations sitting just on the other side, that looks a lot like manipulation. If you looked at the run times at track and field tryouts and there was only Denny with a time anywhere from 5:38 to 5:45, but ten people just a second or two below the cutoff, that looks pretty fishy! That run-timer seems to be up to no good.

The idea of putting this intuition into action comes from McCrary (2008),[43] although there have been improvements to the basic idea since then, with better density estimation and power. The instructions in the code below are for Cattaneo, Jansson, and Ma (2020).[44]

The plan is this:

1. Calculate the density distribution of the treatment variable

2. Allow that density to have a discontinuity at the cutoff

3. See if there is a significant discontinuity at the cutoff

[42] Denny probably had it coming.

[43] Justin McCrary. Manipulation of the running variable in the regression discontinuity design: A density test. *Journal of Econometrics*, 142(2):698–714, 2008.

[44] Matias D Cattaneo, Michael Jansson, and Xinwei Ma. Simple local polynomial density estimators. *Journal of the American Statistical Association*, 115(531):1449–1455, 2020a.

4. Also, plot the density distribution and look at it, and see if it looks like there's a discontinuity

A big discontinuity is not a good sign. It implies manipulation.

WE CAN CODE UP A DENSITY TEST using pre-packaged commands, at least in R and Stata. For Python we will be a bit more on our own. The code will still use the example data from Government Transfers and Political Support.

Before running our code we can ask ourselves about the potential for manipulation here. The government assigned treatment on the basis of not just one characteristic, but a sum of a whole bunch of things. It determined that process before knowing exactly who would qualify (making it hard for the government to manipulate). The whole process wasn't fully obvious (making it hard for people to manipulate), and it would be difficult to know beforehand exactly how close to the cutoff you were (making it hard to know if you should bother trying to manipulate).

So it seems unlikely that this particular study will show evidence of manipulation. But you never know! Let's see. This test uses a slightly expanded version of the data including people who weren't surveyed at the right time to be included in the study, and those outside the bandwidth used in estimation (so we limit to the bandwidth ourselves).

```r
# R CODE
library(tidyverse); library(rddensity)
gt <- causaldata::gov_transfers_density   %>%
    filter(abs(Income_Centered) < .02)

# Estimate the discontinuity
gt %>%
    pull(Income_Centered) %>%
    rddensity(c = 0) %>%
    summary()
```

```stata
* STATA CODE
* If necessary, findit rddensity and install the rddensity package
causaldata gov_transfers_density.dta, use clear download

* Limit to the bandwidth ourselves
keep if abs(income_centered) < .02
* Run the discontinuity check
rddensity income_centered, c(0)
```

As expected, we find no statistically significant break in the distribution of income at the cutoff. Hooray!

Of course, we are more interested in seeing whether our assumption of no manipulation is *plausible* than we are in seeing whether we can reject it to a statistically significant degree.

Does it pass the sniff test? We can check this by plotting the distribution of the (binned) running variable, as in Figure 20.15. It certainly doesn't *look* like there's any sort of discontinuity in the distribution at the cutoff. You can make a very similar graph to this one with `rdplotdensity` in the **rddensity** package in R. In Stata or Python you can follow the steps for the binned scatterplot from before, except instead of getting the mean of the outcome for each bin, instead get the *count of observations* within each bin. Then plot with `twoway bar` in Stata or `seaborn.barplot` in Python.

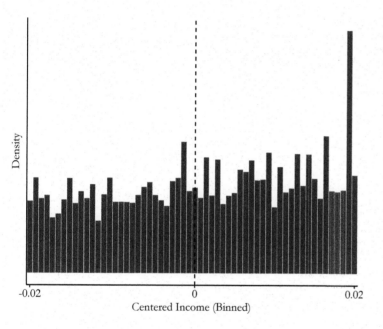

Figure 20.15: Distribution of the Binned Running Variable

20.3 How the Pros Do It

20.3.1 Regression Kink

REGRESSION DISCONTINUITY LOOKS FOR A CHANGE IN THE MEAN OUTCOME AT THE CUTOFF. Why not look for a change in something else? Why not indeed! You could theoretically look for a change in just about anything. Why not a change in the standard deviation of the outcome? Or the median? Or an R^2 of a regression of the outcome on a whole bunch of predictors?[45]

The most common "other thing" to look at, though, is a change in the *slope* of the relationship between the outcome and the running variable. The treatment administered at the cutoff

[45] Each of these would require some careful thought about how exactly to estimate them, but conceptually they should work. I went looking to see if a finance paper has ever used the standard deviation version of regression discontinuity and was surprised not to find one. There's an opportunity for you finance students out there. (Or perhaps I just didn't look hard enough.)

doesn't make the outcome Y itself change/jump—instead, the treatment changes the strength of the relationship between Y and $RunningVariable$. The resulting graph might look something like Figure 20.16—no jump at the cutoff but there's sure *something* happening.

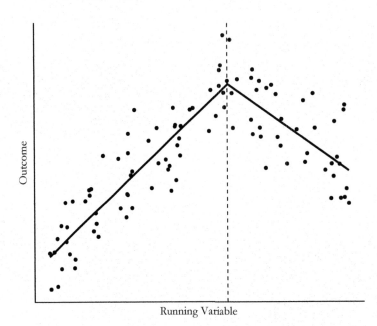

Figure 20.16: A Regression Kink on Simulated Data

WHY MIGHT A TREATMENT CHANGE THE RELATIONSHIP? There are two main ways it might pop up. One might be that a treatment literally changes the relationship. For example, say we were running a regression discontinuity with time as our running variable,[46] and the amount of employment in tech companies as the outcome. Then, at a particular time (our cutoff), a new policy goes into place that encourages more investment in tech companies. We wouldn't see tech employment rise right away (no jump), but we'd expect tech employment to start increasing more quickly than it did before (increase in the slope).

A more common application of regression kink is when *treatment itself has a kink*, like Figure 20.16 but with "Treatment level/rate" on the y-axis instead of "Outcome." Lots of government policies work this way. Take unemployment insurance, for example, which is a common regression kink application. I'll cite Card, Lee, Pei, and Weber (2015) as one example of many, picking it largely because I was already going to cite it for another reason below.[47] They look at data from Austria, where unemployment insurance benefits are 55% of your regular earnings, up to the point where you hit the maximum amount of

[46] In other words, an event study as in Chapter 17. The design mentioned here would be ill-advised as it doesn't really satisfy all the event study assumptions from Chapter 17, but it still works to get the idea across.

[47] David Card, David S Lee, Zhuan Pei, and Andrea Weber. Inference on causal effects in a generalized regression kink design. *Econometrica*, 83(6):2453–2483, 2015.

generosity, and then you get no more. So your regular earnings (running variable) positively affect the amount of unemployment insurance you get (treatment), up to the cutoff, at which point the effect of regular earnings on your unemployment insurance payment becomes zero.

So if you think, for example, that more-generous unemployment benefits make people take longer to find a new job, then you should expect to find a positive relationship between your regular earnings (running variable) and time-to-find-a-new-job (outcome) up to the point of the cutoff, and then it should be flat after the cutoff. That's exactly what they found.

ONE EXAMPLE OF TREATMENT-HAS-A-KINK REGRESSION KINK DESIGN is in Bana, Bedard, and Rossin-Slater (2020).[48] They look at the impact of paid family leave in California. Paid family leave is a policy where pregnant mothers can take an extended paid time off of work so they can, you know, give birth and care for a newborn without having to reply to emails. In California, the state pays 55% of regular earnings up to a maximum benefit amount. So the paid family leave payment (treatment) increases with regular earnings (running variable), until you get to the level of earnings (cutoff) where you're being paid the maximum benefit amount. After that, additional regular earnings don't increase your family leave payment. The setup is very similar to the unemployment insurance case I discussed from Card, Lee, Pei, and Weber (2015), down to the 55% reimbursement rate, coincidentally.

The first thing to check is whether or not they actually see a change in the slope of treatment at the cutoff, and they definitely do. Figure 20.17 shows the average amount of paid family leave received in the sample based on the base-period earnings that their benefits were figured on. Even without a regression line slapped on top of it, that looks like a change in slope to me.

Next we can replace treatment on the y-axis with some outcomes to see whether those outcomes also change slope. If they do, that's evidence of an effect.[49] They look at a number of outcomes in their paper, but I've specifically picked out two: how long the mothers stay on family leave, and whether they use family leave again in the next three years, conditional on going back to work. I've picked these two because you can see an effect on one of them, but no effect on the other, shown in Figure 20.18.

What do we see in Figure 20.18? On the left we see basically nothing. No change in slope at the cutoff at all! This tells us

[48] Sarah H Bana, Kelly Bedard, and Maya Rossin-Slater. The impacts of paid family leave benefits: Regression kink evidence from California administrative data. *Journal of Policy Analysis and Management*, 39(4):888–929, 2020.

[49] Then, you can divide the change in slope you see in the outcome by the change in slope you see in treatment, i.e., run instrumental variables, to get the effect of treatment.

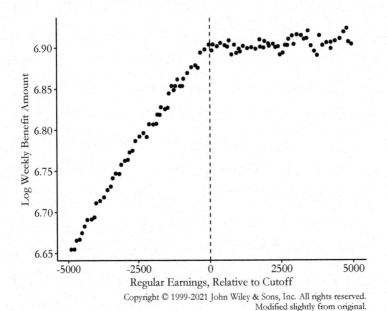

Figure 20.17: Paid Family
Leave Benefits (Treatment)
and Pre-Leave Earn-
ings (Running Variable)
from Bana, Bedard, and
Rossin-Slater (2020)

Copyright © 1999-2021 John Wiley & Sons, Inc. All rights reserved.
Modified slightly from original.

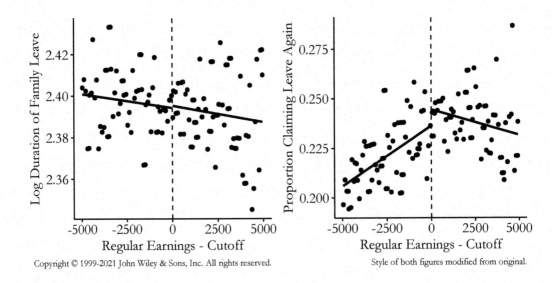

Copyright © 1999-2021 John Wiley & Sons, Inc. All rights reserved.
Style of both figures modified from original.

Figure 20.18: Log Leave
Duration or Proportion
Claiming Leave Again
(Outcomes) and Pre-Leave
Earnings (Running Vari-
able) from Bana, Bedard,
and Rossin-Slater (2020)

that additional family leave payments didn't affect how long the
family leave ends up being—if it did, we'd expect the slope to
change when the payments stopped increasing. In contrast, on
the right, there's a big change in slope at the cutoff. This tells us
that bigger leave payments *do* increase the chances of claiming
family leave again (i.e., having another kid), since the chance

of doing so is increasing while benefits are increasing, but then that stops as soon as the benefit increases stop.

The graph on the right doesn't just change slope though, it *jumps*! What can we make of that? Uh-oh, that might be a cause for concern. Treatment didn't jump, it just changed slope. So if the outcome jumps, rather than just changing slope, that implies that an assumption may be wrong.

Or at least it would be if that's what the original paper actually found. In the original paper there was no jump at all, just a nice change in slope. I just did regular ol' straight-lines-on-either side, but they did some stuff that was a little fancier.[50] In any case, the jump isn't big enough to be convincingly nonzero. I just included it myself to have a reason to talk about this, and to emphasize the importance of gut-checking your results.

[50] Their model doesn't actually allow for the jump, forcing it away. But they include several other details in estimation that make that assumption reasonable.

ACTUALLY PERFORMING A REGRESSION KINK isn't too hard once you have the concept of regression discontinuity down. You're just adding one layer of complexity on top of that. Most of the stuff I already covered still applies here—kernels, bandwidths, polynomials, local regression, and whatnot. The same placebos largely apply—there shouldn't be a change in the slope if you estimate the regression kink model with a control variable as the outcome, for example, and you'll want to make sure the running variable isn't discontinuous at the cutoff. The code doesn't change much—just set the `deriv` (which "deriv"ative of the regression equation do you want) option to 1 in your `rdrobust` code in either R or Stata.

Even the underlying regression equation isn't changing much. The idea is that you're still just estimating a line to the left and right of the cutoff, just like way back in Equation 20.1:

$$Y = \beta_0 + \beta_1(Running - Cutoff) + \beta_2 Treated +$$
$$\beta_3(Running - Cutoff) \times Treated + \varepsilon \quad (20.4)$$

There is sometimes a difference in the equation: some researchers will allow a slope change but *not* allow any sort of jump, i.e., dropping *Treated* by itself from the model, setting $\beta_2 = 0$. Dropping *Treated* helps reduce the influence of noise around the cutoff (and regression kink can use the noise-reducing help), although if there actually *should* be a jump there you wouldn't notice it.[51] The other difference is that now, instead of focusing on the coefficient representing the jump, β_2 (if it's even still in the model), you're instead focusing on the coefficient representing the change in slope, β_3.

[51] You could estimate it once allowing for the jump to see if it's there, and if it's not, estimate it without the jump to improve precision.

Of course, if you're using a setting where you've got a kink because the treatment level has a kink, you'd be using the cutoff as an instrument for treatment just like you did before with fuzzy regression discontinuity.[52]

So is anything else different? Some things, sure. Optimal bandwidth selection procedures are different, for one. You also might be more likely to use use a uniform kernel rather than a triangular one (which conveniently makes calculation way easier— just drop observations outside the bandwidth and go!).[53] See the "uniform" option for `kernel` in `rdrobust`. Also, as you might expect, since we're interested in the slope now instead of just a jump in the outcome, failing to properly model a nonlinear relationship is more of a problem here than for regression discontinuity.[54]

One big difference to watch out for is that the statistical power issue, which was already there with regression discontinuity, gets a lot more pressing. Estimating a change in relationship (slope) precisely takes a lot more observations than estimating a jump in a mean. You're trying to estimate that slope only using the observations you have right around the cutoff. You'd better hope you have a *lot* of observations near that cutoff.

[52] Although now there's an additional level of "fuzziness" to consider—what if the kink in treatment itself isn't perfectly applied? There are notes on this in Card, Lee, Pei, and Weber (2015), cited above. Told you I'd cite it down here.

[53] David Card, David S Lee, Zhuan Pei, and Andrea Weber. Regression kink design: Theory and practice. In *Advances in Econometrics*, volume 38, pages 341–382. Emerald Group Publishing Limited, Bingley, 2017.

[54] Michihito Ando. How much should we trust regression-kink-design estimates? *Empirical Economics*, 53(3):1287–1322, 2017.

20.3.2   Cutoffs Cut Off

SHARP CUTOFFS, FUZZY CUTOFFS, SLOPE-CHANGING CUTOFFS. WHO CAN HANDLE ALL THESE CUTOFFS? I hope it's you because you're about to be buried in cutoffs. Of *course* there's such a thing as a multi-cutoff design.

And if *that* wasn't enough, there are actually *several different kinds* of multi-cutoff designs. Multi multi-cutoff designs. This is exhausting. Let's talk about them, at least briefly.

SOMETIMES, CUTOFFS ARE A MOVING TARGET. A given treatment might have a cutoff that's different depending on where it is, who you are... you name it! This is fairly common. For example, a college admissions office might have one entrance-exam cutoff score for most students, and a different one for students applying for positions on the major sports teams. Or, speaking of exams, in the US if you want a GED (an alternative to a high school degree) you have to take an exam and get above a certain minimum score. But the cutoff score is different for each state. Or maybe you want to use regression discontinuity to analyze elections where third parties are prevalent. In some elections you might need 50.1% of the vote to win, but in others you might be able to win with, say, 42.7%.

Or maybe there is only one cutoff, but it changes over time. The Bana, Bedard, and Rossin-Slater (2020) paper on paid family leave from the last section is an example of this. The quarterly income at which you max out your family leave payments went up over time, from a little under \$20,000 in 2005 to a little over \$25,000 in 2014.

So the cutoff changes by group, or across regions, or across time, but presumably we want to gather all that data together—observations are at a premium in low-powered regression discontinuity, after all. What can we do?

The most common approach is to simply center the data and pretend there's nothing different. As you normally would, estimate your model using $Running Variable - Cutoff$, but simply allow each person's $Cutoff$ to be different depending on their scenario. Easy fix. This is exactly what Bana, Bedard, and Rossin-Slater did.

But is that all we can do? No. Centering the data works, but what exactly it's working *on* is a little different, especially once you consider treatment effect averages.[55] Regular regression discontinuity gives us a local average treatment effect—the average treatment effect among those just around the cutoff. So what do we get if we have multiple cutoffs? It's not exactly the average of those local effects. Instead you get weighted averages of those effects, weighted by how many observations are close to the respective cutoffs. That's a bit more complex than just "average of those at the cutoff," which may impact how useful your results would actually be for policy.

Additionally, by lumping everything together, you're throwing out some useful information. Why settle for some weird average when you can use the different cutoffs to try to get at treatment effects *away* from the cutoff? Perhaps you have group A with a low cutoff for treatment, and group B with a high cutoff. With a few assumptions about how much extrapolating we can do, we can produce an estimate for the treatment effect for people in group A who are far away from the group A cutoff, but near the group B cutoff.[56,57] Now, instead of the multiple cutoffs giving us a treatment effect average the equivalent of mashed potatoes, we get estimates that are even more informative than the straightforward single-cutoff version. Is this french fries in this analogy? I've lost the thread.

[55] Matias D Cattaneo, Rocío Titiunik, Gonzalo Vazquez-Bare, and Luke Keele. Interpreting regression discontinuity designs with multiple cutoffs. *The Journal of Politics*, 78(4):1229–1248, 2016.

[56] Matias D Cattaneo, Luke Keele, Rocío Titiunik, and Gonzalo Vazquez-Bare. Extrapolating treatment effects in multi-cutoff regression discontinuity designs. *Journal of the American Statistical Association*, pages 1–12, 2020b.

[57] For another approach to the same idea, see Bertanha (2020).

Marinho Bertanha. Regression discontinuity design with many thresholds. *Journal of Econometrics*, 218(1): 216–241, 2020.

Conveniently, these approaches to multiple cutoffs make it into the family of regression-discontinuity software packages alongside **rdrobust**, **rddensity**, and **rdpower**, in the form of **rdmulti** in both R and Stata.[58] The **rdmulti** package contains the rdmc function, which can be used to evaluate multi-cutoff settings.[59]

The **rdmulti** package also offers the rdms function, which can be used for *yet another* way in which you can have multiple cutoffs—when everyone is subject to the same cutoffs, but different cutoffs of the same running variable give different kinds or levels of treatment. For example, consider a blood pressure medication, where you're given a low dose if your blood pressure rises into a range of "potential concern," but a strong dose if your blood pressure becomes "dangerously high." The intuition here isn't wildly different from the idea of just running regression discontinuity twice, once at each cutoff.

OF COURSE, THAT'S ALL IF YOUR MULTIPLE CUTOFFS COME FROM ONLY ONE RUNNING VARIABLE. HOW DINKY! OVER HERE, US COOL KIDS HAVE TWO RUNNING VARIABLES. There are a fair number of cases where a treatment is assigned not on the basis of a cutoff of one running variable, but a cutoff on each of two (or more) running variables.

One place this is common is in *geographic* regression discontinuity. In a regression discontinuity based on geography, treatment occurs on one side of the border or another (like in our San Diego/Tijuana example from the start of this chapter). In many applications of geographic regression discontinuity, to cross from an untreated region into a treated one, you have to cross both a certain latitude and a certain longitude. Two running variables![60]

While there are applications of multiple running variables everywhere, one place where this seems to be fairly common is in education.[61] For example, perhaps to progress to the next grade a student needs to meet minimum scores on the math exams *and* on the English exams. Or to get a bonus, a teacher needs their students to meet a certain standard in *both* of the classes they teach. Or a school may be designated as failing to meet state standards only if it falls below multiple performance thresholds.

This is a problem, because now we have, in effect, *infinite* cutoffs. Or at least multidimensional ones. Say you need a 60 in both math and English to go to the next grade. It's not just that there's a cutoff at 60 for math and a cutoff at 60 for

[58] Notice how this "Cattaneo" guy keeps popping up?

[59] The rap-themed causal inference textbook *Causal Inference: The Mixtape* by Scott Cunningham does cover the use of the **rdrobust** and **rddensity** packages, but skips **rdmulti** and its rdmc function, missing the opportunity for an excellent Run-DMC reference.

[60] Keele and Titiunik (2015) discuss the two-running-variable nature of geographic regression discontinuity (which could be a whole section of this chapter on its own), as well as some other features of geographic regression discontinuity. These include the very high chance of manipulation in the running variable (people tend to sort themselves by choice on one side of a border or another), and the fact that crossing a border almost always applies *multiple* treatments, not just one (the effects of crossing from San Diego to Mexico can't be pinned on, say, switching from dollars to pesos. There's plenty going on there).

Luke J Keele and Rocio Titiunik. Geographic boundaries as regression discontinuities. *Political Analysis*, 23(1):127–155, 2015.

[61] Sean F Reardon and Joseph P Robinson. Regression discontinuity designs with multiple rating-score variables. *Journal of Research on Educational Effectiveness*, 5(1):83–104, 2012.

English. There's a cutoff at 60 for math when English is 61, and a cutoff at 60 for math when English is 61.1, and a cutoff at 60 for math when English is 73.7, and a cutoff...

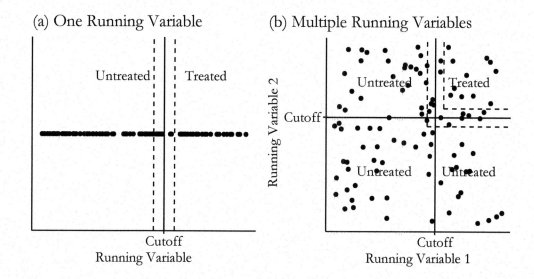

Figure 20.19: A Comparison of Single vs. Multiple Running Variables

The problem is illustrated in Figure 20.19. With a single running variable, it's easy to figure out who you're comparing—those within the bandwidth near the cutoff. But with two running variables, you still want to compare people near the cutoff, but who are you comparing them to? The cutoff is L-shaped! You probably don't want to compare people at the top-left of the L to people in the bottom-right—they're probably not that similar. And if some parts of the L are more populated than others, just going for it may not give you what you want. So you have to figure out how to do an appropriate comparison.

Not only might the groups not be comparable, but the effect of treatment is probably not the same for all parts of the L. Being held back a grade may have very different effects for someone failing math as opposed to someone failing English. So you have to decide *which part(s) of the L you want to estimate the effect at.*

There are a few approaches to estimation in these cases, with Reardon and Robinson (2012) covering five of them. There are also versions, such as the one discussed in Papay, Willett, and Murnane (2011),[62] specifically designed for cases where the treatment you get differs depending on which part of the L you cross—perhaps instead of being held back a grade, kids failing

[62] John P Papay, John B Willett, and Richard J Murnane. Extending the regression-discontinuity approach to multiple assignment variables. *Journal of Econometrics*, 161(2): 203–207, 2011.

math take remedial math while kids failing English take remedial English.

One potential approach, and the one taken by the **rdms** function in **rdmulti**, is to pick specific points on the L and estimate what amounts to a fairly normal regression discontinuity using the other running variable. Do this with enough points on the L and you can get an idea of the effect and how it varies depending on where you cross the border.

20.3.3   When Running Variables Misbehave

REGRESSION DISCONTINUITY IS NICE. IT'S LIKE A VACATION. The treatment variable has few or no arrows heading into it on the causal diagram in the region around the cutoff. Things are simple, results are believable. We're laying back, sipping vodka from a coconut, and idly wondering whether to use a linear model or a 2nd-order polynomial.

Then the rain clouds roll in. Vacation ruined. The running variable isn't behaving.

The main thing that a running variable needs to do in a regression discontinuity, other than be responsible for assigning treatment, is be nice and continuous and smooth, like a gently ebbing tide on the beach.

We've already talked about the problems that can come from manipulation of the running variable. But two other potential issues are the running variable being too *granular*—measured at a very coarse level, or the running variable exhibiting *heaping*—having some values that are suspiciously much more common than others. Either of these can cause problems.

GRANULARITY IS WHEN a variable is measured at a coarse level. For example, if we were measuring annual income, we could say that you earned \$40,231.36 last year, or we could say that you earned \$40,231, or we could say that you earned \$40,000. Or we could even say that you earned "\$40—50,000." Or, heck, we could say "less than \$100,000." These are measurements of your income in decreasing order of granularity.

You can imagine why this might be a problem for regression discontinuity. The whole idea of regression discontinuity is that people just on each side of the cutoff should be pretty easily comparable, with treatment the only real difference between them. That might be pretty believable if the cutoff is \$40,000 and we're comparing you at \$40,231.36 against someone at \$39,990.32. But what if we only have you measured as

$40—50,000? Then we're comparing not only you against the $39,990.32 person. We're also comparing someone who earns $49,999 against someone who earns $30,001. Not so easy to say that treatment is the only thing separating them.

How granular is granular enough? It's subjective, of course, but as long as you can measure the variable precisely enough that you can distinguish who you think is comparable from who you think isn't, you're good to go.

That said, even if you *can* distinguish well enough to do that, but still don't have a variable that's finely measured, you might not be able to properly model the shape of the relationship between the outcome and the running variable *leading up* to the cutoff. If the outcome is increasing with income from $30,000 to $40,000, but you've only got the bin $30—40,000, then you'll get the predicted value at the cutoff wrong, and so your estimate will be wrong too. At the very least, your inability to model that trend will increase your sampling variation.

The real solution here is "find a running variable that's granular enough." But no variable is infinitely granular. So if you're worried about granularity, you may want to pick an estimator that will account for that granularity.

A few papers springing from Kolesár and Rothe (2018) do this.[63] Kolesár and Rothe build better estimates of the sampling variation of the regression discontinuity estimator by making some assumptions about the relationship between the outcome and the running variable *within* that big coarse $30—40,000 bin we can't look inside. One of their estimates assumes that the relationship isn't too nonlinear, and the other assumes that you won't get the shape any *more* wrong than you do at the cutoff. These estimators are available in R in the **RDHonest** package.

One thing Kolesár and Rothe (2018) point out is that you *shouldn't* try to solve the problem by clustering your standard errors on the different values of the running variable. This was standard practice for a long time, but it turns out to be worse in many cases than not addressing the problem at all. Oops! Don't do it.

Non-random heaping is when the running variable seems to be *much more likely* to take certain values than others. Often this can come in the form of rounding. Ask people how old they are and the vast majority of them are going to give you a round number, like "36 years old." However, some of them will be more precise and say "36 years, eight months, and two days." The canonical example is from Almond, Doyle, Kowalski,

[63] Michal Kolesár and Christoph Rothe. Inference in regression discontinuity designs with a discrete running variable. *American Economic Review*, 108(8): 2277–2304, 2018.

and Williams (2010), where they found that baby birth weights were highly likely to be reported as being rounded to the nearest 100 grams.

Either of these processes might result in the distribution of your running variable looking like the simulated data in Figure 20.20 (a). You can see how, every time the running variable hits a certain value (here, a round number), the proportion of observations with that value jumps sky-high. These are the "heaps."

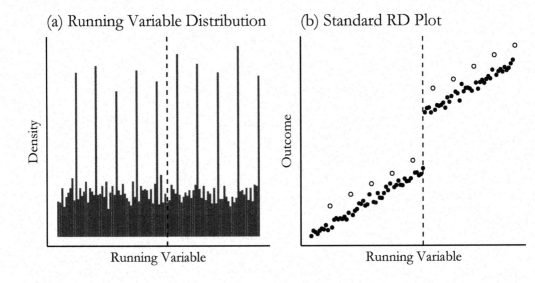

Figure 20.20: Regression Discontinuity Data that Shows Non-Random Heaping

Now, this might not be a huge problem by itself. Having huge numbers of observations at certain points on the x-axis will heavily influence the regression line you fit. But on average, no harm no foul. The real problem comes when the heaping is *non-random*. Notice in Figure 20.20 (b) that not only are the heaps very common, they're also different. The heaps are the non-filled circles, and they have on average *higher* outcome values than the others.

Non-random heaping is likely to occur whenever there's heaping. After all, is it really the same kind of person (or hospital) who rounds their answer as opposed to not rounding it?

With non-random heaping, we have highly influential values of the running variable (because they're way more common than the other values) that are also not representative. This can pull our estimate away from the true effect, and can even flip its sign! You can imagine how this might be especially bad if the cutoff is at one of the heaps—one side of the cutoff would be a heap, and the other side wouldn't. When heaps are inherently different, that's a recipe for bias.

A common approach to this is "donut hole regression discontinuity," where you simply drop observations just around the cutoff so as to clear out heaps near the cutoff. Seems counterintuitive to drop near-cutoff observations in regression discontinuity, which is all about *really wanting* those near-cutoff observations. But that's the problem we face.

However, Barreca, Lindo, and Waddell (2016) examine the problem of heaping and the donut-hole fix.[64] They show that heaping is a problem even when it's *away* from the cutoff. So donut-hole won't fix the problem unless you've gotten rid of the only heap. Instead, they recommend splitting the sample by whether a value is a heap or not, and analyzing each side separately. This gets rid of the problem of there being a clear back door between "being in a heap" and the outcome.

[64] Alan I Barreca, Jason M Lindo, and Glen R Waddell. Heaping-induced bias in regression-discontinuity designs. *Economic Inquiry*, 54(1):268–293, 2016.

20.3.4   Dealing with Bandwidths

HOW FAR AWAY FROM THE CUTOFF CAN YOU GET and still have comparable observations on either side of it? It's an important question, and there's a real tradeoff at play. Pick a bandwidth around the cutoff that's too wide, and you bring in observations that aren't comparable (and rely more on the idea that you've picked the correct functional form), making your estimates less believable and more biased. Pick a bandwidth that's too narrow, and you'll end up estimating your effect on hardly any data, resulting in a noisy estimate. This is the "efficiency-robustness tradeoff." Do you want a precise (efficient) estimate with lots of observations? Well, you're inviting in some bias which will make your estimate less robust to the presence of back doors or poorly-chosen functional form.

What should the bandwidth be? There are three main approaches you can take:

(1) Just pick a bandwidth. This is a pretty common approach. Although perhaps it is becoming less common over time. For this one, you just look at your data and think about how far out you can go without feeling iffy about how far out you've gone. Ideally you can justify why that bandwidth makes sense. Or maybe you just take all the data you have and figure that the range you've got is good enough.

(2) Pick a bunch of bandwidths. This sensitivity test-based approach is also fairly common—both the Fetter (2013) and

the Bana, Bedard, and Rossin-Slater (2020) papers mentioned earlier in this chapter do it (in addition to a little of option 3 below). For this approach, you pick the biggest bandwidth you think makes any sense for the research design, and then you start shrinking it. Each time, estimate your regression discontinuity model.

Once you've got your estimates from each bandwidth, you'd present all the results, potentially in a graph like Figure 20.21, which I made using a very basic non-local linear regression discontinuity with the Fetter (2013) data, varying the size of the bandwidth. We can see the risks of the noise that comes from a too-small bandwidth on the graph. Actually, you can't... I had to cut the tiniest bandwidths (1—3 months) off of the graph because their estimates were so wild, and the confidence intervals so large, that you couldn't really see the other estimates. What you *can* see is that the confidence interval shrinks as the bandwidth expands. Makes sense—a bigger bandwidth means more observations means smaller standard errors.

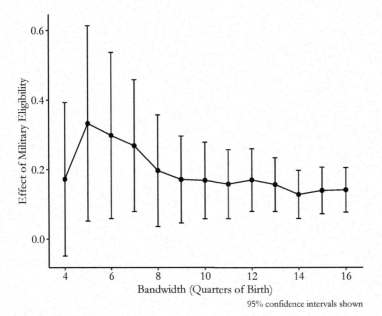

Figure 20.21: Estimating the Effect of Veteran Mortgage Subsidy Eligibility on Home Ownership Rate with Different Bandwidths

Figure 20.21 is largely what you want to see when using this method. The estimate doesn't seem to change that much as I use a wider and wider bandwidth, within reason. So it probably doesn't matter *that* much which one we choose (in the original paper, 12 was the bandwidth in the primary analysis).

(3) Let the data pick the bandwidth for you. Or, rather, pick an objective of "what a good bandwidth looks like" and then use your data to figure out how wide a bandwidth is the best one by that criterion.

There are a few ways to do this. The first, discussed in Imbens and Lemieux (2008),[65] is cross-validation. I introduced cross-validation in Chapter 13. The basic idea is that you're trying to get the best out-of-sample predictive power. So you split your data into random chunks. For each chunk, you estimate your model using *all the other chunks*, and then make predictions for the now out-of-sample chunk you left out. Repeat this for every chunk, and then see overall how well your out-of-sample prediction went. Then, repeat *that* whole process for each potential bandwidth and see which one does the best.

The cross-validation approach was very popular and you still see it often. This is what the Fetter (2013) paper did. However, what we *really* want is good prediction at the cutoff. The standard cross-validation method can give a little too much attention to trying to fit points farther away from the cutoff.

The second approach is the "optimal bandwidth choice" approach, originating in Imbens and Kalyanaraman (2012).[66] This approach takes as its goal getting the best prediction right at the cutoff, just on either side. Of course, we don't know what the true value is right at the cutoff, so we estimate it the best we can.

The processes for several different methods of picking the optimal bandwidth are described in Cattaneo and Vazquez-Bare (2016), which is not too difficult a read as these sorts of papers go.[67] In short, by pursing the goal of getting the best prediction at the cutoff, you end up wanting a bandwidth that is smaller for bigger samples (since you can afford to lose the observations), bigger the more polynomial terms you use (since you can better fit any nonlinearities you introduce by going wider), and beyond that is based on the extent of bias that is introduced from having the wrong functional form, as well as plenty of other details.

The Cattaneo and Vazquez-Bare (2016) paper also points out that using the optimal bandwidth choice approach messes up the standard errors and confidence intervals. The fix for this is to apply a *bias correction*, estimating how big the bias from having the wrong functional form is, and adjusting both the estimate of the effect and the estimate of the variance for it.

They go even further, to talk about bandwidth selection procedures that, instead of trying to get the best prediction at the

[65] Guido W Imbens and Thomas Lemieux. Regression discontinuity designs: A guide to practice. *Journal of Econometrics*, 142(2): 615–635, 2008.

[66] Guido W Imbens and Karthik Kalyanaraman. Optimal bandwidth choice for the regression discontinuity estimator. *The Review of Economic Studies*, 79(3): 933–959, 2012.

[67] Matias D Cattaneo and Gonzalo Vazquez-Bare. The choice of neighborhood in regression discontinuity designs. *Observational Studies*, 3(2):134–146, 2016.

cutoff, try to minimize *coverage error*—the mismatch between the estimated sampling distribution and the actual sampling distribution. This process tends to lead to narrower bandwidths, and is what you'll get as of this writing from running `rdrobust` from the **rdrobust** package in R or Stata.

These bandwidth selection procedures matter a fair amount. Taking Fetter (2013) again as an example, the cross-validation approach in the original paper led to choosing a bandwidth of 12 quarters, which with our method gave an estimate of .170 (.177 using the methods in the original paper). We also saw that the `rdrobust` approach chose a bandwidth of 3.39 quarters, and gave an estimate of .518—very different. And in this case, it really is the bandwidth selection that makes the difference. Forcing `rdrobust` to use the 12-quarter bandwidth brings the estimate to .200, very close to the original paper.

21

A Gallery of Rogues: Other Methods

21.1   It Never Stops

THE TERRIFYING AND DELIGHTFUL THING ABOUT RESEARCH
DESIGN IS THAT THERE'S ALWAYS MORE. New research designs,
new estimation methods, new estimators, new adjustments, new
critiques of old *everything*.[1]

That's what this chapter is about. The twhole wide world
of research design and causality is far too big to cram into one
little book. What you've seen so far are me playing the hits.
Difference-in-differences. Regression discontinuity. Everyone
knows 'em, everyone loves 'em. They're why you bought the
concert ticket. But what about the new album or the b-sides?

It simply wouldn't make sense to spend *too* much time on
new methods. They're untested! It's simply impossible at this
point to know which ones will stand the test of time and which
will fall by the wayside. And even the ones that will be around
a decade from now almost certainly aren't in their final form
yet. The chapter would be immediately outdated.

This chapter, however, *is* about methods that are either new,
or developing, or just super interesting but a bit too

[1] Just think how much of
this book will be outdated
and known as bad advice in
20 years! None of it, actually.   This book is perfect
and eternal.

DOI: 10.1201/9781003226055-21

technologically complex to try to teach you them in this book. I'm not going to go deep into detail on them. No deep dive into caveats and details, no code examples (although I may recommend a few packages you can look into on your own). This chapter is about starting you on a journey of exploration. I'll tell you what's out there. I'll give you an idea of how it works. And if it sounds interesting to you, you'll have some sources to follow up on.

Let's get going.

21.2   Other Templates

In this section, I'll discuss a few promising *template research designs*. That is, they're research designs that have a chance of applying in many different contexts, much like difference-in-differences, instrumental variables, and the other methods I've devoted whole chapters to in the second half of this book. Perhaps in some revision of this book one of these will become a breakout star and get its own chapter and dressing room.

21.2.1   Synthetic Control

SYNTHETIC CONTROL IS THE CLOSEST THING to another chapter that this book got.[2] It's already well-established, popularized in Abadie, Diamond, and Hainmueller (2010),[3] but around before that, and showing up in many places since then. However, it's still enough on the fringe that it wouldn't quite be considered a *standard* part of the toolbox yet, at least not to the degree something like regression discontinuity is. Maybe it will get there.

Synthetic control is *sort of* like a variant of difference-in-differences with matching (Chapter 18). We have the same basic setup as difference-in-differences: there's a policy that goes into effect at a particular time, but only for a particular group.[4] You use data from the pre-treatment period to adjust for differences between the treatment and control groups, and then see how they differ after treatment goes into effect. The post-treatment difference, adjusting for pre-treatment differences, is your effect.

So then what's the difference? A few things:

- Unlike with difference-in-differences, those pre-treatment difference adjustments are not done with regression, but rather by matching. Plus, unlike DID with matching, the purpose

[2] And thank goodness it didn't. Man, this book is long.

[3] Alberto Abadie, Alexis Diamond, and Jens Hainmueller. Synthetic control methods for comparative case studies: Estimating the effect of California's tobacco control program. *Journal of the American Statistical Association*, 105 (490):493–505, 2010.

[4] And in the case of synthetic control, it's specifically *one* group getting treated, not an arbitrary number as in DID. That said, if you have multiple treated groups there are ways of running synthetic control on each of them separately and aggregating the results.

of matching is to eliminate these prior differences, not to account for the propensity of treatment.

- Synthetic control relies on a long period of pre-treatment data.

- After matching, the treated and control groups should have basically no pre-treatment differences. This is often accomplished by including the outcome variable as a matching variable.

- Statistical significance is generally not determined by figuring out the sampling distribution of our estimation method beforehand, but rather by "randomization inference," a method of using placebo tests to estimate a null distribution we can compare our real estimate to.[5]

Synthetic control starts with a treated group and a "donor set" of potential control groups. Using the pre-treatment data periods, it implements a matching algorithm that goes through all the control groups and assigns each of the potential controls a weight. These weights are designed such that the time trend of the outcome for the treated group should be almost exactly the same as the time trend of the outcome for the weighted average of the control group (the "synthetic control" group).

LET'S SEE HOW THIS LOOKS IN DATA FROM ABADIE AND GARDEAZABAL (2003).[6,7] This study looks at an outbreak of violent conflict in the Basque region of Spain in the late 1960s to see what effect the conflict had on economic activity. It compares growth in the Basque region to growth in a synthetic control group. Not too surprisingly, they find that GDP in the region was negatively impacted by the conflict.

Abadie and Gardeazabal match the Basque region to seventeen other nearby regions. The synthetic control algorithm creates weights for each of those seventeen regions in order to make prior trends equal, using information about those prior trends as well as other matching variables including population density, education, and investment levels. Conflict started around 1970, although it didn't really kick into gear for a few years. So matching is done only on the basis of data up to 1969. Cataluna and Madrid turn out to be the strongest matches and get the most weight, with the other regions getting very little weight. These same weights are then applied across the entire time period.

The results of the matching process can be seen in Figure 21.1. As we'd expect, the values of the outcome are very similar in the pre-treatment period for the treated and synthetic control

[5] Specifically, you estimate your synthetic control effect. Then you drop your actual treated observation and cycle through all your control observations, estimating synthetic control estimates for each of them as though they were the treated group. This gives you a null distribution of treatment effects for untreated observations. Finally, check what percentile of the null distribution your actual effect is. If it's far in the tails of the null distribution, that's a good indication your effect wasn't just random chance.

[6] Alberto Abadie and Javier Gardeazabal. The economic costs of conflict: A case study of the Basque country. *American Economic Review*, 93(1):113–132, 2003.

[7] Surely my choice of this data has *nothing* to do with it coming already prepared and ready for use with synthetic control in the R **Synth** package, with code already put together in the help files. Perish the thought.

group. It sure looks like we'd expect them to continue to be the same in the absence of any treatment. Reassuringly, they continue to trend very closely together for a few years after treatment starts, but before the conflict really gets going. We're not actively matching here (nothing's *forcing* the lines to stay together) but they stay together nevertheless in a time period where the treatment intensity was very low. That's good!

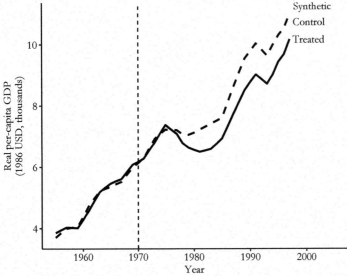

Figure 21.1: Synthetic Control Effect of Conflict on Basque Country Economic Growth from Abadie and Gardeazabal (2003)

Copyright American Economic Association.
Reproduced with permission of the American Economic Review.

Then, the conflict really ramps up. At the same time we see the treated group and the synthetic control start to diverge, with GDP dipping lower for the Basque Country than for its synthetic control. Sure looks like the conflict is having a negative effect!

LIKE ANYTHING, SYNTHETIC CONTROL HAS ITS PROS AND CONS. Its benefits in comparison to difference-in-differences are obvious. It doesn't need to rely on that iffy parallel trends assumption like difference-in-differences did (it sorta *forces* the assumption to be true by its method). It makes the process of selecting a control group a little more disciplined. And because it doesn't rely on regression, it is not as sensitive to functional form issues as difference-in-differences is (and naturally shows you dynamic effects without having to trot out a bunch of inter-action terms). It may well be the "most important innovation in the policy evaluation literature in the last fifteen years," and that's coming from two of the authors of the next method in this chapter.[8]

[8] Susan Athey and Guido W Imbens. The state of applied econometrics: Causality and policy evaluation. *Journal of Economic Perspectives*, 31 (2):3–32, 2017.

Why isn't it more popular than it is, then?[9] It's not quite as widely applicable as difference-in-differences, since it relies on cases where you have access to a lot of pre-treatment data (otherwise the matching quality, or at least your ability to check match quality, will get iffy). It does have a tendency to overfit any noise in the outcome variable when matching—that's no good. Also, the fact that it matches on the outcome variable just *feels* fishy—like using the outcome itself to close back doors is cheating. This is a natural skepticism, and like any method, synthetic control comes with its own set of assumptions that have to hold. But it is important to keep in mind that it's only matching on pre-treatment outcomes, not the post-treatment outcomes where you actually estimate the effect. Plus, other methods like difference-in-differences also use information about pre-treatment outcomes to produce their estimate. They just do so in less obvious ways. Synthetic control isn't pulling a fast one—you've been using pre-treatment outcomes all along.

If you are interested in learning more about synthetic control, I recommend the extended chapter on the method in Scott Cunningham's *Causal Inference: The Mixtape*.[10] Or for a bit more detail, Abadie (2020) covers cases when synthetic control works, and when it doesn't.[11] Synthetic control can be implemented in Stata using the **synth** package, often paired with the **synth_runner** package for easier use. In R there is both the **tidysynth** package, which is a bit easier to use, or **gsynth**, which has more flexibility in how the design works. There is not a mature implementation of the standard synthetic control method in Python, although the **SparseSC** package (installable through Microsoft's GitHub page) provides a related method that applies some machine learning tools in the matching process.

21.2.2   Matrix Completion

LET'S SWING ALL THE WAY FROM ONE OF THE MOST-ESTABLISHED METHODS IN THIS CHAPTER TO ONE OF THE LEAST. Matrix completion is very newfangled as of this writing, making its debut in Athey et al. (2021),[12] and despite that year of publication was around for about five years before that.[13] But it is very promising, and the first of a few causal inference innovations derived from machine learning you'll see in this chapter.

[9] And it is fairly popular—it's almost certainly the next-most-popular template method after the ones I've given their own chapters, despite being much younger.

[10] Scott Cunningham. *Causal Inference: The Mixtape*. Yale University Press, 2021.

[11] Alberto Abadie. Using synthetic controls: Feasibility, data requirements, and methodological aspects. *Journal of Economic Literature*, 59(2), 2021.

[12] Susan Athey, Mohsen Bayati, Nikolay Doudchenko, Guido W Imbens, and Khashayar Khosravi. Matrix completion methods for causal panel data models. *Journal of the American Statistical Association*, pages 1–41, 2021.

[13] Academic publishing is weird. I also happened to see this method presented in a seminar in Los Angeles in maybe 2017 or 2018. As I write this, the coronavirus pandemic is still ongoing. Geez, I miss seminars. I miss classrooms. I even miss Los Angeles, so you know I'm *really* desperate.

Matrix completion works with panel data, where you observe the same units over multiple time periods. In your panel data set, for a given individual (person, firm, country, etc.) in a given time period, that observation is either treated or untreated. We don't just observe whether that observation is treated or untreated, but also we observe their *outcome Y*. Simple enough so far.

But what about that outcome Y? It's not just *any* outcome. It's specifically the outcome that observation gets *conditional on its treatment status*. Say the treatment is that you watched the movie *Home Alone* on Netflix in a particular month, and the outcome is you choosing whether to watch *My Girl*, another Macaulay Culkin movie, on Netflix in that same month when given the chance. If you saw both movies, then your individual outcome is that you did watch *My Girl* that month conditional on having watched *Home Alone*. We have *no idea* whether you would have watched *My Girl* if you hadn't watched *Home Alone*.

We can imagine a matrix (spreadsheet, basically) of outcomes. Each column is a time period, each row is an individual, and each element/cell of the matrix is your outcome. If you watched *My Girl* that month, then your outcome is 1, so that month's column of your row would get a 1. If you didn't it gets a 0.[14]

[14] The matrix completion method is not limited to binary outcomes.

But we don't just have one matrix, we have two. We have one matrix conditional on you getting the treatment (watching *Home Alone*) and a separate matrix conditional on you not watching *Home Alone*. And—here's the kicker—if we're looking in the "treated" matrix in your row, in a month you *didn't* get treated, instead of a 1 or a 0 for your outcome, you get a big ol' "?". We don't know what your outcome would be if you'd gotten treated, since you weren't treated. We'd love to compare your treated self to your untreated self to see how you were affected, but we can't because we never observe both at once.

This whole idea—where the outcome you *would* have gotten is thought of as a missing value we'd love to compare to but can't—is a pretty pure application of the potential outcomes model, which is another way of representing causal relationships besides causal diagrams.

Untreated				Treated			
Ind.	Time1	Time2	Time3	Ind.	Time1	Time2	Time3
1	1	0	1	1	?	?	?
2	?	1	?	2	0	?	1
3	1	?	1	3	?	0	?
4	?	?	?	4	1	1	0

Table 21.1: Outcomes for Treated and Untreated Individuals

One example of these matrices can be seen in Table 21.1. Individual 2 was treated in periods 1 and 3, so we see in the Treated matrix that their outcomes in periods 1 and 3 were 0 and 1, respectively, but period 2 gets a ?. In the Untreated matrix, we *only* see the outcome for period 2—a 1—since that's the only period they're untreated.

MATRIX COMPLETION IS ALL ABOUT FILLING IN ?S to complete the matrix. Specifically, it needs an untreated comparison unit for every treated observation, and so it tries to fill in the ?s in the Untreated matrix. If we can do that, in other words getting our best guess of what *would* have happened to all the treated observations if they'd been untreated, we can compare those outcomes to the ones they *did* get while treated.[15] If you were treated and did see *My Girl*, but we predict that you wouldn't have seen *My Girl* that month if you hadn't been treated, then seeing *Home Alone* had a positive effect on your *My Girl*-watching outcome. Average together the estimated treatment effects over all the treated observations to get an average treatment on the treated.[16]

How can we predict what those ?s would have been? We use the other data in the Untreated matrix. For example, in Table 21.1, we can see that for Individual 3, their Untreated outcome in periods 1 and 3 are both 1, which makes it more likely that the ? in period 2 for Individual 3 would also have been 1 if we'd observed it. Similarly, in period 3, Individuals 1 and 3 both got 1s, making it seem more likely that Individuals 2 and 4 also would have had 1s if they'd been untreated. This is conceptually *sort of* like a two-way fixed effects model with fixed effects for both individual and time. Sort of.[17]

That's the rough idea, but the application is a bit more complex than that. Crucially, it uses regularization, as we talked about in Chapter 13, to improve its prediction powers when filling in the matrix. Also, so far I've talked about using outcomes from other parts of the matrix to fill in ?s, but you could also use covariates to develop something like matching weights that would tell you how useful each *other* cell is in predicting *your* ?. Like with regression and matching, identifying a causal effect using matrix completion requires that the treatment is random conditional on whatever you use to create comparison weights.

If this sounds like an oversimplified description of the method, it is. The technical details do ramp up here quite quickly. But you're about 80% of the way there (and only slightly incorrect) if you think of it as "using regularized regression to predict missing

[15] You could theoretically expand this to fill in values in the Treated matrix, or to use information from Treated to fill in Untreated, but for now the method is focused on scenarios where you only have a few treated observations, so that isn't as useful.

[16] Or you can stop yourself before aggregating fully to look at things like how the treatment effect changes over time, or how varied it is across individuals.

[17] Like two-way fixed effects, matrix completion can be applied wherever a difference-in-difference design could be applied, but it works better than two-way fixed effects when the treatment timing is staggered, for reasons discussed in Chapter 18.

Untreated values, and then comparing the actual Treated values to those predictions in order to get an average treatment on the treated."

APPLICATIONS OF MATRIX COMPLETION ARE FEW AND FAR BETWEEN, with more papers written on improving or understanding the method than using it. That's fair—you don't want to write a whole paper using a new method only to find out a year later something has been discovered and the method doesn't work!

One application, is Wood, Tyler, and Papachristos (2020).[18] The authors in this paper are interested in the use of force by police officers, and whether training on procedural justice methods—which encourage respect, neutrality, and transparency in the course of police work—can reduce the use of force.

The authors look at Chicago, which got thousands of police officers to take the course. Treatment was staggered, with different officers receiving treatment at different times. They then checked whether the officer in question received complaints or used force on the job. Of course, we can't see what amount of complaints they would have gotten, or force they would have used, without the training, but we can use matrix completion to guess.

They find that the training reduced complaints by 11% and the use of force by almost 8%. Not bad! Sure, they don't have a lot of ways for controlling for officer characteristics in regards to who accepts the training. But matrix completion *does* allow them to see that each officer's reductions in complaints and force seems to be timed to when they got the training, rather than just reflecting preexisting changes.

Matrix completion can be implemented in R using the `estimator = "mc"` option in the `gsynth()` function in the `gsynth` package.[19] It is not currently accessible in Stata or Python.

[18] George Wood, Tom R Tyler, and Andrew V Papachristos. Procedural justice training reduces police use of force and complaints against officers. *Proceedings of the National Academy of Sciences*, 117(18):9815–9821, 2020.

[19] Wait, wasn't that the package I suggested for synthetic control? Yep! Turns out matrix completion is a super general idea and synthetic control, difference-in-differences, and a few other designs are actually special cases of matrix completion.

21.2.3   Causal Discovery

OOPS, I LIED. Remember all that business from the first half of this book about needing to use theory and experience to come up with a causal diagram that you could then use for identification? Sure, you could use data to show that you had the wrong diagram—a diagram might imply that two variables are unrelated, so if they're related in data your diagram is wrong—but data could never tell you what the diagram should be in the first place. Or could it?

Causal discovery is the process of using data to develop causal diagrams. How can this be possible? How can we possibly use data to uncover a causal diagram when we know that a given data set must be consistent with *lots* of different causal diagrams? There's no one answer to that, as there are many different algorithms and processes for performing causal discovery—it's more a general idea than a specific method. But let's look at a classic (albeit slow) implementation, the SGS algorithm.[20]

WE START THE SGS ALGORITHM WITH THE SET OF RELE-VANT VARIABLES for our diagram. Let's say we have four variables, just called *A*, *B*, *C*, and *D*.[21]

To build a causal diagram from these variables, we have two tasks ahead of us. First, we need to figure out, for each pair of variables, if there's a direct arrow between them. Second, once we know where the direct arrows are, we need to figure out which direction those direct arrows point.

Let's tackle the first problem. How can we figure out whether there's a direct arrow (of either direction) between two variables? We check for conditional associations in the data. Simple as that. We start with a completely agnostic diagram in Figure 21.2, where every pair of variables is connected (note the lack of arrowheads—we don't know what direction these arrows run).

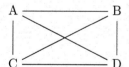

Let's consider that line between *A* and *B*. Should it be there? How can we tell from the data alone? Well, if there's a direct line between *A* and *B* in either direction, then we should see a nonzero relationship between *A* and *B* *no matter what we control for*.[22] So, first we just look at the relationship between *A* and *B*. If that's zero, no direct line! But if it's not zero, try controlling for stuff. Control for *C*. Control for *D*. Control for *C and D*. If any of those sets of controls shuts off the *A* to *B* relationship, that means that the relationship they *do* have is blockable by some set of controls, which wouldn't be possible if there was a direct line between them.[23]

[20] Peter Spirtes, Clark N Glymour, Richard Scheines, and David Heckerman. *Causation, Prediction, and Search*. MIT Press, 2000.

[21] This example was considerably refined after reading Leslie Myint's lecture notes.

Figure 21.2: A Diagram Where Everything is Related to Everything Else

[22] This requires some decision process of what counts as zero. Statistical significance could be one, although there are other reasons besides a true zero that something could be insignificant.

[23] This process assumes that the average effects we're estimating here are representative. For example, if the effect of *A* on *B* is positive in half the sample, but negative in the other half, those effects might cancel out to give an average estimated effect of 0, telling us there's no arrow when there actually is one. We must assume this doesn't happen.

Repeat this process for each pair of nodes. Any relationship that can be set to zero by the use of controls is one more arrow you can knock off.

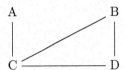

Figure 21.3: A Causal Diagram After Deleting Several Direct Arrows with Causal Discovery

Let's say that we did this whole process and what we ended up with was Figure 21.3. We're already on our way.

NEXT, WE'LL WANT TO FIGURE OUT WHICH DIRECTION THE ARROWS POINT. For this we can head all the way back to Chapter 8 to think about colliders. Remember, if there's a path $X \to Y \leftarrow Z$ on a diagram, where both of the arrows are pointing at the same variable (Y here), then that path is *pre-closed* without any controls, but if we control for the collider, it opens back up.

Let's apply this to Figure 21.3. In particular let's look at the path $A - C - B$. This path could be filled in a few different ways: $A \to C \to B$, $A \leftarrow C \leftarrow B$, $A \leftarrow C \to B$, or $A \to C \leftarrow B$.

What we want to do at this point is look at the relationship between A and B both while controlling for C and not controlling for C.

What could this tell us? If we observe that *without a control for C there's zero relationship between A and B* and also *with a control for C there's a nonzero relationship between A and B*, then there's only one thing it could be—a collider! So if we do find that, it must be $A \to C \leftarrow B$. Now we have some arrows, shown in Figure 21.4.

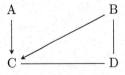

Figure 21.4: A Causal Diagram After Discovering $A \to C \leftarrow B$

But what if we *don't* get that? Let's say we want to finish out our diagram and so check the $C - D - B$ path. This time we get a relationship between B and C whether or not we control for D. This could be consistent with $C \to D \to B$, $C \leftarrow D \leftarrow B$, or $C \leftarrow D \to B$.[24]

[24] How is it consistent with these? Wouldn't controlling for D shut down the relationship in all of these? It would except that we also have the $B \to C$ arrow outside of this particular path.

When we come to a case like this we might still be able to progress—perhaps solving some other part of the diagram would let us determine $D \rightarrow C$ and thus narrow down our options. But also we might have run into an *equivalence class*—a set of diagrams that our data can't distinguish. Sometimes causal discovery can only get us part of the way there. So we'll stay with Figure 21.4. Still, we know more than we did at the start. Maybe other methods, or bringing in some intuition and theory like normal, could help us narrow down the options.

So that's your start in the world of causal discovery. We can ask the data what the causal diagram should be, and then once we have the diagram we can ask the computer how to identify it. Who knows, maybe the whole process will get so effective that we won't even need any people to do causal inference and this whole book will have been a waste of everyone's time.

This description of the algorithm leaves out how causal discovery deals with things like unmeasured variables, and also plenty of other approaches to the problem, some of which use timing to narrow down the potential graphs, or *changes* in the graph itself to figure things out. Not to mention all the massive speed improvements. There's lots of cool stuff in this literature, and some of the algorithms are really ingenious. Several causal discovery algorithms can be implemented in R using the **pcalg** package, or in Python using the **Cdt** package.

21.2.4   Double Machine Learning

A lot of these methods are pretty brain-busting, Let's kick back with something considerably more straightforward and familiar. *Double machine learning* (also known as debiased machine learning) is just another way to close back doors, like regression or matching.[25]

Let's think back to a basic regression setup. We want to know the effect of treatment X on outcome Y. We're in a setting where if we simply regress Y on X, our causal diagram tells us that we won't identify the effect of X on Y. So we need to include some control variables. For simplicity let's just say we only need two control variables to close all the back doors: W and Z. Then, we can identify the effect of X on Y with the regression equation

$$Y = \beta_0 + \beta_1 X + \beta_2 W + \beta_3 Z + \varepsilon \qquad (21.1)$$

[25] Victor Chernozhukov, Denis Chetverikov, Mert Demirer, Esther Duflo, Christian Hansen, Whitney Newey, and James Robins. Double/debiased machine learning for treatment and structural parameters. *The Econometrics Journal*, 21 (1), 2018a.

Simple! But then we remember the whole idea of what *happens* when we include a control variable—we're removing the variation in both X and Y that is explained by the control, and thus the part of their relationship that is explained by the control. This closes any back doors through the controls, hopefully just leaving us with the front-door paths we want. This is how we described controlling for variables in Chapter 4.

As you might expect given this interpretation, we can get the exact same estimate for $\hat{\beta}_1$ as in Equation 21.1 by removing the variation in X and Y explained by the controls ourselves:[26]

1. Regress Y on the controls W and Z, and calculate the residual values of Y, Y^R. Remember, the residual values are the actual value minus the predicted value from the regression, $Y - \hat{Y}$. This residual has removed all parts of Y that can be explained by the controls.

2. Regress X on the controls W and Z, and calculate the residual values of X, X^R. These residuals, too, have been scrubbed of all parts of X explained by the controls.

3. Regress Y^R on X^R.

Here's where the double machine learning comes in. It looks at this list of instructions and asks the question: "uh... why does it have to be regression?"

THE PROCESS FOR DOUBLE MACHINE LEARNING IS SIMPLE. First, predict Y using your set of control variables *somehow* and get the residuals Y^R. Then, predict X using your control variables *somehow*. Finally, regress Y^R on X^R and get your effect.[27]

The *somehow* is where the "machine learning" part comes in. You just use a machine learning algorithm to do the predicting of Y and X using your controls, instead of regression. That's it! It really could be any prediction-based machine learning algorithm. Regularized regression like we talked about in Chapter 13 is one option. As is the random forest, which will get a brief introduction later in this chapter. So is a bunch of stuff you'll need a different, machine learning-focused book to read about. Neural nets, boosted regression, and the fifteen other new ones they came up with while I was writing this sentence.

There's slightly more to double machine learning than just that, but not much—it also employs sample splitting to avoid overfitting. In its basic form, it splits the sample randomly into two halves. Then, it fits its machine learning model using only

[26] These steps are just a regression-based version of the same steps from Chapter 4.

[27] There's also a version of double machine learning that is only slightly different and applies the same trick to instrumental variables estimation.

one half (half A), and then applies that model to get Y^R, X^R, and the resulting $\hat{\beta}_1$ coefficient in the *other* half (half B). Then the halves trade places, fitting a model using B and using that to get residuals in A. Then average the coefficients from your two samples to get an effect.

WHY WOULD YOU WANT TO DO DOUBLE MACHINE LEARNING? The problem with applying machine learning methods to causal inference is that they're usually designed to be good at *prediction*, not inference (statistical or causal). But they are really good at prediction! Double machine learning finds one step in a standard causal inference design that is, inherently, a prediction problem. Then it gives that problem to a machine learning algorithm instead of a linear regression, so machine learning can do what it's good at.[28]

Sometimes, linear regression is just fine. But sometimes it's not. Machine learning methods are much better than linear regression when there are lots and lots and lots of controls (high dimensional data), and also when the true model has lots of interaction terms or highly nonlinear functional forms. In these cases, double machine learning will do a better job fitting all those peculiarities than linear regression. Seems handy to me.

[28] This is also kind of the idea behind matrix completion, and many other machine-learning based causal inference methods.

21.3   Modeling Heterogeneous Effects

IT ALMOST GOES WITHOUT SAYING THAT CAUSAL INFERENCE WOULD BE A LOT EASIER IF EACH CAUSAL EFFECT WERE THE SAME FOR EVERYBODY. If every treatment had the exact same effect on every individual, we wouldn't need to worry about whether our results will generalize to other settings, or about which kind of treatment effect average our estimate has picked up. I could have cut Chapter 10 out entirely. But, alas, we live in a world of heterogeneous effects.

We can't escape heterogeneous effects entirely, but we can give ourselves some tools to face them head on. Why not, instead of treating heterogeneous effects as some sort of nuisance where we have to figure out which average we've just drawn, we just *estimate the distribution of effects?* Then, not only could we take whatever average of that distribution we wanted, but we could toss out the average and look at the distribution overall. Sure, a mean is neat, but why wouldn't we be interested in how the effect varies over the sample? What's the standard deviation of the effect? The effect is bigger or smaller for *who* exactly? Is treatment more effective for old people than young? Poor than rich? French than Russian? And so on.

WE'VE ALREADY DIPPED OUR TOES INTO THE WATER of estimating the heterogeneity in treatment effects. We talked about interaction terms in regression in Chapter 13. That's one way to see how an effect differs across a sample, but it's very limited—you can really only include a couple of interaction terms before your model turns into a poorly-powered impossible-to-interpret slurry.

We discussed *hierarchical linear models* in Chapter 16. These are much better than interaction terms in regression if your goal is understanding the full extent of the variation in an effect across the sample. In a similar way to how regression models an outcome variable as varying based on the values of our predictors, hierarchical linear models model *coefficients* as varying based on the values of predictors. If one of those coefficients is our effect of interest, we'll get a decent idea about how much that effect varies, and who gets the strongest and weakest effects.

While we're at it, let's take a little peek at what else is out there. Research on the estimation of heterogeneous effects has been particularly active since the 2010s. One problem with estimating heterogeneous effects is that there are so many dimensions they *could* vary along. How could we check everything? But you know what's good at checking everything and handling lots and lots and lots of interactions? Machine learning! The estimation of heterogeneous effects may well end up being machine learning's most important contribution to causal inference. There are lots and lots of these approaches. I'll pick only one of these methods—causal forests, for being relatively straightforward to understand, easy to use, and likely to rank among the more popular options. But I want to be clear that there are *zillions* of such approaches, and more on the way.

It's worth pointing out that these methods aren't designed to *identify an effect* in some clever way. These assume that you *already* have a design for identifying an effect. But once you've identified the effect, these can help you see the entire distribution of that effect. You could just as easily use them to see the distribution of a non-causal association.

21.3.1 Causal Forests

WE'RE USED TO THE IDEA OF FITTING A MODEL, AND THEN USING IT TO MAKE A UNIQUE PREDICTION OF THE OUTCOME FOR EACH OBSERVATION. Simply regress an outcome on a bunch of predictors, then plug in the values of someone's predictors to see what prediction the model spits out.

But what if we want to get a unique estimate of *the effect of X on Y* for each observation? That's tougher. We don't have a variable called "the effect of X on Y" in the data, so we can't just do what we did for predicting the outcome. One thing we could do is estimate a model with interaction terms. If we regress the outcome on the treatment as well as the treatment interacted with other predictors, we can plug in the values of someone's predictors to see what the effect of X on Y is for them. The problem with that is that regression simply can't handle a lot of interaction terms, or highly nonlinear interactions. So we end up only using a few predictors to estimate differences in the effect on a pretty basic level.

Enter causal forests![29] Causal forests manage to take the task of estimating a unique effect for each individual, and morph it into a format that works basically the same as predicting the outcome in a certain non-regressiony way. Our ability to handle lots of predictors and high degrees of nonlinearity then works just as well for estimating an individual's effect as it does for predicting their outcome.[30]

How do causal forests manage this? To figure that out, I'll need to start by describing the prediction method that causal forests modify to work for effect estimation—the *random* forest.

RANDOM FORESTS MAKE PREDICTIONS BY SPLITTING THE SAMPLE. Imagine we were trying to predict how many hours of TV you watch per day, and we have only one predictor— whether or not you are married. In that case, it's pretty obvious how we'd make our prediction. We'd split the sample and take conditional means. First, take the non-married people and get their average TV watching—let's say it's 3 hours per day. Then we take the married people and get their average TV watching— 4 hours per day. Now if we're asked to make a prediction, we'd predict 3 hours a day for a non-married person and 4 hours per day for a married person. That's simply the best prediction we can make with that predictor.

How about a different predictor? Instead of having marital status, our only predictor now is age. We can still make a

[29] Stefan Wager and Susan Athey. Estimation and inference of heterogeneous treatment effects using random forests. *Journal of the American Statistical Association*, 113(523):1228–1242, 2018.

[30] In fact, not only *can* it handle lots of predictors, it actually has a hard time when there are only a few (say, 3 or fewer) predictors. If you only want to see heterogeneity across a couple of variables, causal forest is a bad pick.

prediction by splitting the sample in two, but now we have more choice. We could compare people age 6 and below to those 7 and above. Or 7 and below vs. 8 and above. Or 21 and below vs. 22 and above. Or 64 and below vs. 65 and above. Really, we could pick *any* age to split the data at. What's the best prediction we can make? It's whichever one most reduces our prediction error. Often this is measured as the sum of squared residuals. So if the sum of squared residuals when splitting the data by 21-and-below vs. 22+ is 4030.1, but the sum of squared residuals when splitting 64-and-below vs. 65+ is 3901.2, then the split at age 65 is a better one. We wouldn't just pick a few cutoffs to compare, though, we'd try *every* possible cutoff and pick the best (lowest prediction error) option. So we try splitting at 2, and 3, and 4, and so on, all the way up to the oldest people in the sample. If the best cutoff really is at age 65, we use that. If the average below the cutoff is 2.7 and above the cutoff is 4.8, then we predict 2.7 for each person under 65, and 4.8 for each person 65+.

Nothing special so far, and as you can guess, just splitting the sample in two doesn't get you very good predictions. There are still two steps to go before we get to a random forest though.

First is the step of creating a *decision tree*. A decision tree takes this sample-splitting idea and does it repeatedly. Now maybe you have a *bunch* of predictors to work with. Start by finding the *best possible split* for the sample by not just considering each possible cutoff level, but doing so for every predictor variable you have, So you check all the cutoffs possible across marital status, age, race, income, education, etc. etc. etc., and pick the best cutoff for the best variable. *Then* split the sample. Maybe age 64-and-below vs. 65+ does turn out to be the best split overall. Then we can draw a tree describing our split like in Figure 21.5.

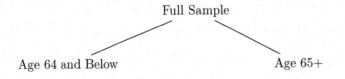

Full Sample

Age 64 and Below   Age 65+

Figure 21.5: A Decision Tree with a Single Split

We're not done! Like I said, splitting the sample in two doesn't get you far. So we split it again. Take just the 64-and-below split and repeat the process of finding the best cutoff for the best variable for *them*. It might even be age again. Let's say it's marital status. Now we head over to the 65+ split and do the process again for them. Let's say the best split for them is income above $65,324 or below. Now our tree looks like Figure 21.6. Our sample is split into four distinct groups: non-married 64-and-unders, married 64-and-unders, 65+ who earn less than $65,324, and 65+ who earn $65,324 or more.

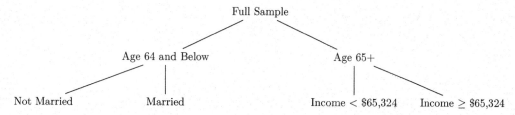

Figure 21.6: A Decision Tree with Two Splits

And we keep going. Keep splitting the sample over and over, finding the best split each time, until the sample we're splitting gets to some definition of "too small," and you stop there. Now we have a decision tree. My prediction for you is whatever the mean is in your split.[31] So if you're a 36-year old Asian married person with a community college degree making $43,225, I make your prediction by following the splits down the tree until I find the small group of people very much like you. Then I take the mean, and that's my prediction.

How do we get from a decision tree to a random forest? Add a bunch of randomness! First off, we bootstrap the whole process, like in Chapter 15, and create a decision tree for each sample. We create that decision tree with a bit of a difference, too: each time we try to split the sample, we don't actually consider *all* the variables. We consider a random subset of the predictor variables instead (choosing a different random sample every time we split), and only look for the best split among those.

Once we're done, we have a whole *bunch* of decision trees, each different and with their own predictions for each individual. The randomness we injected into the process helped make sure the predictions were as independent as possible, so no observation or predictor could be overly influential. The many predictions for each observation can then be averaged together to create a great prediction, taking advantage of the wisdom of crowds from these "independent" predictions.

THAT'S RANDOM FOREST. WHAT IS A CAUSAL FOREST? Causal forests are really just random forests but with a tweak. In a random forest, we chose splits for our data with a goal of minimizing prediction error, because we wanted to predict

[31] There are other options besides taking the mean in your split, but let's keep it simple. Also, notice by the way that it's very easy to use random forest with a categorical or binary outcome. Random forest is often used for "classification" problems, i.e., predicting categories.

the outcome. In a causal forest, we want to see variation in the treatment effect. So instead, we choose splits based on *how different the estimated effect is in each side of the split.*

Let's say we are still looking at TV watching. But now we want to see heterogeneity in the effect of "cord-cutting" (cancelling cable TV and just using streaming services) on TV watch time.

Now when we're considering a split of age 64-and-below vs. 65+, we *estimate the effect of cord-cutting on TV watching* among those 64 and below (however we estimate that effect— controls, instruments, etc.). Then we estimate the same effect among those 65+. We then check how different the effect is. Maybe cord-cutting increases TV watching by .5 hours for those below 65, but decreases it by 1 hour for those 65+. Thats a difference of $1 - (-.5) = 1.5$ in the effect. We do the same thing across all other possible splits and choose the biggest difference.

Then, once the split gets small enough, we stop splitting. The estimate of the effect within your split is what we think the effect is *for you.* Then, as before, we bootstrap the whole thing and limit our choice of splitting variables each time. The overall estimated effect for you is the average across all those bootstrap samples.[32]

THAT'S ABOUT IT! Once the process is done, we have an estimate of the treatment effect for each individual. The method is also capable of estimating *standard errors* for each of those individual effects.[33] Once you have the individual estimates you can ask questions like "whose effects are highest, and whose are lowest?" or investigate the whole distribution of effects.

Causal forests can be estimated using the `causal_forest()` function in the **grf** package in R. In particular the **grf** package uses "honest" causal forests, which just means that the sample is split in half, with one half being used to make tree-splitting decisions, and the other used to estimate effects. In Stata, the `causal_forest` function in the **MLRtime** package can be used to run... the `causal_forest()` function in R, but from Stata.[34] In Python, honest causal forests can be estimated with `econml.grf.CausalForest()`.

[32] There are, for both random and causal forests, some steps I've left out. These are common machine learning "training/holdout" procedures related to using only part of your data to estimate the forest, and then estimating the effect only for the other part.

[33] The neat thing about causal forests, as opposed to some other heterogeneous treatment effect methods in machine learning, is that causal forest estimates have a known sampling distribution, so you can easily use standard errors with them.

[34] https://github.com/NickCH-K/MLRtime

21.3.2 Sorted Effects

CAUSAL FORESTS WERE JUST ONE REPRESENTATIVE from a huge and growing literature on the use of machine learning to estimate treatment effect heterogeneity. Now let's cover a neat little approach that isn't even particularly popular as I write this, but, let's be honest, I like it so it's in the book. That approach is the sorted effects method.[35]

Sorted effects is pretty darn simple. There are plenty of regression-based methods that already allow for treatment effect heterogeneity. Regressions with interactions are one of them. But also, many nonlinear regression models like logit and probit naturally have treatment effect heterogeneity, since the effect can't be that big if your probability of having a dependent value of 1 is near 1 or 0.

Sorted effects simply takes models like that and says "Hey! Why not, instead of reporting average treatment effects with these models, you estimate each individual person's treatment effect, which the model already allows you to do?" Then, once you have that range of estimates, you can present the whole range instead of trying to smush things together into an average. Then, you present the estimates from lowest to highest (i.e., sorted).

This seems... super obvious, to be frank. What makes this its own method? Three things. First, the designers of sorted effects managed to figure out, using bootstrapping, how to estimate the standard errors on these individual effects in an accurate way, and how to handle the noisiness in the tails of the effect distribution, both of which can be difficult. Second, they introduced a few methods for what you can do with the distribution of effects, including comparing *who* is in different parts of the distribution. Who is affected the most and who is affected the least? Third, because this is an approach to treatment effect heterogeneity that lets you summarize that heterogeneity in a single dimension (the effect itself), it allows you to make models with more built-in heterogeneity. A regular model with a whole bunch of interactions would be impossible to interpret. But sorted effects would let you interpret it easily, and also provide standard errors that account for all the interactions.

LET'S SEE WHAT SORTED EFFECTS CAN DO. Let's head back to the data from Oster's study in Chapter 4. In this study, Oster was interested in the effect of a short-lived recommendation for taking vitamin E on the chances that someone would take it. The main thrust of the study is that the people who respond to these recommendations are the ones paying the most attention to their health in general. So when the recommendation is on, it should *look* like vitamin E has all these great health benefits,

[35] Victor Chernozhukov, Iván Fernández-Val, and Ye Luo. The sorted effects method: Discovering heterogeneous effects beyond their averages. *Econometrica*, 86 (6):1911–1938, 2018b.

simply because healthy people are most likely to respond to the new recommendation.

Part of her study was in establishing that people with other great health indicators were most likely to take up vitamin E in response to the recommendation. She had a few ways of doing that descriptively. Let's see what sorted effects can do with a more structured approach.

We're going to regress, in a logit model, the probability of taking vitamin E on being in a during-recommendation period. We'll include as predictors (which we'll also interact with treatment) whether they ever smoked, a standardized score of their exercise level, and a standardized score of their vitamin-taking behavior. The list could reasonably be much longer than that, but let's keep it simple.

The sorted effects method estimates the distribution of effects, as shown in Figure 21.7. The average effect, while it's more precise than any particular effect, leaves out a whole lot of the distribution. There are people in the data estimated to have effects *much* lower and *much* higher than the average. In fact, very few seem to be *at* the average, or even inside its 90% confidence interval. Who is it describing really?

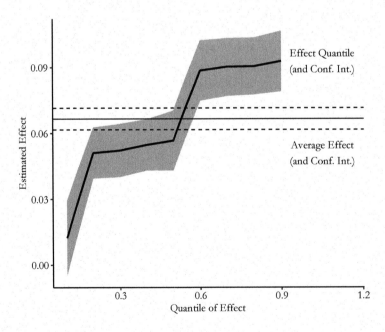

Figure 21.7: Sorted Distribution of Effects of Vitamin E Recommendation on Taking Vitamin E

Sorted effects also lets us see who those people are. Table 21.2 shows the mean and standard deviation of the predictors in the model among those with estimated effects in the top 10% and bottom 10%.

| Effect: | 10% Least Affected | | 10% Most Affected | |
Variable	Mean	SD	Mean	SD
Smoking	1	0	0	0
Exercise Rating Score	-0.422	0.713	1.446	0.193
Vitamin Behavior Score	-1.985	0.387	1.706	0.105

Table 21.2: Characteristics of Those Least and Most Responsive to Health Recommendation for Vitamin E

The differences are pretty stark. Smoking is the biggest tell. Among those who were estimated to be in the 10% least responsive to the vitamin E recommendation, literally 100% of them were smokers. And among the 10% most responsive? 0% were smokers! That's a big difference. Maybe it's not too surprising that people who smoke aren't that responsive to recommendations from health officials, but I didn't expect the difference to be 0 vs. 100![36]

We also see differences for the exercise and vitamin behavior scores. Those who responded the most had considerably higher scores on both of these measures.

These results are pretty strongly supportive of the idea that people who already follow other health recommendations were more likely to follow the vitamin E recommendation, as Oster found. You can imagine, also, how this might be used to see how a given policy intervention might be more helpful to some people than to others.

The sorted effects method can be performed using the **SortedEffects** package in R.

[36] Although keep in mind this is just a demonstration of the method; you shouldn't think of these results as being any sort of conclusive.

21.4   Last But Not Least: Structural Estimation

LET'S FINISH THINGS OUT WITH A LITTLE CHERRY ON TOP. Throughout the first half of this book I focused almost exclusively on attempting to map out the entire data generating process in a causal diagram. Then, we thought about identification in the form of shutting down back-door paths. Then in the second half, I walked that back a bit and have been talking mostly about methods designed to work even if you don't have a firm grasp on the entire data generating process.

The first approach—thinking about the entire underlying model and figuring out from that how to estimate an effect—is known in economics as a *structural* approach.[37] Structural approaches do require some strong assumptions—they only work if you are right about the underlying data generating process.

[37] This is not to be confused with the method "structural equation modeling," which has some similarities with this but is not the same thing.

But if you're wrong about that it's not always clear what other methods actually tell you anyway. So why not take a structural approach?

That's why I devoted the first half of this book, and many sections in the second half, to thinking structurally about your modeling and research design process. If we've covered it already, what's this section on structural estimation about?

This section is about the fact that structural estimation *doesn't stop there*. Using theory to figure out a list of control variables to add to a linear regression is a pretty limited way to do structural modeling. Someone doing structural modeling properly would not just use theory to determine a set of alternative pathways to consider, but would also see what kind of statistical model the theory implies.

FOR EXAMPLE, IF WE HAD A BUNCH OF DATA on object masses and the gravitational force pulling them to other objects. We'd think about the underlying theoretical model, where the gravitational pull between two objects is $F = G\frac{m_1 m_2}{r^2}$, where m_1 and m_2 are the masses of those two objects, r is the distance between them, and G is the gravitational constant.

If we want to know the effect of m_1 on F, A basic approach would recognize that m_2 and r may be alternative explanations for the relationship, and so you might end up running a model like $F = \beta_0 + \beta_1 m_1 + \beta_2 m_2 + \beta_3 r + \varepsilon$. This is not a great model. It ignores the actual way that the variables come together and all that theory you worked on. The effect of m_1 on F *shouldn't* be linear, and we know that from the theoretical equation. Structural estimation would hold on to the theory and would estimate the equation $F = G\frac{m_1 m_2}{r^2}$ directly. It's the process of estimating the *actual model implied by theory*.

As a bonus, once you're done, you know exactly which theoretical parameters relate to which estimated parameters.[38] As another bonus, the structural process shows you exactly how to handle different unobserved theoretical concepts that you can't actually measure.

Because structural models estimate theoretical parameters like G directly also means that it's much easier to answer complex causal questions. Most of this book has been laser-focused on "does X cause Y and how much?" with only a few exceptions. But if you have a structural model, you immediately get access to how *any* variable in the model causes *any other* variable. And how that effect differs in different settings. And how the effect might change if we tried something completely new

[38] Imagine trying to figure out which combination of β_0, β_1, β_2, and β_3 represented G in that linear regression. In a structural model you'd just estimate G directly. You'd know which coefficient represented G because it would be represented by that G you estimated.

that isn't even in the data. You have access to the whole model at that point (assuming, again, you have the right model). You can ask *the model* what's going on, you don't have to wait to see more data.

How can you estimate a whole theoretical model yourself? Structural estimation constructs *statistical* statements out of theoretical ones and uses those to estimate. For example, the gravitational equation I just used isn't perfect; it leaves out quantum effects.[39] So there is some error and we might claim it looks like this: $F = G\frac{m_1 m_2}{r^2} + \varepsilon$. If we are willing to make a claim about the distribution that ε follows, then we can pick values of our parameters (perhaps trying out $G = 3$) and see how unlikely our observed data was to occur given the model. We try a bunch of different parameter values $G = 4, G = 5.3$, and so on, until we find the ones that best fit the data. This is estimating the structural model by maximum likelihood. You can do a similar process with other estimation methods like the generalized method of moments.

Now we come to the reason why I haven't covered this yet. Simply put, the math gets hard! Drawing a causal diagram for your model doesn't cut it—you need to specifically write out, and solve, the mathematical form of your model. Then you need to set up the statistical statements properly—remember, you have to do this all yourself, as the whole idea is that the estimation is unique to your model.[40] Then you can finally estimate your parameters.

So that's why I don't go too deep into structural estimation in this book. However, if you're interested and have the chops for it, I can recommend starting with Reiss and Wolak (2007),[41] which is specifically about structural modeling and estimation in the economic subfield of industrial organization, but does a good job at introducing the general concepts you'd use elsewhere. Galiani and Pantano (2021) do a similar kind of introduction,[42] focusing on labor economics.

[39] And in a social science setting we could easily imagine plenty of things left out of a model.

[40] That said, some mathematical forms do pop up a lot. Multinomial logit estimation comes up so commonly in structural estimation in microeconomics that some people casually think of them as being the same thing.

[41] Peter C Reiss and Frank A Wolak. Structural econometric modeling: Rationales and examples from industrial organization. *Handbook of Econometrics*, 6: 4277–4415, 2007.

[42] Sebastian Galiani and Juan Pantano. Structural models: Inception and frontier. Technical report, National Bureau of Economic Research, 2021.

22

Under the Rug

22.1   The Light, It Burns Us

SO MUCH OF THIS BOOK HAS BEEN DEDICATED TO BEING
CAREFUL. We need to be careful to create good models. We
need to think about the assumptions behind our design. We
need to think about sampling variation and so on and so on and
so on.

A statistician is always prepared to give you something to
worry about, as though you didn't have enough already!

This chapter is all about the assumptions and concerns that
are a part of pretty much *any* causal inference research study,
but which often get ignored or at least brushed aside. Turns
out that we're assuming a lot that we don't even think about.
By the time you're through with this chapter, you'll doubt you
even have a strong enough research design to prove that your
eyes are real.

That said, *someone* has been thinking about this stuff, and
just because it's uncomfortable doesn't make it go away. I'll tell
you about tools to help think about these topics and grapple
with them in regards to your own research.

DOI: 10.1201/9781003226055-22

This is a sort of survey chapter. I won't be going super in-depth on any of these. But I will tell you what they're all about and give you a few places to look in if you're interested in learning more.

Perhaps the scariest thing I can say at this moment is that the list of topics in this chapter is *only a subset of topics that could have easily been included.*

22.2   *How Sure Are You? Model Uncertainty*

IF YOU'VE TAKEN THE TIME TO SIT DOWN AND ACTUALLY TRY TO DESIGN AN ANALYSIS TO ANSWER A RESEARCH QUESTION (as I certainly hope you have by this point), you'll very quickly come to a realization about how many choices you have to make in that process, and how many of those choices you're unsure about. Those arrows in the causal diagram you weren't sure about. The unnecessary control you couldn't decide whether to include or not.

As careful as we like to be about designing our causal diagrams (and using the analyses they imply), we're never going to be certain what the "best" model really is. We might have quite a few reasonable candidates, once we've pinned down the really important parts we're sure about.[1] So we have *model uncertainty* in our analysis. Variation in our estimates comes down not just to sampling variation, but also uncertainty in the question of which statistical model is the right one to estimate.[2]

In many cases, researchers present results from only one preferred model, either ignoring the possibility of model uncertainty or using some criterion, usually based on fit or predictive power, to pick a single best model. Maybe they'll present a few models, checking whether the addition of a few controls that *might* be necessary would change the results. The idea here being that you aren't 100% certain about the causal model—one potential causal model says variable Z must be controlled for, another says it doesn't. If you get similar results controlling it or not, then your result is "robust" to which of the two causal models are right.

Presenting a few hand-picked models is a basic but pretty ad-hoc way to deal with the reality of model uncertainty. Can we do better?

INSTEAD OF TRYING A FEW MODELS WE COULD try *lots* of different models, often differing along the lines of which control variables are included. Then, once you have the results for

[1] Sometimes, even knowing the necessary controls to identify from a causal diagram, even if we're sure about that diagram, leaves some room for interpretation. For example, what if a variable is on a back door, but likely doesn't bias the result *that* much, and controlling for it would make the model way noisier?

[2] This is a bit different from the issue of not knowing which *causal model* is the right one. We talked a bit in Chapter 7 about how to narrow that down. There is some overlap between causal model uncertainty and statistical model uncertainty— for example, both might question whether a certain variable should be controlled for. But the fixes in this section are largely targeted at statistical model uncertainty.

those models, we both have an idea of what degree of model uncertainty we have, and can maybe use that to produce a single estimate.

Let's say that you have a set of variables W_1 that you *know* you have to control for to identify your effect, and a set of variables W_2 that you don't think need to be included, but you either aren't sure or are thinking of including them anyway.

You can estimate every possible model with every possible combination of all the variables in W_2. Or at least a smartly-chosen subset of those models (after all, doing *all* the models means 2^J models if there are J variables in W_2—that grows quickly!). From each model you get an estimate.

Once you have all the models you could show the distribution of all those estimates, which would show in an easy-to-understand way both what a "typical" estimate looks like (perhaps even reporting the mean of the distribution as an average effect), and also how much potential variation in the estimate model uncertainty might be responsible for. If the distribution of the estimates is fairly narrow, then model uncertainty doesn't seem to matter too much and you're good to go. If it's wide, then the results from any particular model should be viewed with the knowledge that it's highly contingent on which controls are included.

ALTERNATELY, YOU COULD GO EVEN MORE PRINCIPLED and use Bayesian model averaging.[3] Bayesian model averaging has been around for a long time. The basic idea is that you don't just plain take the mean of your estimates across all the models. The models should be weighted by how likely they are to be true!

How do we figure out how likely they are to be true? Well, we start by telling *it* how likely we think each is to be true. We pick a "prior probability" for each candidate model. Then, we estimate the models. Based on how well each model explains the data, we update those prior probabilities to get posterior probabilities. Models that fit really well are counted as more likely to be true than the prior, and models that fit poorly are counted as less likely to be true than the prior.

Then, once we have those posterior probabilities, we can use them as weights to average together the results of all of the candidate models, providing us with a good guess about the coefficients while averaging out all the model uncertainty.[4]

DEMONSTRATING OR AVERAGING the distribution of effects across many models is handy, and a great way to deal with uncertain control-variable decisions. Of course, it can't address

[3] Merlise Clyde and Edward I George. Model uncertainty. *Statistical Science*, 19(1):81–94, 2004.

[4] Or at least all the model uncertainty you chose to consider in selecting all the models to average over.

the underlying problem completely—there will always be modeling uncertainties. And our methods generally don't incorporate that kind of uncertainty into our understanding. But we really should know it's there.

22.3   When is The Data Not The Data? Measurement and Validity

WHEN YOU OPEN UP A DATA SET THAT HAS BEEN PREPARED FOR YOU, it's easy to forget the whole process of putting the data together in the first place. And yet!

Before you opened it up on your computer, someone had to produce that data for you. If the data comes from some sort of survey, for example,[5] someone had to decide which questions to ask people and how, what the set of acceptable responses was, and who to sample and when. Then someone had to go out and actually ask the questions and record the answers. Then, the responses come from actual people, not some omnipotent observer of the truth.

Data doesn't fall from the sky. It is produced by humans, or at the very least collected by them. The way in which it is produced affects the way we can use it, and often runs us into problems when the theoretical or statistical model we have in mind relies on assumptions about the data that just aren't true given how it was made.

I already discussed two issues in this vein in Chapter 13. One was the use of sample weights. Our statistical models assume that individuals are sampled randomly, but this is rarely true. Thankfully, many professional surveys provide survey weights that can be used to make the survey sample look more like the general population you'd like to generalize to.[6]

Another issue I brought up in Chapter 13 was measurement error. A lot of the time, variables are simply written down wrong, or reported incorrectly. I can't tell you how many papers I've read trying to solve the problem that people often can't even report their own level of educational attainment correctly. That's the kind of thing you'd think someone would remember!

These two already have their own dedicated sections. What other kinds of concerns arise when we think about the difference between the data we *have* and what we think the data *should be*?

[5] And keep in mind, many large-scale data sources, even government sources, are survey-based on some level. Unemployment rates, for example, involve asking someone whether they're looking for work, at the very least.

[6] Many of the problems that survey designers face about samples being representative are very similar to causal inference problems. This makes sense—they have to ask what causes someone to be more or less likely to be in the sample. Survey weights are a lot like matching weights, where the survey sample is being matched to the population.

22.3.1 Construct Validity

CONSTRUCT VALIDITY ASKS WHAT OUR DATA ACTUALLY MEA-
SURES. What do we want data to do, really? We have a theoretical model that we'd like to estimate some parts of. To
do that, we need to get measurements of the variables in that
model. If we gather some data and that data does a good job
at representing that theoretical concept, we'd say that we have
construct validity.

This is harder than it seems. Let's say that we have a theoretical model and one of the variables in that model is "trust."
That is, how trusting someone is of the world. You can easily
theorize about where this variable fits in your model. You know
where the arrows come in and go out of it.

But how do you measure it? Maybe you just ask people to
rate how much they trust the world on a scale of 1 to 10. Seems
like a good place to start at least.

Someone concerned with construct validity would immediately spot the problems with that plan. Are people good at
self-reporting their own trust levels? Will people understand a
rating scale and what the numbers mean? How likely is someone
to misunderstand the question, or perhaps think that "trust"
means something different than you think it means? Will the
answer to this *even be that closely related* to your idea of trust?
Or will it better represent something like "how important someone thinks trust is"?

One example of this problem at play comes from Hertwig and
Gigerenzer (1999).[7] They look at the classic "Linda problem"
from behavioral economics. In the Linda problem, people are
given a small vignette about someone named Linda. The vignette is written in such a way that suggests that Linda is a
politically active feminist. Then, they are asked which is "more
likely"—that Linda is a bank teller, or that Linda is a bank
teller *and* a feminist.

People routinely say that it's more likely that she's a bank
teller *and* a feminist. However, this breaks the laws of probability. Every bank-teller-and-feminist is a bank teller. So being a
bank teller must be at least as likely as being both. This is taken
as evidence that people make errors in probabilistic judgment.

However, Hertwig and Gigerenzer (1999) point out that this
assumes that respondents are using a probability-based interpretation of the term "more likely." By other interpretations,
their response might be perfectly rational! For example, by

[7] Ralph Hertwig and Gerd
Gigerenzer. The "conjunc-
tion fallacy" revisited: How
intelligent inferences look
like reasoning errors. *Journal of Behavioral Deci-
sion Making*, 12(4):275–305,
1999.

the laws of conversation (not their term), we would be more likely to have given them that feminism-focused vignette in order to demonstrate Linda's feminism, making the bank-teller-and-feminist option more "likely." They also show that, by asking the question in a different way that emphasizes that they really do mean the probability-based interpretation of "more likely," people don't make the error nearly as often.

This problem isn't limited to surveys and experiments. Let's say you want income as a part of your model, and you have complete access to tax data, no surveys involved. So do you use wage income? Or total income? Or family income? Or post-tax income? Which version of income represents the "income" you have in mind for your theory?

What can we do to make sure that we have good construct validity? Well, we can think carefully about what underlying construct our data is likely to actually represent. That's always a good idea. More broadly, we can pay attention to psychometrics. Construct validity is very much their jam. In fact, they'd probably say I'm leaving out a number of other aspects of whether a measure does a good job representing a theoretical concept, such as whether it represents *all* of that concept (content validity), and whether the measure tends to give you the same value when you collect it in the same setting (reliability, i.e., if you measure "trust" for the same person on two different days, do you get more or less the same result?).

I'm not going to attempt to squish the entire field into this section, but one thing they do to help construct validity along is take *multiple* measures, or ask multiple survey questions, that are all intended to represent the same concept. Then, they combine them together with tools like factor analysis or structural equation modeling in order to extract a number representing the underlying theoretical concept that you actually want.[8] The idea here is that every way you ask about your intended theoretical concept is an imperfect representation of what you want, so by looking for the statistical similarities between them you're probably picking up on the real deal. Seems like a good place to start.

[8] Ever taken a survey and it feels like they're asking you the same question in slightly different ways a bunch of times? This psychometric approach is what the survey is going for.

22.3.2   The Observer Effect

THE DOWNSIDE OF DOING SOCIAL SCIENCE IS THAT IT'S THE ONLY FIELD WHERE THE SCIENTIFIC SUBJECTS KNOW THEY'RE SCIENTIFIC SUBJECTS. This leads to another way in which our measurements might fail to represent our theoretical concept—if people modify their behavior knowing that we're watching them.[9] There are a number of names for this, each of which varies slightly in meaning from the others—researcher expectations, researcher demand, hypothesis guessing, Hawthorne effects, observer bias...

For simplicity I'll refer to the whole bundle of concepts as the "observer effect," borrowing a term from physics.[10] The observer effect applies most often in lab experiments, surveys, and polls, although occasionally it pops up in other settings.

There are a bunch of ways in which the *fact that people know we're collecting data on them* might change their behavior, or at least what they say their behavior is:

- They may tell the researcher what they think the researcher wants to hear.

- They may try to guess what the researcher will do with the data and so act in a way that means the data shows what they want it to show (or think it should show).

- They may respond in a way that makes them look good to the researcher.

- They may not like being research subjects, or they may just like causing trouble, and so may actively try to mess up the data.[11]

- They may get self-conscious and self-aware of their responses.

- They may respond to cues from the researcher, who themselves may be consciously or subconsciously trying to get the data to support their hypothesis.

The list goes on. People do stuff for all sorts of reasons.

Unfortunately there's not *too* much you can do after the fact if you think there's likely to be a problem.[12] This is more a problem you need to tackle at the data-collection stage. The study and data collection process itself needs to be designed around the difficult task of getting people to answer questions, often difficult ones, without thinking too much about the fact that they're answering questions.

22.3.3 Processing Data

THEN WE COME TO THE REAL EMBARRASSING CORNER. Perhaps the data collection mechanism has been immaculately

[9] The aforementioned psychometricians might consider this a subset of construct validity.

[10] Borrowing a term from physics and using it for only vaguely related purposes! That's how you know I'm an economist.

[11] Having administered a survey to high school kids myself, I can testify to having seen all kinds of *suuuuuper* hilarious joke answers and penis drawings.

[12] Unless you're willing to make some strong assumptions about the exact way you think it's happening.

designed. The survey is perfectly put together and implemented, the sampling is broad. Then the data comes and you, well... you have some work to do!

It's pretty rare in the real world (that is, outside a classroom) that you are working with a data set that comes to you completely ready to run your analysis on. You generally have to clean and manipulate the data in some way before you can use it.

Therein lies the problem. Just like with model uncertainty, there are often multiple ways to do the data cleaning process. Which observations are likely to be mistakes and should be fixed? Which observations don't apply to your model and should be dropped? How can you code up the free-response survey question into usable categories? How should you define the different variables—for something like education do you keep it at its original value or "bin" it into bigger groups? How big are the bins?

Different people will make these choices in different ways, which can affect the estimates. In Huntington-Klein et al. (2021),[13] we found that multiple researchers separately cleaning the data for the same research project made considerably different decisions, and no two researchers ended up with the same sample size. This affected the estimates—the standard deviation of estimates across different researchers was much bigger than the reported standard errors for each researcher.

[13] Nick Huntington-Klein, Andreu Arenas, Emily Beam, Marco Bertoni, Jeffrey R. Bloem, Pralhad Burli, Naibin Chen, Paul Grieco, Godwin Ekpe, Todd Pugatch, Martin Saavedra, and Yaniv Stopnitzky. The influence of hidden researcher decisions in applied microeconomics. *Economic Inquiry*, 59(3): 944–960, 2021.

That's just differences in data preparation that come down to "reasonable people can disagree on how to do this." Sometimes there are just errors that are hard to catch. One particularly scary version of this popped up in genetics. Geneticists are fond of doing data analysis (or at least data entry) in Microsoft Excel. Excel, however, likes to convert things to dates if it thinks they're dates. And you know some things that look a lot like dates? Certain names of genes.

Analyses of the genetics literature found that up to 20% of all recent papers had errors in them related to Excel reading gene names as dates. Oops! Geneticists actually ended up renaming a lot of date-like terms instead of giving up on Excel.[14]

[14] James Vincent. Scientists rename human genes to stop Microsoft Excel from misreading them as dates. *The Verge*, August 2020.

Data cleaning is a human process just like data collection. The way you do it matters, whether it's an error or just a difference of opinion. Worse, it's the kind of thing where you might not even *notice* that an error has been made until it's far, far too late.

22.4   *Why It Had To Go I Don't Know: Missing Data*

YOU MIGHT THINK THAT, IF YOU HAVE AN OBSERVATION IN YOUR DATA, THEN YOU HAVE THAT OBSERVATION IN YOUR DATA. If only it were that simple! It's pretty darn common to have observations in your data where you have data on *some* of the variables but not others. For example, in a data set of height and weight for a group of people, there might be some people for whom you have measures of both height and weight, some you only have height but not weight, and some you only have weight but not height. When this happens, you have *missing data.*

Missing data can be a real problem. The first, obvious, problem is what the heck we do with observations that have missing data. If we want to regress, say, weight on height, a regression won't allow someone to be in the model if they only have weight *or* height but not both.

The simplest solution is "listwise deletion," where you simply drop everyone from the data if they have missing data for any of the variables in your model. This is the default behavior of most statistical software.

But just because it's the default doesn't mean it's a fantastic idea. Sure, it will let you just run your darn model,[15] but listwise deletion can introduce real problems, even beyond the fact that it shrinks your sample size. Why? It all comes down to *why* the data is missing.

Data can be missing for a number of reasons:

- **Not in Universe (NIU).** A value is NIU if the variable simply doesn't apply to that observation. For example, the variable "number of employees in your small business" *should* be missing for people who don't have a small business.

- **Missing Completely at Random (MCAR).** A value is MCAR if the fact that it's missing is completely unrelated to both the value itself and to the other variables in the model. So in our height and weight example, weight is MCAR if it is equally likely to be missing regardless of someone's true weight and regardless of their height.

- **Missing at Random (MAR).** A value is MAR if the fact that it's missing is unrelated to the value itself *conditional on the other variables.*[16] Weight is MAR if it is equally likely to be missing whether your true weight is higher, lower, or equal to what we'd expect your weight to be given your height.

[15] To be fair, listwise deletion is probably a fine choice if your missing data problem is limited to a pretty tiny portion of your data— it still introduces problems, but less so, and at that point the additional complexity added by some of the more proper methods may hurt more than they help.

[16] MAR is very easy to mix up with MCAR, I know. Sorry, I didn't pick the names!

- **Missing Not at Random (MNAR).** A value is MNAR if the fact that it's missing is related to the missing value itself. For example, weight is MNAR if people with lower actual weights are more likely to have missing weight values in the data.

Listwise deletion works okay for the first two of these: Not in Universe and Missing Completely at Random (MCAR).[17] For Not in Universe missing values, you *want* those observations out of your analysis, since presumably your analysis is trying to model a group that they are not a part of. For MCAR, it's basically a complete fluke that there's missing data somewhere. So dropping an observation with missing data is hardly different from having gathered one less observation in the first place. No problem.

But think about why data might be missing and ask how likely MCAR really is. Why might weight data be missing, for example? People think their weight is personal and don't want to share it? People with particular weights are ashamed of it and don't want to tell a researcher? People lie about their weight and the survey-taker doubts the answer and so doesn't write anything down? If there's a *reason why* the data is missing and it isn't some complete fluke, that reason is probably related to the value of the variable, or at least to one of the other variables in the model.

If the data is Missing at Random (MAR) or Missing Not at Random (MNAR), listwise deletion will do some bad things. In these cases, "Missingness" gets a spot on the causal diagram itself. Using only data without any missing values is like controlling for Missingness. Depending on the exact diagram, this is just asking for collider effects or post-treatment controls. Or just violating your assumptions about random sampling. Your estimates can get biased.[18]

There are some approaches available for dealing with missing data, thankfully. Of course, they bring their own kind of complexity.

22.4.1 Filling in Blank Spots

ONE OBVIOUS WAY TO DEAL WITH MISSING DATA is to fill that data in. There are a host of different methods for *imputing* missing values. That is, trying to figure out what they should be and filling them in.

[17] Although for MCAR, listwise deletion does reduce your sample size.

[18] I should point out, though, that while the other solutions below solve the problem better than listwise deletion if *the additional assumptions they rely on are true*, what if they aren't? Listwise deletion might sometimes still be the least-bad solution to the problem.

To get an idea of how this could work, let's start with a bad way of doing it that is nonetheless popular among the "just let me run my regression please" crowd: mean imputation.

Let's say we have some height and weight data with some missing weight values in Table 22.1.

Height (in.)	Weight (lbs.)
65	120
70	?
66	?
70	180

Table 22.1:   Height and Weight Data with Missing Values

We don't know what the ?s are. So we just take the mean of weight $((120 + 180)/2 = 150)$ and fill in all the missing values with that. This gives us Table 22.2.

Height (in.)	Weight (lbs.)
65	120
70	150*
66	150*
70	180

Table 22.2:   Height and Weight Data with Missing Values Filled in Poorly by Mean Imputation

* Imputed

Now we can run our analysis with all four observations intact. However, this method completely ignores any way in which the rows with missing data might be different. So you can get some weird predictions. Imagine we had a super-tall 84-inch person here with missing weight. We'd predict their weight to be 150 pounds as well. That doesn't seem right.

More sophisticated methods of imputation, which are designed to work for data that is Missing at Random (MAR), take basically the exact same idea but do it better, using the other data that is available to make much better predictions about the missing values.

THE MOST COMMON SOPHISTICATED FORM OF IMPUTATION is probably multiple imputation.[19] In multiple imputation, you (1) fit a model that allows you to predict the variable with missing values using all the other variables in your data,[20] and (2) use that model to predict the missing values.

What kind of model can you fit? Multiple imputation allows for pretty much any kind of model. Traditional approaches have used good ol' regression. However, as with any method where prediction is the real goal, a number of methods have popped

[19] Another method is to use coarsened exact matching, as in Chapter 14, to predict the values of missing data. This requires that the observation with the missing values has some "doppelganger" observations in the data that are exactly the same on a set of matching variables. This approach is far less common but seems interesting for very large data sets.

[20] Note this says "in your data," not "in your model." The idea here is that *all* the variables have information on what the missing value is likely to be.

up based on the use of machine learning algorithms to do the prediction.[21]

At this point, you have a *single* imputation. How do you get from that to *multiple* imputation? You (3) add some random noise to your prediction process (drawing random values of the regression coefficients and then the predicted values, from the sampling distributions you've estimated for those things), (4) predict the value a bunch of different times, with different random values each time.[22] This multiple imputation process improves on the single imputation by not imposing too much certainty on those predictions, which both better reflects the uncertainty in your predictions and makes for better predictions overall.

Now you have multiple predictions for each missing value. Let's say you want to produce ten imputations. You'd end up with ten full data sets, each of them having the missing values filled in slightly different ways. Then, when you want to estimate your model, you estimate it ten different times, once for each imputed data set. Finally, you combine your estimates together according to some estimation-combination equations known as "Rubin's Rule." Now you have your estimate!

Multiple imputation has its ups and downs. It does get you a complete data set that you can work with. However, it also requires that you make a *lot* of decisions in setting it up without a whole lot of guidance, picking a full predictive model for each variable in your data set with missing values, and selecting things like how many imputed data sets to make. Also, because it's such a general method for filling in data, it isn't specialized, or even aware of, the model it's going to be used in, and misses some opportunities there. For that reason, many people prefer the methods I talk about in the next subsection. But multiple imputation still has a real appeal and is intuitively very straightforward.

Now, do keep in mind, this is all intended to work with MAR data (or MCAR I suppose). If missingness is related to the missing value in a way that isn't explained by the other variables, then, well... a method designed to predict those missing values using the other variables won't work well. Are you out of luck if the data is MNAR? Not entirely. You can use multiple imputation with MNAR data. But doing this requires that you explicitly model what you think the missing-data process is so your predictions can account for it. This is really only feasible

[21] Yi Deng, Changgee Chang, Moges Seyoum Ido, and Qi Long. Multiple imputation for general missing data patterns in the presence of high-dimensional data. *Scientific Reports*, 6 (1):1–10, 2016.

[22] A common extension to multiple imputation adds *chained equations*, which is nice when you have multiple variables with missing values. In this method, once you start imputing some values for a variable, you use *those imputed values* to start imputing *other* variables. Then you loop back around to your original variable and repeat the process a bunch of times.

if you have a very strong grasp on why your data is missing. Have fun!

22.4.2   Stronger Without You

MAYBE WE DON'T NEED TO FILL IN THE VALUES. After all, we don't really care about predicting those missing values. That's just a nuisance we have to get out of the way before we can do what we really want, which is estimating our model. What if we had a method that would let us estimate a model even in the presence of missing values? There are actually several ways to do this.

THE FIRST METHOD FOR ESTIMATING A MODEL WITH MISS-ING DATA INCLUDED is also the simplest. Just... model it!

This works especially well when you're dealing with categorical data. Just add "missing value" as its own category. Then you're done. Run the model as normal otherwise.

This is a fairly common approach in the machine learning world. There are some weaknesses, of course. Like the mean imputation method that worked so poorly, this method ignores the differences between different observations with missing data. However, unlike mean imputation, this approach *does* account for the differences between the missing and non-missing observations, and does to some extent model the relationship between missingness and the other predictors.

THE SECOND METHOD IS FULL INFORMATION MAXIMUM LIKE-LIHOOD. Full information maximum likelihood is a way of estimating your *actual* model while letting each observation contribute whatever observed data it has, without requiring it to all be there.

Maximum likelihood in general operates on the idea that, for a given model, you can calculate the probability of a given observation occurring, which we call the likelihood. For example, if your model estimates that a coin has a .6 chance of being heads, then two observations of heads followed by one observation of tails has a likelihood of $.6 \times .6 \times .4 = .144$, i.e., a .144 chance of occurring. Then, you pick the *model* that makes the *data you got* as likely as possible. In our coin example with two observed heads flips and one observed tails, we'd pick whatever $P(Heads)$ value maximized $P(Heads) \times P(Heads) \times (1 - P(Heads))$, which would be $P(Heads) = 2/3$.

Full information maximum likelihood just calculates the probability of a given observation occurring without having to refer to all of the variables together. So if you have data on height and weight, we calculate the probability of observing the particular combination of height and weight you have, and that's your likelihood. But for me we only observe weight. So we'd calculate the probability of observing the weight that I have, and that's my likelihood. Then we combine them together and pick the model that maximizes the likelihood of the whole sample.

The technical details are deeper than that, but that's the main insight here. Full information maximum likelihood figures out how to let an observation contribute only partially, and estimates the model on that basis.

ONE LAST METHOD: THE EXPECTATION-MAXIMIZATION (EM) ALGORITHM. The expectation-maximization algorithm is an estimation method with broader applications than missing data, but missing data is one prominent place where you see it.

The EM algorithm is sort of like a cross between imputation and full information maximum likelihood without really being either. Like imputation, it does "fill in" values, but like full information maximum likelihood, it does so using the model you're estimating. The process of estimating a model with EM is:

1. Use some prediction method to make an initial guess of what the missing values are

2. (The maximization step) use the current version of the filled-in data to estimate your analysis model

3. (The expectation step) use the estimated model to predict the most likely values of the missing data

4. Repeat steps 2 and 3 many times

In other words, we use some filled-in guesses of the missing data to estimate our model. Then, the model itself is used to produce better guesses of the missing values than we had before. The guessing doesn't work like it does in multiple imputation, where each variable is predicted in its own model. Instead, the estimated model from the maximization step is used to figure out the joint distribution of all the variables, and you draw random values from that.

LIKE WITH MULTIPLE IMPUTATION, these methods all are based on using the information you have to infer what you don't. That means that they're all designed to work with data that is Missing at Random (MAR).[23] What if we have data that's Missing Not at Random (MNAR)? Oof. Well, like with multiple

[23] With the slight exception that the first method, simply modeling "missing value" as its own category, will help a little with MNAR data if the true values of missing data tend to be clustered heavily in very few categories—say, if nearly all observations missing an "education" value are PhD holders.

imputation, you can apply any of these methods to MNAR data. However, doing so once again requires you to explicitly model what you think the missing-data procedure looks like. It's a tall order.

22.5   It Lurks! They Call it SUTVA

IT STRIKES FEAR IN THE HEART OF ANYONE DOING CAUSAL INFERENCE. The Stable Unit Treatment Value Assumption, or SUTVA. SUTVA is an assumption that, in simplified terms, means that we know what "treatment" is.

How could we not know what treatment is? There are two ways that concern SUTVA. The SUTVA says that neither of them happens. But if either of them does happen, SUTVA is violated and we have a problem.[24]

1. Treatment means different things for different observations, such that different "treated" individuals didn't really get the same thing. For example, say you want to know the effects of workplace diversity training. But in one workplace that means watching an old VHS and reading a brochure, while in another workplace that means a week-long intensive participatory training program

2. Your outcome is determined in part by the treatment received *by other people*. In other words, there are spillovers. For example, say we want to know the effect of offering tax incentives on a city's economic growth. My neighboring city *does* offer incentives, which makes it grow, and some of that growth comes to *my* city in the form of increased growth as well, whether or not I have tax incentives too

It's not hard to see how either of these could lead to problems, even if we have completely random assignment to treatment.[25] Treatment meaning different things in different cases means that we don't really even know what X is in "the effect of X on Y," so what are we even estimating? Spillovers mean that our treatment *actually* reflects the treatment of many, many people, and picking out just the effect of our own is going to be quite the task.

Even worse, SUTVA violations seem to be just all over the darn place once you start looking for them. When does "receiving treatment" *ever* mean exactly the same thing to everyone who gets it? And when is it ever really the case that what's going on with some people doesn't affect some others?[26]

[24] Guido W Imbens and Donald B Rubin. *Causal Inference in Statistics, Social, and Biomedical Sciences*. Cambridge University Press, 2015.

[25] Learning that there's a causal inference problem that couldn't even be theoretically solved by randomization I'd put on par with learning the truth about certain holiday-related characters as a child.

[26] As I've heard from some very good sources, *we live in a society!*

SO WHAT CAN BE DONE? Unfortunately there aren't exactly "SUTVA fixes" running around. However, there are some things that can be done that allow you to *modify the design itself* to help avoid SUTVA violations.

After all, SUTVA is about not having a stable idea of what treatment is, so... what if we did? Often SUTVA violations can be avoided by just thinking about what treatments we *can* define precisely.

In the case of treatment meaning different things in different cases, one obvious solution, if possible, is to measure not just whether treatment was assigned, but the actual flavor of treatment that was administered. Some experimental designs, for example, will include follow-up measurement that checks in on how faithfully the assigned treatment was followed. Accounting properly for differences in treatment adherence, or just considering different kinds of implementation as actual different treatments, allows the SUTVA violation to be sidestepped.

In the case of spillovers, one common approach is to specify treatment at a broader level. Say we're interested in the effect of tutoring on test scores, but we're worried that *me* being tutored might improve *your* test scores if we're in the same classroom. SUTVA violation! But what if "treatment" is not "I was tutored," but rather "the proportion of students in a classroom who were tutored"? As long as we don't think there are spillovers *between* classrooms we're good to go.

This doesn't necessarily solve all our problems, though. What if the effect we're interested in really does rely on that social interaction? This comes up when the spillovers, peer effects, or social dynamics (the stuff SUTVA said couldn't happen or we'd be in trouble) are *the very thing we're interested in*.

It's not the end of the world. The problems with trying to estimate causal effects in social settings have been known to be real thorny since at least Manski (1993),[27] and the many papers following after it. However, progress has been made in using models that explicitly model the social network going on underneath the data. This is an entire subfield in itself that unfortunately won't get the full treatment in this book, but you can get started with the excellent overview in Frank and Xu (2020).[28]

[27] Charles F Manski. Identification of endogenous social effects: The reflection problem. *The Review of Economic Studies*, 60(3):531–542, 1993.

[28] Kenneth A Frank and Ran Xu. Causal inference for social network analysis. In *The Oxford Handbook of Social Networks*, pages 288–310. Oxford University Press, 2020.

22.6   I Mean Nothing By It: Nonexistent Moments

ONE BIG FAT UNSTATED ASSUMPTION THROUGHOUT THIS EN-
TIRE BOOK IS THAT the tails of our relevant distributions are
neither too big nor too fat. All of the statistical methods we
have used have made the implicit assumption that things like
"the mean" or "the variance" *exist*.

Hold on, how is it possible that the mean or variance could
just *not exist*? If you have a sample of data, you can calculate
the mean and the variance. Nothing's stopping you.

That's true, but those calculations only give you the *sample*
mean and the *sample* variance. It's entirely possible, however, to
have theoretical distributions for which the mean and variance
(technically, the first and second moment) are just plain undefined! When that happens, for one thing, it's harder to figure
out sampling variation, and for another, it's hard to do things
like hypothesis testing when we can't even specify a theoretical
mean to reject. For yet another thing, weird distributions can
lead to noisy data sets that are hard to get a handle on.

One common place where distributions with undefined moments pop up the most is in the case of *fat-tailed* distributions.
These are distributions with a lot of weight in the tails. In a
normal distribution, the probability of an outcome declines very
rapidly as you move away from the mean. But in a fat-tailed
distribution, the probability of an outcome declines very, very
slowly. These tails are fatter than you'd even get in a log-normal
distribution like we discussed in Chapter 3.

Fat-tailed distributions usually pop up in the social sciences
in the context of a *power law distribution*. Under a power law
distribution, most people have pretty small values, but there's
no shortage of people with pretty big values, and there seems to
be no maximum. Plus, the bigger values are *way* bigger. You
can see how this looks in Figure 22.1, which shows an example
of a Pareto distribution, which is one type of power-law distribution. For the scale and shape parameters I've given the
distribution, its mean and variance are undefined.

In Figure 22.1, we see the sharp initial drop that we'd expect from, say, a log-normal distribution. A lot of the values
are extremely tiny. Then it *looks* like it goes to 0. So, nothing
at the high values, right? Wrong! There's an area underneath
that curve; it's just declining *very* slowly. By the time we get
to an x-axis value of 10, there's still 6.7% of the distribution
left to go. Go out ten times further than that and there's still

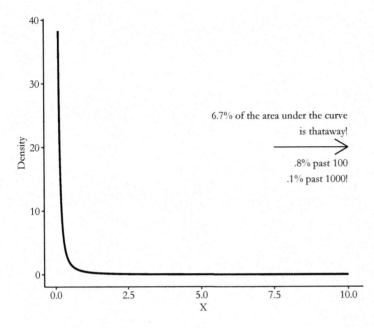

Figure 22.1:   Example of a Power Law Distribution (Pareto scale = .5, shape = .9)

.8% left to go. Nearly a percent of the distribution is *over 100*, even though the majority of the data is below 1. Go ten times further than that and there's still .1% of the distribution left to go! One in a thousand observations will be *1000 times higher* than the median. If .1% doesn't seem impressive to you, a normal distribution with mean (and thus median) 1 and standard deviation 1 has about .00002% of the distribution left to go at only *five* times the median.

This distribution also lets us see what it means for the moments to be undefined. If I generate 100,000 random values from this Pareto distribution and calculate the sample mean and variance, they jump around wildly. With a sample that large, normally-distributed data would produce sample data with the true mean and standard deviation nearly every time. But in the ten samples I try with the Pareto I get sample means anywhere from 9.6 to 38.2 (and if I kept going I'd get means in the hundreds or thousands occasionally), and sample standard deviations from 282 to 4,413.

FAT-TAILED DISTRIBUTIONS IN THE SOCIAL SCIENCES tend to pop up around issues of highly unequal distribution (like income) or anything to do with popularity, where the big tend to be far bigger than the rest. Let's take music as an example, pulling some information about Spotify followers from Wikipedia at the time I write this in April 2021.

The artist with the most followers on Spotify is Ed Sheeran, with 78.7 million followers. The second-most followed artist is

Ariana Grande, with 61.0 million. The #1 spot has a value a full 29% bigger than the #2 spot. By the time we get to #20, Alan Walker,[29] we're down to 27.9 million. The #1 spot is fully 281% as large as the *twentieth* spot. That's an enormous drop over a space of only twenty artists, especially when you consider that there are approximately *7 million* artists on Spotify. I don't have figures on all of them, but I can guarantee you that even though Alan Walker is tiny compared to Ed Sheeran, he is *enormous* compared to the 50th artist on the list, who in turn is enormous compared to the 100th, who in turn... and so on.[30] A massive portion of those 7 million artists probably have no followers at all. *That* is what a fat tail looks like. The big are very big, and there aren't just one or two of them.

Compare that to a distribution that does *not* have a fat tail, like the time it takes someone to run 1500 meters. Hicham El Guerrouj is the world-record holder as of this writing with a time of 3 minutes and 26 seconds. Second place is Bernard Lagat, with 3 minutes and 26.34 seconds. That's a .16% difference between first and second place, a far cry from the 29% difference between Ed Sheeran and Ariana Grande. By the time we get to 20th place we're only up to 3 minutes and 29.46 seconds, 1.7% away from first. Your humble author, not a highly athletic man, can do it in about 6 minutes. In the global population of billions I'm almost certainly worse than 1 billionth place. And yet the ratio of my time to the all-time world record holder is less than 2. But if I started a locally-popular band, released a few tracks, and got a thousand followers on Spotify (which would be pretty darn good), even Alan Walker would still be beating me by a factor of 27,900. Last I checked, 27,900 is a lot bigger than 2!

Fat tails are far from just for music streams. Power law distributions pop up for income, for wealth, for city populations, for word usages, for follower counts on social media, and so on and so on. Further, quite a few variables follow a fairly typical distribution for *most* of the data, but then follow a power law only in the tails. City sizes are a good example of this—the world's biggest city Tokyo is about twice as large as #11, New York City, which is about 50% larger than #22, Guangzhou. It drops off super quickly. But among cities that *don't* have millions of people in them, the distribution is much better-behaved.

So WHO CARES? Different variables have different distributions! So what?

As previously mentioned, the fact that the moments are undefined, but our estimators assume they *are* defined, means that

[29] Coincidentally, the only one in the top 20 I didn't recognize. He is not, as I guessed from the name, a country artist.

[30] One feature of power laws is that the fatness of their tails shows up no matter where you start the distribution. If we looked at the whole distribution of artists, it would be massively unequal. The distribution would be all squished down near 0. Chop off the little artists to, say, the top 10000. It would *still* be massively skewed, and the graph of the distribution would look largely the same as when it was the whole 7 million. Cut to the top 20, as I have here, and it's *still* massively skewed, even within this highly select group of super successful artists.

we're getting something wrong about our estimates of sampling distribution. How wrong? It depends on the setting. Second, these fat-tail distributions are inherently extra noisy. Whether or not you *happen* to get the extra-extra-big observations in your data can really change your estimates.

The most common way of dealing with this problem is with the use of logarithms, as in Chapter 13. If Y follows a power law distribution, then regressing $log(Y)$ on $log(X)$, instead of Y on X, can help turn the relationship into a straight line that OLS can handle.

However, logarithms only go so far. If the tails get fat enough, even the logarithm won't be able to handle it. A great example of this has been people trying to model pandemic spread during the coronavirus. Viruses tend to spread at exponential rates, since each additional infected person has a chance of infecting others, and some "super-spreaders" who come into contact with high numbers of people spread an astonishing amount of the disease. So there's a fat tail.[31] And yet, the number of attempts I've seen that try to use a linear non-fat-tailed model to measure the causal effect of, say, lockdowns, or sunlight, or vaccines, or alternative treatments on coronavirus case rates, throwing at best a logarithm or a polynomial at the problem—let's just say there are a lot of those attempts. And it doesn't work that well.[32]

In these cases there are some methods we can turn to. One is to use *quantile regression*, which is a form of regression that, in short, tries to predict a percentile (often the median) instead of a mean. Because percentiles near the center of the distribution are not sensitive to what's going on in the tail, these can perform better with fat-tailed data.

There are also methods designed to estimate fat tails directly. But this is a deep, deep well of research that gets highly technical. Machine learning techniques have also stepped in to help estimate these models while dealing with the wild values a power-law distribution can throw at you. One approach that is a bit easier to grasp is the application of maximum likelihood,[33] as in the Stronger Without You section of this chapter. Maximum likelihood is all about picking a model that makes the data as likely as possible based on the probability distribution you give it. So, when having it calculate how likely the data is, give it a power-law distribution to work with. It will then pick parameters to help fit the model. Of course, this relies on us having an

[31] Felix Wong and James J Collins. Evidence that coronavirus superspreading is fat-tailed. *Proceedings of the National Academy of Sciences*, 117(47):29,416–29,418, 2020.

[32] One particularly egregious example involved a model from the US White House predicting that daily coronavirus deaths in the United States would drop to 0 by May 15, 2020. This was extremely wrong. The prediction informed policy despite plenty of researchers pointing out the poorly-chosen model, which used a third-order polynomial. The prediction was, of course, way off, with hundreds of thousands more people in the US dying from the virus after May 15.

[33] Heiko Bauke. Parameter estimation for power-law distributions by maximum likelihood methods. *The European Physical Journal B*, 58(2):167–173, 2007.

idea of *which* power-law distribution we're working with. There are many!

22.7   *The Treatment Mystery*

AT THE END OF A LONG AND DEPRESSING CHAPTER, I will lead you to a short and depressing question. It's what I call "the treatment mystery," and it's this:

If the observations with different values of treatment are so comparable, then why did one of them get more treatment than the other?

This whole causal inference exercise is about trying to find observations with different levels of treatment but that are otherwise comparable. Then, the differences we see between them in the outcome should just be due to the treatment. We accomplish this by closing back doors, or isolating front doors.

But if we've really gotten rid of the ways in which different observations are non-comparable, then why did some of them get more treatment than others? You'd think if we'd really accounted for all the reasons they're different, there wouldn't be any difference in treatment, either.

A great example of this is in twin studies. There are a lot of research questions that want to know the effect of education on something—wages, civic participation, etc. There are a lot of back doors into education, however, with things like family background, genetics, demographics, personality, and so on affecting both educational attainment and other outcomes. One way to deal with this problem is by using *identical twins*.

By comparing one identical twin who has more education to another twin who has less, you are by necessity making a comparison between people with the same family background, genetics, demographics, and probably a lot of similarities on things like personality. Seems like a slam dunk for all those back doors being closed! And so there are quite a few studies using this design. Some of the early ones are collected in Card (1999).[34]

[34] David Card. The causal effect of education on earnings. *Handbook of Labor Economics*, 3:1801–1863, 1999.

But then the question rears its head again. If those twins are really so identical, then why does one of them have more education than the other? Is it the result of some random outside force, as the research design would imply, or is there some real difference between the twins that led one to get more education? If it's the latter, then that same difference that led to more education may also lead to better outcomes in some other way. The back door returns.

Sometimes this mystery isn't such a mystery. If the process leading to treatment is really well-known (as in an experiment or regression discontinuity), then there's no real problem. But if it's not, it's something we have to grapple with.

Bibliography

Alberto Abadie. Using synthetic controls: Feasibility, data requirements, and methodological aspects. *Journal of Economic Literature*, 59(2), 2021.

Alberto Abadie and Javier Gardeazabal. The economic costs of conflict: A case study of the Basque country. *American Economic Review*, 93(1):113–132, 2003.

Alberto Abadie and Guido W Imbens. On the failure of the bootstrap for matching estimators. *Econometrica*, 76(6): 1537–1557, 2008.

Alberto Abadie, Alexis Diamond, and Jens Hainmueller. Synthetic control methods for comparative case studies: Estimating the effect of California's tobacco control program. *Journal of the American Statistical Association*, 105(490):493–505, 2010.

Alberto Abadie, Susan Athey, Guido W Imbens, and Jeffrey M Wooldridge. When should you adjust standard errors for clustering? Technical report, NBER, 2017. NBER Working Paper No. 24003.

Daron Acemoglu and Simon Johnson. Disease and development: The effect of life expectancy on economic growth. *Journal of Political Economy*, 115(6):925–985, 2007.

Anna Aizer and Joseph J Doyle Jr. Juvenile incarceration, human capital, and future crime: Evidence from randomly assigned judges. *The Quarterly Journal of Economics*, 130(2): 759–803, 2015.

Theodore W Anderson and Herman Rubin. Estimation of the parameters of a single equation in a complete system of stochastic equations. *The Annals of Mathematical Statistics*, 20(1):46–63, 1949.

Michihito Ando. How much should we trust regression-kink-design estimates? *Empirical Economics*, 53(3):1287–1322, 2017.

Joshua D. Angrist. Lifetime earnings and the Vietnam era draft lottery: Evidence from Social Security administrative records. *The American Economic Review*, pages 313–336, 1990.

Joshua D. Angrist and William N. Evans. Children and their parents' labor supply: Evidence from exogenous variation in family size. *American Economic Review*, 88(3):450–477, 1998.

Joshua D. Angrist and Alan B. Keueger. Does compulsory school attendance affect schooling and earnings? *The Quarterly Journal of Economics*, 106(4):979–1014, 1991.

Joshua D. Angrist and Alan B. Krueger. Instrumental variables and the search for identification: From supply and demand to natural experiments. *Journal of Economic Perspectives*, 15 (4):69–85, December 2001.

Joshua D. Angrist and Jörn-Steffen Pischke. *Mostly Harmless Econometrics: An Empiricist's Companion*. Princeton University Press, 2008. ISBN 978-1-4008-2982-8.

Susan Athey and Guido W Imbens. The state of applied econometrics: Causality and policy evaluation. *Journal of Economic Perspectives*, 31(2):3–32, 2017.

Susan Athey, Mohsen Bayati, Nikolay Doudchenko, Guido W Imbens, and Khashayar Khosravi. Matrix completion methods for causal panel data models. *Journal of the American Statistical Association*, pages 1–41, 2021.

Chris Auld. Breaking news! `https://twitter.com/Chris_Auld/status/1035230771957485568`, 2018. Accessed: 2020-02-20.

David H Autor. Outsourcing at will: The contribution of unjust dismissal doctrine to the growth of employment outsourcing. *Journal of Labor Economics*, 21(1):1–42, 2003.

Michael A. Bailey. *Real Econometrics*. Oxford University Press, 2019.

Andrew C. Baker, David F. Larcker, and Charles C. Y. Wang. How much should we trust staggered difference-in-differences estimates? Technical report, Social Science Research Network, 2021.

Sarah H Bana, Kelly Bedard, and Maya Rossin-Slater. The impacts of paid family leave benefits: Regression kink evidence from California administrative data. *Journal of Policy Analysis and Management*, 39(4):888–929, 2020.

Alan I Barreca, Jason M Lindo, and Glen R Waddell. Heaping-induced bias in regression-discontinuity designs. *Economic Inquiry*, 54(1):268–293, 2016.

Timothy J Bartik. *Who Benefits from State and Local Economic Development Policies?* WE Upjohn Institute for Employment Research, 1991.

Erich Battistin, Agar Brugiavini, Enrico Rettore, and Guglielmo Weber. The retirement consumption puzzle: Evidence from a regression discontinuity approach. *American Economic Review*, 99(5):2209–2226, 2009.

Heiko Bauke. Parameter estimation for power-law distributions by maximum likelihood methods. *The European Physical Journal B*, 58(2):167–173, 2007.

Andrew Bell and Kelvyn Jones. Explaining fixed effects: Random effects modeling of time-series cross-sectional and panel data. *Political Science Research and Methods*, 3(1):133–153, 2015.

Marc F Bellemare and Jeffrey R Bloem. The paper of how: Estimating treatment effects using the front-door criterion. Technical report, Working Paper, 2019.

Marc F Bellemare and Casey J Wichman. Elasticities and the inverse hyperbolic sine transformation. *Oxford Bulletin of Economics and Statistics*, 82(1):50–61, 2020.

Marinho Bertanha. Regression discontinuity design with many thresholds. *Journal of Econometrics*, 218(1):216–241, 2020.

Marianne Bertrand, Esther Duflo, and Sendhil Mullainathan. How much should we trust differences-in-differences estimates? *The Quarterly Journal of Economics*, 119(1):249–275, 2004.

Jay Bhattacharya and William B Vogt. Do instrumental variables belong in propensity scores? Technical report, NBER, 2007. NBER Working Paper No. t0343.

Bernard S Black, Parth Lalkiya, and Joshua Y Lerner. The trouble with coarsened exact matching. *Northwestern Law & Econ Research Paper Forthcoming*, 2020.

Sören Blomquist and Matz Dahlberg. Small sample properties of LIML and jackknife IV estimators: Experiments with weak instruments. *Journal of Applied Econometrics*, 14(1):69–88, 1999.

David E Bloom, David Canning, and Günther Fink. Disease and development revisited. *Journal of Political Economy*, 122(6): 1355–1366, 2014.

David E Broockman. Black politicians are more intrinsically motivated to advance blacks' interests: A field experiment manipulating political incentives. *American Journal of Political Science*, 57(3):521–536, 2013.

John M Brooks and Robert L Ohsfeldt. Squeezing the balloon: Propensity scores and unmeasured covariate balance. *Health Services Research*, 48(4):1487–1507, 2013.

Stephen J Brown and Jerold B Warner. Using daily stock returns: The case of event studies. *Journal of Financial Economics*, 14(1):3–31, 1985.

Matias Busso, John DiNardo, and Justin McCrary. New evidence on the finite sample properties of propensity score reweighting and matching estimators. *Review of Economics and Statistics*, 96(5):885–897, 2014.

Jing Cai, Alain De Janvry, and Elisabeth Sadoulet. Social networks and the decision to insure. *American Economic Journal: Applied Economics*, 7(2):81–108, 2015.

Marco Caliendo and Sabine Kopeinig. Some practical guidance for the implementation of propensity score matching. *Journal of Economic Surveys*, 22(1):31–72, 2008.

Brantly Callaway and Pedro HC Sant'Anna. Difference-in-differences with multiple time periods. *Journal of Econometrics*, 2020. Forthcoming.

Sebastian Calonico, Matias D Cattaneo, Max H Farrell, and Rocio Titiunik. Regression discontinuity designs using covariates. *Review of Economics and Statistics*, 101(3):442–451, 2019.

Susan Camilleri and Jeffrey Diebold. Hospital uncompensated care and patient experience: An instrumental variable approach. *Health Services Research*, 54(3):603–612, 2019.

David Card. The impact of the Mariel boatlift on the Miami labor market. *ILR Review*, 43(2):245–257, 1990.

David Card. The causal effect of education on earnings. *Handbook of Labor Economics*, 3:1801–1863, 1999.

David Card, David S Lee, Zhuan Pei, and Andrea Weber. Inference on causal effects in a generalized regression kink design. *Econometrica*, 83(6):2453–2483, 2015.

David Card, David S Lee, Zhuan Pei, and Andrea Weber. Regression kink design: Theory and practice. In *Advances in Econometrics*, volume 38, pages 341–382. Emerald Group Publishing Limited, Bingley, 2017.

Matias D Cattaneo and Gonzalo Vazquez-Bare. The choice of neighborhood in regression discontinuity designs. *Observational Studies*, 3(2):134–146, 2016.

Matias D Cattaneo, Rocío Titiunik, Gonzalo Vazquez-Bare, and Luke Keele. Interpreting regression discontinuity designs with multiple cutoffs. *The Journal of Politics*, 78(4):1229–1248, 2016.

Matias D Cattaneo, Michael Jansson, and Xinwei Ma. Simple local polynomial density estimators. *Journal of the American Statistical Association*, 115(531):1449–1455, 2020a.

Matias D Cattaneo, Luke Keele, Rocío Titiunik, and Gonzalo Vazquez-Bare. Extrapolating treatment effects in multi-cutoff regression discontinuity designs. *Journal of the American Statistical Association*, pages 1–12, 2020b.

Julia Chabrier, Sarah Cohodes, and Philip Oreopoulos. What can we learn from charter school lotteries? *Journal of Economic Perspectives*, 30(3):57–84, 2016.

John C Chao and Norman R Swanson. Consistent estimation with a large number of weak instruments. *Econometrica*, 73(5):1673–1692, 2005.

Victor Chernozhukov, Christian Hansen, and Martin Spindler. Post-selection and post-regularization inference in linear models with many controls and instruments. *American Economic Review*, 105(5):486–90, 2015.

Victor Chernozhukov, Denis Chetverikov, Mert Demirer, Esther Duflo, Christian Hansen, Whitney Newey, and James Robins. Double/debiased machine learning for treatment and structural parameters. *The Econometrics Journal*, 21(1), 2018a.

Victor Chernozhukov, Iván Fernández-Val, and Ye Luo. The sorted effects method: Discovering heterogeneous effects beyond their averages. *Econometrica*, 86(6):1911–1938, 2018b.

Richard C Chiburis, Jishnu Das, and Michael Lokshin. A practical comparison of the bivariate probit and linear IV estimators. *Economics Letters*, 117(3):762–766, 2012.

Carlos Cinelli and Chad Hazlett. Making sense of sensitivity: Extending omitted variable bias. *Journal of the Royal Statistical Society: Series B (Statistical Methodology)*, 82(1):39–67, 2020.

Merlise Clyde and Edward I George. Model uncertainty. *Statistical Science*, 19(1):81–94, 2004.

Thomas Coleman. Causality in the time of cholera: John Snow as a prototype for causal inference. *Available at SSRN 3262234*, 2019.

J Michael Collins and Carly Urban. The dark side of sunshine: Regulatory oversight and status quo bias. *Journal of Economic Behavior & Organization*, 107:470–486, 2014. DOI: 10.1016/j.jebo.2014.04.003.

Timothy G Conley, Christian B Hansen, and Peter E Rossi. Plausibly exogenous. *Review of Economics and Statistics*, 94 (1):260–272, 2012.

Jason Connor and Wayne Hall. Thresholds for safer alcohol use might need lowering. *The Lancet*, 391(10129):1460–1461, 2018.

Christopher Cornwell and William N Trumbull. Estimating the economic model of crime with panel data. *The Review of Economics and Statistics*, 76:360–366, 1994.

Scott Cunningham. *Causal Inference: The Mixtape*. Yale University Press, 2021.

Jamie R Daw and Laura A Hatfield. Matching and regression to the mean in difference-in-differences analysis. *Health Services Research*, 53(6):4138–4156, 2018.

Clément De Chaisemartin and Xavier d'Haultfoeuille. Two-way
fixed effects estimators with heterogeneous treatment effects.
American Economic Review, 110(9):2964–96, 2020.

Rajeev H Dehejia and Sadek Wahba. Propensity score-matching
methods for nonexperimental causal studies. *Review of Economics and Statistics*, 84(1):151–161, 2002.

Yi Deng, Changgee Chang, Moges Seyoum Ido, and Qi Long.
Multiple imputation for general missing data patterns in the
presence of high-dimensional data. *Scientific Reports*, 6(1):
1–10, 2016.

Daniel K Fetter. How do mortgage subsidies affect home ownership? Evidence from the mid-century GI bills. *American
Economic Journal: Economic Policy*, 5(2):111–47, 2013.

Carlos A Flores and Alfonso Flores-Lagunes. Partial identification of local average treatment effects with an invalid instrument. *Journal of Business & Economic Statistics*, 31(4):
534–545, 2013.

Kenneth A Frank and Ran Xu. Causal inference for social network analysis. In *The Oxford Handbook of Social Networks*,
pages 288–310. Oxford University Press, 2020.

Wayne A Fuller. *Measurement Error Models*, volume 305. John
Wiley & Sons, 2009.

Sebastian Galiani and Juan Pantano. Structural models: Inception and frontier. Technical report, National Bureau of
Economic Research, 2021.

Markus Gangl. Partial identification and sensitivity analysis.
In *Handbook of Causal Analysis for Social Research*, pages
377–402. Springer, 2013.

Gapminder Institute. Gapminder. https://www.gapminder.
org/, 2020. Accessed: 2020-03-09.

Andrew Gelman. You need 16 times the sample size to estimate
an interaction than to estimate a main effect. Statistical Modeling, Causal Inference, and Social Science, 2018.

Andrew Gelman and Jennifer Hill. *Data Analysis using Regression and Multilevel/hierarchical Models*. Cambridge University Press, 2006.

Andrew Gelman and Guido W Imbens. Why high-order polynomials should not be used in regression discontinuity designs.
Journal of Business & Economic Statistics, 37(3):447–456,
2019.

Charles E. Gibbons, Serrato Juan Carlos Suárez, and Michael B.
Urbancic. Broken or fixed effects? *Journal of Econometric
Methods*, 8(1):1–12, 2019.

Dan Goldhaber, Cyrus Grout, and Nick Huntington-Klein.
Screen twice, cut once: Assessing the predictive validity of
applicant selection tools. *Education Finance and Policy*, 12
(2):197–223, 2017.

Paul Goldsmith-Pinkham, Isaac Sorkin, and Henry Swift. Bartik instruments: What, when, why, and how. *American Economic Review*, 110(8):2586–2624, 2020.

Andrew Goodman-Bacon. Difference-in-differences with variation in treatment timing. Technical report, National Bureau
of Economic Research, 2018.

William H Greene. *Econometric Analysis*. Pearson Education
India, 2003.

Beth Ann Griffin, Daniel F McCaffrey, Daniel Almirall, Lane F
Burgette, and Claude Messan Setodji. Chasing balance and
other recommendations for improving nonparametric propensity score models. *Journal of Causal Inference*, 5(2), 2017.

Jinyong Hahn. On the role of the propensity score in efficient semiparametric estimation of average treatment effects.
Econometrica, 66(2):315–331, 1998.

Jens Hainmueller. Entropy balancing for causal effects: A multivariate reweighting method to produce balanced samples in
observational studies. *Political Analysis*, 20(1):25–46, 2012.

Scott Hankins, Mark Hoekstra, and Paige Marta Skiba. The
ticket to easy street? The financial consequences of winning
the lottery. *Review of Economics and Statistics*, 93(3):961–
969, 2011.

James J Heckman and Edward J Vytlacil. Local instrumental variables and latent variable models for identifying and
bounding treatment effects. *Proceedings of the National
Academy of Sciences*, 96(8):4730–4734, 1999.

Ralph Hertwig and Gerd Gigerenzer. The "conjunction fallacy" revisited: How intelligent inferences look like reasoning errors. *Journal of Behavioral Decision Making*, 12(4):275–305, 1999.

Keisuke Hirano, Guido W Imbens, and Geert Ridder. Efficient estimation of average treatment effects using the estimated propensity score. *Econometrica*, 71(4):1161–1189, 2003.

Daniel G Horvitz and Donovan J Thompson. A generalization of sampling without replacement from a finite universe. *Journal of the American Statistical Association*, 47(260):663–685, 1952.

Nick Huntington-Klein, James Cowan, and Dan Goldhaber. Selection into Online Community College Courses and their Effects on Persistence. *Research in Higher Education*, 58(3): 244–269, 2017.

Nick Huntington-Klein, Andreu Arenas, Emily Beam, Marco Bertoni, Jeffrey R. Bloem, Pralhad Burli, Naibin Chen, Paul Grieco, Godwin Ekpe, Todd Pugatch, Martin Saavedra, and Yaniv Stopnitzky. The influence of hidden researcher decisions in applied microeconomics. *Economic Inquiry*, 59(3):944–960, 2021.

Stefano M Iacus, Gary King, and Giuseppe Porro. Causal inference without balance checking: Coarsened exact matching. *Political Analysis*, 20(1):1–24, 2012.

Guido W Imbens and Karthik Kalyanaraman. Optimal bandwidth choice for the regression discontinuity estimator. *The Review of Economic Studies*, 79(3):933–959, 2012.

Guido W Imbens and Thomas Lemieux. Regression discontinuity designs: A guide to practice. *Journal of Econometrics*, 142(2):615–635, 2008.

Guido W Imbens and Donald B Rubin. *Causal Inference in Statistics, Social, and Biomedical Sciences*. Cambridge University Press, 2015.

Luke J Keele and Rocio Titiunik. Geographic boundaries as regression discontinuities. *Political Analysis*, 23(1):127–155, 2015.

Judd B Kessler and Alvin E Roth. Don't take "no" for an answer: An experiment with actual organ donor registrations. Technical report, National Bureau of Economic Research, 2014.

Gary King and Richard Nielsen. Why propensity scores should not be used for matching. *Political Analysis*, 27(4):435–454, 2019.

Heather Klemick, Henry Mason, and Karen Sullivan. Superfund cleanups and children's lead exposure. *Journal of Environmental Economics and Management*, 100:1022–1089, 2020.

Michal Kolesár and Christoph Rothe. Inference in regression discontinuity designs with a discrete running variable. *American Economic Review*, 108(8):2277–2304, 2018.

Michal Kolesár, Raj Chetty, John Friedman, Edward Glaeser, and Guido W Imbens. Identification and inference with many invalid instruments. *Journal of Business & Economic Statistics*, 33(4):474–484, 2015.

Sagar P Kothari and Jerold B Warner. Econometrics of event studies. In *Handbook of Empirical Corporate Finance*, pages 3–36. Elsevier, 2007.

David L Lee, Justin McCrary, Marcelo J Moreira, and Jack Porter. Valid t-ratio inference for IV. *arXiv preprint arXiv:2010.05058*, 2020.

Moshe Lichman. UCI machine learning repository, 2013.

Colin Mahony. Effects of dimensionality on distance and probability density in climate space. The Seasons Alter, 2014.

Marco Manacorda, Edward Miguel, and Andrea Vigorito. Government transfers and political support. *American Economic Journal: Applied Economics*, 3(3):1–28, 2011.

Charles F Manski. Identification of endogenous social effects: The reflection problem. *The Review of Economic Studies*, 60 (3):531–542, 1993.

Michael A. Martin. *Wild Bootstrap*, pages 1–6. John Wiley & Sons, Ltd., 2017. ISBN 9781118445112.

Justin McCrary. Manipulation of the running variable in the regression discontinuity design: A density test. *Journal of Econometrics*, 142(2):698–714, 2008.

Bruce D McCullough and Hrishikesh D Vinod. Verifying the solution from a nonlinear solver: A case study. *American Economic Review*, 93(3):873–892, 2003.

Edward Miguel, Shanker Satyanath, and Ernest Sergenti. Economic shocks and civil conflict: An instrumental variables approach. *Journal of Political Economy*, 112(4):725–753, 2004.

Ronald T Milam, Marc Birnbaum, Chris Ganson, Susan Handy, and Jerry Walters. Closing the induced vehicle travel gap between research and practice. *Transportation Research Record*, 2653(1):10–16, 2017.

Douglas C Montgomery, Cheryl L Jennings, and Murat Kulahci. *Introduction to Time Series Analysis and Forecasting*. John Wiley & Sons, 2015.

Thomas A Mroz. The sensitivity of an empirical model of married women's hours of work to economic and statistical assumptions. *Econometrica*, 55(4):765–799, 1987.

Yair Mundlak. On the pooling of time series and cross section data. *Econometrica*, 46(1):69–85, 1978.

Aviv Nevo and Adam M Rosen. Identification with imperfect instruments. *Review of Economics and Statistics*, 94(3):659–671, 2012.

Emily Oster. Data and code for: Health recommendations and selection in health behaviors: Replication files. Nashville, TN: American Economic Association [publisher], 2020. Ann Arbor, MI: Inter-university Consortium for Political and Social Research [distributor], 2020a.

Emily Oster. Health recommendations and selection in health behaviors. *American Economic Review: Insights*, 2(2):143–60, 2020b.

John P Papay, John B Willett, and Richard J Murnane. Extending the regression-discontinuity approach to multiple assignment variables. *Journal of Econometrics*, 161(2):203–207, 2011.

Paulo MDC Parente and JMC Santos Silva. A cautionary note on tests of overidentifying restrictions. *Economics Letters*, 115(2):314–317, 2012.

James M Patell. Corporate forecasts of earnings per share and stock price behavior: Empirical test. *Journal of Accounting Research*, 14(2):246–276, 1976.

Judea Pearl. *Causality*. Cambridge University Press, Cambridge, MA, 2nd edition, 2009.

Judea Pearl and Dana Mackenzie. *The Book of Why: The New Science of Cause and Effect*. Basic Books, New York City, New York, 2018.

Dimitris N Politis and Halbert White. Automatic block-length selection for the dependent bootstrap. *Econometric Reviews*, 23(1):53–70, 2004.

Patrick A Puhani. The treatment effect, the cross difference, and the interaction term in nonlinear "difference-in-differences" models. *Economics Letters*, 115(1):85–87, 2012.

Sean F Reardon and Joseph P Robinson. Regression discontinuity designs with multiple rating-score variables. *Journal of Research on Educational Effectiveness*, 5(1):83–104, 2012.

Peter C Reiss and Frank A Wolak. Structural econometric modeling: Rationales and examples from industrial organization. *Handbook of Econometrics*, 6:4277–4415, 2007.

Jessica Wolpaw Reyes. Environmental policy as social policy? The impact of childhood lead exposure on crime. *The BE Journal of Economic Analysis & Policy*, 7(1), 2007.

Paul R Rosenbaum and Donald B Rubin. Constructing a control group using multivariate matched sampling methods that incorporate the propensity score. *The American Statistician*, 39(1):33–38, 1985.

Jonathan Roth. Should we adjust for the test for pre-trends in difference-in-difference designs? *arXiv preprint arXiv:1804.01208*, 2018.

Pedro HC Sant'Anna and Jun Zhao. Doubly robust difference-in-differences estimators. *Journal of Econometrics*, 219(1): 101–122, 2020.

Heather Sarsons. Rainfall and conflict: A cautionary tale. *Journal of Development Economics*, 115:62–72, 2015.

Sebastian Schneeweiss, Jeremy A Rassen, Robert J Glynn, Jerry Avorn, Helen Mogun, and M Alan Brookhart. High-dimensional propensity score adjustment in studies of treatment effects using health care claims data. *Epidemiology*, 20 (4):512, 2009.

Justin M. Shea. *wooldridge: 111 Data Sets from "Introductory Econometrics: A Modern Approach, 6e" by Jeffrey M. Wooldridge*, 2018. URL https://CRAN.R-project.org/package=wooldridge. R package version 1.3.1.

John Snow. *On the Mode of Communication of Cholera*. John Churchill, 1855.

Peter Spirtes, Clark N Glymour, Richard Scheines, and David Heckerman. *Causation, Prediction, and Search*. MIT Press, 2000.

James H Stock and Motohiro Yogo. Testing for weak instruments in linear IV regression. In *Identification and Inference for Econometric Models: Essays in Honor of Thomas Rothenberg*, pages 80–108. Cambridge University Press, Cambridge, 2005.

Liyang Sun and Sarah Abraham. Estimating dynamic treatment effects in event studies with heterogeneous treatment effects. *Journal of Econometrics*, 2020. Forthcoming.

Monica Taljaard, Joanne E McKenzie, Craig R Ramsay, and Jeremy M Grimshaw. The use of segmented regression in analysing interrupted time series studies: An example in pre-hospital ambulance care. *Implementation Science*, 9(1):1–4, 2014.

Zhiqiang Tan. Bounded, efficient and doubly robust estimation with inverse weighting. *Biometrika*, 97(3):661–682, 2010.

James Vincent. Scientists rename human genes to stop Microsoft Excel from misreading them as dates. *The Verge*, August 2020.

Stefan Wager and Susan Athey. Estimation and inference of heterogeneous treatment effects using random forests. *Journal of the American Statistical Association*, 113(523):1228–1242, 2018.

Frank Windmeijer, Helmut Farbmacher, Neil Davies, and George Davey Smith. On the use of the LASSO for instrumental variables estimation with some invalid instruments. *Journal of the American Statistical Association*, 114(527): 1339–1350, 2019.

Justin Wolfers. Did unilateral divorce laws raise divorce rates? A reconciliation and new results. *American Economic Review*, 96(5):1802–1820, 2006.

Felix Wong and James J Collins. Evidence that coronavirus superspreading is fat-tailed. *Proceedings of the National Academy of Sciences*, 117(47):29,416–29,418, 2020.

Angela M Wood, Stephen Kaptoge, Adam S Butterworth, and 239 more. Risk thresholds for alcohol consumption: Combined analysis of individual-participant data for 599,912 current drinkers in 83 prospective studies. *The Lancet*, 391(10129):1513–1523, April 2018. DOI: 10.1016/S0140-6736(18)30134-X.

George Wood, Tom R Tyler, and Andrew V Papachristos. Procedural justice training reduces police use of force and complaints against officers. *Proceedings of the National Academy of Sciences*, 117(18):9815–9821, 2020.

Jeffrey M Wooldridge. Instrumental variables estimation of the average treatment effect in the correlated random coefficient model. In *Modelling and Evaluating Treatment Effects in Econometrics*. Emerald Group Publishing Limited, Bingley, 2008.

Jeffrey M Wooldridge. *Econometric Analysis of Cross Section and Panel Data*. MIT press, 2010.

Jeffrey M Wooldridge. *Introductory Econometrics: A Modern Approach*. Nelson Education, 2016.

Jeffrey M Wooldridge. Two-way fixed effects, the two-way Mundlak regression, and difference-in-differences estimators. Technical report, SSRN, 2021.

Cheng Xu. *Essays on Urban and Environmental Economics*. PhD thesis, The George Washington University, 2019.

Qingyuan Zhao and Daniel Percival. Entropy balancing is doubly robust. *Journal of Causal Inference*, 5(1), 2017.

Zhong Zhao.   Using matching to estimate treatment effects:
  Data requirements, matching metrics, and Monte Carlo evidence. *Review of Economics and Statistics*, 86(1):91–107,
  2004.

Index

2SLS, 480

always-takers, 489
autocorrelation, 235, 238
autoregression, 430
average treatment effect, 145, 325
average treatment on the treated, 147, 325
average treatment on the untreated, 147, 325

back door, 121
back-door stuff, 162
bad path, 121
balance, 307
bandwidth, 282, 289, 506
between variation, 383, 385
bias, 184
binary variable, 21
bivariate probit, 500
block bootstrap, 376
bootstrap, 240, 370
bootstrap standard errors, 314, 319, 573

calculations of data, 36
categorical variable, 21
causal forests, 569
causal inference in action, 386
causality, 87, 89
charter schools, 131
classical measurement error, 257
closing multiple doors, 165
cluster bootstrap, 376
clustered standard errors, 239, 242, 380, 447

collider, 124, 140
collinearity, 253
compliers, 489
conditional average treatment effect, 147
conditional distribution, 47
conditional independence assumption, 302
conditional mean, 49
conditional median, 49
conditional standard deviation, 49
construct validity, 583
continuous variable, 20
control function, 500
control groups, 166
controlling for a variable, 59
convergence, 38
count variable, 20
cross-validation, 263, 553
cutoff, 506

d-separation, 126
data, 36
data generating process, 67, 102
data mining, 12, 14
data science, 12
data transformation, 34
debiased machine learning, see double machine learning
decision tree, 570
defiers, 489
density, 24
difference-in-differences, 435
direct effect, 96, 98, 121
distance matching, 271

distribution, 22
double machine learning, 565
doubly robust, 318, 463
Durbin-Wu-Hausman test, 399, 494
dynamic treatment effects, 454, 455

error term, 181, 400
errors-in-variables, 256
estimates, 37
event study, 407, 435
evidence, 10
exclusion restriction, 474
exogenous variation, 133, 183
expectation-maximization algorithm, 592
explaining away results, 12
explaining why, 13, 16

factor analysis, 255
false negative rate, 188
false positive rate, 188
fixed effects, 381, 447
for loop, 344
forbidden regression, 499
forcing variable, see running variable
forecasting, 429
frequency table, 22
frequency weights, 249
front door, 121
front door method, 139
full information maximum likelihood, 591
fuzzy regression discontinuity, 510, 544